Lecture Notes in Physics

The Editorial Policy for Proceedings

The series Lecture Notes in Physics reports new developments in physical research and teaching – quickly, informally, and at a high level. The proceedings to be considered for publication in this series should be limited to only a few areas of research, and these should be closely related to each other. The contributions should be of a high standard and should avoid lengthy redraftings of papers already published or about to be published elsewhere. As a whole, the proceedings should aim for a balanced presentation of the theme of the conference including a description of the techniques used and enough motivation for a broad readership. It should not be assumed that the published proceedings must reflect the conference in its entirety. (A listing or abstracts of papers presented at the meeting but not included in the proceedings could be added as an appendix.)

When applying for publication in the series Lecture Notes in Physics the volume's editor(s) should submit sufficient material to enable the series editors and their referees to make a fairly accurate evaluation (e.g. a complete list of speakers and titles of papers to be presented and abstracts). If, based on this information, the proceedings are (tentatively) accepted, the volume's editor(s), whose name(s) will appear on the title pages, should select the papers suitable for publication and have them refereed (as for a journal) when appropriate. As a rule discussions will not be accepted. The series editors and Springer-Verlag will normally not interfere with the detailed editing except in fairly obvious cases or on technical matters.

Final acceptance is expressed by the series editor in charge, in consultation with Springer-Verlag only after receiving the complete manuscript. It might help to send a copy of the authors' manuscripts in advance to the editor in charge to discuss possible revisions with him. As a general rule, the series editor will confirm his tentative acceptance if the final manuscript corresponds to the original concept discussed, if the quality of the contribution meets the requirements of the series, and if the final size of the manuscript does not greatly exceed the number of pages originally agreed upon. The manuscript should be forwarded to Springer-Verlag shortly after the meeting. In cases of extreme delay (more than six months after the conference) the series editors will check once more the timeliness of the papers. Therefore, the volume's editor(s) should establish strict deadlines, or collect the articles during the conference and have them revised on the spot. If a delay is unavoidable, one should encourage the authors to update their contributions if appropriate. The editors of proceedings are strongly advised to inform contributors about these points at an early stage.

The final manuscript should contain a table of contents and an informative introduction accessible also to readers not particularly familiar with the topic of the conference. The contributions should be in English. The volume's editor(s) should check the contributions for the correct use of language. At Springer-Verlag only the prefaces will be checked by a copy-editor for language and style. Grave linguistic or technical shortcomings may lead to the rejection of contributions by the series editors. A conference report should not exceed a total of 500 pages. Keeping the size within this bound should be achieved by a stricter selection of articles and not by imposing an upper limit to the length of the individual papers. Editors receive jointly 30 complimentary copies of their book. They are entitled to purchase further copies of their book at a reduced rate. As a rule no reprints of individual contributions can be supplied. No royalty is paid on Lecture Notes in Physics volumes. Commitment to publish is made by letter of interest rather than by signing a formal contract. Springer-Verlag secures the copyright for each volume.

The Production Process

The books are hardbound, and the publisher will select quality paper appropriate to the needs of the author(s). Publication time is about ten weeks. More than twenty years of experience guarantee authors the best possible service. To reach the goal of rapid publication at a low price the technique of photographic reproduction from a camera-ready manuscript was chosen. This process shifts the main responsibility for the technical quality considerably from the publisher to the authors. We therefore urge all authors and editors of proceedings to observe very carefully the essentials for the preparation of camera-ready manuscripts, which we will supply on request. This applies especially to the quality of figures and halftones submitted for publication. In addition, it might be useful to look at some of the volumes already published. As a special service, we offer free of charge LATEX and TEX macro packages to format the text according to Springer-Verlag's quality requirements. We strongly recommend that you make use of this offer, since the result will be a book of considerably improved technical quality. To avoid mistakes and time-consuming correspondence during the production period the conference editors should request special instructions from the publisher well before the beginning of the conference. Manuscripts not meeting the technical standard of the series will have to be returned for improvement.

For further information please contact Springer-Verlag, Physics Editorial Department V, Tiergartenstrasse 17, D-69121 Heidelberg, FRG

U. G. Jørgensen (Ed.)

Molecules in the Stellar Environment

Proceedings of IAU Colloquium No. 146
Held at Copenhagen, Denmark, May 24-29, 1993

Springer-Verlag
Berlin Heidelberg New York
London Paris Tokyo
Hong Kong Barcelona
Budapest

Editor

U.G. Jørgensen
Niels Bohr Institute
Blegdamsvej 17
DK-2100 Copenhagen

ISBN 3-540-57747-5 Springer-Verlag Berlin Heidelberg New York
ISBN 0-387-57747-5 Springer-Verlag New York Berlin Heidelberg

Printed in Germany

Printing and binding: Druckhaus Beltz, Hemsbach/Bergstr.
SPIN: 10080311 55/3140-543210 - Printed on acid-free paper

Preface

The region of a star where the spectrum is formed is called the atmosphere. It consists of the photosphere (where most of the visual and infrared light comes from), the chromosphere (a somewhat hotter layer above the photosphere where, in cool stars, most of the ultraviolet light is produced), and sometimes a corona (the very hot outermost region). Stars with an "average" (effective) temperature in the photosphere of less than around 4000 K are called "cool". The low luminosity cool stars include the M dwarfs, some white dwarfs, and the brown dwarfs. All these are stellar types that often can be as old as the universe, and they can provide us with important and detailed information about things as different as the conditions in the early universe, the unseen mass in the galaxies, and of course stellar structure and evolution itself. The high luminosity cool stars are called red giants. They are often larger than the Earth's orbit around the Sun, and a wind of gas and dust from their uppermost layers can be so strong that the mass of our Sun would be expelled in less than 100,000 years. These stars continuously enrich the interstellar medium with material from the nucleosynthesis deep in their interior. New stars and planets forming from the interstellar clouds are therefore built "of the ashes of the red giants". In fact about half of all the elements that surround us come from this process (2/3 of all the elements heavier than iron, and a few of the lighter and more abundant ones too; the rest of the elements come from supernovae). Nevertheless, there is no theory capable of explaining how the material is blown out of the red giant stars. Part of the wind-material can gather temporarily in shells around the stars. Dense shells of gas and dust around red giants are known as circumstellar envelopes. All these phenomena are collectively referred to in this book as the stellar environment. It has not yet been possible to construct a unified model for this part of a star.

The gas in cool stars is dominated by small molecules (like CO, H_2O, HCN) which slowly grow into bigger molecules (like carbon chains and polyaromatic hydrocarbon rings) and grains (like silicon carbide and graphite dust) in the outflowing gas. A major obstacle to constructing a unified model for the stellar environment in cool stars has been the severe lack of data for the molecular physics and chemistry that prevail at the high temperatures and/or low pressures of the stellar environment. It is the aim of the present book to review the status and knowledge of the theory, observations, and experiments relating to both the astrophysical and the chemical aspects of this problem, and to stimulate work in the area where data and progress are most needed. One problem that any scientist entering this field will immediately discover is that "the two sides" – the chemistry and the astrophysics – each speak their own language. It is this problem which in the first chapter of the book is described as "the two villages". Each chapter in the book has therefore been aimed at presenting both the basic concepts and the underlying theory, together with the most recent knowledge and progress. This aim is described in the first chapter as creating "a bridge between the villages" – i.e., the ability to pass knowledge from one of the communities to the other.

I am thankful to all the authors for their patience with me when I returned their manuscripts for the n'th time claiming that we need more "stones on the bridge" in order to "make it easy to pass from one of the villages to the other" – i.e., in order to make the concepts from one of the disciplines understandable for scientists and students from the other discipline. It is my hope that the book will be used as such a bridge between molecular chemistry and stellar astrophysics.

The idea of the book (and the IAU Colloquium 146 upon which it is based) arose from discussions between Georges Graner, Mike Seaton, Rainer Wehrse and myself during a fruitful meeting on atomic and molecular opacities in Venezuela in the summer of 1991. The project was well received by the host institutes in Copenhagen (the Niels Bohr Institute and Nordita), and the financial wherewithal was secured by generous support from the International Astronomical Union (on recommendations by commissions 14, 29, and 36), the Danish Natural Science Research Council, the Augustinus Foundation, the Carlsberg Foundation, and Knud Højgaard's Foundation. I gratefully acknowledge this support, as well as the work done by the scientific organizing committee (consisting of Georges Graner, David L. Lambert, Björn O. Roos, Takashi Tsuji, Rainer Wehrse, and myself) and the local organizing committee (consisting of Ben Mottelson, Bernard Pagel, Peter Thejll, and myself). Particular thanks are due to Peter Thejll without whose continuous effort the organization of the conference would have been impossible.

Copenhagen, November 1993

Uffe Gråe Jørgensen

Contents

Stellar Photospheres and Molecules — A View From The Bridge

David L. Lambert

Department of Astronomy, University of Texas, Austin, Texas 78712 USA

1 Introduction

The allegory alluded to in the subtitle of my review came to mind in the early morning hours when I was battling insomnia after a successful observing run at our McDonald Observatory. Similar bouts with insomnia are more widely classified as "jetlag," a condition probably too familiar to many of the readers!

Imagine two villages connected by a bridge spanning a deep, fast-flowing river. In my village on the south bank live the astronomers. Across the bridge live the chemical physicists and the physical chemists who are engaged in determining by all possible theoretical and experimental means the structure and spectra of molecules (and atoms). The astronomical community can be sorted into various clans of whom I consider here only the stellar spectroscopists and, in particular, the small subclan fascinated by quantitative spectroscopy who to pursue their trade must establish and maintain excellent communications with the village on the other side of the bridge. The necessity for contacts between the two villages has grown with the development of the tools of the stellar spectroscopists' trade: e.g., the advances of telescopes and spectrographs with respect to coverage of the electromagnetic spectrum and sensitivity, and the theoretical developments in the construction of model stellar atmospheres and the computation of synthetic stellar spectra. The recent decades of astronomical advances have been matched by exciting years in the village across the bridge: e.g., tunable lasers have spawned a renaissance in molecular (and atomic) spectroscopy, and theoretical work on molecular structure and spectra has advanced with the awesome growth of computers. Yet, a curious paradox may be sketched. These welcome and remarkable developments in the two villages have made it more difficult to maintain effective communications across the bridge. How can I possibly exploit fully the successes of MRD-CI calculations and appreciate their limitations? My counterparts across the bridge will echo my question: What experiments or quantum calculations might be usefully made in support of the astronomers' programs ATLAS, MARCS, SOSMARCS, and their relatives?

This review is one attempt to sketch the needs of stellar spectroscopists for basic molecular data. If we are to deserve collaborations with chemists and physicists, it is helpful if the astrophysical problems are of fundamental importance. The "stellar environment" presents a rich array of such problems. The stellar spectrum is our observational tool for understanding the physical structure and chemical composition of the stellar atmosphere from the deepest photospheric layers to the outermost limits of the stellar wind. The data primarily but not exclusively relating to the chemical composition are then applied to questions in stellar evolution and nucleosynthesis. Information gathered star-by-star is then used to tackle questions such as "How do stars evolve?" "How has the Galaxy evolved?" "How, when, and where were the chemical elements synthesized?" These are fundamental questions.

Here, I follow this introduction with a sketch of asymptotic giant branch (AGB) stars which are an especially interesting class of red giants whose environments can only be analysed when full attention is given to the molecular component of the stars' spectra. After explaining the structure and significance of these stars, I give examples of the basic molecular data needed in the analysis of stellar spectra, and I draw in large measure on personal examples. This review is directed primarily at the villagers across the bridge. In the last chapter, I return to my allegory and offer a few (obvious) remarks on how communication across the bridge might be made more effective.

2 AGB Stars: A Key Site of Nucleosynthesis

2.1 Evidence of Nucleosynthesis

Supernovae of Type II and Type Ia are widely recognized as major sites of nucleosynthesis. Type II SN are the terminal phase for massive stars ($M > 8M_\odot$). Type Ia SN are commonly supposed to result when two white dwarfs coalesce to form an unstable compact object whose mass violates Chandrasekhar's limit for the maximum mass supportable by degenerate electrons. The AGB stars make major and distinctive contributions to stellar nucleosynthesis. In particular, these stars contribute elements that are not obviously synthesized by the SN: lithium, fluorine, and many heavy nuclides. AGB stars have initial masses in the range of about 1 to 8 $M_\odot$; stars less massive than about 1 $M_\odot$ have not yet exhausted their hydrogen supply, and stars more massive than 8 $M_\odot$ evolve to die as SN and do not attain the structure of an AGB star.

The role of the AGB stars was demonstrated long before the structure of the stars was understood even in bare outline. Merrill (1952) provided this demonstration with his discovery of absorption lines of Tc I in blue spectra of a class of red giants known as S stars. Technetium was the signal of internal nucleosynthesis because this element has no stable isotopes. All isotopes are short-lived in the astronomical sense: e.g., ^{99}Tc, the longest lived at low temperatures, has a half-life of about 200,000 yr., which is several orders of magnitude shorter than the time taken by the progenitors of typical S stars to exhaust their supply of hydrogen and evolve to become AGB stars. Relative to normal (M-type) stars of similar temperature, spectra of S stars also show an enhancement of other heavy elements such as Zr and La; the latter enhancements are readily revealed on low dispersion spectra from strengths of ZrO, LaO, and other bands. Merrill's observational evidence for the synthesis of heavy nuclides was paired a few years later with the theoretical idea that synthesis occurs through successive captures of neutrons using abundant iron-group nuclei as the seeds for this synthesis. This is termed the *s*-process, where *s* denotes "slow" which in turn implies that unstable neutron-rich nuclides will, in general, β-decay rather than experience another neutron capture. The operation of the *s*-process was described in the classic paper on nucleosynthesis by Burbidge et al. (1957). An excellent recent introduction is offered by Käppeler et al. (1989). Key unanswered questions about the *s*-process in AGB stars remain: What is the neutron source? What is the neutron density at the *s*-process site? Do models of AGB stars reproduce the observed pattern of *s*-process enrichments in individual AGB stars and in the mixture of stellar ejecta represented by unevolved stars of various ages and in the well-characterized solar system (carbonaceous chondrites)? What mass range among the AGB stars are the leading contributors of *s*-process nuclides to the Galaxy?

2.2 The Structure of AGB Stars

Stars do not become red giants until the initial H in their core is burnt to helium. Exhaustion of H terminates the main sequence phase of a star's life: main sequence lifetimes are $t_{\rm MS} \propto (M/M_\odot)^{-3}$, where the solar lifetime is about 10 Gyr. The remaining lifetime of a star is 10 percent or less of $t_{\rm MS}$. Unable to generate energy by nuclear fusion, the He core contracts and concomitantly the H-rich outer layers expand and cool; the star becomes a red giant. The He core ignites eventually and He is burnt to C via the 3α-process ($3^4{\rm He} \rightarrow ^{12}{\rm C}$ + energy). The upper mass limit for AGB stars is set by the requirement that the C-O core after He exhaustion be supported against gravitational collapse by the pressure of degenerate electrons. This mass limit is about 8 $M_\odot$.

Stars more massive than 8 $M_\odot$ ignite their C-O cores and proceed to a Type II SN. The lower mass limit for AGB stars of (presently) slightly less than 1 $M_\odot$ is set by the age of the Galaxy ($t_{\rm gal} \simeq$ 10–16 Gyr.) and the long main sequence lifetimes of low mass stars; obviously, $t_{\rm MS}(M) \simeq t_{\rm gal}$ sets the lower mass limit for AGB stars.

AGB stars are a study of sharp contrasts. The C-O core of about 1 $M_\odot$ has a radius approximately equal to that of the Earth; the mass density is about 10^6 g cm^{-3}. The outer radius of the AGB star may exceed the radius of the Earth's orbit around the Sun. Even with one to several solar masses the mean density of the H-rich envelope is very low. The H-rich envelope is almost completely convective. This has important consequences for observers because, if mixing occurs between He-shell and the base of the envelope, convection will carry the products of nucleosynthesis into the spectroscopically accessible atmosphere. The Tc discovered by Merrill shows that such mixing can occur.

AGB stars must live by nuclear energy. Two potential sources exist: the thin He-shell around the degenerate C-O core, and the H-rich layers just exterior to the He-shell. As shown long ago by Schwarzschild & Härm (1967), this double-shell source is unstable in the sense that neither the H-burning layers nor the He-shell can burn uninterruptedly. Calculations show that the two nuclear sources burn approximately out of phase. For the majority of the time, H burns at the base of the H-rich envelope. Helium, the ash of H-burning, is added to the He-shell. After an interval of 10^4–10^5 yr., compression and heating of the He-shell causes it to ignite. On ignition, the He-shell is almost fully convective for a brief interval ($\sim 10^2$ yr.). Following this "thermal instability" or "thermal pulse," He is burnt quiescently at the base of the thin (now radiative) shell. After a short period, He-burning can no longer be sustained, H-burning resumes, and the cycle of H-burning, the thermal instability, and He-burning is repeated. The cycle for intermediate mass AGB stars is illustrated in Fig. 1. The number of cycles experienced by an AGB star depends on several factors and is, in nature, likely to be severely determined by the mass loss rate, a factor ill-determined from observations and not yet predictable from theory. AGB evolution appears to terminate in a rapid and severe shedding of the H-rich envelope known as the ' 'superwind."

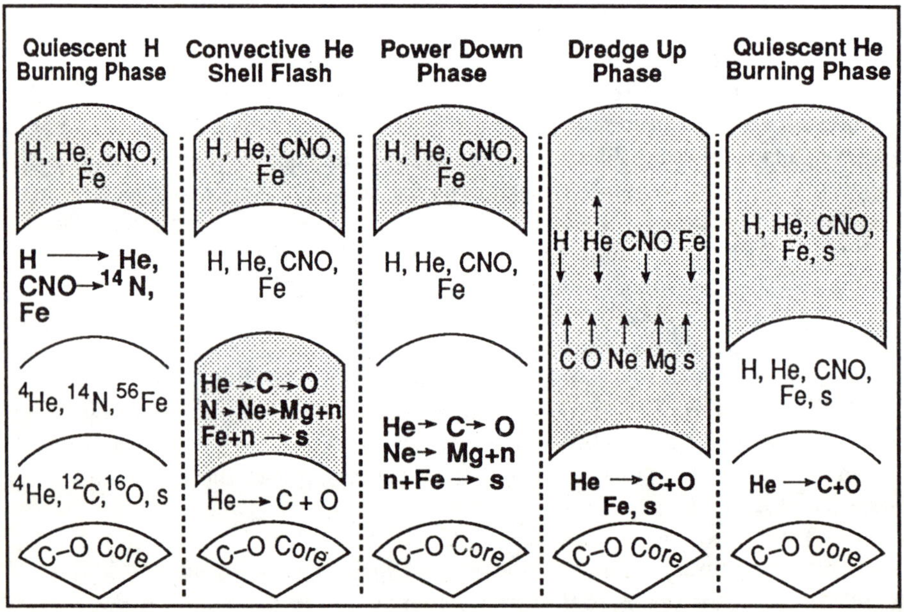

Fig. 1. Nucleosynthesis and dredge-up in an intermediate mass AGB star. This is a highly schematic representation that is not to scale in either the radial distance or mass coordinate, but instead emphasizes the ordering of dominant nuclear processes. The screened areas represent convective zones. The outer and H-rich convective zone extends all the way to the surface. Regions of nuclear burning are shown in bold type. (After Iben 1985.)

2.3 The Third Dredge-Up

The nuclear reactions comprising H-burning, He-burning, and accompanying reactions change the composition of the interior. These changes are observable only if the deep convective envelope can penetrate into the regions exposed to nuclear reactions. This penetration, if it occurs, leads to a phenomenon known as "the third dredge-up." Figure 2 depicts a successful dredge-up. (The first dredge-up occurs in all stars as they first become red giants or supergiants following the termination of the main sequence. The second dredge-up is predicted to occur in intermediate mass stars after exhausting of their He core as blue supergiants and on the return to the domain of the red giants). The S and other red giants are incontrovertible evidence that the third dredge-up occurs in nature; the *s*-process, as I sketch below, accompanies He-burning. The theoretical evidence for the third dredge-up is mixed!

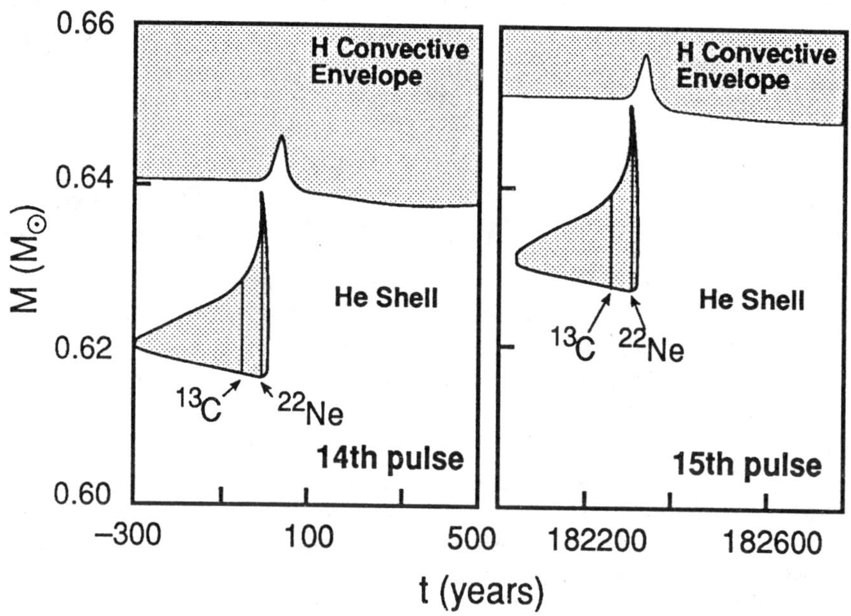

Fig. 2. Mixing regions as a function of time and activation of neutron sources during two thermal pulses in a low mass AGB model. Fully convective zones are denoted by screened areas. The times at which the ^{13}C and ^{22}Ne neutron sources are activated are indicated by the vertical lines within the convective shells (after Hollowell & Iben 1989.)

An opportunity for the third dredge-up exists during the brief interval when He burns quiescently. At this time the base of the convective envelope will descend into the top of the He-shell. For the observer, the critical issue is the depth of penetration into the He-shell. Freshly synthesized ^{12}C with a little ^{16}O and the products of the *s*-process were distributed over the lower parts of the He-shell by the convection associated with the thermal pulse. If the convective envelope can penetrate into these parts, now quiescent and not convective, ^{12}C and *s*-process products will be added to the envelope and so to the atmosphere. A successful third dredge-up is sketched in Figure 2. Repeated applications of the third dredge-up increase steadily the C-to-O ratio and the *s*-process enrichment of the atmosphere. The classical S stars such as those discovered by Merrill to contain Tc have a higher C/O ratio than less evolved red giants. When the C/O ratio exceeds unity, C-containing molecules (e.g., C_2, CN, HCN) dominate the spectrum of these "carbon" stars. Spectra of the S stars and their less evolved counterparts, the M stars, are dominated in the optical region by bands of oxides (e.g., TiO, ZrO, LaO). The switch at C/O = 1 from oxides to C-containing molecules is engineered by the CO molecule. The He abundance of the envelope is increased even if the dredge-up does not penetrate the layers that experienced the thermal instability. In contrast to C and *s*-process elements, the He enrichment of the envelope is quite minor. The reason for this is that C and *s*-process elements in the He-shell are overabundant by a factor of about 1,000 relative to their initial abundances in the envelope, but He is overabundant by the modest factor of 10 (by number of atoms). This is fortunate because the He abundance of the atmosphere is not directly measurable and, hence, it is reassuring to suppose the He abundance is not altered significantly by the third (or the preceding) dredge-up. Of course, the possibility exists that He-rich atmospheres may go undetected until almost all the H has been converted to He or removed!

A successful third dredge-up of ^{12}C and *s*-process elements was demonstrated theoretically for the more massive AGB stars by Iben (1975—see the comprehensive review papers Iben [1985, 1991]). In some quarters, these stars of about 3 to 8 solar masses are called intermediate mass (IM) stars. The *s*-process is driven by the neutron source ^{22}Ne (α, n) ^{25}Mg which is ignited in the He-shell at temperatures $T \gtrsim 3 \times 10^8$K. In the third dredge-up, the convective envelope skims off the top of the layers exposed to He-burning and the *s*-process. These products of nucleosynthesis are then mixed throughout the envelope including the surface. At the next thermal pulse, the remainder of the exposed layer in the He-shell is re-heated and re-exposed to neutrons; this re-exposure is characterized by an overlap factor often denoted by r. This sequence of re-exposures leads in a natural way after a few pulses to the dredged-up material being characterized by an exponential distribution of exposures to neutrons (Ulrich 1973). Such a distribution characterizes the solar/meteoritic abundances of *s*-process elements heavier than about Rb (the "main" *s*-process elements). In early calculations (Iben 1975) it was suggested that the ^{22}Ne neutron source and the predicted overlap factors led to a solar-like distribution of *s*-process elements in the envelopes of these massive AGB stars. Furthermore, the numbers of these stars and the mass of their envelopes which are ejected before the stellar cores die as white dwarfs appeared in combination to suggest that these stars are the Galaxy's principal supplier of the "main" *s*-process elements (Truran & Iben 1977). Unfortunately, this conclusion is now in doubt, in part because the competition to the ^{22}Ne neutron source from ^{22}Ne (α,γ) ^{26}Mg was probably underestimated. This serves as a reminder that quantitative stellar spectroscopists in pursuit of the sites of stellar nucleosynthesis look across the river to nuclear physicists as well.

In the low mass ($M \lesssim 3\ M_\odot$) AGB stars, the temperatures in the He-shell are too low during the thermal pulses for the ^{22}Ne neutron source to be very effective. The principal source is most likely to be $^{13}C(\alpha,n)^{16}O$. Questions remain as to how the necessary ^{13}C supply is generated; protons from the base of the convective envelope are mixed or diffused into the top of the He-shell during the long interpulse period of H-burning. Then $^{12}C(p,\gamma)^{13}N(e^-\nu)^{13}C$ provides the ^{13}C which is later ingested into the convective regions of the He-shell when He-burning resumes. Thanks to the paucity of protons, little ^{13}C is converted by proton capture to ^{14}N, as occurs in H-burning when protons are abundant. At present, the ^{13}C abundance at He-burning is uncertain theoretically. As Figure 2 implies, the ^{22}Ne source may be ignited following the ^{13}C source in low mass AGB stars, but the neutron flux from the ^{22}Ne source is relatively weak. Note that a comparable uncertainty about the initial ^{22}Ne concentration in the He-shell of an IM AGB star does not exist, because the ^{22}Ne abundance is set by the initial abundance of C, N, and O (and ^{22}Ne, a minor species) which by H-burning is converted to ^{14}N, and prior to He-burning ^{14}N is converted by 2 α-captures to ^{22}Ne.

The reader is invited to read some recent papers to see how the issue of ^{13}C seeding of the He-shell is approached in theoretical calculations: e.g., Boothroyd & Sackmann (1988), Hollowell & Iben (1989), Busso et al. (1992). To the observer there would seem to be two potentially observable consequences of the degree of ^{13}C seeding. One, a high seeding means a high neutron exposure at each pulse and relatively higher production ratio of the heavier to the lighter *s*-process elements (say La to Sr), as long as the overlap factor is unchanged. Second, the neutron density during *s*-processing is dependent on the rate at which ^{13}C is ingested into the He-burning convective layers. A high ^{13}C abundance tends to drive the *s*-process at a high neutron density. I discuss below the prospects for diagnosing the neutron density from analysis of the surface enrichments of key nuclides.

2.4 Hot Bottom Burning in Very Luminous AGB Stars

In the preceding look at nucleosynthesis in AGB stars, the focus was on the He-shell in which the ^{12}C and the *s*-process elements are made and whose presence in the atmosphere of an AGB star is often demonstrable from low resolution qualitative spectroscopy. H-burning at the base of the convective envelope also provides for nucleosynthesis. This contribution is potentially spectacular in two principal ways: an excess of C over O may be eliminated as ^{12}C is converted to ^{14}N and a carbon star is reconverted to a now N-rich S star, and a reservoir of ^{3}He is converted to ^{7}Li to create super Li-rich stars that are quite possibly a major source of Li in the Galaxy.

H-burning at the base of the convective envelope is a central feature of AGB evolution except in the short intervals when the He-shell is ignited. Although these episodes of H-burning convert some ^{12}C to ^{14}N and consume ^{3}He, no ^{7}Li is then or later injected into the convective envelope; the convective envelope does not penetrate the H-burning layer. The reason is very simple: the ^{7}Li is destroyed by protons as fast as it is produced. Also, the conversion of ^{12}C to ^{14}N is constrained by the fact that the mass in the H-burning layer is finite.

As an AGB star evolves to higher and higher luminosities, a structural change is predicted to occur. The H-burning layer and the convective envelope join so that products of nucleosynthesis in the H-burning layer are swept to the surface and fresh "raw materials" are converted into the layer for nuclear processing. This episode is referred to by various phrases: I shall label it "hot bottom burning" or HBB—see Uus (1973) and Scalo et al. (1975) for an early description, and Blöcker & Schönberner (1991) and Sackmann & Boothroyd (1992) for more recent calculations of evolution and/or accompanying nucleosynthesis.

The possibility that ^{7}Li may be synthesized in red giants was recognised by Cameron & Fowler (1971), who proposed the "^{7}Be-transport mechanism" in which ^{7}Li is created via the reaction chain $^{3}He(\alpha,\gamma)^{7}Be(e^{-},\nu)^{7}Li$, where the ^{3}He is a product of H-burning in the main sequence phase and the ^{7}Li (and ^{7}Be) is transported by convection to cooler layers to avoid destruction by protons; ^{7}Li destruction occurs very rapidly at the temperatures necessary to induce conversion of ^{3}He to ^{7}Be. The ^{3}He was created by the pp-chain of H-burning. ^{3}He production occurs *very* slowly because the initial reaction $p(p,e^{+}\nu)^{2}H$ proceeds via the weak interaction; the nuclear reaction $^{2}H(p,\gamma)^{3}He$ is much faster. ^{3}He is produced in significant quantities only in main sequence stars. Calculations show that, as the star evolves to become a red giant, the ^{3}He produced avoids destruction over a thick zone outside the H-burning core and later the H-burning shell. It is this reservoir of ^{3}He that is available to the HBB AGB star for processing to ^{7}Li. Since the reservoir is finite, the rise of the surface ^{7}Li abundance of an HBB AGB star may be followed by a decline as the ^{7}Li is destroyed on its return to the H-burning layer at the base of the convective envelope and replenishment is impossible owing to exhaustion of the ^{3}He reservoir. I comment below on Li-rich AGB stars and the role of molecular spectra in the analysis of atomic (Li I) lines.

3 Structure and Spectra of Molecules and Stars

3.1 Why Study Molecules in Stellar Photospheres?

To the easily intimidated, the association of atoms into molecules results in a spectrum of awful complexity. This complexity must be faced by the observer curious about AGB stars. Indeed, molecular spectra offer novel diagnostics of AGB behaviour not provided by the atomic lines. In this section, I shall note briefly the unique contributions of molecular spectra to the analysis of photospheric structure and composition and then turn to discussion of the data on molecular structure and data needed to effect the analyses of AGB stars. (I shall not discuss the chromospheres and circumstellar shells of AGB stars, see reviews by Olofsson, Hinkle, and by Sedlmayr, this volume, on that subject). Not unexpectedly, my emphasis is on the roles of molecules in determining photospheric chemical compositions in order to address questions concerning stellar evolution and nucleosynthesis. Applications of molecules to the observational definition of the physical properties of stellar photospheres are important but are not covered in detail. Molecular spectra provide information on the chemical composition and the physical structure of the photosphere. These two aspects are coupled, as is well shown by Jørgensen (this volume). I shall not discuss this coupling except to remark that the molecular line blanketing, which is dependent on chemical composition, is a principal determinant of the (theoretical) structure of an AGB star's photosphere. Specification of the blanketing calls for much of the molecular data needed in the analysis of photospheric spectra.

Molecular spectra may be used to estimate the temperature and pressure of the photosphere. In practice, their use as barometers is rare. Here, one seeks to use molecules in an analogous way to the use of atomic and ionic spectra (say, Fe I and Fe II lines) with the Saha ionization equilibrium to derive the stellar surface gravity; for example, one might use the infrared OH rotation-vibration lines with the forbidden [O I] 6300 and 6363Å lines and dissociation equilibrium to derive the surface gravity of an AGB star (the assumption of hydrostatic equilibrium links the gas pressure affecting the dissociation equilibrium to the surface gravity). In practice, this application of molecular spectra is very difficult for AGB stars because the necessary available atomic lines are few in number and often severely blended. Applications are possible for warmer red giants where molecular lines are less numerous. In AGB stars, one might envision applications of diatomic with triatomic molecules as barometers, say OH and H_2O in O-rich stars or CN and HCN in C-rich stars.

Spectroscopists have long used the measured equivalent widths of lines of an atom (or ion) to derive an excitation temperature (T_{exc}) for that species. If LTE is valid, this T_{exc} is obviously a weighted mean of the kinetic temperature over the layers contributing to the lines' formation. In recent years, comparison of the measured T_{exc} and predictions from model atmospheres has been used to derive a star's effective temperature (T_{eff}). Space prohibits a thorough critique of the atomic lines as a stellar thermometer. While valuable and valid estimates of T_{eff} are provided for many types of stars, there are often severe limitations to this application of atomic lines. A majority of the limitations are rooted in fundamentals of atomic structure and spectra.

Use of atomic or molecular lines as a thermometer providing T_{exc} and, thence, T_{eff} requires that there be available an adequate collection of weak lines sampling a range in excitation potential, possessing accurately known *gf*-values, and drawn from levels whose populations show no significant departures from LTE. The latter constraint is cited because, with few exceptions, predictions of non-LTE populations for atoms in stellar atmospheres have not advanced beyond a level that I call "semi-quantitative" and do not permit the T_{exc}(NLTE)–T_{eff} relation to be predicted. This is especially true for cool stars and those atomic species frequently adopted as thermometers (e.g., Fe I, Ti I). My list of qualitative conditions could be translated to quantitative conditions embracing the number of lines, the range in excitation potential, the accuracy of the *gf*-values, and the precision of the stellar equivalent widths.

My intent is not to explore the quantitative accuracy of T_{exc} estimates, but to point out that molecules offer inherent advantages over atoms. Molecules are more practical thermometers than atoms. In particular, the rotational structure of a given band of an electronic or rotation-vibration transition provides a set of lines whose relative f-values may often be fixed to satisfactory accuracy from a rotational line strength (Hönl-London factor) and a quantity describing the rotation-vibration interaction. The range in excitation potentials, which controls the sensitivity of the thermometer, may be extended by combining bands provided that their relative f-values are known. Preliminary theoretical studies indicate that the rate constants for rotational and vibrational excitation and de-excitation through H, He, H_2, and e collisions dominate the statistical equilibrium of the ground and, possibly, the excited vibrational states of a molecule resident in the photosphere of a red giant (Thompson 1973; Hinkle & Lambert 1975); i.e., the local excitation and kinetic temperatures are identical or almost so.

By contrast, atomic spectra are far less attractive thermometers. An atomic line list spanning an adequate range in excitation potential will necessarily involve different terms which are most probably connected by strong permitted radiative transitions and, hence, potentially subject to non-LTE effects. A satisfactory line list may also span a broader wavelength interval than the molecular alternative. In general, the available supply of accurate atomic f-values will be found wanting. Then, in the case of cool stars, the supply of atomic lines will be depleted because several to many potential lines will be masked by strong molecular absorption (the high density of molecular lines will also mean that a molecular thermometer may be compromised by the paucity of unblended weak molecular lines).

As a final note on molecules and the atmospheric structure, I mention the use of molecules as probes of the photospheric velocity field—see, for example, the analyses by Tsuji (1988) of the CO 2.3μm rotation-vibration lines in the spectra of M giants showing the radial velocity to vary systematically with a line's excitation energy. These shifts and line asymmetries show that "a pack of low excitation CO lines should be originating in an extra molecular component, not yet expanding at large velocity, but well distinct from the photosphere" (Tsuji 1988). As Tsuji points out, this zone revealed by the CO lines may be the environment in which dust formation occurs and mass loss is initiated by radiation pressure on the grains.

To date, the contributions of observed molecular lines to our understanding of the chemical composition exceed the insights obtained into the physical structure of AGB stars. I divide these prelusive remarks into two parts: isotopic ratios and elemental abundance ratios.

Three concise statements suffice to show why molecular lines in stellar spectra are analysed for the isotopic abundance ratio of one or more of the molecule's constituent atoms:

1. Often, the isotopic ratios address astrophysical questions of great import,

e.g., stellar nucleosynthesis of the chemical elements.

2. Often, ratios of high accuracy are obtainable with limited knowledge of the atmospheric structure and a partial description of the basic molecular data, e.g., the $^{12}C/^{13}C$ ratio may be extracted from CN lines without a precise value for the molecule's dissociation energy whose persistent uncertainty afflicts all attempts to use the CN lines as monitors of either the C or the N elemental abundances.
3. In general, the detectable molecules in stellar spectra provide isotopic ratios for those atoms for which the isotopic atomic wavelength shifts are too small to be resolved in stellar spectra. This fortunate circumstance arises because the nuclear masses have a direct influence on a molecule's rotation-vibration energy levels.

A combination of isotopic *with* elemental abundance ratios is often essential in addressing the key astrophysical questions. In quantitative analyses of warm/solar stars, molecular spectra may supplement the data on elemental abundance ratios provided by atomic spectra. In the cooler stars such as the inhabitants of the AGB, however, molecular spectra may be the primary or even the sole source of elemental abundance ratios. Certainly, for the CNO trio whose abundances reflect the presence of H- and He-burning products in the stellar atmosphere, molecules are the key to the abundances. Note too that a simultaneous analysis of two or more indicators may be required for a single elemental abundance. Consider, for example, the C-rich AGB stars. Carbon monoxide and N_2 largely control the partial pressures of free C, N, and O and, hence, of molecules containing these elements. While the CO lines are a direct measure of the O abundance $\epsilon(O)$, the C_2 lines reflect the difference $(\epsilon(C) - \epsilon(O))^2$ and, hence, the C abundance is set by the combination of CO and C_2 lines. An obvious extension of this to CN shows that the N abundance requires a simultaneous fit to CO, C_2, and CN lines. A corollary is obvious: errors in the basic data (f-values and dissociation energies) for a molecule may affect the abundance of more than one element, and the effects may not be equal. For example, an error in the f-value of the C_2 lines contributes the following errors:

$$\Delta \log(\epsilon(C) - \epsilon(O)) \sim 0.5 \Delta \log f(C_2)$$

$$\Delta \log \epsilon(N) \sim \Delta \log f(C_2)$$

where the CN lines are the source of the N abundance and "all" of the N is supposed to be associated into N_2 (i.e., $p(N_2) \propto \epsilon(N)$ and $p(N)^2 \propto p(N_2)$, hence $\epsilon(N) \propto p(CN)^2$ where p denotes a partial pressure).

The spectrum of an AGB star may provide novel information for the molecular spectroscopist because the star's photosphere is a spectacular light source whose composition, column density, and combination of low pressure and moderate temperatures cannot be realized in the laboratory. As evidence, I cite our discovery of ZrS bands in 7,000–10,000Å spectra of S stars (Lambert & Clegg 1980) ahead of laboratory detections of these bands. The necessity for laboratory work is well shown by the recent work on TiS using a King-type furnace (Jonsson et al. 1992). The TiS bands from 1.05 to 1.35μm were shown to account quite well for bands previously attributed (Hinkle et al. 1989) to ZrS on the basis of wavelengths predicted by *ab initio* calculations (Langhoff & Bauschlicher 1988). Analysis of high resolution stellar spectra will now be needed to show if the stellar band are a mix of TiS and ZrS lines. Before this can be done, the predicted infrared bands of ZrS will have to be detected in the laboratory.

3.2 Molecules as Irritants and Stimulants

Civil servants in the villages on either side of the bridge would assuredly suppose that the interests of the astronomers in one village and of the physicists and chemists across the river in the other village would be efficiently served if the astronomers compiled a list of their needs for basic molecular data, and if the physicists and chemists prepared a catalogue of available data with a supplement describing products that could be made on special request. Then, messengers would be detailed to meet at the mid-point of the bridge and to exchange the list and the catalogue. In my experience such an impersonal exchange is rarely fruitful. Therefore, I decline to offer a list of current needs for basic molecular data. I elect to present a few examples from our recent pursuit of AGB stars where molecular spectra either interfered with the primary atomic line in the investigation ("molecules as irritants") or the molecular lines were themselves the key to a fundamental question of stellar evolution and nucleosynthesis ("molecules as stimulants"). It will be clear that readily available molecular data are often inadequate for the purposes of the stellar analyses.

Lithium Synthesis by HBB AGB Stars. Lithium production by very luminous AGB stars with an H-burning occurring at the base of the convective envelope may be a major contributor to the lithium budget of the Galaxy. Our recent observations of AGB stars in the Small and Large Magellanic Clouds showed, as predicted, that Li-production is confined to the most luminous AGB stars (Smith & Lambert 1989, 1990; Smith et al. 1993). This result could not be demonstrated from the well-known and brighter examples of super Li-rich AGB stars in our Galaxy because their distances and, hence, their luminosities are ill-determined. For the Magellanic Clouds, our nearest neighbours among galaxies, many independent and generally concordant methods have been applied to the measurement of their distances. As the distances of the AGB stars in the Clouds are known, reliable luminosities may be estimated. In order to assess the role of these AGB stars as lithium producers, we need to estimate the Li abundance of their photospheres from analysis of the Li I lines (molecules containing Li—e.g., LiH—are unlikely to be present in detectable quantities).

Inspection of Fig. 3 shows the obvious presence of a bandhead of the TiO $\gamma(1-0)R_1$ band. The Li I 6707Å resonance doublet is in a region crossed by TiO lines from, for example, the R_2 and R_3 branches of the $\gamma(1-0)$ band. This interference from TiO lines must be taken into account. TiO lines are of such a high density in this spectral region that lines overlap to simulate a quasi-continuous opacity that depresses the apparent continuum level. A few stronger

TiO lines may also be blended with the Li I 6707Å feature, and they too must be taken into account. Today, the standard tool for such an abundance analysis is the synthetic spectrum in which a comprehensive list of atomic and molecular lines is combined with a model photosphere and the predicted spectrum matched to the observed spectrum.

An example of a fit of a synthetic spectrum to an observed spectrum is shown in Figure 4. The reader may find the details of this analysis in Plez et al. (1993). Here, I wish to draw attention to the depression of the local continuum: the relative flux near the Li I feature is approximately 0.4, where the continuum flux in the absence of lines is at 1.0. The 60% depression of the continuum is due to TiO lines which populate this region at a line density of several hundred lines per Å.

Many of these lines have not been observed in laboratory spectra, but their wavelengths are predicted from molecular constants and their oscillator strengths estimated from a few available radiative lifetimes (see Schamps this volume). Indeed, except through selective laser excitation the individual lines will not be observable in laboratory spectra; they will overlap to form quasi-continuous emission or absorption just as their stellar counterparts do. This example shows clearly that stellar demands may outstrip the present capabilities of laboratory spectroscopy. And TiO is a molecule extensively investigated in the laboratory!

Correct representation of the TiO blanket is acute for lithium because in many red giants, the 6707Å feature is the sole evidence for lithium. In the case of iron and other abundant elements with rich atomic spectra, one may select the spectral region so that molecular line blanketing is minimized. Happily, the Li abundance of these Li-rich stars is sufficiently high that a subordinate line at 8126Å is seen in a region of weak blanketing; the local continuum is depressed by only 5 to 10% and not 60% as at 6700Å. Since the 8126Å Li I line is weak, it is important to identify completely the few contributing TiO (and other) blending lines. The available laboratory spectroscopy appears adequate for this purpose (Fig. 5).

Rubidium and Neutron Densities in Thermally Pulsing He Shells. As a second example of "molecules as irritants" I cite the case of the Rb I 7800Å resonance line and the use of the Rb abundance to measure the neutron density in the thermally pulsing He-shell that drives the *s*-process. Rubidium shares with its fellow alkali lithium the characteristic that it is of low abundance and the neutral atom has a low ionization potential. Such characteristics mean that generally only the strongest resonance lines are potentially detectable. For Rb I these lines are at 7800 and 7947Å. In spectra of O-rich AGB stars, the 7947Å line is irretrievably blended. By thorough accounting of the molecular lines (mostly TiO) and a blending Si I line it is possible to extract the Rb abundances from the 7800Å line. Our early work on Rb was done on Barium giants (Tomkin & Lambert 1983; Smith & Lambert 1984; Malaney & Lambert 1988) which are considerably warmer than the AGB stars. Analyses of Barium giants are relevant to studies of AGB stars because these giants are created when

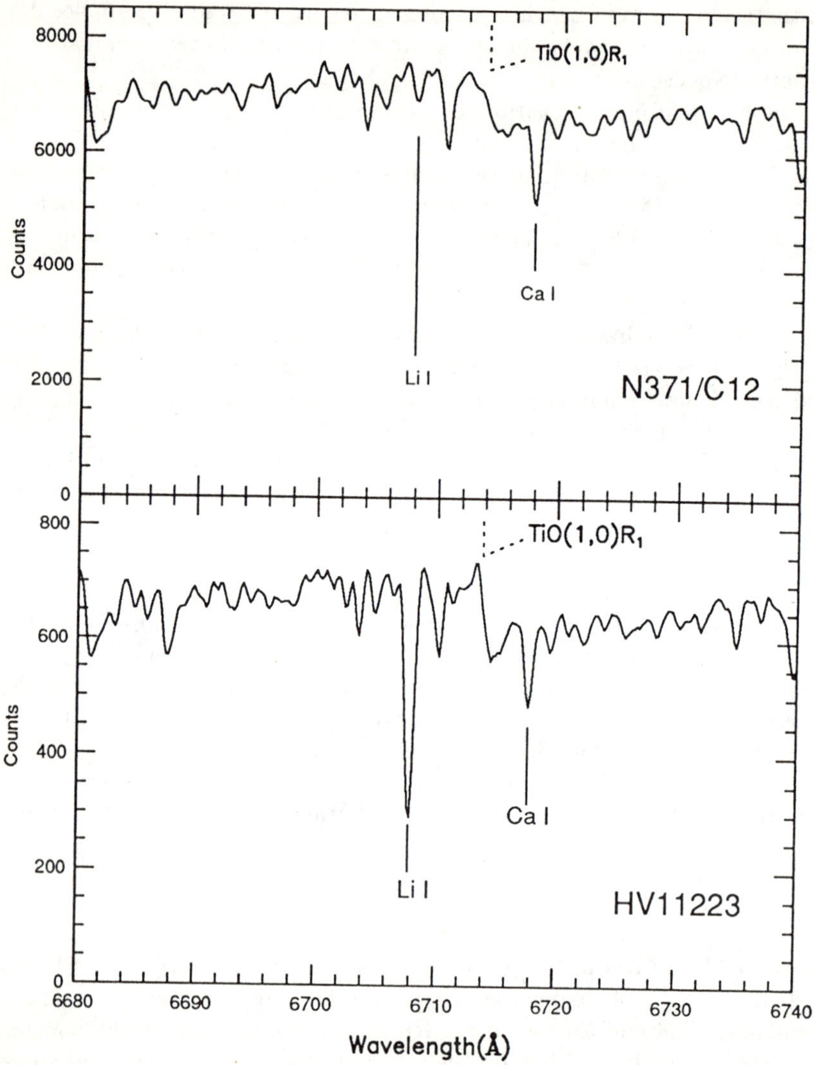

Fig. 3. Spectra of an AGB star HV11223 and a supergiant N371/C12 in the small Magellanic Cloud. N371/C12 is a cool massive star with a temperature and surface gravity similar to that of the AGB star. Note the general similarity of the spectra; the TiO (1–0)R_1 band and the Ca I line have quite similar strength in the two stars. The one striking difference is the strength of the Li I 6707Å feature in the AGB star.

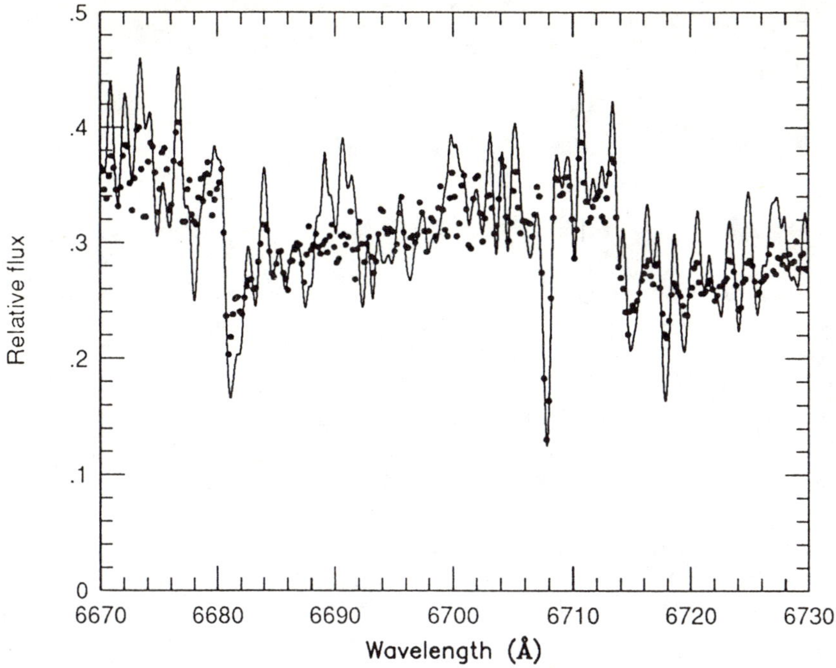

Fig. 4. Observed (dots) and calculated (line) spectrum of the AGB star HV1963. Two TiO 1–0 R bandheads are clearly seen. Note the continuum in the absence of lines is at a relative flux of 1.0.

an AGB star transfers large amounts of mass to a companion to convert it to the Barium star. The C-O core of the AGB star remains as a white dwarf in orbit around the Barium star. Since the Barium giant and its main sequence predecessor have warm photospheres, molecular contamination of the spectra is greatly reduced permitting analyses that are impossible for AGB stars. Recently, we have completed analyses of Rb in a sample of O-rich AGB stars.

Why is Rb of especial interest? To answer this question we must examine the path taken by the *s*-process along the valley of nuclear stability. In the limit that the neutron density N(n) driving the *s*-process tends to zero, the *s*-process takes a unique path; I ignore here the possibility that the temperature and electron density may influence the path. The uniqueness results from the fact that the first n-rich unstable nuclide of an element decays rather than captures an additional neutron. At the neutron densities expected in the thermally pulsing He-shell, however, an additional neutron capture may occur before an unstable nuclide decays. This competition between neutron capture and (usually) β-decay creates a potential branchpoint in the *s*-process path. For each branchpoint, there is a critical value of the neutron density $N(n)_c$: if $N(n) \ll N(n)_c$, the unstable nuclide at the branchpoint decays, but if $N(n) \gg N(n)_c$, the nuclide experiences a neutron capture. This difference between the low and high neutron density

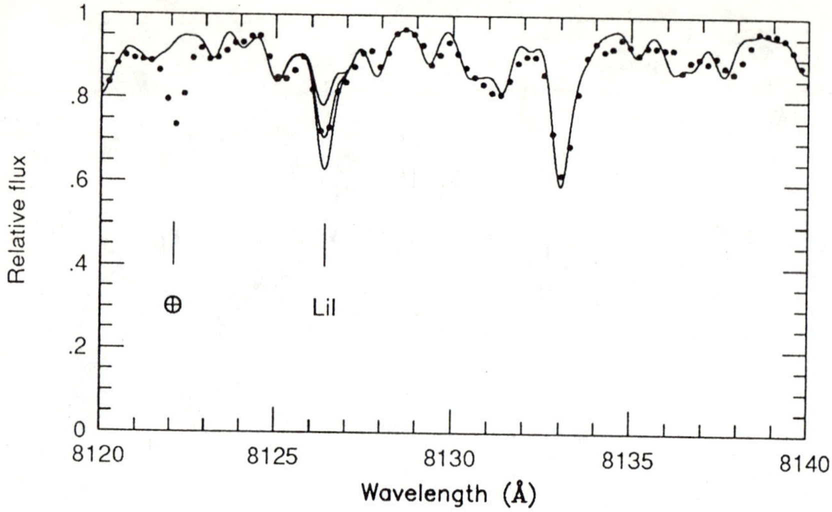

Fig. 5. The observed (dots) and synthetic spectrum (line) of HV1963 near the Li I 8126Å line. Synthetic spectra are shown for three Li abundances: the best fit and ±0.5 dex from that fit. The observed feature at 8122Å is due to terrestrial water vapor.

paths affects the yields from the *s*-process of nuclides immediately following the branchpoint. Then, there is the prospect of measuring the photospheric abundances of the affected nuclides and, hence, inferring the neutron density that directed the *s*-process. Just two branches are available for use with AGB stars:

1. the branch at unstable ^{85}Kr that determines the Rb abundance relative to its neighbours Sr and Y.
2. the branch at unstable ^{95}Zr that controls synthesis of the stable isotope ^{96}Zr.

Here, I comment further on ^{85}Kr. The interested reader may find discussion of ^{95}Zr and of the detection of isotopic bandheads of ZrO in S stars in Smith (1988) and Lambert (1988).

Operation of the ^{85}Kr branchpoint, which is illustrated in Fig. 6. , is complicated slightly by the fact that ^{85}Kr has a short-lived state which reduces the effectiveness of the branch run by ^{85}Kr with a long-lived ground state. A benefit resulting from the complication is that the influence of ^{85}Kr on the *s*-process path is sensitive to the duration of the neutron burst in the thermal pulse. The potential branch at ^{86}Rb plays a minor role. Beer & Macklin (1989) provide accurate measurements of the ^{85}Rb and ^{87}Rb neutron capture cross-sections and discuss operation of the ^{85}Kr branchpoint in AGB models.

Exploitation of the branchpoint for AGB stars must rely on its effect on the Rb abundance. In the limit $N(n) \rightarrow 0$, the *s*-process path is through ^{85}Rb, ^{86}Rb to ^{86}Sr. In the high density limit, the path runs through ^{86}Kr to ^{87}Rb and ^{88}Sr. Thus, the elemental abundances of Rb relative to Sr or Y is sensitive to neutron densities around $N(n)_c$. Thanks to the fact that the neutron capture cross-sections of ^{85}Rb and ^{87}Rb (see Fig. 6) differ by a factor of about 20, the Rb/Sr and Rb/Y ratios vary by more than an order of magnitude between the low and high N(n) limits.

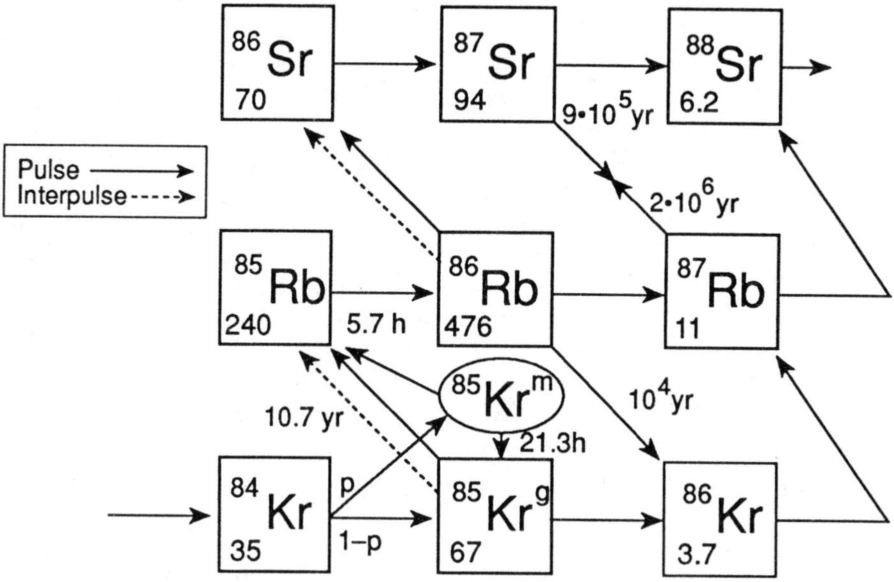

Fig. 6. Synthesis of Kr, Rb, and Sr isotopes during a thermal pulse (solid line) and in the interpulse phase (dashed lines) when the ^{85}Kr and ^{86}Rb which are built up decay. The population P of the isomeric state in ^{85}Kr is generated by neutron capture on ^{84}Kr ($P = 0.49 \pm 0.06$). The time scales for ^{87}Rb and ^{87}Sr β-decay during the pulse are too long to be of any significance for the synthesis. The branching of ^{86}Rb to ^{86}Kr can be neglected too. (After Beer & Macklin 1989.)

Our analyses of Rb in Barium giants and directly in O-rich AGB stars (Lambert & Smith 1993) have shown no case indicative of a high N(n). Beer & Macklin (1989) predicted Rb/Sr ratios for the neutron pulses expected for intermediate mass AGB stars where $^{22}Ne(\alpha, n)^{25}Mg$ is the neutron source. These pulses have a high N(n) ($\gtrsim 3 \times 10^9 cm^{-3}$) and a rapid termination of s-process and a Rb/Sr ratio is predicted in excess of that observed. Beer and Macklin show that pulses with a peak N(n) $\sim 10^8$ to 10^9 cm^{-3} and pulse lengths $\Delta t \simeq 10$ yr. are required to fit the observations. Pulses of this kind have been engineered for low mass AGB models where $^{13}C(\alpha, n)^{16}O$ is the leading neutron source. We identify the stars examined by us for Rb as low mass AGB stars and are continuing to search for intermediate mass AGB stars rich in Rb (principally ^{87}Rb) and ^{96}Zr.

MgH Bands and Mg Isotopic Ratios. In the chase for the Li and Rb abundances of AGB stars, the TiO molecule is an "irritant" (ZrO serves as a stimulant when it is used to derive the abundance of ^{96}Zr). Here, I comment on the role the MgH $A^2\Pi - X^2\sum^+$ green system plays in the identification of the neutron source driving the *s*-process in the He-shell's thermal pulse. To the observer, MgH in this example is a "stimulant."

Two neutron sources are likely candidates: $^{22}\mathrm{Ne}(\alpha,n)^{25}\mathrm{Mg}$ for $T \simeq 3 \times 10^8$K and $^{13}\mathrm{C}(\alpha,n)^{16}\mathrm{O}$ for $T \simeq 1.5 \times 10^8$K. The higher ignition temperature for $^{22}\mathrm{Ne}(\alpha,n)$ means that it runs the *s*-process in intermediate mass AGB stars. $^{13}\mathrm{C}(\alpha,n)$ is the primary source of neutrons in low mass AGB stars. The ^{22}Ne source cannot easily run the *s*-process and elude the stellar spectroscopist because it leads to a non-solar mix of the Mg isotopes in the He-shell and, after dredge-up of *s*-process products, in the photosphere of an IM-AGB (Truran & Iben 1977; Scalo 1978; Malaney 1987); in a typical case, the photospheric ratios of ^{24}Mg:^{25}Mg:^{26}Mg may be 1:4:4 for a factor of 10 enrichment of *s*-products at the surface instead of the star's initial ratios (presumed close to solar) of 8:1:1. The MgH lines provide an opportunity to measure the isotopic ratio. Operation of the ^{13}C neutron source makes no significant alteration to the Mg isotopic ratio, and the additional amount of ^{16}O synthesized in the He-shell is negligible relative to that from He-burning through $^{12}\mathrm{C}(\alpha,\gamma)^{16}\mathrm{O}$.

Examination of MgH lines in about 15 Barium stars and several MS and S stars (see Lambert 1991 for references) shows none with the ^{25}Mg and ^{26}Mg enrichments expected from the ^{22}Ne source. The conclusion is that the examined stars are low mass AGB stars. It will be of interest to search for MgH lines in spectra of the Magellanic Cloud's Li-rich AGB stars whose masses according to pulsational calculations are high enough for the ^{22}Ne source to run the *s*-process. Perhaps the test should be applied to the immediate progenitors of these stars (i.e., AGB star not yet experiencing HBB), because there is a possibility that excess ^{25}Mg and ^{26}Mg in the convective envelope are removed by protons: $^{25}\mathrm{Mg}(p,\gamma)^{26}\mathrm{Al}$ and $^{26}\mathrm{Mg}(p,\gamma)^{27}\mathrm{Al}$. It appears, however, that significant Li-production (and preservation) occurs only for temperature in the hot bottom too cool to destroy ^{25}Mg and ^{26}Mg. Conversely, if ^{25}Mg and ^{26}Mg are destroyed in very hot bottomed envelopes, Li is not added to the envelope and the atmosphere in detectable amounts.

C, N, O Abundances and Infrared Spectroscopy. Stellar spectroscopists were long confined to the spectral regions detectable with photographic plates. Exploration of nucleosynthesis occurring in AGB stars was severely compromised by the lack of access to infrared spectra. One may cite three principal reasons for exploring the infrared (here, 1.0 to 5μm):

- The AGB stars are cool and, hence, emit most of their flux in the infrared.
- The line density in infrared spectra may be lower than in visual spectra.
- The infrared is essential to the determination of the C, N, and O elemental and isotopic abundances that monitor the mixing of H- and He-burning products into the atmosphere.

The third point deserves amplification in the message carried across the bridge from the astronomical village to the other side. To highlight the need for infrared spectroscopy, I shall compare the indicators of the C, N, and O abundances available for the G–K giants and the AGB stars.

At the photospheric temperatures of G–K stars, molecules mark but do not deface the visual spectrum. C, N, and O abundances are obtainable from visual-to-near infrared ($\lambda \lesssim 9000$Å) spectra. Lines that have been used include the C_2 Swan system's $\Delta v = 0$ and -1 bands, the CN Red system $\Delta v = 2$ to 4 bands, and the [O I] 6300 and 6363Å lines (see for example, Lambert & Ries 1981; Kjærgaard et al. 1982). Infrared lines could (and should) also be used: the CO rotation-vibration (or rovibrational for short) $\Delta v = 2$ and 3 lines, the CN Red system's $\Delta v = 0$ and -1 bands, and the OH rovibrational $\Delta v = 1$ and 2 bands.

In the case of the O-rich AGB stars, the C, N, and O indicators in the visual are lost in a wealth of molecular lines (the CN Red system is often detectable among the other lines). Fortunately, the infrared remains a source of C, N, and O indicators. In addition to the list cited above for G–K giants I would add the NH rovibrational lines. In the coolest stars, one should add the H_2O rovibrational lines.

For C-rich AGB stars, the infrared is similarly of paramount importance. The principal indicators are the CO rovibrational $\Delta v = 2$ and 3 lines, the CN Red system $\Delta v = -2$ and -3 lines, the C_2 Phillips $\Delta v = -2$ and Ballik-Ramsay $\Delta v = 0$ lines, the CH and NH rovibrational $\Delta v = 1$ lines and, in the cooler stars, the rovibrational lines of HCN and C_2H_2.

Since the atmosphere of an AGB star is so extensive, the stellar spectra provide a more complete representation of these various transitions than has been obtained in the laboratory. In addition, the absolute *gf*-values of the transitions are needed. Much new data have been supplied in recent years, principally from calculations. But, as the discussion of CN (Costes & Naulin this volume) see below too—indicates, some basic astrophysical needs remain unfulfilled. Laboratory infrared quantitative spectroscopy and quantum chemistry calculations of infrared transitions on astrophysically relevant molecules are to be encouraged.

4 What Does the Village Store Need?

This review, written to intrigue the villagers in the physical-chemical community across the river, began with a description of the contributions of AGB stars to nucleosynthesis and then turned to portraits of selected problems in which stellar spectroscopists have drawn on molecular spectra in pursuit of answers. In these pursuits, the incompleteness of the available data on the molecules of interest may become apparent and, in some cases, may severely limit the precision of the answers provided to the astrophysical questions. In the final section, I comment on two topics that deserve continuing attention by the physical-chemical community. The astrophysical village store needs to be supplied with the "products" that I describe now.

4.1 The Dissociation Energy of CN

A red giant's photospheric N abundance is a tracer of the photosphere's contamination with the product's of H-burning. Recall that the CNO-cycle H-burning reactions turn H to He as they convert "all" initial C, N, and O nuclei into ^{14}N nuclei, i.e., H-burning creates N-rich material. Evidence of this material at the surface of red giants is predicted and observed following the first and second dredge-ups. A HBB AGB star necessarily achieves further N enrichment. A full characterization of the N enrichment demands accurate measurement of the photospheric N abundance.

For all red giants except the very coolest, the CN molecule is effectively the sole indicator of the N abundance (of course, the N abundance cannot be obtained from CN alone but from combinations of CN with other C-containing molecules and even O indicators, as the C and O abundances are linked through CO molecule formation). The NH $A-X$ 0–0 band near 3360Å is unattractive as an N indicator because the red giants emit little flux in the ultraviolet and the spectrum is very crowded with lines. In the coolest red giants, the NH rovibrational lines near 3μm may be exploited. In short, knowledge of a giant's N abundance depends heavily on the reliability of the basic data for CN. One item required is the molecule's dissociation energy D_0(CN). Our village store needs a "new improved" product, as I shall now sketch.

I consider the extreme case of the cool carbon stars where nitrogen atoms are almost fully associated in N_2 molecules, and O is similarly associated into CO molecules. Then, the number density of CN molecules is related to the elemental C, N, and O abundances (here denoted by ϵ) as follows:

$$n(\mathrm{CN}) \propto (\epsilon_c - \epsilon_o)\epsilon_N^{\frac{1}{2}}$$

where $(\epsilon_c - \epsilon_o)$ has to be determined from C_2 lines. This relation shows that $\log \epsilon_N \propto 2 \log n(\mathrm{CN})$. The dissociation energy enters into the breakdown of n(CN) between the N abundance and the temperature:

$$n(\mathrm{CN}) \propto 10^{D_0^0 \theta}$$

where $\theta = 5040/T$ and the temperature dependent partition functions are ignored. For cool carbon stars $\theta \simeq 2$. Then, an uncertainty ΔD_0^0 (in eV) contributes an abundance error $\Delta \log \epsilon_N \simeq 4 \Delta D_0^0$ where, for example, $\Delta D_0^0 \simeq 0.25$ eV results in $\Delta \log \epsilon_N \simeq \pm 1$.

Since the total conversion of solar C, N, and O to N changes the N abundance by about 1.2 dex, it is clear that an uncertainty of ± 1 dex severely limits our ability to detect even large additions of H-burnt material to a red giant's photosphere. The problem is alleviated for warmer red giants such as the K giants, which have experienced the first dredge-up because the N_2 molecules are not a major sink of N and the temperatures are warmer. Nonetheless, the need for an accurate value of D_0 remains acute. The need can be met to some extent by the stellar spectroscopists' skill in devising differential methods of analysis whose results may be nearly independent of the adopted value of D_0(CN).

To answer the question "What is the "best" value of D_0(CN)?", a search of the literature would be made by the astrophysical villager. Fortunately, Huang et al. (1992) have listed published estimates of the heat of formation of CN (ΔH_f(CN)) obtained from 1962 to 1990 from a wide variety of experiments including analysis of solar CN absorption lines and *ab inito* quantum calculations. Note that $D_0\mathrm{CN} = \frac{1}{2}D_0^0(\mathrm{N}_2) + \mathrm{L(C)} - \Delta\mathrm{H}_f(\mathrm{CN})$, where L(C) is the heat of sublimation of graphite and $D_0(\mathrm{N}_2) = 9.760 \pm 0.005$eV and L(C) = 7.353 ± 0.02eV are established values (see JANAF Tables, Chase et al. 1984) Inspection of Huang et al.'s list shows the experimental values range from $D_0(\mathrm{CN}) = 7.66 \pm 0.05$eV to 7.90 ± 0.08eV. Although the range is less than my example of ± 0.25eV, it nonetheless is greater than the errors assigned to these (and most other entries) in the list.

Happily, the more recent entries do appear to be converging. Huang et al. (1992) estimate $D_0(\mathrm{CN}) = 7.72 \pm 0.03$eV from photodissociation of C_2N_2. Costes et al. (1990—see also Costes & Naulin this volume) obtained $D_0(\mathrm{CN}) = 7.77 \pm 0.04$eV from analysis of the energy of the CN molecules produced from C + NO reactions.

The temptation to adopt $D_0(\mathrm{CN}) = 7.75 \pm 0.04$eV is probably strong. However, not only do some experiments suggest a lower value but *ab initio* calculations empirically corrected from calculations on similar molecules with well determined dissociation energies also suggest lower values: $D_0(\mathrm{CN}) = 7.53$eV (Larsson et al. 1983, as quoted by Huang et al. 1992) and $D_0 = 7.65 \pm 0.06$eV (Bauschlicher et al. 1988).

Recalling the case of the cool carbon stars ($\Delta \log \epsilon_{\mathrm{N}} \approx 4\Delta D_0^0(\mathrm{eV})$), one would like to get $\Delta D_0^0 \lesssim 0.05$eV so that the N abundance uncertainty from D_0 (CN) alone is less than 50%. The suggestion that the recent experiments have reached this precision is not supported by refined *ab initio* calculations. Are the experiments correct? Are the *ab initio* calculations less reliable than estimated? Can a spectroscopic method be devised to provide—finally— an accurate value of D_0(CN)?

Analyses of stellar CN lines use the red ($A - X$) system to the almost total exclusion of the violet ($B - X$) system. These analyses require both an estimate of D_0(CN) and the f-values of the appropriate red system bands. As for D_0(CN), estimates of the f-values have shown a dismayingly large scatter. Larsson (this volume) reviews the status of measurements and *ab initio* predictions of the radiative lifetimes of the vibrational levels of the $A^2\Pi$ state. The "best" experimental lifetimes ($\tau_{v'}$) including a laser-induced fluorescence experiment (Lu et al. 1992) give $\tau_{v'}$-values that are consistently shorter than the most recent *ab initio* predictions (e.g., Bauschlicher et al. 1988): the difference amounts to about 30% at $v' = 0$ but a factor of 3 for $v' = 7$. This discrepancy too deserves continued scrutiny by experimentalists and theoreticians. The variation of $\tau_{v'}$ with v' or equivalently the run of band f-values across the red system may be well determined from analysis of the solar CN absorption lines Sauval et al. (1994).

4.2 Excitation by H Atoms?

The assumption of Local Thermodynamic Equilibrium (LTE) so simplifies the analysis of stellar spectra and the construction of model stellar atmospheres that there is an understandable reluctance to abandon it in favor of the less restrictive assumption of Steady State Statistical Equilibrium or non-LTE. Under LTE, all local properties (excluding the radiation field) assume the values attained in thermal equilibrium at the local kinetic temperature. The power of LTE is that it reduces greatly the atomic and molecular data needed to construct model stellar atmospheres and to employ those atmospheres in the analysis of stellar spectra. In particular, LTE restricts the kind of data needed to lists of classified lines, their *gf*-values, and the photoionization (photodetachment) cross-sections for bound-free (and free-free) transitions contributing to the continuous opacity.

Departures from LTE *must* exist. Then, the LTE assumption that level populations may be calculated from the Boltzmann-Saha equation is dropped, and level populations are obtained from a series of rate equations that in turn are coupled to equations of transfer. If N(i) is the population of state i, the rate equation for N(i) is

$$\frac{dN(i)}{dt} = 0 = -\sum_i N(i)(R_{ij} + C_{ij}) + \sum_j N(j)(R_{ji} + C_{ji}).$$

The level is considered to be depopulated by radiative processes (rate constants R_{ij}) and collisional processes (rate constants C_{ij}). The level i is repopulated by radiative and collisional processes from other levels j. The set of levels includes various stages of ionization; for example, a non-LTE study of Fe I should consider radiative and collisional coupling of Fe I and Fe II levels and, hence, an adequate treatment will require a model Fe atom and a Fe^+ ion. In cool stars, certain atoms and molecules have to be treated together.

It is obvious that introduction of non-LTE calls for a greater range of atomic and molecular data. In particular, collisional excitation, ionization, and dissociation cross-sections are needed. Calculation of the radiation fields that enter into the rates R_{ij} will also call for more atomic and molecular data than was necessary in LTE; for example, the ultraviolet radiation field is unimportant to the calculation of a model atmosphere of a cool star because little flux is carried by the ultraviolet, but the ionization or excitation of an atom in non-LTE may be significantly affected by radiative rates involving the ultraviolet.

In cool stars, hydrogen atoms outnumber electrons by a wide margin with the ratio being roughly the abundance of the easily ionized metals. For the upper layers of the solar photosphere, $n(\mathrm{H})/n(\mathrm{e}) \sim 10^4$ where electrons are contributed by Mg, Si, and Fe principally. Of course, the ratio $n(\mathrm{H})/n(\mathrm{e})$ is even larger in metal-poor stars. One expects that all particles—electrons and H atoms—are describable by a single kinetic temperature. If, then, the fraction of particles with energies in excess of an excitation energy $E_{\mathrm{UL}} = E^A_{\mathrm{U}} - E^A_{\mathrm{L}}$ is the same for electrons and hydrogen atoms, why is it that excitation of A is generally assumed to occur through e on A collisions and not through collisions between A and the far more numerous H atoms? It will strike the observant novice as curious that neglect of the H on A collisions is almost never justified.

The rate constant for collisional excitation $A_L \rightarrow A_U$ is an integral of a velocity dependent cross-section $\sigma(v)$ over the Maxwellian velocity distribution; I write this rate constant as $\langle\sigma(v)v\rangle$. Certainly, the H atoms' contribution relative to that of the electrons is reduced by the ratio of the thermal velocities, i.e., $(m_e/m_H)^{\frac{1}{2}} \simeq 1/43$. There is also a general argument to suggest that the cross-section for excitation by H atoms is likely to be less than that for excitation by electrons.

The general argument was given by Massey (1949) and was discussed recently by Anderson (1981). A collision occurs with a characteristic time scale $t \sim r/v$, where r is the effective range and v is the relative velocity. The transition in the perturbed atom corresponds to a frequency $v \sim \Delta E/h$. Classically and, also, quantum mechanically, the expectation is that the cross-sections for excitation and de-excitation will be small unless $1/t \sim v$ or $tv \sim 1$. It is instructive to examine this limit:

$$tv = \frac{r}{v}\frac{\Delta E}{h} = \frac{4r\,\Delta E}{\sqrt{T}}\text{(electrons)}$$
$$= \frac{170r\,\Delta E}{\sqrt{T}}\text{(hydrogen atoms)}.$$

In these formulae, the effective range is given in Å units and the energy ΔE in eV. In the limit $tv \rightarrow 0$, the cross-section will decrease from a maximum near $tv \lesssim 1$, but it will remain significantly large for small tv values. On the other hand, the cross-section is expected to be very small for $tv \gg 1$. A typical optical transition has $\Delta E \sim 3$eV so that with $T \sim 5000$K and $r \sim 1$Å, tv $\sim 1/6$ for an electron collision and ~ 7 for a hydrogen atom collision. Then, the cross-section for an inelastic collision with an H atom is expected to be much smaller than for an inelastic collision with an electron. The effective range of the H-atom collision is possibly less than that of the e-atom collision. In an atom, fine-structure and some term-to-term transitions correspond to small ΔE. In a molecule, the rotational and vibrational transitions correspond to small excitation energies: $\Delta E_{\text{rot}} \sim 0$–$0.05$ eV and $\Delta E_{\text{vib}} \sim 0.2$–$0.4$ eV for typical molecules. When ΔE is much less than 1eV, the product $tv < 1$ and excitation by hydrogen (also helium) atoms must be considered.

The total rate appearing in the equations of statistical equilibrium is the product of $\langle\sigma v\rangle$ and the density of projectiles: can the small $\langle\sigma v\rangle$ expected for optical transitions ($\Delta E \gtrsim 1$eV) from collisions with H atoms be offset by the large ratio $n(\text{H})/n(\text{e})$?

In a series of non-LTE calculations on Li I, Fe I, and other atoms, Holweger and colleagues have included the inelastic collisions with H atoms where the rate constants were taken from a generalization of a "modified classical Thomson formula" (Drawin 1968, 1969), which was expected to yield "an order of magnitude estimate of collisional excitation and ionization cross-sections." The Thomson formula alluded to is that for ionization of atoms by electrons. The reader interested in the employed rate constants and the empirical evidence from analysis of absorption lines in stars for the efficacy of collisions with H atoms is referred to Lambert (1993).

In view of the fact that the rate constants for the H-atom collisions are based on a rudimentary estimate adapted from Thomson's (1912) classical estimate for collisional ionization by electrons, it is surely of interest to test the rudimentary estimate against experiments and quantum mechanical predictions. Such a test seems especially worthwhile when it is recalled that Thomson's estimate predicts cross-sections for collisional ionization by electrons that are considerably larger than the real cross-sections near threshold (Seaton 1962). What follows is a brief discussion of a theoretical estimate for excitation of the Li I $2s^2S \rightarrow 2p^2P$ resonance (6707Å) transition and experimental measurements for excitation of the (Na D) transition.

Theoretical estimates of excitation through H-atom collisions must consider the H-atom pair as forming a temporary molecule and, hence, a set of molecular potentials. The LiH molecule is a relatively simple molecule as far as contemporary quantum chemistry is concerned. As a stellar spectroscopist with no more than a superficial knowledge of quantum chemistry, I would suppose that excitation rates by H-atom collisions can now be estimated with fair precision. Allen & Dickinson (private communication) have estimated the rate constants for excitation of the Li I $2p$ level from the ground $2s$ level by H atoms. Allen & Dickinson's estimates are substantially smaller than expected from Drawin's formula. The quantum mechanical ($\equiv$ true?) rate is 100–200 times smaller at the temperatures of interest $T \sim 2000$–6000 K than the rates adopted by Steenbock and Holweger (1984). If the lower rate for $2s \rightarrow 2p$ excitation proves typical of rates for other transitions, Steenbock & Holweger's calculations show that H-atom collisions are not competitive with the electron-atom collisions in determining the departures from LTE in the excitation and ionization equilibrium of atoms.

Although it seems clear that additional quantum mechanical calculations are feasible for at least simple atoms, experimental confirmation of these results and particularly measurements on transitions not so readily amenable to theoretical attack would be of interest. One key problem is the production of a low-energy beam of H atoms. Since $kT \sim 0.2$–0.6 eV for cool stellar atmospheres, the rate constants under stellar conditions are usually dominated by the cross-section within 1eV of the threshold for excitation or ionization. Unfortunately, "no reliable hydrogen atom beam source for the 1–10 eV range is known" (Fleck et al. 1991). For energies above about 10eV, photodetachment of an H^- beam provides a useful H-atom beam. Such a source was used to measure the emission cross-section for the Na I $3p \rightarrow 3s$ transition for Na colliding with H atoms of energies from 15–1000 eV. The results shown in Figure 7 are not exactly equivalent to the $3s \rightarrow 3p$ excitation cross-section because the collisions with H atoms can lead to population of higher Na I levels that populate the $3p$ level through cascades. Certainly, the emission cross-section is an upper limit to the desired excitation cross-section for $3s \rightarrow 3p$.

The cross-section estimated from Drawin's expression (see Lambert 1993) is shown in Fig. 7. In the energy range of overlap with experiment, the Drawinian

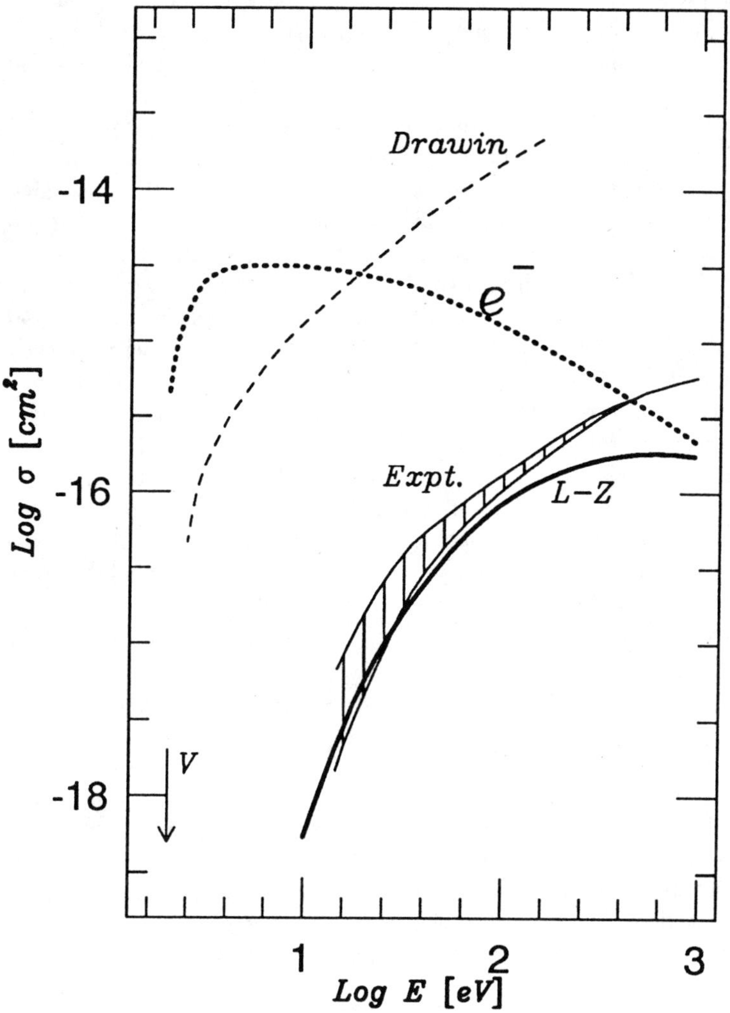

Fig. 7. Collision cross-sections for excitation Na I $3s^2S \rightarrow 3p^2P$ as a function of the laboratory energy of the projectile (H or e⁻). Experimental estimates (upper and lower error bounds) for excitation by H atoms are shown for $E_{\rm H} \sim$ 10–1000eV by the hatched area (Fleck et al. 1991). Theoretical estimates for excitation by H atoms include the Landau-Zener model (see line marked L-Z), and the Drawinian prediction (Lambert 1993). The experimental cross-section for electron excitation is shown by the curve labelled e⁻.

estimates are a factor of about 100 larger than the experimental results. The cross-section for electron excitation is also shown in Fig. 7. Near threshold, it is apparent that the cross-section of H-atom excitation is likely to be 4 or more orders of magnitude smaller than that of electron excitation. In view of the lower thermal velocities of the H atoms, it appears that the electrons will be greatly favoured in a comparison of the overall rates of excitation, i.e., $n_e\langle\sigma v\rangle_e$ vs. $n_{\rm H}\langle\sigma v\rangle_{\rm H}$.

A realistic calculation of the cross-section requires access to accurate potential energy curves for the NaH molecule. Fleck et al. (1991) examined computed $^1\Sigma^+$ potential curves. The $H^- + Na^+$ ion pair potential introduces avoided crossings of potential energy curves. A collision pair can make a transition between potential energy curves at an avoided crossing. Fleck et al. (1991) use the Landau-Zener (LZ) model to estimate the $3s - 3p$ emission cross-section and obtain a fair fit to the experimental results. Extrapolation of the LZ predictions to near threshold give $\sigma \sim 10^{-21}$ cm^2. Fleck et al. (1991) report that they also undertook quantum mechanical scattering calculations which gave a much larger cross-section near threshold ("The origin of this discrepancy is not yet understood."). Since the rate constants at the low temperatures of the photospheres where H-atom collisions are of potential importance are determined by the near-threshold cross-sections, it is critical that the theory be accurate in this region. Fleck et al. (1991) conclude that "the curve crossing mechanism is the essential mechanism" and "reliable quantum mechanical scattering calculations are required for the energy region between the threshold and the range of the present experimental data." This combination of conclusions make it unlikely that a general recipe can be found for the H-atom excitation (and ionization) cross-sections. It is to be hoped that one or two atoms of astrophysical interest can now be examined in detail and the rate constants for H–atom collisions applied to a non-LTE study.

The Massey criterion suggests that atoms seem likely to be effective in causing excitation between rotational and vibrational levels of electronic states of molecules. H, H_2, and He are the prospective contenders in the competition with electrons. There is laboratory data on H_2 and He and a very limited amount on H as the cause of inelastic collisions with molecules. But much of this data involves molecules of little or no interest to the stellar spectroscopist.

5 Information, Languages, and Communication

In this review, I have attempted to set out why a particular group of cool stars—the AGB stars—are today of burning interest to observers and theoreticians in the astronomical village on the south bank of the river. I hope my fellow villagers will excuse the omissions and generalities that I was forced to make. They exist for several reasons but primarily because I was trying to beguile the villagers on the north bank into working with us in solving the fundamental questions presented by the stars: How and why do they evolve? and how are the chemical elements synthesized?

By sketching a few topics in the spectroscopy of AGB stars, I trust that I have shown how "information" drawn from the northern village is needed by those in the southern village who wish to extract results ("information") about stellar evolution and nucleosynthesis from stellar spectra. There is a corresponding flow in the opposite direction; for example, the laboratory molecular spectroscopist may find previously unreported transitions in spectra of cool stars.

The reasons for and benefits of exchange of information over the bridge are, I think, abundantly clear. In an era of instant communication, it would seem churlish to complain that not all is well. I shall complain! In order for the communication between the villages to be optimally effective, we and you should speak the same language. Often, we do not. Therefore, I will end this review by urging that we should train our students to become acquainted with the topics studied across the river. Astronomy students (and their professors) should welcome exposure to the state of the current art of quantum chemistry and to the beautiful variety of new techniques for experimental spectroscopy. I hope that we may meet on another occasion to enjoy "En udsigt fra Broen."

References

Anderson, N. 1981, *Comments At. Mol. Phys.*, **10**, 133

Bauschlicher C.W. Jr., Langhoff S.R., Taylor P.R., 1988, *Astrophys. J.*, **332**, 531

Beer H., Macklin R.L., 1989, *Astrophys. J.*, **339**, 962

Blöcker T., Schönberner D., 1991, *Astron. Astrophys.*, **244**, L43

Boothroyd A.I., Sackmann I.-J., 1988 *Astrophys. J.*, **328**, 653 and 671

Burbidge G.R., Burbidge E.M., Fowler W.A., Hoyle F., 1957, *Rev. Mod. Phys.*, **29**, 54

Busso M., Gallino R., Lambert D.L., Raiteri C.M., Smith V.V., 1992, *Astrophys. J.*, **399**, 218

Cameron A.G.W., Fowler W.A., 1971, *Astrophys. J.*, **164**, 111

Chase M.W., Davies C.A., Downey J.R., Frurip D.J., McDonald R.A., Syverud A.N. (eds.), 1984, *JANAF Thermochemical Tables, 3rd edition*

Costes M., Naulin C., Dorthe G., 1990, *Astron. Astrophys.*, **232**, 270

Drawin H.W., 1968, *Z. Phys.*, **211** 404

Drawin H.W., 1969, *Z. Phys.*, **225** 483

Fleck I., Grosser J., Schecke A., Steen W., Voigt H., 1991, *J. Phys. B*, **24**, 4017

Hinkle K.H., Lambert D.L., 1975, *Monthly Notices Roy. Astron. Soc.*, **170**, 447

Hinkle K.H., Lambert D.L., Wing R.F., 1989, *Monthly Notices Roy. Astron. Soc.*, **238**, 1365

Hollowell D., Iben I. Jr., 1989, *Astrophy. J. (Letters)*, **340**, 966

Huang Y., Barts S.A., Halpern J.B., 1992, *J. Phys. Chem.*, **96**, 425

Iben I. Jr., 1975, *Astrophys. J.*, **196**, 525

Iben I. Jr., 1975, *Astrophys. J.*, **196**, 549

Iben I. Jr., 1991, *Astrophys. J. Suppl.*, **76**, 55

Jonnson J., Launila O., Lindgren B., 1992, *Monthly Notices Roy. Astron. Soc.*, **258**, 49p

Käppeler F., Beer H., Wisshak K., 1989, *Repts. Prog. Phys.*, **52**, 945

Kjærgaard P., Gustafsson B., Walker G.A.H., Hultquist L., 1982, *Astron. Astrophys.*, **115**, 145

Lambert D.L., 1988, in *The Impact of Very High S/N Spectroscopy on Stellar Physics*, ed. G. Cayrel de Strobel & M. Spite, p. 563
Lambert D.L., 1991, in *Evolution of Stars–the Photospheric Abundance Connection*, ed. G. Michaud & A.V. Tutukov, p. 299
Lambert D.L., 1993, *Phys. Scripta*, T47, 186
Lambert D.L., Clegg R.E.S., 1980, *Monthly Notices Roy. Astron. Soc.*, **191**, 367
Lambert D.L., Ries L.M., 1981, *Astrophys. J.*, **248** , 228
Lambert D.L., Smith V.V., 1993, *in preparation*
Langhoff S.R., Bauschlicher C.W. Jr., 1988, *J. Phys. Chem.*, **89**, 2160
Larsson M., Siegbahn P.E.M., Ågren H., 1983, *Astrophys. J.*, **272**, 369
Lu R., Huang Y., Halpern J.B., 1992, *Astrophys. J.*, **395**, 710
Malaney R.A., 1987, *Astrophys. J.*, **321**, 832
Malaney R.A., Lambert D.L., 1988, *Monthly Notices Roy. Astron. Soc.*, **235**, 695
Massey H.S.W., 1949, *Repts. Prog. Phys.*, **12**, 248
Merrill P.M., 1952, *Astrophys. J.*, **116**, 21
Plez B., Smith V.V., Lambert D.L., 1993, *Astrophys. J. in press*
Sackmann I.-J., Boothroyd A.I., 1992, *Astrophys. J.*, **392**, L71
Scalo J.M., 1978, *Astrophys. J.*, **221**, 627
Scalo J.M., Despain K.H., Ulrich R.K., 1975, *Astrophys. J.*, **196**, 805
Schwarzschild M., Härm R., 1967, *Astrophys. J.*, **246**, 251
Seaton M.J., 1962, in *Atomic and Molecular Processes*, ed. D.R. Bates, 1962, Chapter 11
Sauval A.J., Blomme R., Grevesse N. 1994, in *Poster Session Proceedings of IAU Coll. 146,* eds. P. Thejll & U.G. Jørgensen, p. 107
Smith V.V., 1988, in *The Origin and Distribution of the Elements*, ed. G. J. Mathews, p. 535
Smith V.V., Lambert D.L., 1984, *Publ. Astron. Soc. Pacific*, **96**, 226
Smith V.V., Lambert D.L., 1990a, *Astrophys. J. (Letters)*, **36**, L69
Smith V.V., Lambert D.L., 1990b, *Astrophys. J. Suppl.*, **72**, 387
Smith V.V., Lubowich D., Plez B., Lambert D.L., 1993, *in preparation*
Thompson R.I., 1973, *Astrophys. J.*, **181**, 1039
Thomson J.J., 1912, *Phil. Mag.*, **23**, 449
Tomkin J., Lambert D.L., 1983, *Astrophys. J.*, **273**, 722
Truran J., Iben I. Jr., 1977, *Astrophys. J.*, **216**, 797
Tsuji T., 1988, *Astron. Astrophys.*, **197**, 185
Ulrich R.K., 1973, in *Explosive Nucleosynthesis*, eds. D.N.Schramm&D.W.Arnett, p.139
Uus U., 1973, *Soviet Ast.*, **17**, 196

Dominating Molecules in the Photospheres of Cool Stars

Uffe Gråe Jørgensen

Niels Bohr Institute, Blegdamsvej 17, DK-2100 Copenhagen, Denmark

1 Introduction

Molecular data bases (including very different numbers of molecular lines, and based on computations spanning a wide range in quality) have been used in the construction of models of the stellar atmospheric structure, and effects of various molecules on the stellar structure and evolution have been described in the literature. A systematic study showing which molecules are of importance in which types of stars is, nevertheless, still lacking, and the line lists used in various model computations necessarily reflect what was at hand at the time of computation rather than a systematic effort to solve the molecular opacity problem. A pioneering investigation was presented in two extensive papers by Tsuji (1964, 1973) where the foundation of the reaction scheme used in most models today was laid down. Later Tsuji (1986) has given a review of the existing work on molecules known in stars, and Johnson (1986) and Gustafsson (1989) have touched upon the same subject in reviews covering a broader range of stellar aspects. Jørgensen (1992) has recently summarized the existing molecular data (and discussed the methods used) available for model atmosphere construction.

In the present review I attempt to present a systematic analysis of the effect of the most abundant molecules in cool stars, and to touch upon what can be the most important missing molecular data for further progress in the field. For this purpose, I have computed about 150 new model atmospheres where various molecules have been included in, or excluded from, the adopted opacity in a systematic way. The models are based on a recent version (Jørgensen et al. 1992) of the Marcs code (Gustafsson et al. 1975) with molecular opacities (described further in section 4) from CO, CN, C_2, TiO, HCN, C_2H_2, C_3, and H_2O considered in the computations. The code, with the input data used, is particularly tuned for computations of giant stars (stars with low gravity), and the emphasis in the present review will therefore be on red giants, but dwarf models will be computed for comparison purposes and for evaluation of trends. Whereas most of the results for the giant stars presented in this review are as reliably as is possible today, the results for the dwarfs ($\log(g) = +4.5$) are to be considered as more qualitative in nature. A thorough description of the dwarf stars and opacity problems particular to these are described elsewhere in this book by Borysow, Liebert, Scholz & Wehrse, and by Tsuji.

A stellar atmosphere is the outer, observable layer of a star – as opposed to the invisible stellar interior. The photosphere is the (lower) layer of the atmosphere where most of the visible (and infrared) part of the stellar spectrum is produced, and where the temperature decreases outward from the star. Often the *atmosphere* is used synonymously with the *photosphere* because only very few models exist for the layers above the photosphere. The effective temperature, $T_{\rm eff}$, is an "average" temperature of the photosphere, but there are both (deeper) layers in the photosphere with temperatures much higher than $T_{\rm eff}$ and (higher) layers in the photosphere with temperatures somewhat below $T_{\rm eff}$. Also the surface gravity (g, here referred to in cgs units), the metallicity Z (the mass fraction of the elements heavier than helium) and the C/O ratio are of vital importance for the atmospheric structure. The basic stellar parameters in the models computed here consist of various combinations of $T_{\rm eff} = 4500$, 3500, 3100, 2800, 2500 K; $\log(g) = -0.5, +0.5, +1.5, +4.5$; C/O = 0.43, 0.86, 0.92, 0.95, 0.98, 1.0, 1.01, 1.02, 1.05, 1.35, 2.0, and $Z = Z_\odot$ (i.e., the solar abundance except for carbon).

2 Stellar Evolution into the "Molecular Regime"

The majority of stars end their life in the red giant phase where the outer parts expand and cool, while the interior slowly transforms into a hot, compact object (a white dwarf) of material that will slowly cool and undergo no further nuclear processing. While the stars are red giants they will, for the first time in their life, mix nuclear processed material from the interior to the surface (see e.g., Iben & Renzini 1983 and Lambert this volume). The cooler and more luminous they get the more of this material will be blown into space (e.g., de Jager et al. 1988), and in this way the gas and dust clouds, that new stars and planets are created from, will slowly be more and more enriched with nuclear processed material from cool red giants. Today it is even possible to trace some of the material that our own solar system was created from back to various types of red giants (e.g., Anders & Zinner 1993).

The Sun will become a red giant in about 7 billion years from now, and during about 200 million years the Sun will develop from a phase where it is 3500 K (and has 1000 times its present luminosity) to a final stage another 5 times more luminous and with $T_{\rm eff} \approx 2500$ K (Jørgensen 1991). It is in this interval of effective temperature that molecules dominate the structure of the stellar atmosphere, and I will here call it the "molecular regime" of stellar evolution. In stars warmer than 3500 K, the atomic lines become of increasing importance relative to the molecular lines, although a proper study has never been performed of where exactly the transition between the "atomic regime" and the "molecular regime" is (see Seaton this volume). There are also stars assumed to be cooler than 2500 K, but we have no reliable models of them yet. Dust may be an important opacity source in such stars (see Alexander & Ferguson and Sedlmayr, this volume). Stars tend to lose the bulk of their mass to the interstellar space while they are in the molecular regime – perhaps even triggered by the radiative pressure on the molecules (Jørgensen & Johnson 1992). The Sun is expected to lose 50% of its present mass while 2500 K $<T_{\rm eff}<$ 3500 K.

At far lower luminosities there is another important group of stars within the same temperature interval. They are the M dwarfs and the brown dwarfs. They spend basically all of their lifetime in the molecular regime. The brown dwarfs may be so numerous that they contribute significantly to the "missing" mass in the universe, and the group of M dwarfs house some of the oldest stars in existence. Also the coolest white dwarfs – remnants of a higher mass generation of stars formed in the very early days of our Galaxy – are in this temperature interval (see Liebert this volume and Thejll this volume). Analysis of the spectrum of these more compact objects involve problems that are very different from those found in the red giants, but common for all the stars in the molecular regime is that the structure in the layers that we can observe is dominated by molecular opacities. Further progress in our understanding of their role in the evolution of the universe is therefore dependent on development in our knowledge of the molecules that dominate their atmospheric opacity.

3 Which Molecules Dominate the Stellar Opacity?

In stars of solar abundance hydrogen and helium accounts for 98% of the mass, but the only stable molecule formed from these atoms is H_2. For all cool stars, H_2 is the most abundant molecule throughout the atmosphere. If H_2 was dipole active in its ground state, the study of molecular absorption in stars would be completed by the study of molecular hydrogen. Molecules made up of combinations of H, C, N, and O (in the Sun, CNO accounts for 86% of the atoms heavier than He), or even more rare elements, are much less abundant in the stellar atmosphere. However, they have so much stronger absorption coefficients that their resulting opacity in the visual and near infrared spectral region (where cool stars transmit their energy) is orders of magnitude bigger than that of H_2. For example, in a red giant of 3500 K and solar elemental composition there are 1 million H_2 molecules for each TiO molecule, but still TiO dominates the opacity.

The second most abundant molecule in nearly all cool stars is CO, because of its very high dissociation energy and because of the high cosmic abundance of carbon and oxygen. In the Sun, oxygen accounts for 55% of all the atoms heavier than helium, and carbon for 52% of the rest (Anders & Grevesse 1989). In contrast to the Earth's atmosphere, CO is much more abundant than CO_2 in red giants, which is an effect of the extremely low pressure in their atmospheres. For the same reason molecules like NH_3 and CH_4, that are very abundant in planetary atmospheres, do not seem to form in red giants. For the less studied "high-pressure" stars like brown dwarfs and cool white dwarfs these molecules may be important (see Tsuji this volume).

CO has its vibration-rotation spectrum in a region (1-5 μm) where cool stars emit a considerable fraction of their energy, but although its absorption coefficient is much higher than that of H_2 (i.e., the quadrupole transitions) in the same region, its absorption is still weak compared to other molecules made of H, C, N, O and heavier elements, and particularly its absorbtion is discrete (i.e., concentrated to relatively few but strong lines) so that the stellar flux can escape the star between the lines. This effect drastically decreases the importance of CO compared to what one might have guessed based on its high partial pressure and relatively high integrated absorption coefficient. Only in situations where there are no or only few other molecules, the effect of CO may be substantial. In the Sun, where the temperature throughout the atmosphere is high compared to the stars we discuss in this review, CO will be the only molecule of importance, and its effect is presumably to form clouds of cool gas in an otherwise warm solar atmosphere, just as it would cool and contract the entire atmosphere of a red giant if it was the only molecule to form in such stars, as we will see in the following sections.

In a number of cool stars (but not all, as will be seen below) the group of the most abundant molecules after CO, consists of N_2, CS, and OH. N_2 is dipole inactive in its ground state, and can (just as H_2) be excluded from the opacity computations (but must obviously be taken into account in the molecular equilibrium computations). CS has a discrete absorption structure, just as CO, and usually a partial pressure 2 to 3 orders of magnitude lower than that of CO. Its opacity has been included in recent model atmosphere computations by "the Indiana group" (Brown et al. 1989) based on a line list of 35000 lines compiled by Piñeiro et al. (1987), but it has never been included in the models computed with the Marcs program or in the "Kurucz-models" (Kurucz 1979) based on the "Smithsonian data base" (Kurucz this volume). The effect of this molecule on the model structure has not been described in the literature, but it is an opacity that deserves much more attention. CS is also abundant in circumstellar envelopes and in the stellar wind (see Hinkle this volume and Olofsson this volume). OH is very abundant in certain oxygen-rich red giants, and it is often observed as a maser, too. It is an illustrative example of a very abundant molecule where big confusion about its role still exists. It has never been included in computations with the Marcs code, it is listed with 330 000 lines (although most in the UV where they will be without effect for cool stars) in the Smithsonian data base, and it has been included with less than 1% of this number of lines in the Indiana group computations. No study of its role for the atmospheric structure exists. Other molecules that may have exceptionally high partial pressures in some model atmospheres include SiO, H_2O, and NO.

The molecules presumed to have the biggest effect on the model structure are generally those in the fourth most abundant group (after H_2, CO, and N_2/CS/OH) and their partial pressures are typically, say, 7 orders of magnitude below that of H_2 – a tiny, tiny fraction of the total gas pressure. Due to their rarity the balance between them is easily shifted, and therefore small errors in the adopted molecular absorption coefficient can result in big adjustments of the model structure, via adjustments in the molecular equilibrium. This is one of the reasons why high accuracy and completeness is required in the molecular data. If the electronic transitions of CO (and/or H_2) had been in the spectral region where cool stars emit the bulk of their energy, the situation would have been very much simpler (and of course also relatively dull).

An extreme example of the effect of incompleteness in the molecular data can be examined by setting certain of the molecular absorption coefficients to zero (more realistic examples will be considered in a later section). Consider first a carbon star model with $T_{\rm eff} = 3500$ K and only continuum sources in the opacity. This model turns out to be completely unaffected by the possible inclusion of the opacity of polyatomic molecules (HCN, C_2H_2, and C_3). A similar model including continuum sources, and now also the opacity of the diatomics CO, C_2 and CN, expands with 25% in size (and the gas pressures in the surface layers decrease with a factor of 30) if the opacitiy from the polyatomics is introduced. The large difference in the way the two models react to the inclusion of the opacity from the polyatomics, is due to the cooling effect of CO, C_2, and CN, which gives rise to a much larger partial pressure of the polyatomic molecules in the model which considers also opacity from CO, C_2, and CN. The lesson from this is clear: it is in principle not possible to say which molecules will dominate the opacity before all molecular opacities have already been included. The inclusion of additional opacity sources (or improved completeness in the line list of already included opacities) may change the atmospheric structure in such a way that other molecules suddenly become dominant.

Also due to the small partial pressure of the "opacity-strong" molecules, small changes in the most abundant atoms can cause very big changes in the molecular equilibrium. The ratio that is most important for such changes is the C/O ratio (but also the Si/S ratio can effect the spectrum drastically), and it causes the distinction between K- or M-type (oxygen-rich) red giants, S-type stars, and carbon stars, where different molecules will dominate the opacity, and which will therefore be discussed individually below.

During the late phases of stellar evolution carbon and s-process elements are created in the stellar interior and mixed into the visible layers. In the Sun (where no nuclear processed material has been mixed to the surface) C/O = 0.43, which is therefore assumed to be the "starting condition" for "normal" stars. When enough carbon is mixed to the surface during the red giant phase, the ratio C/O approaches unity, and the chemical balance shifts drastically from oxygen bearing molecules to carbon bearing molecules, because either all the carbon or all the oxygen is bound in CO. The molecules that dominate the structure in oxygen and in carbon-rich stars are therefore completely different from one another. In high metallicity environments, like our own galaxy, oxygen-rich red giant stars are much more common than carbon stars, but in metal deficient environments the opposite is true (Richer & Westerlund 1983, Jørgensen & Johnson 1992).

A good feeling about which molecules should be expected to be found in stellar atmospheres can be obtained by looking at the dissociation energies of various molecules. Those with the highest dissociation energies will form first and thereby use up all of the available material of the least abundant atom in the molecule. Some of the diatomic molecules with highest dissociation energies

Table 1. Dissociation energy (from Tsuji (1964) except for CN from Costes et al. (1990) and TiO from Colket (1984)) of some important molecules, and the elemental abundance (in units of atoms per 10^6 Si atoms; from Anders & Grevesse (1989)) in the solar system (numbers in parenthesis are exponents of 10).

molecule:	CN	HCN	C_2	C_2H	C_2H_2	C_3	C_3H	C_4	C_5	CH
D_0(eV):	7.77	4.94	6.12	5.72	4.95	7.74	4.92	5.21	7.41	3.47
molecule:	MgH	CaH	SiC	SiC_2	CH_4	NH	NH_3	NO	CO	CO_2
D_0(eV):	1.97	1.75	4.5	8.7	4.38	3.8	4.52	6.50	11.09	5.45
molecule:	H_2O	OH	SiO	TiO	TiO_2	VO	ZrO	ScO	YO	LaO
D_0(eV):	5.10	4.40	8.1	7.95	6.7	6.4	7.8	6.9	9.0	8.15

element:	H	C	N	O	Mg	Si	Fe	S
abundance:	2.79(10)	1.01(7)	3.13(6)	2.38(7)	1.07(6)	1.00(6)	9.00(5)	5.15(5)
element:	Ca	Ti	V	Sc	Zr	Y	La	
abundance:	6.11(4)	2.4(3)	2.93(2)	3.42(1)	1.14(1)	4.64	4.46(−1)	

are listed in Table 1. More complete lists can be found in Tsuji (1964, 1973), Huber & Herzberg (1979), and in Chase et al. (1984; the JANAF tables).

4 The SCAN data base of molecular lines

The original computations with the Marcs code (Gustafsson et al. 1975) included about 50 000 (observed) lines from various diatomic hydrides, and were tuned particularly for computation of models in the range of $T_{\rm eff}$ approximately between 3700 K and 6000 K. In more recent computations of cool star models, these lines are no longer included, but in stead a more complete line list, which excludes some of the original hydrides, is used. This fact does not reflect a knowledge of, say, CH, MgH and NH being unimportant, but rather the recognition of the overwhelming importance of completeness in the line list of the molecules. Inclusion of a few lines for an additional molecule may give the false impression that this molecule is "considered". The most complete computations with the Marcs code, as concerns the molecular opacities, are those of Jørgensen et al. (1992) for the carbon-rich models, and those of Plez et al. (1992) and Jørgensen (1993) for the oxygen-rich stars.

These models include the opacities of CO and C_2 from Querci et al. (1974) and the opacities of CN, C_3, C_2H_2, HCN, TiO and H_2O from what we now call the SCAN data base. These opacities are produced and described by Jørgensen et al. (1985) for HCN, by Jørgensen et al. (1989) for C_3, by Jørgensen & Larsson (1990) for CN, by Jørgensen (1993) for TiO, and by Jørgensen & Jensen (1993) for H_2O. The TiO opacity represents an updating and extension of the original list by Krupp et al. (1978). The lists for CN, HCN, C_3 and for H_2O are based on full ab initio computations of the dipole moment and potential energy surfaces by use of the CASSCF theory (see Langhoff & Bauschlicher this volume and

Malmqvist this volume). The potential energy surface and the solution of the Schrödinger equation for the H_2O computations have been improved by use of the MORBID theory (see Jensen this volume), and a similar update for the HCN molecule is in progress. The opacity of C_2H_2 is constructed from observational data and some simplified assumptions about the unobserved combination bands based on experience with other molecules (unpublished work, but the method has been described in some detail by Jørgensen 1990). A SCAN line list for CH will soon be available too (Jørgensen & Larsson, in preparation), and work on FeH, SiC_2, CaH, and other small molecules is under consideration. In total the SCAN data base contain about 60 million lines (identification, frequency, excitation energy, and line strength) from 7 of the most important molecules in stellar atmospheres, and many more will hopefully be added in the future.

One of the most important experiences we have got from our construction of the SCAN data base is that completeness of the line lists used in the opacity construction is of overwhelming importance. With the strongly increased computer capacity over the last, say 10 years, it is today affordable, for small molecules, to compute the complete set of lines between basically all eigenstates with energies up to very close to the dissociation energy. Most data in use have a cut off at some arbitrary energy level, or (v,J)-level, chosen to make the computation and the data handling affordable in terms of computing time and required data storage space. For this reason it must be expected that there is quite a lot of work ahead in re-computing the data for a number of diatomic and other small molecules where some data already exist.

In our computation of the "red" system of CN (Jørgensen & Larsson 1990, see also Larsson this volume) we introduced, as a measure of the completeness of the line list, the ratio of the partition functions obtained by summation of all included eigenstates and by summation of all possible states, respectively. The method has been developed further in our work on H_2O (Jørgensen & Jensen 1993, Jørgensen et al. 1994). In our work on CN we found that 99.9% of the partition function was included in our computation for temperatures up to 5 000 K. About 40% of the missing 0.1% of the population was due to exclusion of vibrational levels with $v>30$ (up to the dissociation energy corresponding to $v \approx 60$), and about 60% was due to the exclusion of transitions with lower levels in the third electronic state (the B level). Experiments with use of the SCAN-CN data, have shown 10 times higher opacity in the visual region compared to previous estimates of the CN "red" opacity.

In our work on CN, all lines with J less than a high value, J_{max}, was included on the tape, but for H_2O, where space demands a more careful evaluation, we computed the line intensity at a reasonably high temperature and included a line on the list only if its intensity was above a given pre-specified value. Ideally, lines with such high values of J should be included so that the energy of the level (v,J_{max}) is (for each v) equal to the dissociation energy. In practical terms, it can be problematic to include the rotational levels up to the dissociation energy, because the molecular constants often are insufficiently known, and because it involves unreasonably many lines (of doubtful effect). Nevertheless, the inclusion of high values in J is less critical than the inclusion of high values in v, because the integrated band intensity is often known, and one then just has to make sure that the sum of all the intensities of all lines is equal to the integrated band intensity within a specified accuracy. This method was used by Jørgensen et al. (1985, 1989) in the computation of the absorption coefficient of HCN and C_3, respectively. The vibrational transition moment was computed between all the vibrational eigenstates, without consideration of rotation but in a full quantum mechanical treatment. In our original use of the (simplified) vibrational transition moments for HCN (Eriksson et al. 1984), we estimated how much of the integrated band intensity was missing when computing to a given (affordable) value of J_{max} and then distributed the rest as a continuum opacity, but with the increased computer capacity of today we simply compute each band by including sufficiently high values of J till a given fraction (e.g., 99.9%) of the band intensity is included. Typically such computations lead to about 10 million lines per molecule, which can reasonably be stored on tapes and handled on disks, but with a fast computer the line list can also be generated directly from the band list (for linear molecules) and stored in the form of an opacity sampling (OS) or an opacity distribution function (ODF) within a few CPU hours of computing.

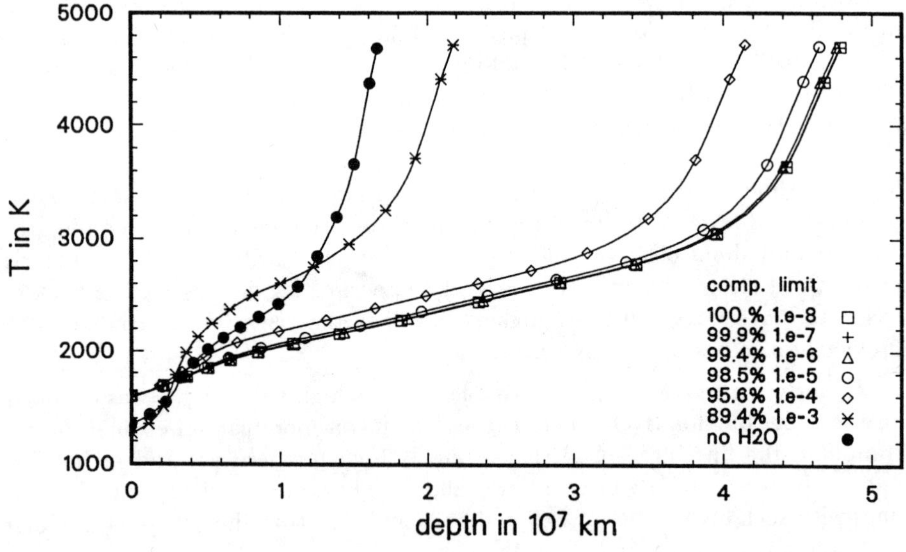

Fig. 1. The geometrical size of an oxygen-rich model atmosphere of $T_{eff} = 2800\,K$, $\log(g) = -0.5$, C/O = 0.43 when various degrees of completeness in the opacity of the water molecule is considered.

For asymmetric top molecules, such as H_2O, having both J and K rotational quantum numbers, the total number of lines easily becomes billions. The handling of this big number of lines is both difficult in computing time and disk space, and is, fortunately, without much meaning. On the other hand the computation of line intensities from the vibrational intensities is much more time consuming for an asymmetric top molecule (like H_2O) than for a linear molecule (like HCN), and it is not affordable to re-compute the lines each time a new OS or ODF is desired (hundreds or thousands of CPU hours are needed on a medium size computer depending on the required accuracy). Therefore the effect of still weaker lines on the stellar atmospheric structure has been explicitly studied for H_2O by including only lines stronger than a given threshold limit, $S_{\rm lim}$, which was tested as both 10^{-3} km/mol, 10^{-4} km/mol, 10^{-5} km/mol, etc to 10^{-8} km/mol (Jørgensen et al. 1994). The value of $S_{\rm lim}$ necessary for convergence in the model structure depends not only on the basic stellar parameters (a lower value of $S_{\rm lim}$ is required for high gravity objects and/or lower values of $T_{\rm eff}$), but also on the quality of the other opacities included. For a test model of $T_{\rm eff}$ = 2800 K without TiO, the difference in model temperature in the surface layer when $S_{\rm lim}$ (for water) was 10^{-6} and when it was 10^{-8}, was 20 K, whereas the difference was only 6 K for corresponding models with TiO included. The more complete the total set of opacities of the molecules is, the less important it is whether the opacity for one of the molecules is computed to the highest possible accuracy or not. The more recent opacities in the SCAN data base are accompanied by a completeness analysis which gives an impression of how big a fraction of the opacity is included. This is important not only for the model structure, but also for the continuum drawing in spectral analysis.

5 Oxygen-Rich Giant and Dwarf Stars

The oxygen-rich stars are the most common cool stars in the solar neighbourhood, and therefore also the most well studied. Several of the bright stars we see with the naked eye on the night-sky are cool oxygen-rich red giants. The molecular opacity of these stars is dominated by H_2O, TiO, and CO. CO and TiO are important opacity sources throughout the "molecular regime", but H_2O is the dominant molecular opacity source for the cooler giants, whereas CN, in oxygen-rich stars, only affects the structure of the stars warmer than $T_{\rm eff}$ = 3500 K. For example, the partial pressure of CN is 7 orders of magnitude below that of TiO, at $\tau_{\rm Ross} = 10^{-2}$ in a $T_{\rm eff}$ = 2500 K model of solar abundance, whereas it is as abundant as TiO for a $T_{\rm eff}$ = 3500 K model. For a similar model of $T_{\rm eff}$ = 4500 K, CN is the most abundant molecule in all layers below $\tau_{\rm Ross} = 10^{-4}$, but at those temperatures atomic lines and continuum sources may dominate the atmospheric opacity, and the relative contribution of molecules and atoms has not been thoroughly studied yet. The ratio between the partial pressures of CO and TiO is almost constant, P(CO)/P(TiO) $\approx 10^4$, throughout the atmosphere, as is also seen in Fig. 2. The ratio between the partial pressure of H_2O and that of TiO (or CO), on the other hand, increases rapidly for models cooler than $T_{\rm eff} \approx 3000\,{\rm K}$. At $T_{\rm eff} = 3100\,{\rm K}$ and $\log(g) = -0.5$, the partial pressure of H_2O equals that of TiO in the surface ($\log \tau_{\rm Ross} = -5.5$), but already for $T_{\rm eff} = 2800\,{\rm K}$ the partial pressure of H_2O in the surface is nearly 4 orders of magnitude bigger than that of TiO.

When compared to a model where only the continuum opacities are included, the effect of CO is mainly to cool the surface layers and to contract the whole atmosphere. The effect of TiO alone is to cool the surface layers (and to create a back-warming in the deeper layers) and the effect of H_2O is mainly to expand the atmosphere. In combination the molecules may, however, react in a more complex manner, and TiO is usually considered a heating agent (for the surface layers too) because models with continuum sources, CO, CN, and H_2O included in the opacity generally are cooler than models including also TiO. The effect of these molecules, individually and together, on the structure of a typical red giant is shown in Fig. 2.

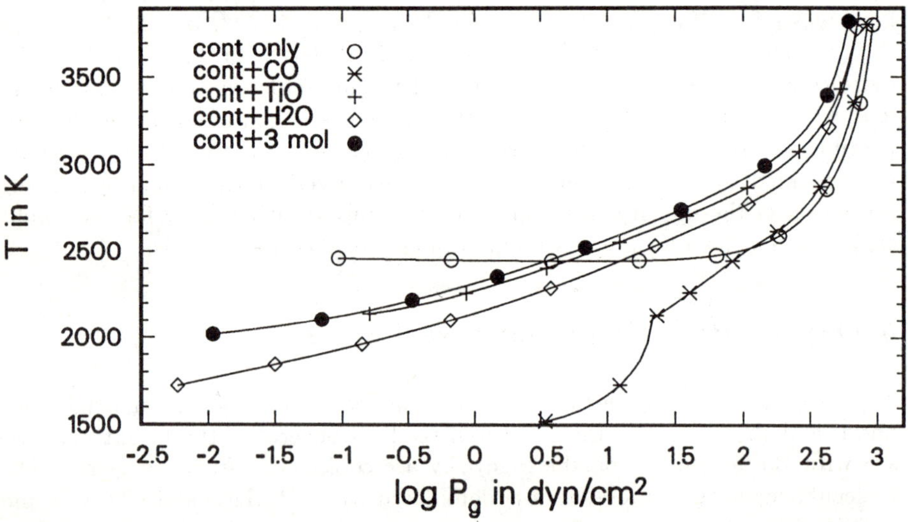

Fig. 2. Effects of CO, TiO, and H_2O on the model structure (temperature versus gas pressure) of a typical oxygen-rich giant ($T_{\rm eff} = 3100\,{\rm K}$, $\log(g) = -0.5$, and $C/O = 0.43$). The tickmarks indicate increasing optical depths in steps of $\log \tau_{\rm Ross} = 0.75$ from the value $\log \tau_{\rm Ross} = -5.5$ in the surface layer.

For the giants, the C/O ratio will increase after the helium shell flash phase when freshly produced ^{12}C is mixed to the surface. This will gradually decrease the amount of available free oxygen, after the formation of CO, and will eventually change the model structure and spectral appearance completely as C/O approaches and pass unity. At slight increases of C/O the effect on the model structure is nevertheless negligible. Even for $C/O = 0.87$ the gas pressure in the surface of a model with $T_{\rm eff} = 3100\,{\rm K}$ is decreased with less than 0.3 dex (caused

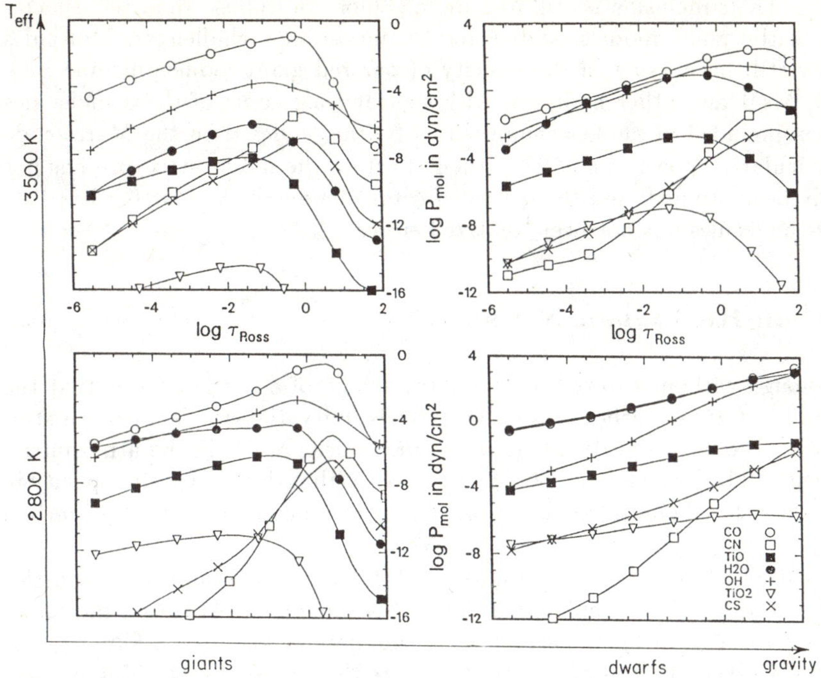

Fig. 3. The partial pressure of various molecules in the atmosphere of four oxygen-rich stars in the "molecular regime". The two left panels show typical giants ($\log(g) = -0.5$) and the two panels to the right in the figure show typical M dwarfs ($\log(g) = +4.5$). The two uppermost panels are for $T_{\rm eff} = 3500$ K and the two lower panels are for $T_{\rm eff} = 2800$ K.

by the decreased formation of H_2O) and the surface temperature increased with only 15 K, compared to a corresponding model with C/O = 0.43.

For stars of higher surface gravity (M dwarfs, cool white dwarfs, and brown dwarfs) CO, TiO, and H_2O are still abundant, but the relative importance of water increases dramatically compared to the two other molecules. Effects of the molecular abundance in oxygen-rich models of different temperature and gravity are shown in Fig. 3. It is, for example, seen that at $T_{\rm eff} = 2800$ K and $\log(g) = +4.5$ (an M dwarf), the partial pressure of H_2O equals the partial pressure of CO throughout the atmosphere. Even at $T_{\rm eff} = 3500$ K it equals that of CO from $\log \tau_{\rm Ross} = -3$ to -1. CN and H^- are of less importance in the high gravity objects than in the giants. As expected, the more complex molecules increase in importance for the high gravity objects. For example the partial pressure of TiO_2 is only 3 to 4 orders of magnitude below that of TiO in the $\log(g) = 4.5$ models compared to up to 8 orders of magnitude below that of TiO in the warmer low gravity models. Also CaOH, CaH, FeH, CO_2, CH_4, and collision induced absorption are candidates of being important opacity sources in the cool high gravity objects, and some of them are described in detail in other chapters of this book. Their inclusion is still in a more exploratory phase than the opacity sources in the giant models, and seems to impose new challenges. Molecules that may still be missing in the opacity of our red giant models include SiO, CS, OH, MgH and other hydrides. Although at least some of these molecules have been included in models computed with other codes than the Marcs code (see particularly Brown et al. 1989), their effect on the model structure has not explicitly been studied, and the degree of completeness of the existing line lists for these molecules has not been analyzed either.

6 Carbon-Rich Giant Stars

When enough carbon is mixed into the atmosphere of a red giant star that the number ratio C/O of carbon to oxygen exceeds unity, the star is a carbon star. Since CO is the most tightly bound of the molecules present in the atmosphere, all oxygen will be locked up in CO. TiO and H_2O (that dominates the opacity in the oxygen-rich red giants) are therefore almost completely absent in the carbon stars.

The carbon stars are more rich in the molecular variety than the oxygen-rich stars, due to the higher complexity of the carbon chemistry. At least 6 molecules are known to be vital for the structure of cool carbon stars (CO, CN, C_2, C_3, HCN, and C_2H_2), and it seems that even more than these 6 molecules are necessary for a good description of the photospheric opacity in the carbon stars. CH, CS, NH, C_2H, C_3H, SiC_2, and other molecules are likely candidates for this "missing" opacity. Typically, the introduction of the opacity from molecules (diatomics or polyatomics) in cool carbon stars where otherwise only continuum sources were considered, gives rise to a cooling of the surface layers of about 1000 K, as is seen in Fig. 4 where the effect of various molecules on the model structure of two representative carbon stars are shown. As is seen from the figure, the diatomic molecules give rise to a cooling and contraction (increase of the gas pressure for given optical depth) of the atmosphere. The polyatomic molecules, on the other hand, again expands the atmosphere compared to the structure where only diatomics are included, and the final effect of all the molecules together is therefore to change drastically the temperature versus gas pressure structure, but to produce a less pronounced change in the gas pressure versus optical depth structure. In older models where no polyatomic molecules were included in the molecular opacity of the carbon stars, the temperature versus gas pressure structure (which is particularly relevant for the predicted spectrum, and therefore also for determination of the abundance, gravity, etc.) were actually further from the right structure than models with no molecules included in the opacity at all, as is seen from Fig. 4, whereas the less sensitive temperature versus optical depth structure were predicted better in the models with diatomic molecules than in the models completely without molecular opacities.

For the cooler carbon stars, polyatomic molecules dominate the opacity fairly deep into the atmosphere whereas diatomics have the highest partial pressure all the way to the surface in our warmer models. In a model with $C/O = 1.05$

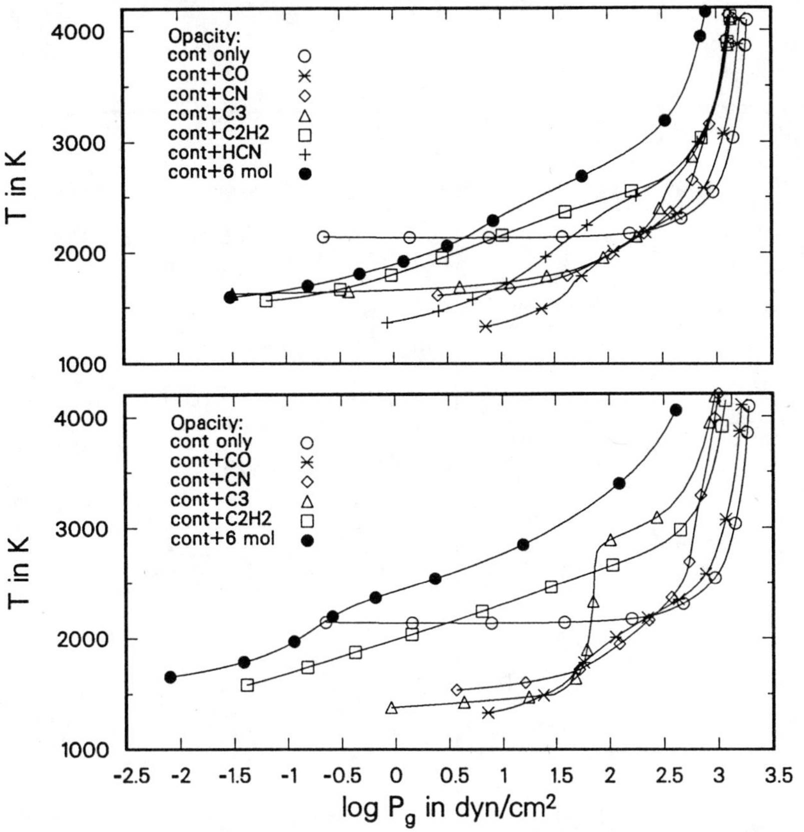

Fig. 4. Effects of CO, CN, C_3, HCN and C_2H_2 on the model structure of two typical carbon-rich red giants of $T_{\rm eff} = 2800$ K and $\log(g) = -0.5$, with C/O = 1.05 (upper panel) and 2.0 (lower panel), respectively. Tickmarks as in Fig. 2. The response from inclusion of C_2 (not shown) is very similar to that of CN.

and $T_{\rm eff} = 2500$ K, HCN is the most abundant of the 6 molecules included in the opacity (apart from CO which has a very low absorption coefficient) from the surface and to the layer where $\tau_{\rm Ross} \approx 1$. For a corresponding $T_{\rm eff} = 3500$ K model, CN dominates the molecular pressure (apart from CO) throughout the atmosphere. For the warmer stars, CS has a partial pressure higher than any of the 5 (CO excluded) through a substantial part of the atmosphere (particularly for the carbon stars with C/O only slightly above 1, and for low values of the gravity), but its opacity has not been included in the computations presented here (or in any other computations based on the Marcs code). As for CO, the band systems of CS are not as rich as those of TiO, C_2 and CN, but due to the high partial pressure, its effects certainly deserves further investigation.

The increasing excess of carbon resulting from successive He-shell flashes, will leave more and more free carbon (after the formation of CO molecules) to form carbon-rich molecules. For this reason the C_2 band strengths are traditionally

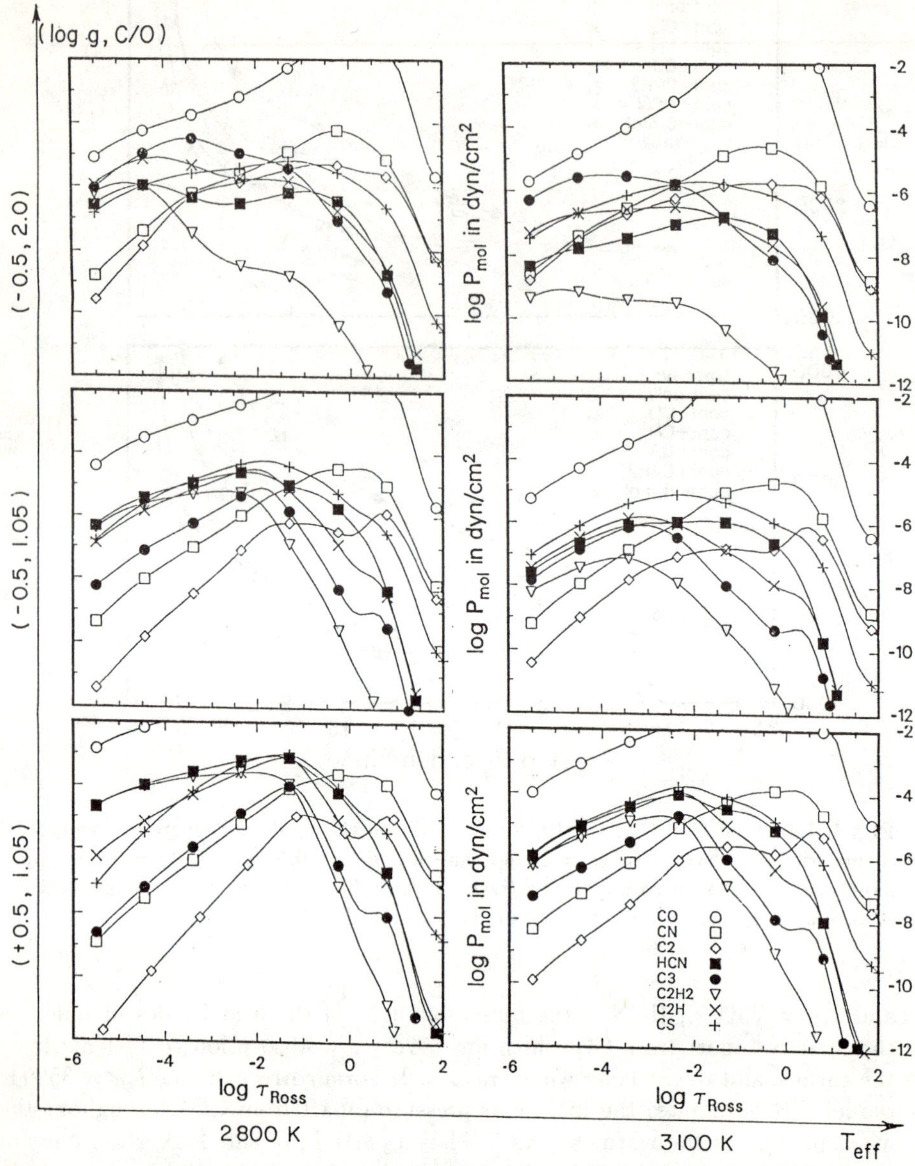

Fig. 5. The partial pressure of various molecules in the photosphere of six representative carbon stars of $T_{\rm eff} = 2800\,\rm K$ (left panels) and $T_{\rm eff} = 3100\,\rm K$ (right panels). From top to bottom the stars have (log(g), C/O) = (−0.5, 2.0), (−0.5, 1.05), and (+0.5, 1.05), respectively. Tickmarks as in Fig. 2.

used as an index for the C/O ratio (this is the j in the $C_{i,j}$ classification scheme of carbon stars; see, e.g., Yamashita 1972). For a model with $T_{\rm eff}$ = 3100 K the partial pressure of C_2 at $\tau_{\rm Ross} = 10^{-2}$ is 100 times smaller than that of CN when C/O = 1.05, but it is less than a factor of 10 below CN for C/O = 2.0. In the upper photosphere where the small polyatomic molecules form, the sensitivity to adjustments in the model structure is nevertheless even bigger, and our models show the abundance of C_3 to be considerably more sensitive to the C/O ratio than the abundance of C_2, but the bands due to C_3 are predominantly in the less studied infrared spectral region, whereas the C_2 bands are in the visual region accessible by use of CCDs and photographic plates. When C/O increases from C/O = 1.05 to C/O = 2.0 for a $T_{\rm eff}$ = 3100 K model, the partial pressure of C_3 increases with a factor of 100 more than the partial pressures of C_2, as is readily seen in Fig. 7 (in the next section). Also C_2H and C_2H_2 are strongly increasing functions of the C/O ratio, whereas NH, CN, and HCN are remarkably unaffected by changes in the C/O ratio, and the partial pressure of CS even decreases slightly when C/O increases. Among the polyatomics, C_3 has its main influence for $T_{\rm eff} \approx 3100$ K and for high C/O ratio and low gravity, whereas C_2H_2 increase in importance for lower temperatures and/or high C/O. For C/O close to unity, the partial pressure of C_2H_2 and C_3 are both relatively low, and HCN is therefore the dominating polyatomic opacity source for the "carbon-poor" carbon stars. At C/O approaching 2 (which is a very high C/O ratio at least for carbon stars in the solar vicinity) the partial pressures of both C_3 and C_2H are considerably higher than those of HCN and C_2H_2. It imposes a serious problem for reliably modelling of the carbon-rich carbon stars that the monochromatic absorption coefficient of C_2H is still unknown in a form useful for astrophysics (see Peyerimhoff this volume), and therefore considered to have zero opacity in all existing models.

The general effect on the model structure when increasing the $T_{\rm eff}$ and keeping the other fundamental parameters constant, is to expand and heat the surface layers. The same trend appears when C/O is increased.

Carbon stars with higher gravity (dC, CH, and Ba stars) than AGB red giant carbon stars can be understood as formed from mass transfer in binary systems. Exploratory models have been constructed for their atmospheres (Gass et al. 1988 and Kipper 1992, see also Liebert this volume). Molecules like CH_4, CaH, FeH and/or others might be important opacity candidates in such stars. Figure 5 gives an impression about which direction the changes will take if models of the higher gravity carbon stars are to be constructed. The higher the gravity the more likely will the formation of larger molecules be. For example the abundance of C_2H_2 relative to C_2H or relative to HCN will be an increasing function of gravity, as is also easily seen in Fig. 5, and generally the effect of polyatomic molecules will increase relative to the diatomics when the gravity increases. It is also seen that the traditional measures of the carbon abundance, C_2 (and C_3), are dependent on gravity too, in the sense that their partial pressures relative to CN and CH, or HCN, is a decreasing function of gravity, which of course impose a problem to direct use of the two dimensional carbon star classification scheme

(C_{ij}, with j based on the C_2 band intensities, and with i based on the Na D atomic line intensities which are also gravity dependent) to the dwarf stars.

7 S-Type Stars

Beside the oxygen-rich and the carbon-rich red giant stars, there are also the less common S-type stars. These are classified from having strong ZrO, VO, LaO, and YO spectral bands. The name S stars is historical from the time the stars were classified into the spectral groups A, B, C, D, ..., but it happens to be so, that the spectra of the S stars are also rich in s-process elements. Let $N(X)/N_6(Si)$ be the number of atoms X per 10^6 Si atoms. For solar composition the number of oxygen atoms per 10^6 Si atoms, after the CO formation, is $[N(O)-N(C)]/N_6(Si) = 1.37\,10^7$. Si is the most abundant atom that can form diatomic oxide with a dissociation energy higher than that of TiO. Since $N(Ti)/N_6(Si)$ is as low as 2400, there is 10 times more free oxygen atoms than Si and Ti atoms to form SiO and TiO. When C/O increases and more oxygen is bound into CO, $[N(O)-N(C)]/N_6(Si)$ decreases and approaches 1 for $C/O = 0.96$. At this value of C/O the amount of TiO is therefore no longer determined by the abundance of Ti, but by the amount of available free oxygen. The two dominant oxygen-rich molecular absorbers (TiO and H_2O) will therefore decrease in amount, which will cause the atmosphere to shrink, which will in turn cause the gas pressure to increase. Since both ZrO, LaO, and YO have higher dissociation energies than TiO, the partial pressure of these molecules will, to a start, increase. The result will be that the intensity of the bands of these molecules will increase strongly compared to the intensity of the TiO bands. When ZrO bands are visible the star is called an S star, and the point when the intensity of the ZrO and the TiO bands are equal is expected to correspond to $C/O \approx 0.96$ (Jaschek & Jaschek 1990). It is generally assumed that there is an evolutionary sequence from the oxygen-rich to the S-type to the carbon-rich red giants, although details in this simple scenario have several times been criticized (e.g., Iben & Renzini 1983, Lundgren 1988).

From the point of view of molecular opacities the S-type stars are an interesting challenge, because both TiO and H_2O are disappearing gradually, and the carbon-rich molecules are not yet formed. This gives rise to a "molecular zoo" those members have not yet been fully identified, and many molecules may act together to define the opacity. To judge from the spectra, the elemental abundances, and the dissociation energies (see Table 1), ZrO, LaO, YO, VO, FeH, MgH, CaH, ZrS, and SiS may all play a role for the atmospheric structure of S stars. Only few of these molecules have yet been studied in the sense that reasonably complete line lists exist. For the important opacities of ZrO, YO, and FeH, an extensive analysis has been performed as part of the Berkeley program (see Davis this volume), and a line list has been constructed for ZrO and YO, which shows the ZrO to be the most important heir (it has many lines in several electronic systems, and high f-values) of TiO when the abundance of TiO decreases.

We can presumably learn about a number of qualitative effects in the S-type star atmospheres from studying models based on only the molecular opacities included in the carbon-rich and oxygen-rich models (CO, CN, C_2, TiO, HCN, C_3, C_2H_2, H_2O), although the effects for C/O close to 1, of course can only be taken as indicative. From such models the first surprise is maybe to discover how relatively insensitive the M-type atmosphere is to changes in the C/O ratio until C/O approaches very close to unity. When C/O change from 0.43 (the solar value) to 0.95 the gas pressure and temperature in the surface change with 0.5 dex and 75 K, respectively, and the physical size of the atmosphere decreases with only 15% for a test-model with $T_{\mathrm{eff}} = 3100\,\mathrm{K}$ and $\log(g) = -0.5$. This reflects that there is still more free oxygen after CO and SiO formation than there is elemental Ti (for solar elemental abundances). The effect is comparable to the effect of excluding (from the opacity in M-type giants) those of the weak lines of the H_2O rotation-vibration system that makes up only 4% of the H_2O absorption coefficient (see Fig. 1). Models based on an absorption coefficient of H_2O with 4% uncertainty will consequently not be able to distinguish the physical properties of an M-type giant with C/O = 0.43 from an S-type giant with C/O = 0.95 (but the chemical properties are of course calculated better despite of this).

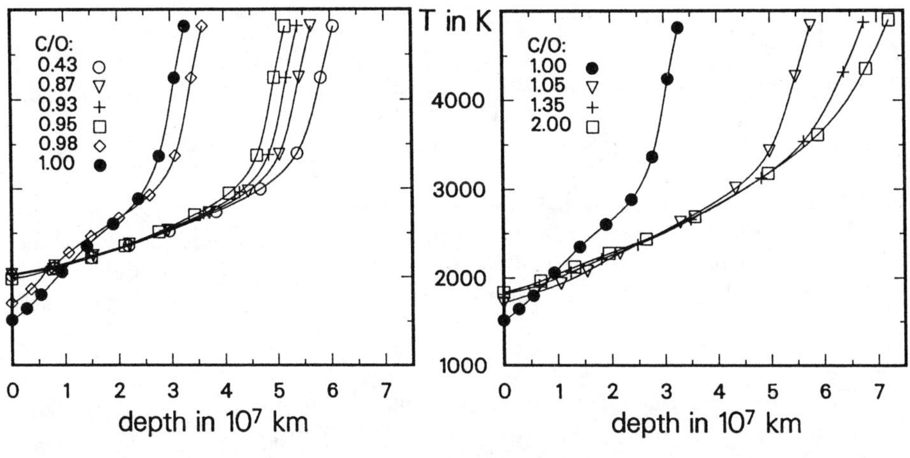

Fig. 6. The geometrical size of the atmosphere of red giant models of $T_{\mathrm{eff}} = 3100$, $\log(g) = -0.5$, and various C/O ratios ranging from C/O = 0.43 to C/O = 2.0 (as indicated with the legend). It is seen that when C/O increase beyond the Solar value, the atmosphere first shrinks because of the disappearance of TiO and H_2O, until the star is an S star with C/O $\approx$ 1 (left panel), but then rapidly increases in size again when C/O > 1 (right panel) and carbon-rich molecules (CN, C_2, HCN, C_2H_2, C_3) form.

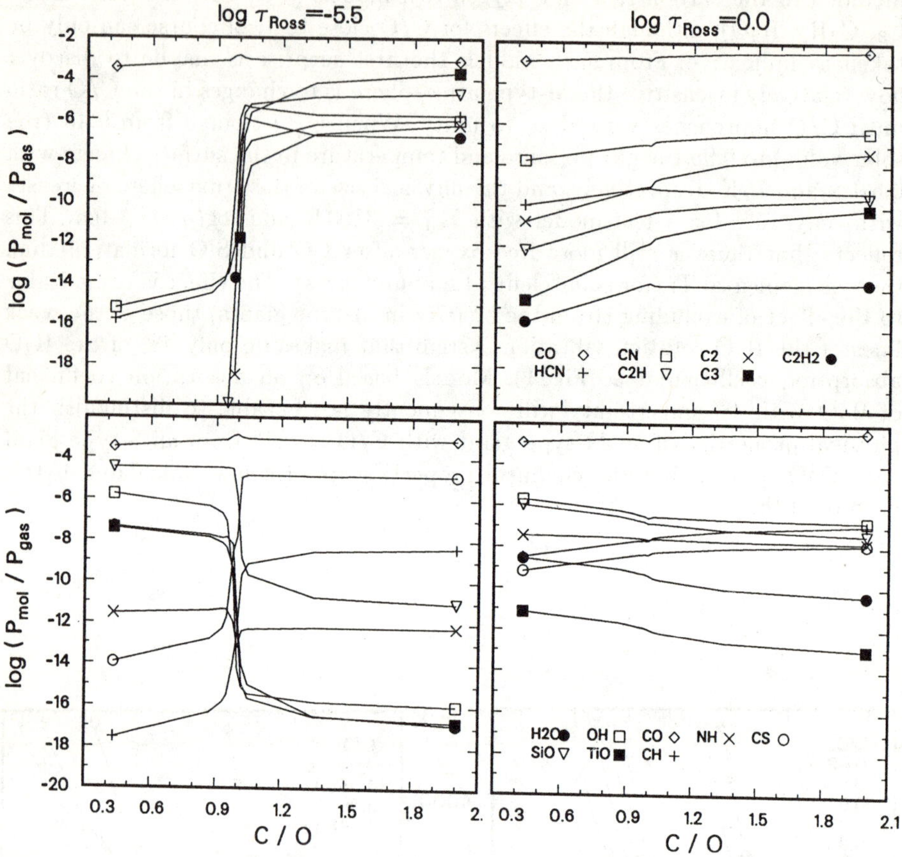

Fig. 7. The changes in the partial pressure of various molecules is large when C/O change from below to above 1, particularly in the upper photosphere (left panel) and less pronounced in the lower part of the photosphere where the molecules that are traditionally used as spectroscopic indicators of the C/O ratio, are formed (right panel).

When changing the C/O ratio from 0.95 to 1.0 there occurs a dramatic readjustment of the model structure. For a $T_{\rm eff}$ = 3100 K model, the surface cools with 500 K, and the size of the atmosphere is reduced with nearly a factor of two, as is seen in Fig. 6. This effect may of course partly be due to lacking knowledge of the opacities of the molecules that show strong features in the spectrum, but qualitatively the effect is likely to be correct, because the strong contraction basically is due to the lack of polyatomic molecules in S-type stars – from an observational as well as from a theoretical point of view. There is a dramatic change in the relative partial pressure of the oxygen and the carbon bearing molecules when C/O increases from 0.95 to 0.98, as is seen in Fig. 7. The number-ratio of H_2O and HCN molecules changes with 11 orders of magnitude in a red giant model atmosphere with $T_{\mathrm{eff}} = 3100\,\mathrm{K}$ and $\log(g) = -0.5$ when C/O changes from 0.95 to 1.0. At C/O = 0.95 the partial pressure of H_2O in the surface layers is 5 orders of magnitude higher than that of HCN. At C/O = 0.98 it is only 1 order of magnitude above, and at C/O = 1.0 it is 6 orders of magnitude below HCN. From C/O = 1.0 to C/O = 2.0 the partial pressure of HCN stays approximately constant, while those of the more carbon-rich molecules increase rapidly. For example the partial pressure of C_3 increases with nearly 7 orders of magnitude. It is these enormous changes in the partial pressures of the polyatomic molecules that are responsible for the big changes in the surface gas pressure, and in the size, of the red giants with C/O ≈ 1, and these effects are therefore relatively independent of the inclusion of the proper diatomic S-type opacities (ZrO, LaO, YO, FeH, and VO). The main changes in the partial pressures of the diatomic molecules included in our computations are a strong decrease in TiO when C/O approaches and pass unity, and a corresponding increase in the amount of CN and CH. The ratio between TiO and CN (or CH) changes with 7 orders of magnitude when C/O increases from 0.95 to 1.0 (and this change must be expected to increase when ZrO, VO, YO, and LaO are introduced into the opacities, because they use some of the oxygen otherwise available for TiO).

8 Summary

Molecules are particular important in the photosphere of stars in the effective temperature interval from about 3500 K to 2500 K (and below). This temperature regime include the cool red giants as well as the M dwarfs, brown dwarfs, and cool white dwarfs. I have described the SCAN data base of 60 million molecular lines from CN, TiO, CH, HCN, C_2H_2, C_3, and H_2O, and analyzed the effect of the opacities constructed from these data (and data from CO and C_2) on models of cool stellar atmospheres. 150 new models were computed by use of the Marcs code and with the molecular input data varied in a systematic way in order to understand the effect of the individual molecules. Typical changes in the model structure of a cool red giant when molecules are included into the opacity is a cooling in the surface layer of about 1000 K. The diatomic molecules cause a contraction of the atmosphere, whereas a typical effect of polyatomic molecules is to expand the atmosphere considerably. Marked changes have occurred during recent years in the obtained model structure, and this will probably be the case for some more years when additional and better molecular data become available, and particularly for the S stars and the dwarfs, the models are still exploratory. The strong interaction between the molecules was stressed, and potential candidates for "missing" opacities were suggested.

References

Anders E., Grevesse N., 1989, *Geochim. Cosmochim. Acta*, **53**, 197

Anders E., Zinner E., 1993, *Meteoritics*, **28**, 490

Brown J.A., Johnson H.R., Alexander D.R., Cutright L.C., Sharp C.M., 1989, *Astrophys. J. Suppl.*, **71**, 623

Chase M.W., Davies C.A., Downey J.R., Frurip D.J., McDonald R.A., Syverud A.N. (eds.), 1984, *JANAF Thermochemical Tables, 3.edition*

Colket M.B., 1984, *J. Quant. Spectrosc. Radiat. Transfer*, **31**, 7

Costes M., Naulin C., Dorthe G., 1990 *Astron. Astrophys.*, **232**, 270

de Jager C., Nieuwenhuijzen H., van der Hucht K. A., 1988, *Astron. Astrophys. Suppl.*, **72**, 259

Eriksson K., Gustafsson B., Jørgensen U.G., Nordlund Å., 1984, *Astron. Astrophys.*, **132**, 37

Gustafsson B., Bell R.A., Eriksson K., Nordlund Å., 1975, *Astron. Astrophys.*, **42**, 407

Gass H., Liebert J., Wehrse R., 1988, *Astron. Astrophys.*, **189**, 194

Gustafsson B., 1989, *Ann. Rev. Astron. Astrophys.*, **27**, 701

Huber K.P., Herzberg G., 1979, *Constants of Diatomic Molecules*, Van Nostrand, N.Y.

Iben I., Renzini A., 1983, *Ann. Rev. Astron. Astrophys.*, **21**, 271

Jaschek C, Jaschek M., 1990, *The Classification of Stars*, Cambridge Univ. Press

Johnson H.R., 1986, In: *The M-type stars*, H.R.Johnson&F.R.Querci (eds.), NASA SP-492, p.323

Jørgensen U.G., 1990, *Astron. Astrophys.*, **232**, 420

Jørgensen U.G., 1991, *Astron. Astrophys.*, **246**, 118

Jørgensen U.G., 1992, *Rev. Mexicana Astron. Astrof.*, **23**, 49

Jørgensen U.G., 1993, *Astron. Astrophys.*, in press

Jørgensen U.G., Almlöf J., Gustafsson B., Larsson M., Siegbahn P., 1985, *J. Chem. Phys.*, **83**, 3034

Jørgensen U.G., Almlöf J., Siegbahn P., 1989, *Astrophys. J.*, **343**, 554

Jørgensen U.G., Jensen P., 1993, *J. Mol. Spectrosc.*, **161**, 219

Jørgensen U.G., Jensen P., Sørensen G.O., 1994, In: P.Thejll&U.G.Jørgensen (eds.), *Poster Session Proceedings of IAU Coll.146*, Copenhagen University, p. 51

Jørgensen U.G., Johnson H.R., 1992, *Astron. Astrophys.*, **265**, 166

Jørgensen U.G., Johnson H.R., Nordlund Å, 1992, *Astron. Astrophys.*, **261**, 263

Jørgensen U.G., Larsson M., 1990, *Astron. Astrophys.* **238**, 424

Kipper T., 1992, *Baltic Astronomy*, **1**, 181

Krupp B.M., Collins J.G., Johnson H.R., 1978, *Astrophys. J.*, **219**, 963

Kurucz R.L., 1979, *Astrophys. J. Suppl.*, **40**, 1

Lundgren K, 1988, *Astron. Astrophys.*, **200**, 85

Piñeiro A.L., Tipping R.H., Chackerian C.Jr., 1987, *J. Mol. Spec.*, **125**, 91

Plez B., Brett J.M., Nordlund Å, 1992, *Astron. Astrophys.*, **256**, 551

Querci F., Querci M., Tsuji T., 1974, *Astron. Astrophys.*, **31**, 265

Richer H.B., Westerlund B.E., 1983 *Astrophys. J.*, **264**, 114

Tsuji T., 1964, *Ann. Tokyo Astron. Obs.*, **9**, no. 1

Tsuji T., 1973, *Astron. Astrophys.*, **23**, 411

Tsuji T., 1986, *Ann. Rev. Astron. Astrophys.*, **24**, 89

Yamashita Y., 1972, *Ann. Tokyo Astron. Obs.*, **13**, no. 3

Opacity Problems in Cool Low Mass Stars

Michael Scholz and Rainer Wehrse

Institut f. Theoret. Astrophysik der Universität Heidelberg, Im Neuenheimer Feld 561, D–69120 Heidelberg, Germany

1 Introduction

In this contribution we want to discuss M star atmospheres and their dependence on molecular opacities. A star belongs to the spectral class M if its optical and infrared spectrum shows strong bands of TiO and numerous strong metal lines so that for wavelengths < 4000 Å there is "hardly any flux left" (Jaschek & Jaschek 1987). M stars cover a very large range in luminosity: M dwarfs are the intrinsically faintest stars, whereas M giants and supergiants reach luminosities that are among the highest known. General properties of these objects are given in Table 1 (after Schmidt-Kaler 1982).

A good knowledge of M star spectra is of crucial importance for the understanding of the properties of the Milky Way (and other galaxies as well) for the following reasons (see also Lambert, this volume, and Liebert, this volume):

- M dwarfs have a space density of about 0.1 pc^{-3} and therefore constitute about 80 % of the total number of stars in the solar neighborhood. Although their individual masses are very low, their large number even let them dominate the mass distribution.
- Since the luminosity of stars varies approximately with mass $\propto M^{-3.2}$, M dwarfs evolve very slowly. Therefore, M dwarfs of very different ages can be found and the abundances of metals in their atmospheres reflect in most cases (possible exceptions may be the carbon dwarfs, see Green et al. 1992, and Liebert, this volume) the abundance distribution of the Galaxy at the time of their birth. In other words: from the understanding of the M dwarfs we can get a full account of the chemical evolution of the Milky Way.
- M giants return large amounts of heavy elements to the interstellar medium where the enriched gas is used for a subsequent generation of stars.
- Due to their high luminosity M giants often dominate the integrated (infra)red light of galaxies.

Unfortunately, spectrum formation is extremely complicated in these stars so that we are only now starting to obtain models that can be used for spectral analysis. Up to very recently, understanding was also hampered by the fact that reliable and reasonably high resolution spectra in the region of maximum flux were not available. The progress in infrared detector technology in the last years made it possible that now even for extremely faint stars digital spectra with resolutions sufficient for detailed analysis are available (G. Rieke, private communication). Another jump is expected with the ISO satellite to be launched 1995.

In the next section we describe in some detail the temperature and pressure stratifications in these stars and comment on the geometrical extension of the giants. Section 3 is devoted to a discussion of consequences of errors in the opacities, in particular we address the run-aways in giant atmospheres that may occur when water vapor forms and we point to the fact that the convection in the dwarfs extends from very shallow layers to the center so that errors in the outermost parts propagate through all parts of the object. In Section 4 we review the present agreements and disagreements between synthetic and observed spectra. Based on this discussion, in Section 5 a list of most urgently needed molecular data for future improvements of M star models is presented.

Table 1. Approximate physical parameters of M stars (after Schmidt-Kaler 1982); note that the luminosities of dwarfs are lower by a factor ≈ 2 after Kirkpatrick et al. (1993)

	dwarfs	giants/supergiants
luminosities $/L_\odot$	10^{-3}—8×10^{-2}	3×10^{2}—3×10^{5}
radii $/R_\odot$	2×10^{-1} —6×10^{-1}	4×10^{1}—2×10^{3}
masses / $M_\odot$	0.06—0.5	1—24
eff. temperatures (K)	2600 —3800	2600—3800
log gravity (cgs units)	4.6 —4.9	−1—1

2 Structure of M star atmospheres

In Fig. 1 we show the pressure distributions in some giants and in dwarfs as a function of the radial coordinate r: giant atmospheres are in many cases geometrically extended (both in relative and absolute terms) and their pressures are relatively high in spite of the low gravity; in fact, the pressures are comparable to those in the solar photosphere (see Vernazza et al. 1981). This is a consequence of the fact that the absorption coefficients are very small (e.g. H^- absorption is very weak due to the lack of electrons) and therefore photons can traverse large column densities. In contrast, the atmospheres of dwarfs are extremely thin (typically a few hundred kilometers compared to a radius of 1 to 4×10^5 kilometers) so that a comparison of different cases is hardly possible in this representation.

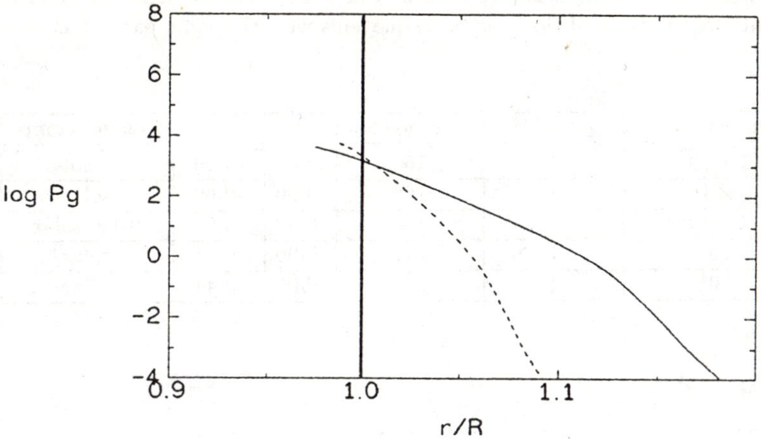

Fig. 1. Gas pressure distributions in M stars plotted as function of the normalized radial coordinate. The thick vertical line indicates the pressures in dwarfs, the broken curve refers to a giant model with parameters: M=1 $M_{\odot}$, L = 500 $L_{\odot}$, R= 83 $R_{\odot}$, the full curve to a giant model with M=1 $M_{\odot}$, L = 2000 $L_{\odot}$, R = 166 $R_{\odot}$. All models have solar composition. Note the compactness of the dwarf atmospheres compared to the large geometrical extension of giant models, in particular if they have high luminosity.

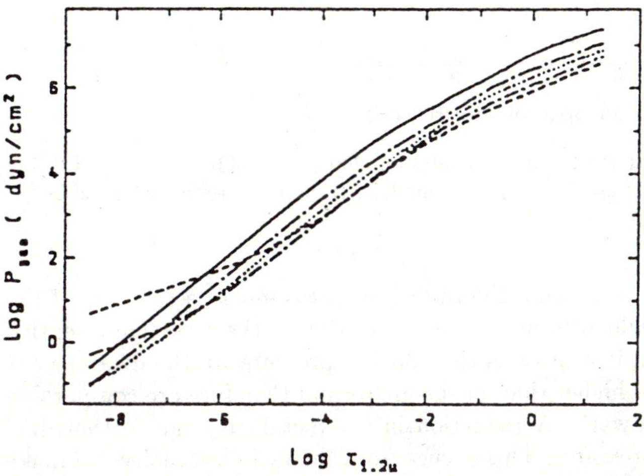

Fig. 2. Typical gas pressure distributions in M dwarfs (after Allard 1990) plotted as a function of the optical depth at 1.2 μm. The curves refer to models with effective temperatures of 2000 (—), 2500 (-.-), 2750 (...), 3000 (-.-.) and 3250 (- - -) K, solar metallicity, and $\log g = 5$.

Table 2. Parameters of giant models (see Bessell et al. 1989). The models X350H and X350C represent two solutions of the structure equations with the same parameter set (see text).

model	luminosity/$L_\odot$	mass/$M_\odot$	radius/$R_\odot$	T_{eff}	log g	chem. comp.
ZZ300	2×10^3	1	166	3000	-0.00	3× solar
Z300	2×10^3	1	166	3000	-0.00	solar
WW300	2×10^3	1	166	3000	-0.00	0.1× solar
X350C	10^4	1	273	3500	-0.44	solar
X350H	10^4	1	273	3500	-0.44	solar

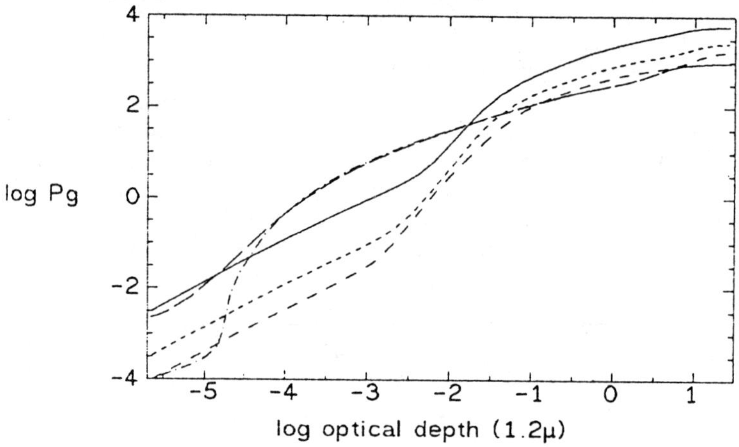

Fig. 3. Same as Fig. 2 but for the giant models ZZ300 (- - -), Z300 (– – –), WW300 (—), X350C (- . -) and X350H (— —). The model parameters are given in Table 1.

Therefore, we give in Figs. 2 and 3 the run of the pressures as a function of the optical depth at 1.2μ (which in most cases is similar to the Rosseland depth). From this representation it is obvious that the gas pressures in the dwarfs are 2 to 5 orders of magnitude higher than in the giants and therefore are comparable to those in cool white dwarfs. A reduction in the metallicity can further dramatically increase the pressures. This ensues from the fact that molecules make up the opacity and that their abundance depends usually quadratically on the metallicity.

The corresponding temperature distributions (Figs. 4 and 5) also show significant differences between dwarfs and giants: in the outer (optically thin) parts the dwarfs have to a very good approximation a constant temperature whereas for the giants it drops $\propto 1/\sqrt{r}$ as a consequence of the geometrical extensions

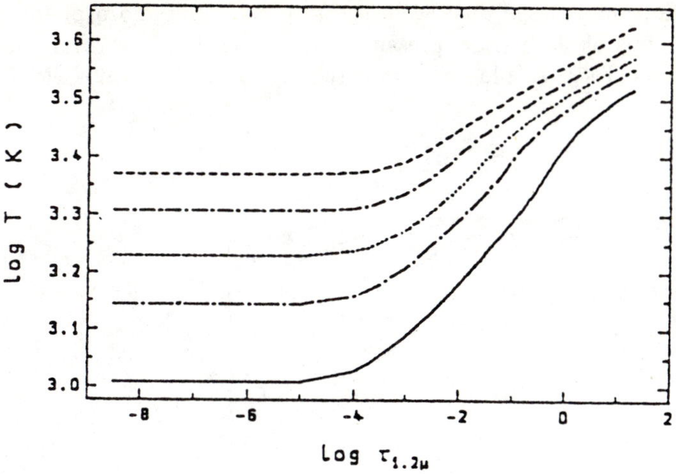

Fig. 4. Same as Fig. 2 but for the temperature distribution

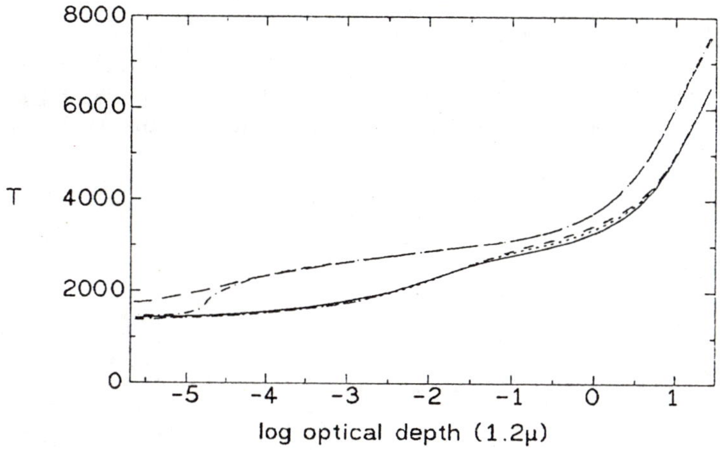

Fig. 5. Same as Fig. 3 but for the temperature distribution

of the atmospheres (in principle, the same is true for dwarfs; however, here the density decrease is so rapid at practically constant r that this effect is not seen in the spectra). On the other hand, in the inner parts, the temperatures of giants rise very fast since all energy is transported by radiation, whereas for the dwarfs convection reduces the temperature gradient strongly. Therefore, for identical effective temperature the giants' temperatures are lower in the outer atmospheric layers and much higher in the deeper layers than the temperatures for dwarfs.

These pressures and temperatures imply that the photospheric extinction in M stars is essentially determined by molecular absorption (whereas in so-called infrared stars significant amounts of dust seem to form inside the photosphere). Figs. 6 - 11 give the runs of the partial pressures (normalized to the total gas pressure) of some molecules that play an important role for the atmospheric structure.

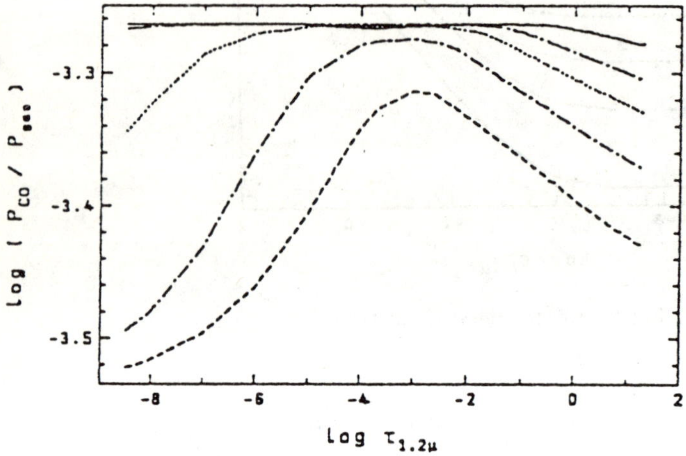

Fig. 6. Partial pressure of the CO molecule (normalized to the total gas pressure) as a function of the optical depth at 1.2 μ in dwarf atmospheres (from Allard 1990).

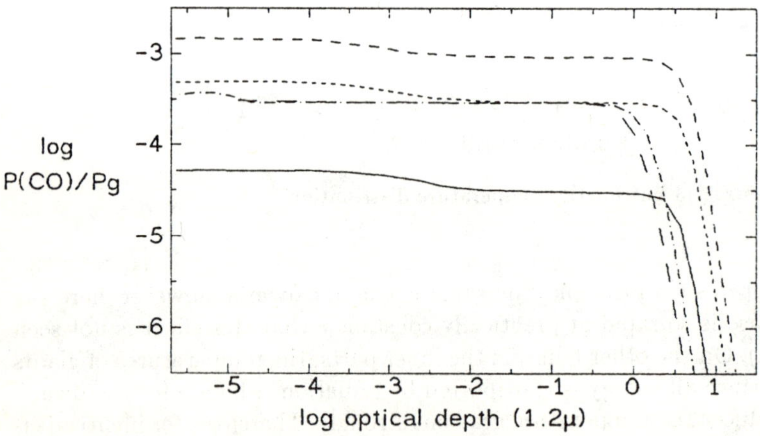

Fig. 7. Same as Fig. 6 but for the giants of Fig. 3

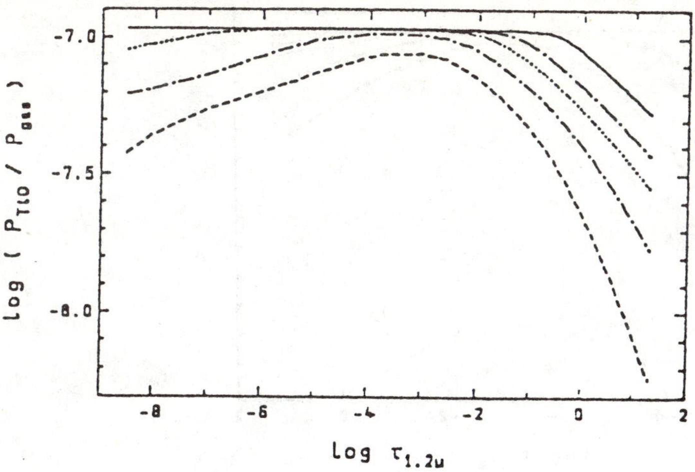

Fig. 8. Same as Fig. 6 but for TiO.

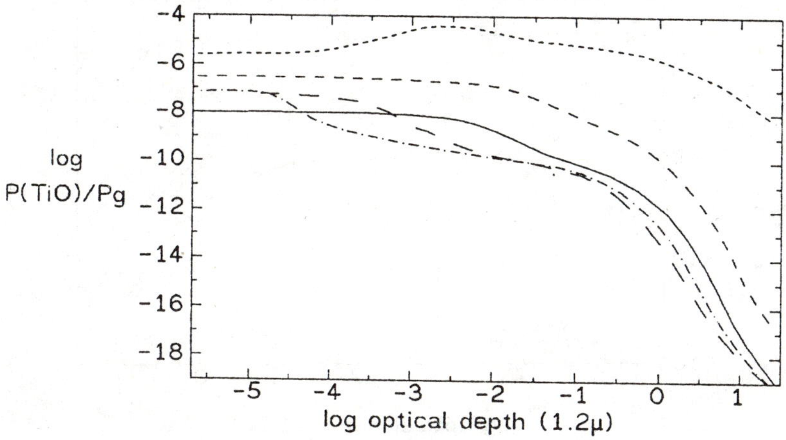

Fig. 9. Same as Fig. 7 but for TiO

3 Effects of incomplete and incorrect opacities

3.1 First order estimates of direct and indirect effects

Due to the low temperatures (and high pressures in dwarfs) in the atmospheres of M stars errors in the dissociation energies and/or temperature fluctuations lead to serious changes in the equilibrium constants and therefore often in the particle concentrations. For example, uncertainties of 0.5 eV for a dissociation energy of 5 eV and of 50 K for a temperature of 1600 K imply a change in

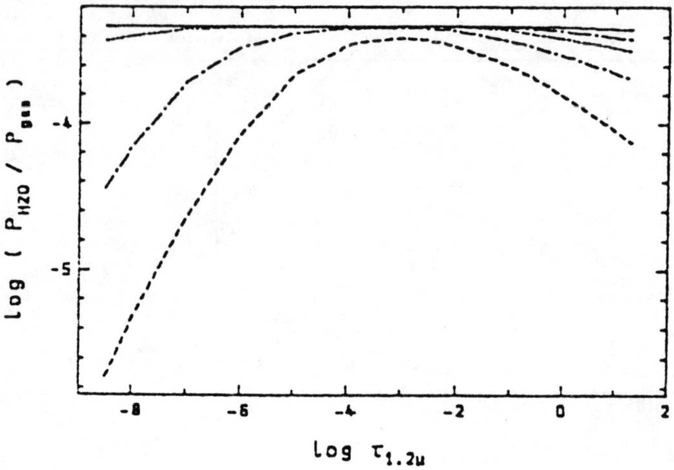

Fig. 10. Same as Fig. 6 but for H_2O

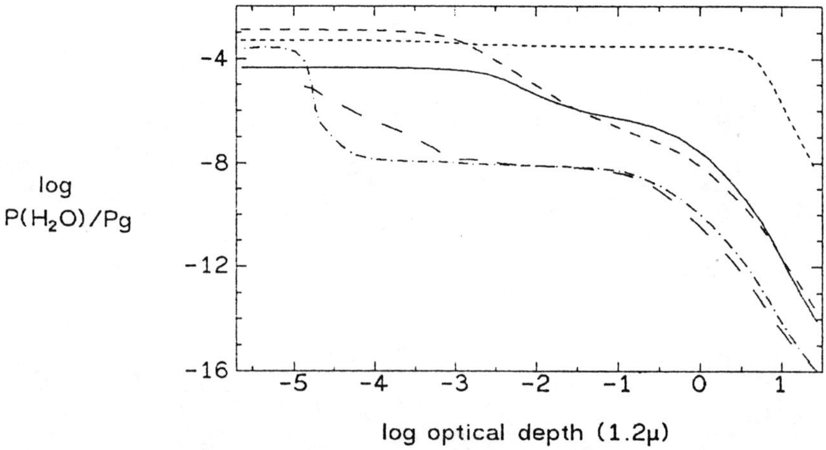

Fig. 11. Same as Fig. 7 but for H_2O

the equilibrium constant of a factor 20 (!) corresponding to an uncertainty in the density of a given species up to the same factor (see Costes & Naulin, this volume, Eq. (2)). Note, however, that in a spectral analysis the molecular data are considered to be fixed laboratory quantities and the temperatures are to be adjusted to such values that the band strengths and therefore the molecular concentrations are fitted.

The consequences of incorrect and/or incomplete opacities can easily be estimated if we consider the radiative transfer equation (or radiative transport equation) which reads in its conventional form for spherical media (Mihalas 1978)

$$\cos\theta\frac{\partial I_\lambda}{\partial r} + \frac{\sin^2\theta}{r}\frac{\partial I_\lambda}{\partial\cos\theta} = (\kappa_\lambda + \sigma_\lambda)(S_\lambda - I_\lambda) \tag{1}$$

(r = radial coordinate, θ = polar angle, S_λ = source function = ratio of emission to extinction). After discretization in space and angle it reduces to a linear system of algebraic equations that can be written in matrix notation for wavelength λ

$$\mathbf{A}_\lambda \mathbf{I}_\lambda = \mathbf{B}_\lambda, \tag{2}$$

or

$$\mathbf{I}_\lambda = \mathbf{A}_\lambda{}^{-1}\mathbf{B}_\lambda \tag{3}$$

where the intensity vector $\mathbf{I}_\lambda$ contains as components the specific intensities for all depths and angles, $\mathbf{A}_\lambda$ describes the spatial and angular couplings between the intensity components, and $\mathbf{B}_\lambda$ gives the photon sources. $\mathbf{A}_\lambda$ and $\mathbf{B}_\lambda$ depend linearly on the absorption and scattering coefficients per unit volume, κ_λ and σ_λ, resp. Therefore an uncertainty or modification of κ_λ or σ_λ implies that

$$\begin{aligned} \mathbf{A}_\lambda &\rightarrow \mathbf{A}_\lambda + \Delta\mathbf{A}_\lambda \\ \mathbf{B}_\lambda &\rightarrow \mathbf{B}_\lambda + \Delta\mathbf{B}_\lambda. \end{aligned} \tag{4}$$

As a consequence, the intensity vector is changed to first order by

$$\Delta\mathbf{I}_\lambda = \Delta\mathbf{A}_\lambda{}^{-1}\Delta\mathbf{B}_\lambda - \mathbf{A}_\lambda{}^{-1}\Delta\mathbf{A}_\lambda\mathbf{I}_\lambda \tag{5}$$

Equation (5) gives the direct effects that are maximal if the norm $||\mathbf{A}_\lambda||$ is minimal as is well known (see Unsöld 1958). They are relevant only at the considered wavelength. The indirect effects that concern the whole spectrum result from a change in the temperature distribution $\Delta\mathbf{T}$ induced by the opacity changes. For a static configuration in radiative equilibrium they can be estimated from the requirement that the total flux has to be the same at all depths. If we write the flux vector

$$\mathbf{F}_\lambda = \mathbf{g}\mathbf{A}_\lambda{}^{-1}\mathbf{B}_\lambda \tag{6}$$

(the matrix $\mathbf{g}$ contains the weights for the angle integration) the temperature modification due to the change in the opacity is

$$\begin{aligned} \Delta\mathbf{T} \approx \quad & \tfrac{\pi}{4\sigma_s}\Big(\mathbf{g}\textstyle\int_0^\infty \Big[-\mathbf{A}_\lambda{}^{-1}\frac{\partial\mathbf{A}_\lambda}{\partial\mathbf{T}}\mathbf{A}_\lambda{}^{-1}\Delta\mathbf{B}_\lambda + \mathbf{A}_\lambda{}^{-1}\frac{\partial\Delta\mathbf{A}_\lambda}{\partial\mathbf{T}} \\ & -\mathbf{A}_\lambda{}^{-1}\frac{\partial\mathbf{A}_\lambda}{\partial\mathbf{T}}\mathbf{A}_\lambda{}^{-1}\Delta\mathbf{A}_\lambda\mathbf{I}_\lambda + \mathbf{A}_\lambda{}^{-1}\frac{\partial\Delta\mathbf{A}_\lambda}{\partial\mathbf{T}}\Big]\,d\lambda\Big)^{-1}\textstyle\int_0^\infty \Delta\mathbf{F}_\lambda d\lambda \end{aligned} \tag{7}$$

(σ_s indicates the Stefan-Boltzmann constant). The expression shows that for the indirect changes the *temperature derivatives* of $\mathbf{A}_\lambda$ and $\mathbf{B}_\lambda$ as well as of $\Delta\mathbf{A}_\lambda$ and $\Delta\mathbf{B}_\lambda$ are of particular importance, i.e. that a species like H_2O with a strong temperature dependence of its concentration has a much stronger effect on the stratification than a molecule that hardly varies with $\mathbf{T}$.

Since the dwarfs have convection zones that extend to very close to the surface the radiative temperature gradient $\nabla_r = (d\ln T/d\ln P)_r$ plays a central role for the atmospheric structure. A similar calculation as above shows that an additional opacity changes ∇_r not only directly but also via its first and *second* temperature derivative.

3.2 Opacity induced temperature ambiguities in M giant atmospheres

The considerations so far have been completely general. However, we will discuss now two effects that are particular to M stars as far as we know:

If in giant atmospheres a molecule forms that can absorb effectively at infrared wavelengths carrying high fluxes (so that it cools via a picket-fence type cooling mechanism due to a near-surface absorber; see e.g. Mihalas 1978) and that has an increasing concentration with decreasing temperature, there may be a positive feedback: the molecule starts to form, it cools the upper atmosphere somewhat, the monochromatic absorption of the molecule increases, the layers of relevant monochromatic optical depth move outward resulting in an even lower temperature so that still more molecules form that in turn cool even further etc. The process only stops when equilibrium is reached and no more absorbing particles are formed. In actual calculations the feedback can be initialized by the iteration to radiative equilibrium and can lead for one given parameter set to two temperature distributions that fulfil all structure equations. An example is given by the models X350H (without H_2O cooling) and X350C (with H_2O cooling) in Fig. 3 (see also the results of detailed numerical experiments by Jørgensen, this volume).

3.3 The radiative layer and the structure of M dwarfs

The second effect occurs in the M dwarfs and is a consequence of the fact that these stars are fully convective except for a tiny layer very close to the surface (mean optical depth often smaller than 0.01): errors in the opacity lead to errors in the radiative gradient ∇_r and these in turn lead to errors in the optical depth, i.e. at the pressure and temperature, where convection starts. This implies that the "wrong adiabate" is entered. Since the real temperature gradient is very close to the adiabatic one for the high pressures of M dwarfs it means that such an error is hardly damped out and therefore the whole stellar structure may be erroneous. Note also that additional errors may be introduced by discretisation schemes for the radiative transfer equation (as e.g. the Feautrier scheme, see Mihalas 1978) that are of first order only close to the surface.

4 Comparison of synthetic and observed M star spectra

After the pioneering dwarf models of Mould (1976) progress has recently been made by the PhD theses of Kui (1991) and Allard (1990; see also Allard et al. 1992) that include both the relevant bands and proper radiative transfer. Allard also includes some metal line blanketing. For warmer objects quite satisfactory agreements with observations could be obtained (Leinert et al. 1990, Kirkpatrick et al. 1993) in most wavelength ranges; in particular, for Gl 866 also a very good fit for the NaI D lines could be obtained. By selecting better transition probabilities Allard (1994) was able to get in addition very good fits to infrared spectra of very cool objects (except for wavelength regions where water vapor dominates) and therefore eliminates the problem of too flat flux gradients found by Kui. However, it should be pointed out that these comparisons were – with the exception of line profiles for Gl 866 – all made for data of rather low resolution. When high resolution observations for dwarfs become available, new problems have to be expected.

For giants, comparisons between synthetic and observed spectra also show satisfactory agreement at most effective temperatures over wide portions of the spectrum, but various significant discrepancies remain. A comprehensive review of recent achievements in interpreting red giant spectra is given by Jørgensen 1991. For the Mira variables among the late M type giants, spectrum synthesis is still in a pioneer stage (see Bessell & Scholz 1989; Scholz 1992) since the structure of the atmospheres is determined by the propagation of shock fronts.

5 Data needed for future improvements of synthetic M star spectra

From the description given above it should be evident that still much work is needed for a full understanding of M star atmospheres. Regarding the opacities the main needs concern to TiO, H_2O, FeH, VO;

- complete line lists (so that the sum rules are well fulfilled) and corresponding transition probabilities that are accurate to better than 10%;
- accurate thermodynamical data, in particular partition functions; they should be in a form that the first and second temperature derivatives could be calculated accurately;
- data on line broadening.

In addition, it should be mentioned that for the modeling of very cool and luminous giants as Mira stars it is necessary to have highly accurate ($\lambda/\Delta\lambda \approx 3 \times 10^5$) relative line positions since the atmospheres of these objects are no longer in hydrostatic equilibrium but pulsate and therefore the correlation of line positions has a strong effect on the temperature distribution and the emitted spectrum (as can be seen from the $\partial I/\partial\lambda$ term in the transfer equation for the comoving frame, see Scholz 1992). Furthermore, all M dwarfs seem to possess coronae that radiate strongly in the X ray range (Th. Fleming, private communication). Thus it can be expected that quite a high X ray flux impinges on the matter of the photosphere and changes significantly its thermodynamical state. In particular, the level populations are very far from their LTE values (see Johnson, this volume). If we want to model this effect we need enormous amounts of photoionization, photoexcitation and collisional excitation cross-sections as well.

References

Allard F., 1994, In: P.Thejll & U.G.Jørgensen (eds.), *Poster session proceedings of IAU Coll. 146*, Copenhagen Univ., p. 1

Allard F., 1990, Ph.D. thesis, Heidelberg University, Heidelberg
Allard F., Scholz M., Wehrse R., 1992, *Rev. Mexicana Astron. Astrof.*, **23**, 203
Bessell M.S., Brett J.M., Scholz M., Wood P.R., 1989, *Astron. Astrophys. Suppl. Ser.*, **77**, 1
Bessell M.S., Scholz M., 1989, in: Evolution of Peculiar Red Giants, Johnson, H.R., Zuckerman, B., eds., Cambridge Univ. Press, p. 67
Green P., Margon B., Anderson S.F., MacConnell D.J., 1992, *Astrophys. J.*, **400**, 659
Jaschek C., Jaschek M., 1987, The classification of stars, Cambridge Univ. Press
Jørgensen U.G., 1991, in: The Infrared Spectral Region of Stars, Jaschek C., Andrillat Y., eds., Cambridge Univ. Press, p. 165
Kirkpatrick J.D., Kelly D.M., Rieke G.H., Liebert J., Allard F., Wehrse R., 1993, *Astrophys. J.* **402**, 643
Kui R., 1991, Ph.D. thesis, Australian National University, Canberra
Leinert Ch., Haas M., Allard F., Wehrse R., McCarthy D.W., Jahreiß H., Perrier Ch., 1990, *Astron. Astrophys.*, **236**, 399
Mihalas, D., 1978, Stellar Atmospheres, Freeman
Mould J.R., 1976, *Astron. Astrophys.*, **48**, 443
Schmidt-Kaler Th., 1982, in: Landolt-Börnstein, Vol. VI/2b, Schaifers, K., Voigt, H.H., eds., Springer, §4.3
Scholz M., 1992, *Astron. Astrophys.* **253**, 203
Unsöld A., 1958, Physik der Sternatmosphären, Springer
Vernazza J.E., Avrett E.H., Loeser R., 1981, *Astrophys. J. Suppl.*, **45**, 635

Molecular Opacities in Cool Dwarf Stars

James Liebert

Steward Observatory, University of Arizona, Tucson AZ 85721, USA

1 Introduction

The term dwarf stars identifies objects of small radius in the Hertzsprung–Russell (H–R) Diagram, but encompasses more than one phase of stellar evolution. The M dwarfs (type dM) populate the main sequence at the low mass end; these are the coolest core hydrogen-burning stars. They belong generally to the Galactic disk, or Population I, have relatively small space motions with respect to the Sun, and have similar metallicities to the Sun (although perhaps only within a factor of several). In particular, this means that the abundance of oxygen is always greater than that of carbon. The M subdwarfs (sdM) are the Population II counterparts, showing low metallicities and high space motions. Because they have smaller radii, they define a main sequence at lower luminosity than the M dwarfs for a given temperature. Hence the term subdwarf.

A third type of dwarf star is the white dwarf. This is an object which has ended its nuclear-burning life, and has contracted into an electron-degenerate configuration the size of the Earth, most likely with a core composed of the products of helium-burning – carbon and oxygen. Sometimes called degenerate dwarfs, these objects fall along a diagonal cooling track in an H–R Diagram. They range in temperature (luminosity) from higher than 100,000 K (above solar luminosity) to perhaps 4,000 K ($< 10^{-4}\ L_{\odot}$), depending on how long the object has been cooling at a constant radius.

Likewise, there may be objects called brown dwarfs, too low in mass to initiate hydrogen burning. They radiate first by the release of half the energy of gravitational contraction, may initiate limited cycles of nuclear-burning for a finite period, after which they enter a cooling track for hydrogen-rich configurations; they have roughly ten times the radii of carbon-oxygen degenerate dwarfs.

Finally, in the last two decades, a fifth category has been recognized – the carbon dwarf. This is a low mass main sequence star whose carbon abundance exceeds the oxygen abundance. Normally, in the dM–sdM atmospheres the carbon is locked up in CO and a few other molecules. In the dwarf carbon (dC) stars, it is the oxygen that is locked up in CO while the spectra reveal a rich chemistry of carbon-based molecules.

In the following sections, I will review the properties of the spectra and the analyses for each of these categories except for the white dwarfs (see Thejll this volume).

2 M Dwarf and Subdwarf Stars

There has been relatively little attention given to studying the spectra and atmospheres of the dM stars, less so even than for M giants. With relatively low temperatures and high pressures, these atmospheres are fertile ground for molecule formation. Given the modest amount of quantitative data for molecules more complex than diatomic, one might fear that our ability to match observations with synthetic spectra may be quite limited for this type of star. Indeed, up until recently, very few relevant model atmosphere calculations even existed, and researchers routinely estimated the temperatures of these stars by fitting colors (and/or spectra) to blackbodies. As we shall see, the energy distributions of these stars are very unlike that of a blackbody.

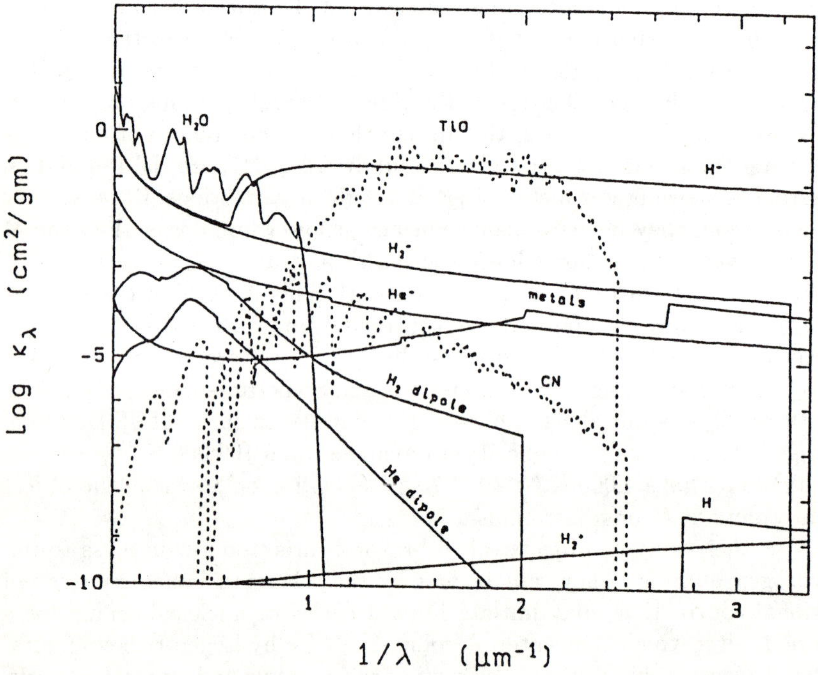

Fig. 1. Opacities calculated for various molecular and atomic constituents plotted against wave number for a 3000 K, log g = 5, solar composition model from Allard (1990). Note the prominent role of H_2O, TiO and H^-.

Fortunately, in the last few years the situation has changed dramatically for the better. This is primarily due to two PhD dissertations – by Allard (1990) at the University of Heidelberg, and by Ruan (1991) at Mt. Stromlo Observatory and the Australian National University. Allard (1994) adds the exclamation point to some of the following comments concerning her dissertation and some subsequent work. Unfortunately, neither Allard's nor Ruan's dissertation have been published.

The molecular (and atomic) opacities important in M dwarf stars are well illustrated in Fig. 1, taken from Allard's (1990) thesis, for a model at 3,000 K, log g = 5, and solar composition. Two of the dominant opacity sources are the TiO molecule in the optical and H_2O in the infrared. Dominant continuous opacity sources include H_2^- in the infrared and H^- in the optical.

For low metallicity stars (the subdwarf M), the pressure-induced dipole opacities due (see Borysow this volume) due to H_2-H_2 and H_2-He can dominate the infrared opacity, while H^- is strongest in the optical. Molecules involving heavier elements are almost completely restricted to hydrides – MgH, CaH being the strongest in the optical spectrum.

What can we say about the accuracy of these models and the fits of synthetic to observed spectra? Certainly, a number of improvements to the physics have been made in recent years. It appears that these models now can yield improved estimates of the effective temperature ($T_{\rm eff}$) of low mass stars.

Kirkpatrick et al. (1993a) made the first comprehensive attempt to fit a sequence of M dwarf spectroscopic standards, from early M spectral types at $\geq$ 3500 K to as low as 2700 K. Two examples of these fits are shown as Fig. 2. The overall energy distribution is fit rather well, though the quantitative fits to TiO and other band strengths are only fair, and become worse at shorter wavelengths.

Since all of these near neighbors of the Sun have excellent trigonometric parallax measurements, the luminosities of the stars can be determined rather

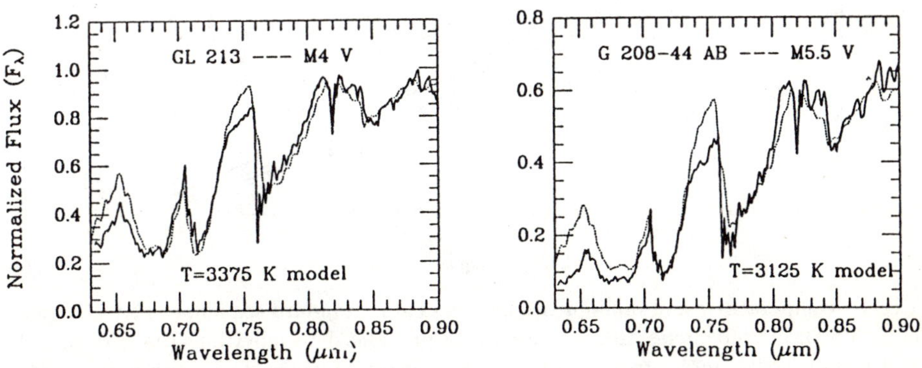

Fig. 2. Two examples of fits to optical spectra of middle M dwarfs using Allard models, from Kirkpatrick et al. (1993a).

accurately once the temperatures are estimated. The derived positions in a Hertzsprung-Russell (log L vs. log $T_{\rm eff}$) diagram can then be compared to the predictions of stellar interiors models for main sequence stars of very low mass.

When this comparison was made using $T_{\rm eff}$ values estimated by fitting blackbodies, the observed stars appeared usually to have smaller $T_{\rm eff}$ at a given luminosity – or alternatively a larger radius – than predicted by the interiors models (see Burrows & Liebert 1993, Fig. 19). This implied either that calculated atmospheric fits were wrong, the interiors models were wrong, or that many low luminosity stellar objects in the solar neighborhood were in fact substellar mass "brown dwarfs."

Using the Allard atmosphere models, however, Kirkpatrick et al. (1993a) find essentially a $T_{\rm eff}$ scale for the sequence of M dwarfs which is in fair agreement with the interior models – see Fig. 3. This implies that the atmospheric and interiors models are consistent with each other and with the assumption that the observed objects are hydrogen-burning stars (M dwarfs). However, the agreement was still poor for the lowest mass stars with $T_{\rm eff} \leq 3{,}000$ K.

Improvements made since this 1994 paper (see Allard 1994) include notably the addition of FeH bands near one micron in the spectrum and the treatment of

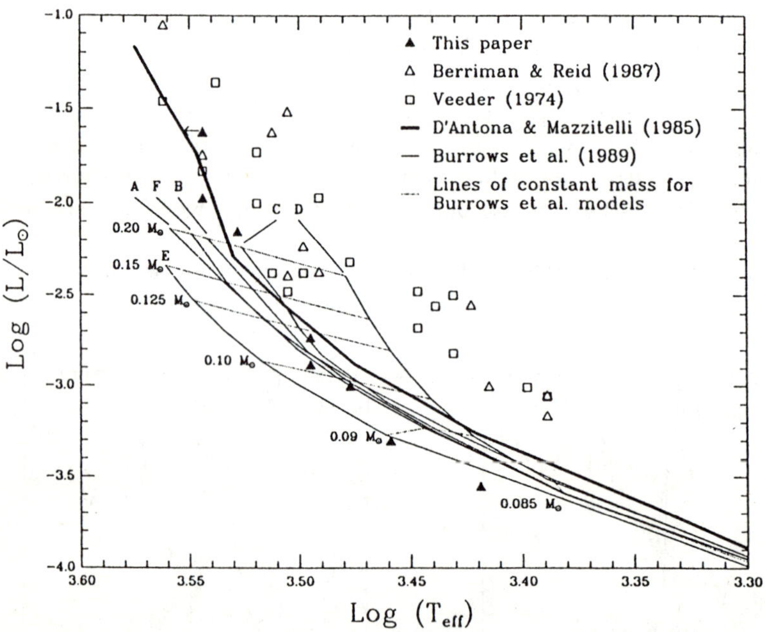

Fig. 3. Temperatures and luminosities for the M star standard sequence of Kirkpatrick et al. (1993a) (filled circles) are compared with previous determinations (open symbols) and various stellar interiors models (curves) for main sequence stars, with mass points labelled.

the pressure-induced dipole opacities by Lenzuni et al. (1991). Now the synthetic spectrum fit to one of the lowest luminosity stars in the Kirkpatrick et al. (1993a) is much improved for $T_{\rm eff}$ of 2,800 K; the fit has now been extended to include observations to 3.5 μ by Jones et al. (1993) (see also review by Miller et al. in this volume). The complicated band spectrum of H_2O remains the important piece which does not fit well.

Can we now be confident of the atmospheric and stellar paramaters derivable for low mass stellar spectra? Such an attitude would be premature. The models predict far too much flux in the visual and bluer wavelengths, which must tell us that there are still many opacity sources to be added to the models. A similar problem persists with attempts to fit subdwarf M stars with models having drastically reduced metallicities. The missing components undoubtedly are primarily molecular in nature, and for the subdwarfs AlH may be particularly important at blue wavelengths. But we are off to a good start, finally, with M dwarf stars.

3 Brown Dwarf Candidates

Not a single object has yet been discovered outside of our solar system (but not including evolved, binary systems) that can unambiguously be proclaimed a substellar object. However, this situation is arguably due to the fact that it is so difficult to prove that a given candidate is a brown dwarf, and does not necessarily mean that substellar objects do not exist. Indeed, numerous candidates exist, and more are being found at a rapid rate. A detailed review of the candidates and the methods of discovery is given in Burrows & Liebert (1993).

Here I wish to discuss only two of the candidates – among the few for which more than a little spectrophotometric information exists. The first, GD165B, is a spatially-resolved companion to a white dwarf star and clearly has a much lower temperature than any visible object yet known (Becklin & Zuckerman 1988). A poor-quality optical spectrum (Kirkpatrick et al. 1993b) appears to differ qualitatively from those of the least luminous known stars – see Fig. 4 taken from that paper, Davis (this volume) discuss the possibility that a band system of VO, expected to grow stronger at cooler $T_{\rm eff}$, may explain the enhanced absorption near 8500Å found by Kirkpatrick et al. (1993b).

The observations are not yet accurate enough for the $T_{\rm eff}$ of GD165B to be estimated, although the infrared colors suggest it is at least several hundreds of degrees cooler than vB10, for example. Still, the implied luminosity is nearly 10^{-4} $L_{\odot}$, and is consistent with high opacity main sequence models near the stellar mass limit of 0.075–0.08 $M_{\odot}$. GD165B could be a hydrogen-burning star, but then very likely it would be the lowest mass star known. A second brown dwarf candidate for which considerable (and very confusing!) information exists is G29–38B, another potential companion to a white dwarf discovered by the same group (Zuckerman & Becklin 1987). I say "potential companion" because what these authors found was that the white dwarf has a spatially-unresolved infrared

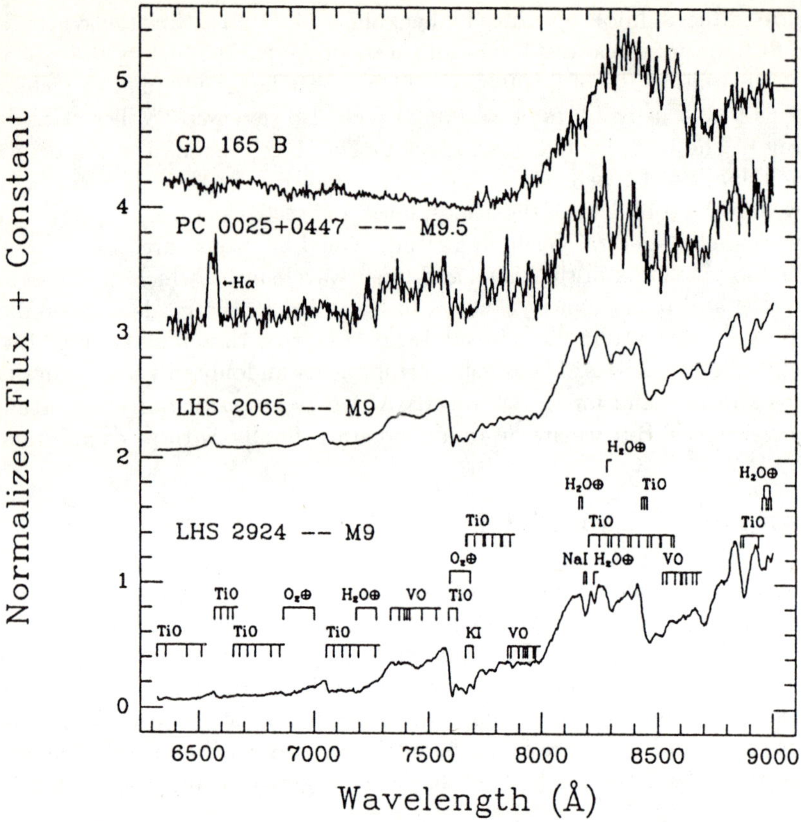

Fig. 4. Spectra of four of the coolest M dwarf stars known, the top of which, GD165B, is discussed in the text. Molecular band features are labelled for LHS2924 (from Kirkpatrick et al. 1993b).

excess, which they initially suggested might be a brown dwarf companion having a $T_{\rm eff}$ of about 1,200 K and a considerable luminosity of 5 x $10^{-5}L_{\odot}$, implying a radius not too different from that expected for a brown dwarf. (Brown dwarfs from $0.001M_{\odot}$ to stars of $0.08M_{\odot}$ basically should have a radius similar to that of Jupiter similar to that of Jupiter which is $0.1R_{\odot}$). These and subsequent authors have detected the excess out to wavelengths as long as ten microns, and the flux at these long wavelengths is far in excess of that expected from the above $T_{\rm eff}$. Further attempts to spatially resolve the infrared source from the white dwarf have failed, and infrared spectra show no evidence for spectral features. Hence, it has been proposed that the excess is due to some kind of dust shell around the white dwarf. However, it should be noted that the white dwarf has a cooling age of order 10^9 years, and may show small amplitude radial velocity variations consistent with having a companion of substellar mass. This complicated situation is reviewed more extensively by Zuckerman (1992), from which more detailed references to the work on G29–38 may be found.

Numerous brown dwarf candidates are apparently being found as members of young clusters, including some imbedded in giant molecular clouds and other regions of recent star-formation – see Zinnecker et al. (1993) and Burrows & Liebert (1993).

I wish to promote the bias that the cup is best considered half full in the widening search for brown dwarfs. To be sure, the optimist must acknowledge that there have been several false claims, missteps, and some null results. But the biggest problem is how we can recognize a brown dwarf, if we find it, from an energy distribution or spectrum. This may require the calculation of spectra and energy distributions for objects with $T_{\rm eff}$ of 1,000–2,000 K. Such atmospheres would have some properties intermediate between those of Jovian planets and low mass stars, as explored by Lunine et al. (1986) and by Tsuji (this volume). Needless to say, different molecular opacities will be important in trying to calculate an emergent spectrum in this intermediate temperature range. Substellar atmospheres may be a challenging new subject in the decades to come.

4 Dwarf Carbon Stars

The first carbon dwarf, listed in the Lowell Observatory proper motion catalogue as G77–61, was discovered by Dahn et al. (1977) amongst the few hundred nearby stars with trigonometric parallax measurements in the U.S. Naval Observatory program. For more than ten years, this 13th magnitude star remained unique among low mass dwarf stars in showing strong C_2 and CH with weak CN, instead of the normal TiO features. This implied that most oxygen was tied up in CO molecules, instead of most of the carbon. At $M_{bol} \sim +9$, it has the luminosity of an early M or late K dwarf, a bit more than $10^{-2} L_{\odot}$, and $T_{\rm eff}$ near 4,000 K.

Its existence posed a problem for stellar evolution: While G77–61 has halo-like space motions and apparently a very low abundance of heavy elements (Gass et al. 1988), Population II stars consistently show higher O/C ratios than stars of higher metallicity. Thus it is very unlikely that such a star could form out of material with C/O > 1. The normal explanation for luminous carbon stars is that these are asymptotic giant branch (AGB) objects with carbon-oxygen cores powered by both hydrogen- and helium-burning shell sources (see Lambert this volume). Due to mixing processes associated with thermal flashes of the helium-burning shell, some carbon from the stellar core gets dredged up into the outer envelope and stellar atmosphere, inverting the O/C ratio.

However, it has been known for a long time that there are carbon stars too low in luminosity to be in such an advanced phase of evolution. The subgiant CH stars, and related types of stars, have generally turned out to be binary; this permits the explanation that a now-unseen companion transferred material to the now-visible star when in the carbon AGB phase. And, sure enough, G77–61 turned out to be a 243-day spectroscopic binary; its likely white dwarf companion is too cool for detection, even with the International Ultraviolet Explorer (IUE) Observatory (Dearborn et al. 1986).

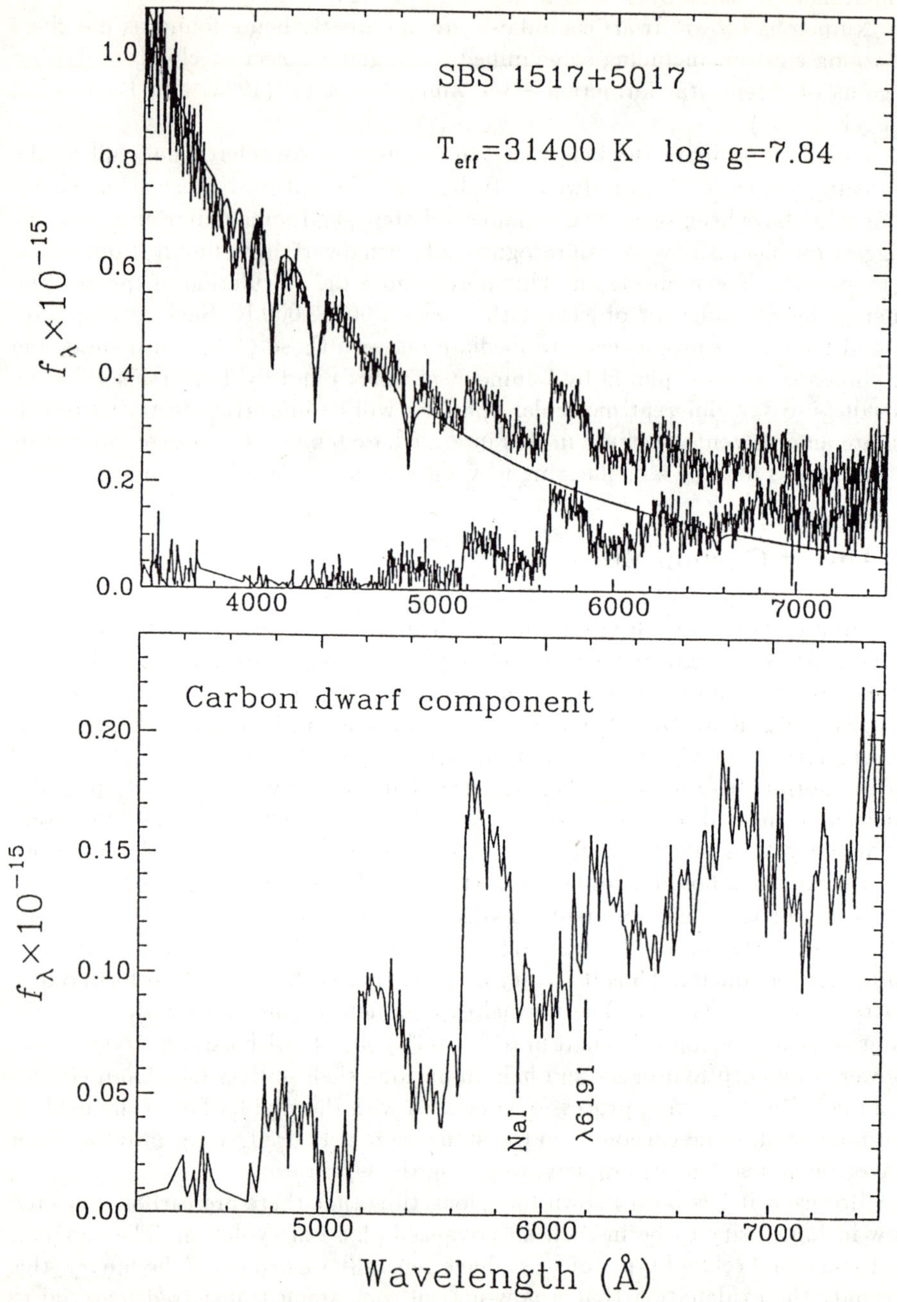

Fig. 5. (a) (top) composite spectrum of DA + dC components; (middle curve) best fitting white dwarf atmosphere model; (bottom) resultant spectrum of carbon dwarf star, as described in text; (b) carbon dwarf spectrum on an expanded scale.

In the last few years, new discoveries have swelled the number of dwarf carbon stars to ten (cf. Green, Margon and MacConnell 1992; Warren et al. 1993). They all appear to be remarkably similar in T_{eff} and luminosity to G77–61. Most, though apparently not all of these, show the kinematics of halo or at least old disk stars, though none of the new discoveries have abundance analyses. It is not known whether all are binaries.

In Fig. 5, I illustrate the spectra of the dwarf carbon stars with the latest discovery – called SBS 1517+5017 from the Russian Second Byurakan Survey (Markarian & Stepanian 1983). This is one of two known cases where a white dwarf companion is actually visible (Liebert et al. 1994). The blue side of this composite optical spectrum (top tracing of the figure) is dominated by the broad hydrogen lines of the DA white dwarf, the hot component of the system. This component can be modelled accurately with an atmosphere having T_{eff} = 31,300 K and log g= 7.84 and pure hydrogen component; the synthetic spectrum from this model (middle curve) can then be subtracted from the composite spectrum, leaving the bottom tracing – the resultant spectrum of the dwarf carbon component of the binary, which dominates longward of 5000Å. The Swan bandheads of C_2 degrading to shorter wavelengths are clearly visible near 5100Å and 5600Å.

The atmospheric modelling of G77–61 by Gass et al. (1988) is the only attempt thus far to analyze a carbon dwarf; this study must be regarded as exploratory. The formation of carbon-based molecules is greatly enhanced at these high pressures compared with the atmospheres of AGB carbon stars, so the opacity set in this study, must be inadequate. The amount of opacity in AGB carbon stars increases with decreasing wavelength, and there is a strong indication that the same is true for dwarfs. While the fits to the red part of the spectrum in the Gass et al. analysis are adequate, far too much flux is predicted at shorter wavelengths, which must be a direct consequence of the missing opacity. It is obvious that much work remains to be done in order to produce accurate models of this rare but interesting new type of dwarf star.

Acknowledgments. It is a pleasure to thank F. Allard, R. Wehrse and T. Tsuji for illuminating discussions, and F. Allard for some results in advance of publication. The author acknowledges financial support from the National Science Foundation through grant AST 92-17961.

References

Allard F., 1990, *Ph.D. thesis, Univ. Heidelberg*

Allard F., 1994, In: P.Thejll & U.G.Jørgensen (eds.), *Poster session proceedings of IAU Coll. 146*, Copenhagen Univ., p. 1

Becklin E.E., Zuckerman B., 1988, *Nature*, **336**, 656

Burrows A., Liebert J., 1993, *Rev. Mod. Phys.*, **65**, 301

Dahn C.C., Liebert J., Kron R.G., Spinrad H., Hintzen P.M., 1977, *Astrophys. J.*, **216**, 757

Dearborn D.S.P., Liebert J., Aaronson M., Dahn C.C., Harrington R.S., Mould J.,Greenstein J.L., 1986, *Astrophys. J.*, **300**, 314

Gass H., Liebert J., Wehrse R., 1988, *Astron. Astrophys.*, **189**, 194

Green P., Margon B., MacConnell J., 1992, *Astrophys. J. Let.*, **380**, L31

Kirkpatrick J.D., Kelly D.M., Rieke G.H., Liebert J., Allard F., Wehrse R., 1993, *Astrophys. J.*, **402**, 643

Kirkpatrick J.D., Henry T.J., Liebert J., 1993, *Astrophys. J.*, **406**, 701

Jones H.R.A., Longmore A.J., Miller S., Tennyson J., 1993, *Astrophys. J.*, in preparation

Liebert J., Schmidt G.D., Lesser M., Stepanian J.A., Lipovetsky V.A., Chaffee F.H., Foltz C.B., Bergeron P., 1994, *Astophys. J.*, in press

Lunine J.I., Hubbard W.B., Marley M.S., 1986, *Astrophys. J.*, **310**, 238

Markarian B.E., Stepanian J.A., 1983, *Astrophyzica*, **20**, 21

Ruan P., 1991, *PhD Dissertation, Australian National University*

Warren S.J., Irwin M.J., Evans D.W., Liebert J., Osmer P.S., Hewett P.C., 1993, *Mon. Not. R. Astr. Soc.*, **261**, 185

Zinnecker H., McCaughrean M.J., Wilking B.A., 1993, In: *Protostars and Planets III*, eds. E. Levy, J.I. Lunine, Univ. of Arizona Press, Tucson, in press

Zuckerman B., 1992, In: *Planets Around Pulsars*, ASP Conf. Series vol. 36, eds. J.A. Phillips, S.E. Thornett, S. Kulkarni, ASP, Provo UT, p. 303

Zuckerman B., Becklin E.E., 1987, *Nature*, **330**, 138

White Dwarfs

Peter Thejll

Niels Bohr Institute, Blegdamsvej 17, DK-2100 Copenhagen, Denmark

1 White dwarfs as the end product of most stellar evolution

It is the intention of this review to explain what white dwarfs are and why it is interesting to study them, and why the H_2^+ molecule is of special interest.

The evolution, from start to finish, of a star of mass less than about 2 solar masses ($M_\odot$), can roughly be summarized as follows:

- A cloud of gas contracts from the interstellar medium until hydrogen ignites at the center and a *main sequence* (MS) star forms. H is transformed to He and the MS phase continues until H is exhausted in the stellar core.
- H continues burning in a shell outside the He core while the core contracts. He "ashes" are added to the core, and a *red giant* star is formed as the envelope expands. The star evolves up the Red Giant Branch (RGB) (i.e. it becomes more and more luminous and the surface cools).
- Towards the end of the RGB phase, mass-loss from the upper layers increases until helium to carbon burning in the core ignites suddenly under degenerate conditions – this is called the *Helium Flash* (HF). The HF terminates the RGB evolution, and therefore also the mass-loss and the growth of the stellar core.
- The star readjusts its structure and the He-core burns steadily on the *horizontal branch* (HB) (a phase of nearly-constant luminosity) until fuel is exhausted in the He-core.
- Then the C/O core contracts anew and the expansion of the envelope, and the growth of the core, during He-shell burning, mimics RGB evolution but relatively little mass is added to the core this time.
- The second ascent of the giant branch (the so-called Asymptotic Giant Branch, or AGB) continues with increased mass loss towards the end
- Rapid detachment of a considerable fraction of the remaining envelope and the hot core takes place, sometimes observable as the *Planetary Nebulae* (PN) phase.
- The PN is dispersed as the core contracts to a white dwarf (WD).

– The WD cools for a long time, as internal kinetic energy and latent heat is released.

For stars of mass above 2 $M_\odot$ the evolution is different and will not be touched on here. A good, thorough introduction to stellar evolution can be found in Collins (1989) Kippenbahn & Weigert (1991).

White dwarfs represent the end-stage of evolution for 90% of all stars. The remaining 10% evolve into neutron stars or black holes; they undergo rapid MS evolution followed by a supernova explosion. Because most stars end up as white dwarfs one can learn much about stellar evolution history by studying them – the origins of WD properties are to be sought in the preceding evolution stages, as well as in basic physics.

The equation of state for degenerate matter determines when the ignition of the degenerate helium core takes place: this event essentially sets the mass of white dwarfs – if the flash is delayed the core grows larger and if the flash is early a smaller He-core is formed (Jørgensen & Thejll 1993). Also, mass-loss increases towards the end of the RGB ascent so that the point of termination of the ascent sets upper limits to the eventual WD envelope mass. The mass-loss events that follow during the AGB ascent and during the PN phase trim the envelope mass to its final WD value. The nature of the mass-loss mechanisms is not well understood, but empirical fits to observations of red giant star mass-loss (RGB stars and AGB stars) exist (Reimers 1977, and Niuewenhuijzen & de Jaeger 1990).

Residual burning of the H-shell in the upper layers throughout the evolution also deplete the amount of hydrogen left. Theory predicts that something like 10^{-4} $M_\odot$ of H will be left when the star finishes its AGB and PN phases. Thus, the total mass and detailed structure (envelope-mass, core mass, chemical composition of the layers, etc.) of white dwarfs depend on the physics of degenerate electrons, nuclear forces (fusion), atomic physics (opacities) and thermodynamics (structure, and conditions for convection).

After the PN has been dispersed the hot bare core of C/O contracts until the electrons are degenerate yet again and no further contraction is possible. Cooling for the rest of the evolution of the Universe follows. This means that *WD cooling can be used to measure the age of the star-forming regions of the Galaxy.* Cooling is rapid at first (e.g. tables of D'Antona & Mazzitelli 1990), with the temperature reaching half its maximum value in just one million years, but as the object cools the rate of cooling slows down. In about 10^{10} years it is expected that 3000K will be reached – in other words: it may be possible to find white dwarfs as old as the Galaxy itself. This method of measuring the age of the Galaxy (Wood 1992) is an important alternative to the other methods such as globular cluster chronology. No important nuclear reactions take place in white dwarfs once the degenerate state has been reached so the only source of energy for the flow of radiation is kinetic energy of the ions, and latent heat released in the crystallization of the ions that occurs towards the end. The rate of WD cooling is governed by the transparency of the surface layers. The interior is isothermal and the rate at which radiation leaks through the photosphere is set by the opacity. As the cooling rate slows down with lower T_{eff}, the WDs spend most of their life while cool, i.e. where molecular opacities are important - that is why it is worthwhile investing significant efforts in calculating the best possible molecular opacities for white dwarfs.

2 White dwarf characteristics

One important characteristic of white dwarfs is that the majority have more or less the same mass – 30% of the DAs have masses between 0.5 and 0.55 $M_\odot$(Bergeron et al. 1992). As stars of masses from near 1 $M_\odot$ up to about 8 $M_\odot$ all manage to become white dwarfs it is clear that mass-loss is needed in the preceding stages – and mass loss that is carefully tailored so as not to be too large nor too small.

The radiation we can see from white dwarfs comes from the photosphere - a layer only about 500 m thick. By analyzing spectra it is possible to determine T_{eff}, $\log(g)$ and the composition – using line profiles and/or photometry. At an early stage of WD research it became evident that WDs fall into 2 major groups: Those with H dominated photospheres (75% of the WDs), called DAs, and those dominated by He, called DBs, plus a few intermediary cases (DABs and DBAs) as well as stars contaminated by elements heavier than He (DQs: showing C and/or C_2 features, DZs: showing Ca and other "metal" features), and the DC that have only very faint depressions in their continua.

Explaining the origins of the abundance differences is a major goal of WD research as a successful explanation may have important bearing on earlier evolution stages: Because the gravitational field is so large, elements of different mass separate quickly, H floating on top of He etc. – this means that a DB may not contain any H at all and that perhaps such a lack can be explained by events in earlier stages of evolution where layers of the envelope was shed on the RGB, AGB and in the PN phase. Other possible explanations for this problem are discussed and reviewed by Shipman (1988, 1992).

As the WDs become cold it is observed that the relative abundance of DAs decreases – i.e. that photospheric H is absent. This could mean that any H present in cooling DAs is mixed with underlying He in a process governed by the efficiency of convection in He and the layer thickness of H left from the previous evolutionary stages – but unless the layer of H is thinner than about $10^{-10}M_\odot$, convective mixing cannot "mix away" the H. This is an important clue to the previous stages of stellar evolution: Mass loss before the WD stage is so tuned that the condition for the maximum H-layer thickness is met in some cool DA WDs.

Further clues to the internal structure of WDs can be had from observing oscillations of WDs. Astereoseismology interprets non-radial g-mode pulsations (Bradley 1993), seen on the stellar surface as brightness variations. Such analyses have the potential for revealing the stellar mass, the interior structure, the rotation of the star and the rate of cooling which is related to the composition of the core, and are thus important tools for studying the interior of WDs.

DBs as well as DAs oscillate, as do the much hotter PG1159-class objects (Winget et al. 1985). The oscillations are driven by mechanisms in which the opacity of the gas is coupled to its ionization-state so that the opacity depends on the pressure and temperature and these in turn depend on the radiation field. In H rich envelopes the oscillations take place near 10,000K, while the instability strip lies somewhere between 20,000 to 25,000K in the He-rich DBs, and even higher for oxygen (the PG1159's). Oscillations become feasible when a convection zone has properties suitable for sustaining the cycle of partial ionization in each of these elements. Therefore the physics of convection becomes linked to the study of oscillations.

This coupling has been utilized in attempts to test current convection theories, such as the mixing-length approach (Fontainc 1984). Such theories contain rather arbitrarily chosen parameters that in principle can be tuned by requiring them to have such values that the predicted temperatures of instability strips coincide with the observed temperatures. Here accurate knowledge of stellar photosphere opacities enter the picture, because accurate temperatures can only be had from observations if the opacities that are used for calculating line profiles and continuum fluxes are accurate. One study of the accurate modelling of DB photospheres, and the impacts on mixing length theory calibrations, is that by Thejll et al. (1991).

For the DAs one important source of opacity, until recently not well modelled, is the ultraviolet absorption of H_2 and H_2^+. In the DAs surrounding the DA instability strip this molecule shows strong absorption features near the Lyman-α line – so-called satellites of Lyman-α. At the relevant temperatures the flux in DAs drops very quickly on the blue side of Lyman-α anyway but the flux dominates the luminosity of the stars in the range from Lyman-α into the 2000-3000 Å range. Figure 1 shows an example of such an object, observed with the *International Ultraviolet Explorer* satellite.

If we understand cool white dwarf opacities well we have a better chance of calculating their cooling rate, and hence the age of the Galaxy by counting the space density of cool white dwarfs. This gives one of the strongest motivations for studying molecular opacities in white dwarfs.

The presence of molecules in white dwarf photospheres is not always just determined by having the appropriate temperature range for the existence of the molecules in question. An example are the DQ WDs – these are helium rich cool white dwarfs (i.e. DBs) that have a convection zone in the envelope that reaches so deep into the stellar interior that the diffusion-distributed tail of the central carbon core is "dredged-up" to the photosphere and becomes visible at temperatures from 15,000K and down (e.g. Thejll et al. 1990). At these temperatures carbon shows CI and CII lines as well as – at the lower temperatures – molecular C_2 bands, such as the Swan bands, which also must be modelled under the extreme pressure conditions present in WDs.

Magnetic fields are observed in some, mainly cool ($T_{\rm eff} < 10000$K), white dwarfs and the effects on atomic as well as molecular absorption features must then be included in the analysis – see Bues & Karl (1994). Bues (1991) discusses

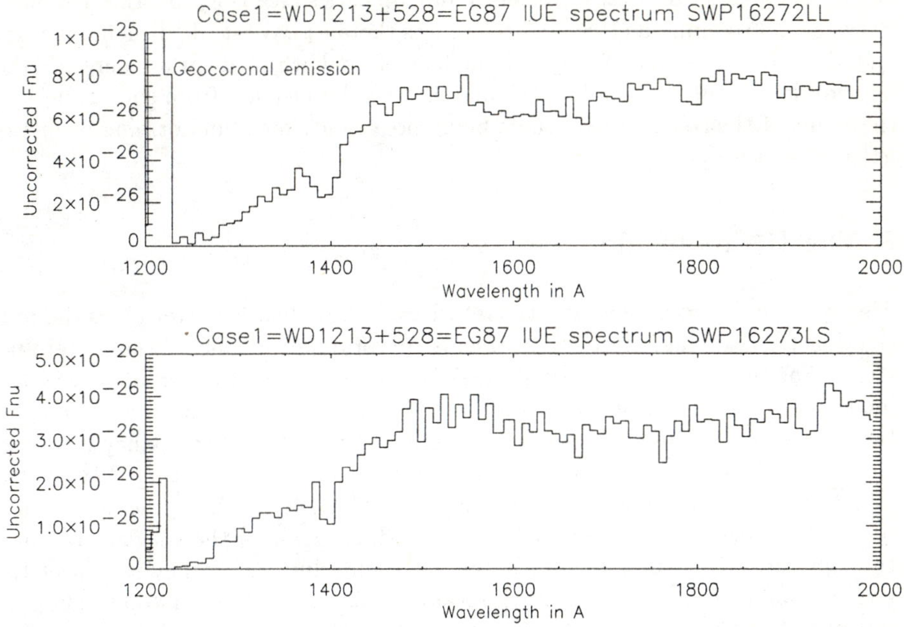

Fig. 1. Examples of ultraviolet spectra, observed by Sion et al. (1984) of a DA white dwarf showing H_2^+ absorption near 1400 Å. The DA is Case 1 and the figure headers gives other names as well as the numbers of the two IUE spectra shown. The upper spectrum was taken using the large aperture of the IUE camera and the lower with the small, hence the differences in the size of the geocoronal emission feature in the center of the Lyman-α-line. The spectra have been smoothed for display purposes, and some of the non-photospheric features removed, such as cosmic ray hits. The spectra have not been otherwise corrected for instrument artifacts or interstellar reddening. It seems that only the 1400 Å feature is clearly present. Other features, near 1600 Å and 2000Å, due to H_2 and higher systems have been identified by Allard & Koester (1992) in this object, but very careful reductions of the data are required to bring out these fainter features with certainty.

the WD G99-37 which shows strong C_2 and CH absorbtion features as well as CI lines. Strong molecular features from other species are rare in WDs, and the non-observability of CN and CH features are typically used to set upper limits to element abundances (e.g. Zeidler-K.T. et al. 1986).

Strong features from "metals" such as Ca, Mg, Si and Fe in DZ WDs become more and more apparent as the stars cool, perhaps because of accretion onto cool stars from the interstellar medium – as the radiation pressure decreases accretion becomes possible. Zeidler-K.T. et al. (1986) shows spectra and analyses of typical DZ WDs.

At low temperatures it is also possible to form collision induced systems like H_2-H_2 and H_2-He that have continuum opacities (see Borysow this volume). Borysow & Frommhold (1989 and references therein) have studied such opacities and they are important for the modelling of cool white dwarfs where all the information about T_{eff}, and $\log(g)$ may have to be gleaned from such things as the shape of the continuum – there being no, or only few, line-forming elements left in the atmosphere.

3 The H_2^+ molecule

The opportunity to understand the interior of white dwarfs better gives the motivation for studying any opacity that is important near one of the instability strips. For the case of the oscillating DBs one opacity-related problem was the inconsistency that existed between the UV-flux based T_{eff} determinations and those based on analysis of optical line profiles of HeI. This discrepancy has been cleared up (Thejll et al. 1991). For the cool DAs and the oscillating DAs (called "ZZ Ceti" stars) the focus is on the H_2 and H_2^+ molecules and the special absorption features present near the maximum-flux region of the spectra. Because these features are strong and are in the maximum-flux region they represent the largest source of opacity that, until recently, was not well understood. The reason uncertain opacities are of concern to interpreters of spectra is not just that regions of the spectrum may not be analyzable, but that strong features block the radiation and cause back-warming of the atmosphere so that the structure of the photosphere is changed and the form of the spectrum in *other* regions change and may become unanalyzable, in the worst cases. Being strong, the UV features of H_2, H_2^+, and also H_3^+, therefore require understanding for satisfactory modelling of cool DAs and ZZ Ceti stars.

These "quasi-molecules", formed during atom-atom or atom-ion collisions, give rise to line-like features, called satellites of the Lyman-α line, predicted theoretically in the 1970's and observed first, outside laboratories, in the 1980's by the IUE satellite in DA WDs (Greenstein 1980). Stewart et al. (1973) calculated crossections of the satellites due to H_2^+, using nearest-neighbour, quasi-static-approximations and this work was used to identify the features at 1400 Å in IUE spectra by Nelan & Wegner (1985), Holm (1985) and Koester (1985), simultaneously. At temperatures above 15000K the H_2^+ satellite at 1400 Å is stronger than the feature at 1600 Å due to H-H collisions. Sando & Wormhoudt (1973) calculated classically the expected absorption of the H_2 feature. While both the Stewart et al. (1973) and the Sando & Wormhoudt (1973) work served well to identify the features they did not give satisfactory fits to real spectra.

Allard et al. (1991) have lately modelled the many Lyman-α satellites in a comprehensive study of atom-atom and ion-atom interactions for binary, and higher, systems, and today this work (Allard & Koester 1992) gives the best fits to observed DA spectra. The work rests on the use of square-well potentials

(Allard 1978), which permits analytical calculation of the correlation function, and an appropriate density expansion that allows for the inclusion of effects in the far wings, which are important for the broad Lyman-α line.

Museth et al. (1994) have recently calculated full non-adiabatic quantum mechanical expressions for the crossection of H_2^+ avoiding some of the simplifications of the work by Allard et al. (1991).

4 Summary

I have described how WDs fit in to the stellar evolution picture and what questions they may help answer – e.g. the details of nucleosynthesis, mass-loss in the giant stages, and the age of the Galaxy. Notably astereoseismology of WDs can give information about internal structure to such a high degree of accuracy that understanding the "satellite" UV absorption features near Lyman-α becomes of great importance for bringing into agreement the photospheric analysis, and the things revealed by ZZ Ceti analysis. Recent advances in the modelling of H_2^+, and other quasi-molecular absorbers, have the potentials for achieving this in the near future.

For those seeking further insight into the state of the art in white dwarf research, the biennial proceedings of the WD workshops can be recommended: Wegner (1988), Vauclair & Sion (1990) and Barstow (1992).

Acknowledgments. Research support from the Carlsberg Foundation is gratefully acknowledged.

References

Allard N., 1978, *J. Phys. B.*, **11**, 1383

Allard N., Kielkopf J., 1991, *Astron. Astrophys.*, **242**, 133

Allard N., Koester D., 1992, *Astron. Astrophys.*, **258**, 464

Barstow M. (ed.), *White Dwarfs*, 1992, Kluwer Academic, NATO ASI Series vol. 403

Bergeron P., Saffer R., Liebert J., 1992, *Astrophys. J.*, **394**, 228

Borysow A., Frommhold L., 1989, *Astrophys. J.*, **341**, 549

Bradley P.A., Winget D.E., Wood M.A., 1993, *Astrophys. J.*, **406**, 661

Bues I., 1991, In: G. Vauclair, E.M. Sion (eds.), *White Dwarfs*, Kluwer Academic, NATO ASI Series vol. 336

Bues I., Karl L., 1994 In: P.Thejll & U.G.Jørgensen (eds.), *Poster Session Proceedings of IAU Coll. 146*, Copenhagen University, p. 20

Collins G.,W. II, 1989, *The fundamentals of stellar astrophysics*, W.H. Freeman Co., New York

D'Antona F., Mazzitelli I., 1990, *Annual Reviews of Astronomy and Astrophysics*, **28**, 139

Fontaine G. Tassoul M., Wesemael F., 1984, *Proc. of the 25th Liege International Astrophysical Colloquium*, Liege: Université de Liege, p. 328

Greenstein J., 1980, *Astrophys. J.*, **241**, L89

Holm A.V., Panek R.J., Schiffer III F.H., Bond H.E., Kemper E., Grauer A.D., 1985, *Astrophys. J.*, **289**, 774

Jørgensen U.G., Thejll P., 1993, *Astron. Astrophys.*, **272**, 255

Kippenbahn R., Weigert A., 1991, *Stellar Structure and Evolution*, Springer Verlag A&A Library, Berlin

Koester D., Weidemann V., Zeidler-K.T. E.-M., Vauclair G., 1985, *Astron. Astrophys.*, **142**, L5

Museth K., Linderberg J., Billing G., Thejll P., 1994, In: P.Thejll & U.G.Jørgensen (eds.), *Poster Session Proceedings of IAU Coll. 146*, Copenhagen University, p. 82

Nelan E.P., Wegner G., 1985, *Astrophys. J.*, **289**, L31

Niuewenhuijzen H., and de Jaeger C., 1990, *Astron. Astrophys.*, **231**, 134

Reimers D., 1977, *Astron. Astrophys.* , **61**, 217

Sando K., Wormhoudt J.C., 1973, *Phys. Rev. A.*, **7**, 1889

Shipman H. L., 1988, In G. Wegner (ed.), *White Dwarfs*, p. 220, Springer Verlag: LNP 328

Shipman H. L., 1992, In M. Barstow, *White Dwarfs*, p. 555, Kluwer Academic, NATO ASI Series, vol. 403

Sion E.M., Wesemael F., Guinan E.F., 1984, *Astrophys. J.*, **279**, 758

Stewart J.C., Peek J.M., Cooper J., 1973, *Astrophys. J.*, **179**, 983

Thejll P.A., Shipman H.L., MacDonald. J., 1990, *Astrophys. J.*, **361**, 197

Thejll P., Vennes S., Shipman H.L., 1991 *Astrophys. J.*, **370**, 355

Vauclair G., Sion. E.M., (ed.) 1990, *White Dwarfs*, Kluwer Academic, NATO ASI Series vol. 336

Wegner G. (ed.), 1988, *White Dwarfs*, Springer Verlag, Berlin, Lecture Notes in Physics vol. 328

Winget D.E., Kepler S.O., Robinson E.L., Nather R.E., D. O'Donoghue, 1985, *Astrophys. J.*, **292**, 606

Wood M. A., 1992, *Astrophys. J.*, **386**, 539

Zeidler-K.T. E.-M., Weidemann V., Koester D., 1986, *Astron. Astrophys.*, **155**, 356

Astrophysical Applications of Approximate Methods for Molecular Opacities - from stars to substellar brown dwarfs

Takashi Tsuji

Institute of Astronomy, The University of Tokyo, Mitaka, Tokyo, 181 Japan

1 Introduction

For studies of radiative processes in stellar environment, knowledges on opacities are essential, and molecular line opacities play a dominant role in many cases where temperatures are roughly below 4,000K. Accurate treatments of molecular line opacities require high precision line data for millions (and up to billions?) of lines. Recent progress in theoretical as well as in experimantal molecular physics made it possible to realize such an approach, and dedicated efforts are being done to developing extensive databases for polyatomic (Jørgensen in this volume) as well as for diatomic (Kurucz in this volume) molecules. These accurate databases should be of fundamental importance in our understanding of astronomical observations that are progressing rapidly both in scope and in accuracy.

Once the database includes so many lines, it can be regarded as a statistical sample, and actual use of such a database generally relies on statistical treatments such as the opacity distribution function (ODF) method or the opacity sampling (OS) method. Although these treatments are more or less approximate in nature, they are regarded to be reasonably accurate and used as the standard methods for line opacities. Then, by approximate methods for molecular opacities in this review, we restrict ourselves to such treatments as the band model methods in which detailed line-by-line databases are not referred to at all. In fact, it is known that simple analytical expressions for approximate molecular line opacities can be formulated with only a few spectroscopic constants including the band f-values (Sect.2).

One major role of approximate opacities has been to explore new problems in which molecular opacities play a major role, yet accurate opacity data are not yet available. For example, when infrared observations in the 1960's revealed a wealth of molecular bands such as of H_2O and CO in cool stars for the first time, it was immediately recognized that the molecular opacities play a dominant role in determining the atmospheric structures of cool stars, by the use of simple straight mean (Tsuji 1967) or harmonic mean (Auman 1967a) opacities. Then, somewhat improved methods such as the regular band model (Tsuji 1969) or more sophisticated random band model (Alexander et al. 1989) have been introduced. Apart from such exploratory purposes, approximate opacities are quite useful in practical applications so far as they are used with their limitations in mind. For example, the band model opacities may suffice for determining the thermal structure of stellar atmospheres for many practical purposes, while they may be not accurate enough for analyzing the stellar emergent spectra in general (Sect.3).

Although accurate molecular opacities are being developed for some molecules of astrophysical interests, explorations of model atmospheres in which molecular opacities play a dominant role may still be pursued. One good example of such a new exploration may be the case of brown dwarfs. In fact, at the very high density and low temperature environment to be expected in the atmospheres of brown dwarfs, some new opacity problems will appear. For example, methane (CH_4) opacity will play an important role but, as accurate opacity for CH_4 is still not available, approximate opacity will play a major role in exploring brown dwarf atmospheres (Sect.4).

Now, even if exploration of model atmospheres in the classical sense may come near the end by the brown dwarf atmospheres, this fact does not imply that exploration of stellar atmospheres themselves have also finished. It is true that the classical model atmospheres may now come to near the completion by the use of accurate opacities, and it is no doubt that this is an important achievement by itself. At the same time, such accurate classical models revealed the limitation of the classical atmospheres themselves. Especially, recent observations revealed various dynamical and non-thermal phenomena in the atmospheres of cool stars. As the classical model atmosphere is essentially a static equilibrium model, it is beyond the scope of the classical models to handle these phenomena. Certainly, for better understanding of stellar atmospheres, it is needed to understand the dynamical structure through the sub-photospheric convective zone to the outer atmosphere in details. For analyses of radiative processes together with hydrodynamics in the atmospheres of cool stars, molecular opacities will play an important role. Here again, the simplicity of approximate molecular opacities should be very attractive, since it will give more room for other thinking in an exploratory stage (Sect.5).

By the reasons outlined above, there may be sufficient reasons why we pay some attention to approximate methods for molecular opacities, even in the welcome circumstances at present when accurate molecular opacities are being developed.

2 Basic Formulation of Band Model Methods

2.1 Regular and Random Band Models

The band model method for molecular opacities was originally introduced for treating atmospheric transmission in meteorology (e.g. Goody 1964). The purpose was to provide a relatively simple representation of the mean value of the spectral transmission for a spectral mesh including many spectral lines. Also, the band model method has been applied for other purposes such as analyses of laboratory gas emissivities. Probably, the band model method is somewhat analogous to the classical theory of curve-of-growth familiar to astronomers: the band model method is to evaluate a growth of transmission (or emissivity) under the presence of multiple lines, while the curve-of-growth method is to analyze a growth of absorption for a single line. Since the radiative transfer problems in stellar atmospheres include much more complicated cases than just to obtain transmissions, the major techniques of the band model method cannot directly be applicable to astrophysical problems, just as the classical theory of curve-of-growth cannot to modern problems of spectral line formations.

For astrophysical purpose, what is useful may be the basic idea of how to model the spectral absorption coefficients for a large number of lines. For this purpose, we consider two models: one is the regular model and the other is the random model. The regular band model consists of an array of lines of equal intensity spaced at equal intervals, and such a model is known as the Elsasser model. This model can be characterized by two parameters - the mean absorption coefficient ($\bar{k}$) and the mean line separation ($\bar{d}$) - for each spectral mesh of appropriate interval, once the spectral line shape is defined. The case of the Voigt profile has been formulated by Golden (1969) and the result is known as the Voigt-Analogue Elsasser Band Model (VAEBM) . It is straight forward to obtain the ODF for this model, since the same spectral shape of the same intensity repeats at the equal intervals. Roughly speaking, the ODF for this model is a Voigt profile itself truncated at the mid-point between the two adjacent lines, plus the contributions of line wings of nearby lines in the array.

On the other hand, the random band model consists of randomly spaced lines of different intensities. For example, the probability $P(S)dS$ that a line has an intensity between S and $S + dS$ can be assumed as

$$P(S) = exp(-S/S_o)/S_0, \tag{1}$$

where S_0 is the mean line intensity. Also, if a sufficiently large number of lines are randomly placed in a spectral mesh, the probability that there is a gap of x is given by

$$p(x) = exp(-x/\bar{d}), \tag{2}$$

where $\bar{d}$ is the mean line separation (Goody 1964). Recently, such a model has been applied to generate the ODF for H_2O by Alexander et al. (1989). For this purpose, synthetic spectrum is generated, in which the line position of the next line can be determined by Eq.(2) using a random number generator and the strength of this line by Eq.(1) again using a random number generator. Then, the OS or the ODF method can be applied to this somewhat fictitious spectrum. This is an interesting attempt, but requires as much effort as to generate a more realistic spectrum based on a line-by-line database. It is to be noted, however, that this model is also characterized by two parameters - the mean absorption coefficient $\bar{k}$ and the mean line separation $\bar{d}$ (note that $S_0 = \bar{k}\bar{d}$).

2.2 Band Model Parameters

Rotation-Vibration Bands of Simple Molecules For linear (and spherical top) molecules, the rotational term value is given by

$$F_v(J) = B_v J(J+1), \tag{3}$$

with the rotation constant B_v. Then, line positions for the P- and R-branches are given by a single formula as

$$\omega = \omega_{v'v''} + (B_v' + B_v'')m + (B_v' - B_v'')m^2, \tag{4}$$

where $\omega_{v'v''}$ is the wavenumber of the band origin, $m = -J$ for the P-branch and $m = J+1$ for the R-branch. This equation can be solved for m as

$$m = [-(B_v' + B_v'') \pm \sqrt{(B_v' + B_v'')^2 + 4(B_v' - B_v'')(\omega - \omega_{v'v''})}]/2(B_v' - B_v''). \tag{5}$$

The integrated line intensity for a single rotational line can be given by

$$S_{v'v''}(m) = \frac{\pi e^2}{m_e c^2} f_{v'v''} \frac{\omega}{\omega_{v'v''}} |m| F(m) exp[-\frac{hc}{kT}\{G_v'' + B_v''|m(m-1)|\}]$$
$$(1 - e^{-hc\omega/kT})/Q_r(T)Q_v(T), \tag{6}$$

where $f_{v'v''}$ is the vibrational f-value, $F(m)$ is the Herman-Wallis factor, G_v'' is the term value of the lower vibrational level, $Q_r(T)$ and $Q_v(T)$ are the rotational and vibrational partition functions, respectively, and other notations have their usual meanings. $S_{v'v''}(m)$ can also be understood as $S_{v'v''}(\omega)$ by means of Eq.(5).

The line separation in the (v', v'') band is given by

$$d_{v'v''}(m) = \frac{d\omega}{dm} = (B_v' + B_v'') + 2(B_v' - B_v'')m. \tag{7}$$

By applying Eq.(5), this can also be written (for each of the P- and R-branches) as

$$d_{v'v''}(\omega) = \sqrt{4(B_v' - B_v'')(\omega - \Omega_{v'v''})}, \tag{8}$$

where $\Omega_{v'v''}$ is the wavenumber of the (v', v'') band head and is given by

$$\Omega_{v'v''} = \omega_{v'v''} - (B_v' + B_v'')^2/4(B_v' - B_v''). \tag{9}$$

For the Q-branch (perpendicular band), the line position is given by

$$\omega = \omega_{v'v''} + (B_v' - B_v'')m(m+1), \tag{10}$$

where $m = J$. As for the P- and R- branches, the line separation is given by

$$d_{v'v''}(\omega) = \sqrt{4(B_v' - B_v'')[\omega - \omega_{v'v''} + (B_v' - B_v'')/4]}. \tag{11}$$

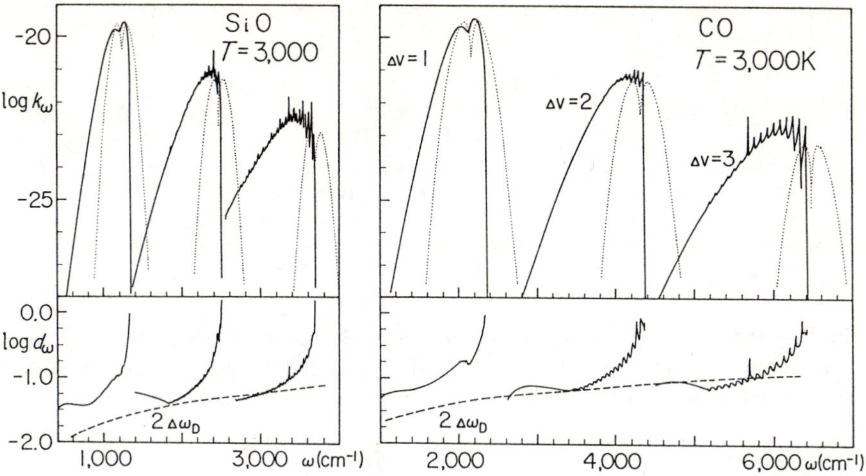

Fig. 1. The absorption cross-sections $\bar{k}$(cm^2/molecule) and the mean line separations $\bar{d}$(cm^{-1}), averaged over 10cm^{-1}, for SiO and CO at T =3,000K. The results of the rigid rotator harmonic oscillator (RRHO) model are shown by the dotted lines. The full Doppler widths corresponding to $v_{\rm mic}$=3km s^{-1}are indicated by the dashed lines.

Note that $\Omega_{v'v''}$ in Eq.(8) is replaced by $\omega_{v'v''}$ in Eq.(11), if we neglect a small term in Eq.(11) by utilizing the fact that $(\omega - \omega_{v'v''}) \gg (B_v^{'} - B_v^{''})/4$.

Now, we follow the standard definitions of the band model parameters. First, the weak line parameter is defined by

$$(\frac{S}{d}) = \sum_{v'v''}[\frac{S_{v'v''}(\omega)}{d_{v'v''}(\omega)}], \tag{12}$$

where the summation should be extended to all the bands that are contributing at ω. The resulting (S/d) is nothing but the straight mean absorption coefficient and can be denoted as $\bar{k}(\omega)$. Second, the strong line parameter is defined by

$$(\frac{S^{1/2}}{d}) = \sum_{v'v''}[\frac{S_{v'v''}(\omega)^{1/2}}{d_{v'v''}(\omega)}]. \tag{13}$$

Then, we defined

$$(\frac{1}{\bar{d}}) = (\frac{S^{1/2}}{d})^2/(\frac{S}{d}), \tag{14}$$

which provides the effective mean line separation $\bar{d}(\omega)$. As $1/d_{v'v''}$ is the number of lines of the (v', v'') band in 1cm^{-1}, a more naive definition of $\bar{d}$ may be

$$(\frac{1}{\bar{d}}) = \sum_{v'v''}\frac{1}{d_{v'v''}(\omega)}. \tag{15}$$

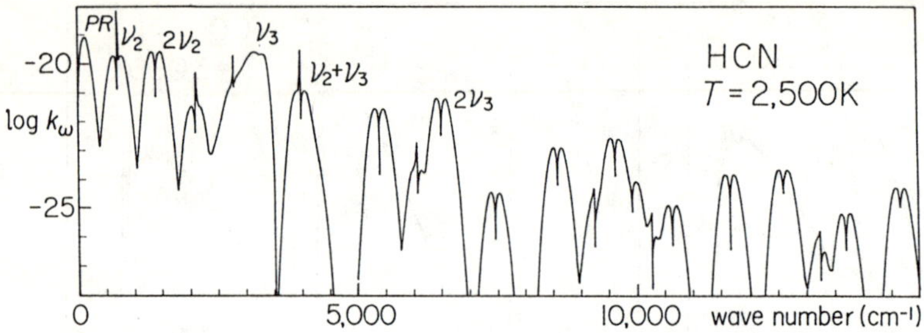

Fig. 2. The absorption cross-sections $\bar{k}$(cm^2/molecule) of HCN for 28 band systems including the pure rotation band, at T =2,500K. The RRHO model is applied, except for the ν_3-band system for which about 1,000 bands are considered band-by-band.

In this case, however, we must introduce a cutting criterion, since $\bar{d}$ may be unrealistically small if we consider very weak bands in the summation.

If we introduce the rigid-rotator-harmonic-oscillator (RRHO) model, the straight mean absorption coefficient defined by Eq.(12) can be reduced to a simple analytical formula and it is no longer necessary to do summations band-by-band (e.g. Tsuji 1984). This model provides a large simplification especially in the case of polyatomic molecules, but the shape of the absorption coefficient as a function of wavenumber cannot be represented very well if the effects of non-rigidity and anharmonicity are important (see Fig.1).

As examples of diatomic molecules, the band model parameters are shown in Fig.1 for CO and SiO. Also shown are the cases of the RRHO model, which appear to be a poor approximation to the detailed band-by-band approach especially for overtone bands. Also, it is shown that the mean line separations are generally larger than the full Doppler widths for turbulent velocity of 3km s^{-1}. As an example of polyatomic linear molecules, the straight mean absorption cross-sections for HCN are shown in Fig.2. The mean line separations are generally smaller than the Doppler widths for appropriate turbulent velocity as shown in our previous analysis (Tsuji 1984), where only a limited number of band systems were included. Here, we also included the pure rotation transitions, which are already important in the mid-infrared at high temperatures.

Another molecule of potential importance at high density and low temperature environment is CH_4, since CO is no longer the most stable molecule, and most carbon atoms will be in CH_4 at temperatures below 1,000K and at high densities, even in the oxygen-rich environment. As CH_4 is a spherical top molecule, its line positions can be given by Eq.(4) for rotation-vibration transitions, and the formulation of the band model outlined above can also be applied to CH_4. Only slight modification is necessary to accomodate the fact that the statistical weight of a rotational level with the quantum number J is $(2J+1)^2$ rather than $2J+1$ (Herzberg 1945). The resulting straight mean absorption cross-sections of CH_4 are shown in Fig.3 together with those of H_2O based on the experimental data by Ludwig (1967). It is interesting to notice that CH_4 bands nicely cover

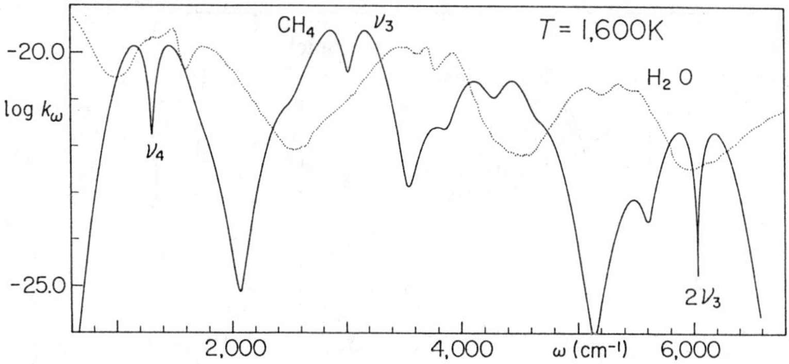

Fig. 3. The absorption cross-sections $\bar{k}$(cm^2/molecule) of CH_4 at T =1,600K, including 14 band systems by the RRHO model, and those of H_2O (Ludwig 1967) for comparison.

the regions that are least absorbed by H_2O. Thus, the infrared spectral region is completely opaque if these two species exist together and such a situation will actually be realized in the atmospheres of brown dwarfs (Sect.4).

Electronic Bands of Diatomic Molecules More or less similar consideration can be applied to electronic bands of diatomic molecules. With an approximation that no distinction is made between the P-, Q-, and R-branches, the straight mean absorption cross-section has been given by Golden (1967) as

$$k_{v'v''}(\omega) = \frac{\pi e^2}{m_e c^2} f_{v'v''} \frac{1}{|B_v' - B_v''|} exp[-\frac{hc}{kT}\{G_v'' + \frac{B_v''}{B_v' - B_v''}(\omega - \Omega_{v'v''})\}]$$
$$g_e'' exp(-\frac{hc}{kT}T_e'')(1 - e^{-hc\omega/kT})/Q_r(T)Q_v(T)Q_e(T), \qquad (16)$$

where $Q_e(T)$ is the electronic partition function, g_e'' and T_e'' are the statistical weight and the term value of the lower electronic state, respectively. If the electronic dipole moment is independent of the internuclear distance, the band oscillator strength $f_{v'v''}$ can be given by

$$f_{v'v''} = f_e \frac{\omega_{v'v''}}{\omega_{00}} q_{v'v''}, \qquad (17)$$

where ω_{00} is the wavenumber of the band origin of the (0,0) band, f_e is the electronic oscillator strength defined at ω_{00}, and $q_{v'v''}$ is the Franck-Condon factor.

In consistency with Golden's formulation, we use Eq.(8) for the line separation throughout but, if the P-, Q-, and R-branches are present, actual line separation must be 3 times smaller than that given by Eq.(8). Further, the line separation may still be smaller due to the Λ-type doubling and due to the splitting by the spin-orbital interaction. Thus, we finally have

$$d_{v'v''}(\omega) = \sqrt{4(B_v' - B_v'')(\omega - \Omega_{v'v''})}/3(2S+1)(2 - \delta_{\Lambda,0}), \qquad (18)$$

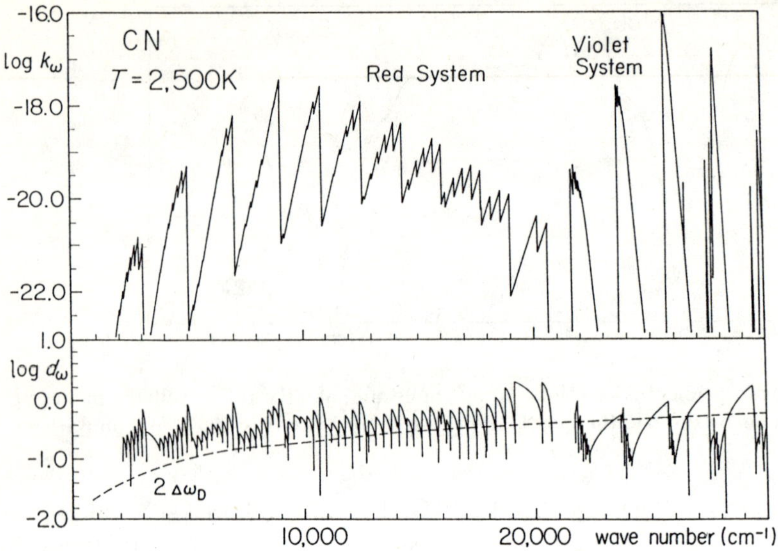

Fig. 4. The absorption cross-sections $\bar{k}$(cm^2/molecule) and the mean line separations $\bar{d}$(cm^{-1}), averaged over 10cm^{-1}, for the CN Red and Violet systems at T =2,500K. The full Doppler widths corresponding to $v_{\rm mic}$=3km s^{-1}are indicated by the dashed lines.

where $2S+1$ is the spin multiplicity, Λ is the component of the electronic angular momentum along the internuclear axis, δ is the Kronecker's delta function. The band model paremeters based on this model are shown in Fig.4 for the CN Red and Violet systems.

A formulation free from some of the simplifications used in Eq.(16) is given by Zeidler-K.T. & Koester (1982), who treated the P-, Q-, and R- branches separately.

2.3 Band Model Opacities for Astrophysical Applications

Now, we have the two parameters - the straight mean absorption cross-section $\bar{k}_i(\omega)$ and the mean line separation $\bar{d}_i(\omega)$ - for each molecule that is important as an opacity source. Then, we evaluate the absorption ccoefficient per gram of stellar material by solving the chemical equilibrium for given temperature and gas pressure. We now apply Eqs.(12)-(14), which were applied to different bands of the same species in Sect.2.2, to different species that give non-negligible contribution at a given spectral mesh [note that $S_{0i}(\omega) = \bar{k}_i(\omega)\bar{d}_i(\omega)$ for i-th species]. The results are the integrated mass absorption coefficient and the average mean line separation for the given spectral mesh, chemical composition, temperature, and gas pressure. Then, the ODF can easily be estimated within the framework of the VAEBM, by which line broadenings due to turbulence and collision are considered at each step of integration of the model atmosphere.

In this way, once the two parameters - $\bar{k}_i(\omega)$ and $\bar{d}_i(\omega)$ - for each molecule are prepared, an approximate ODF can easily be estimated for any chemical composition, temperature, gas pressure, and turbulent velocity during the construction of a model atmosphere, and it is not necessary to prepare large tables of opacities covering of all the parameter space. Such simplicity and flexibility are the most attractive features of the VAEBM even at the loss of accuracy as compared with the ODF based on a detailed line-by-line database. By the way, the ODF based on a line-by-line database is not necessarily very accurate when several different species are contributing in the same spectral mesh, since the line that gives the maximum absorption is not necessarily due to the same species at different depths in stellar atmosphere. As is well known, the ODF method cannot consider such an effect, and the OS method is more consistent in this regard. On the other hand, an efficient method for calculating composite ODF's from the ODF for each individual opacity source has been developed by Saxner & Gustafsson (1984), and it is possible to change the chemical composition without recalculation of ODF's in the construction of model atmospheres. Such a flexibility is certainly an advantage of the ODF method over the OS method.

3 Applicability and Limitation of Approximate Opacities

The band model opacities can further be simplified if the mean line separation is comparable to or smaller than the line width. In this case, the line absorption coefficient can be treated as a quasi-continuous absorption and is given by the straight mean $\bar{k}(\omega)$ itself. Line opacity based on this assumption is referrred to as the Just Overlapping Line Approximation (JOLA). It is in this most simple form that the molecular line opacities such as of H_2O, CO, and OH have first been introduced in stellar opacities (Tsuji 1966), although it has been recognized that such an approximation may not necessarily be valid for stellar environment in general. Neverthless, such a simple opacity played an important role in the exploratory stage of constructing model atmospheres of cool stars (Sect.1). Also, harmonic mean was considered for H_2O (Auman 1967b), but the reason for using the harmonic mean for atmospheric problems was not very clear.

Further, the JOLA found some useful applications. For example, the cooling effect of CO in the atmospheres of cool stars has been examined in detail by this simple technique by Johnson (1973). A case where the JOLA may actually be valid can be found in the atmospheres of cool white dwarfs, since molecular lines are completely smeared out by the large pressure broadening at the high densities of white dwarf atmospheres. Thus, successful applications of the JOLA have been done to model the atmospheres of C_2 white dwarfs (Grenfell 1974; Wehrse 1975). Also, it is by the JOLA that the collision-induced absorption (CIA) of H_2 has first been introduced in the actual computation of stellar model atmospheres (Tsuji 1969). In fact, the JOLA is well justified in this case, since individual lines are highly broadened because of the short time of the intermolecular interaction inducing the dipole moment and hence completely smeared out.

The next simple approximation - the regular band model - was introduced

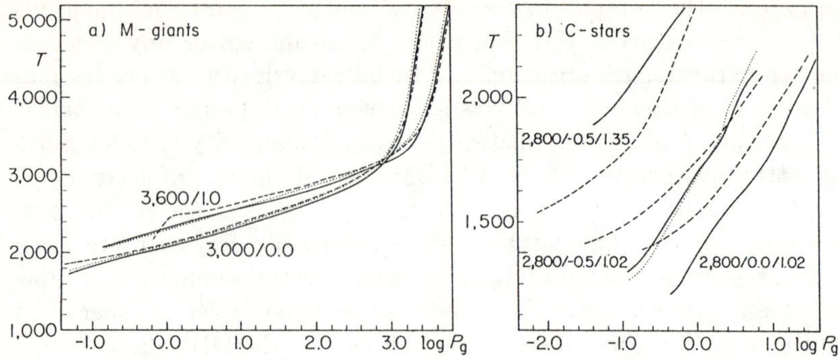

Fig. 5. a) Model atmospheres of oxygen-rich giant stars based on the VAEBM (solid lines) and on the JOLA (dotted lines) are compared with those based on the OS method (Brown et al. 1989; dashed lines). $T_{\rm eff}$/log g are indicated for each model, and other input parameters are $v_{\rm mic}$=1km s^{-1}and the solar composition. **b)** Model atmospheres of carbon stars based on the VAEBM (solid lines) and on the JOLA (dotted lines) are compared with those base on the OS method (Jørgensen et al. 1992; dashed lines). $T_{\rm eff}$/log g/ (C/O)-ratio are indicated for each model and $v_{\rm mic}$=2km s^{-1} throughout.

first for modelling the atmospheres of M dwarfs (Tsuji 1969). Then, the VAEBM was applied to estimating the Rosseland mean opacities of the oxygen-rich mixture (Tsuji 1971), and to constructing model atmospheres of red giant and supergiant stars (Tsuji 1976; 1978). In this VAEBM approach, the band model parameters ($\bar{k}_i$ and $\bar{d}_i$) of 9 molecules including H_2O, CO, OH, CN, CH, TiO, MgH, CaH, SiH (and H_2 CIA as a quasi-continuum) were given for 72 spectral meshes between 0.3 and 30 μm (hereafter be referred to as VAEBM72). Here, the band model parameters for H_2O were based on the experimental data by Ludwig (1967). At the same time, more elaborate treatments such as the ODF method (e.g. Querci et al. 1974; Gustafsson et al. 1975) or the OS method (e.g. Johnson et al. 1980) have been preferred in general, and the band model opacities have not necessarily been used extensively. Recently, however, an interesting attempt to apply the random band model to H_2O opacity sampling has been given by Alexander et al.(1989).

Generally, the degree of success of the approximate opacities as applied to model atmospheres can be tested by comparing with the results based on more elaborate opacities (also see Carbon 1979; Jørgensen 1993). Such a test, however, is by no means easy, since it is difficult to find model atmospheres with the same input physics and data except for the treatments of opacities. As an example, we compare in Fig.5a the models of red giant atmospheres by our VAEBM72 with those by the OS method (Brown et al. 1989) based on the H_2O opacity by Alexander et al.(1989). Inspection of the results reveals that the VAEBM provides model atmospheres that are very similar to those based on the OS method. Even the thermal structures of the model atmospheres based on the JOLA are not much different from those based on the VAEBM as shown in Fig.5a or from those based on the OS opacities as shown by Alexander et al.(1989). As another example, we compare in Fig.5b the model atmospheres of carbon stars based on the OS method (Jørgensen et al. 1992) with those based on our new VAEBM opacities in which band model parameters of CO, CN, C_2, HCN, and C_2H_2 are prepared for 194 spectral meshes between 0.2 and 45 μm (hereafter be referred to as VAEBM194). Here, the agreement in temperatures is generally within 100K and differential effects due to the gravity and the C/O ratio show reasonable agreement, between the two models based on the different opacities.

A rationale for the result that the temperature structure can be relatively well determined by the approximate opacities can be found in the fact that the thermal structure of stellar atmosphere is decided by the constraint of radiative equilibrium - namely,

$$\int_0^\infty \kappa_\nu B_\nu d\nu = \int_0^\infty \kappa_\nu J_\nu d\nu, \tag{19}$$

where κ_ν is the absorption coefficient including line absorption, B_ν is the Planck function, and J_ν is the mean intensity. That is, the radiative energy absorbed by a volume element (right hand side) must balance with the energy emitted (left hand side), at radiative equilibrium. Now, as

$$\int_0^\infty \kappa_\nu B_\nu d\nu \approx \sum_{i=1}^{n} B_{\nu_i} \int_{\Delta\nu_i} \kappa_\nu d\nu, \tag{20}$$

if the frequency interval $\Delta\nu_i$ is small enough so that the Planck function is nearly constant, the volume emissivity does not depend on whether κ_ν is given by the straight mean, by the band models, by the ODF, or by the detailed OS, so far as the integrated absorption coefficient is the same in each frequency interval. Although J_ν depends on the details of κ_ν, this effect is also largely tempered in the product $\kappa_\nu J_\nu$ in the integrand of the right hand side of Eq.(19). Thus, the temperature that is determined by Eq.(19) may depend only weakly on the details of κ_ν. In deeper photospheric layers, backwarming effect may be over-estimated by the JOLA which over-estimates the blocking of radiation by lines, but this is largely relaxed by the band model approach such as the VAEBM.

Next problem is how well the approximate opacities can predict such quantities as emergent fluxes. It is interesting to know that even the JOLA predicts the emergent fluxes blanketed by H_2O within 20% of those based on the OS method for cool giant stars (Alexander et al. 1989), and this may be because the line density of polyatomic molecules such as H_2O is so large that the lines are actually overlapping. Such a favourable condition is generally not met for diatomic molecules (see Figs.1 & 4), and we examine the case of CO in the region of the fundamental VR bands. First, for a model atmosphere of an M dwarf star ($T_{\rm eff}$=2,400K, log g = 5.0, and $v_{\rm mic}$=1.0km s^{-1}), emergent fluxes based on our VAEBM72, on the JOLA, and on the detailed OS method are shown by the solid, dotted, and dashed lines, respectively in Fig.6a, together with the continuum flux by the thin solid line. The OS flux is evaluated with the sampling interval of 0.2cm^{-1} based on our $^{12}C^{16}O$ line database ($J \leq 150$ and $v \leq 15$),

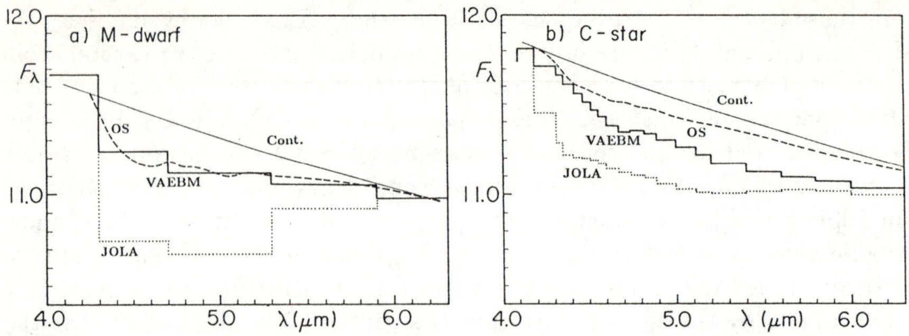

Fig. 6. Predicted emergent fluxes based on the VAEBM (solid lines) and on the JOLA (dotted lines) are compared with those base on the detailed line-by-line OS method (dashed lines), in the region blanketed by the CO fundamental. Predicted continuous flux is also shown by the thin solid line. **a)** For a model atmosphere of M dwarf star with ($T_{\rm eff}$/log $g/v_{\rm mic}$)=(2,400K/5.0/ 1.0km s^{-1}). **b)** For a model atmosphere of carbon star with ($T_{\rm eff}$/log $g/v_{\rm mic}$)=(3,000K/0.0/3.0km s^{-1}).

and then averaged over interval of 30cm^{-1} each. It is evident that the emergent flux based on our VAEBM72 agrees well with the accurate OS flux while that based on the JOLA shows unacceptably large deviation from the true flux. Second test is for a carbon star ($T_{\rm eff}$=3,000K, log g = 0.0, and $v_{\rm mic}$=3.0km s^{-1}) with our VAEBM194, and the results are shown in Fig.6b. Here, the emergent flux based on the VAEBM194 does not agree so well with the accurate OS flux, although it provides a large improvement over that based on the JOLA. The discrepancy between the flux based on the VAEBM194 and the OS flux is larger at the longer wavelength side of the CO band where $\bar{d}$ based on Eq.(14) is also smaller (see Fig.1). This may indicate that the estimation of $\bar{d}$ based on Eq.(14) is not necessarily correct. On the other hand, $\bar{d}$ of the VAEBM72 was based on Eq.(15) with an appropriate cutting criterion (Tsuji 1976) and the result appears to be fine. Thus, how to estimate the mean line separation is crucial in the band model approach and this problem should be examined further.

Finally, it is to be remembered that the JOLA has played some roles in exploring effects of sphericity in the model atmospheres of oxygen-rich (super)giants (e.g. Watanabe & Kodaira 1978; Schmid-Burgk et al. 1981) as well as of cool carbon stars (Scholz & Tsuji 1984). The major physical effects have mostly been confirmed by the recent works based on the OS method both for the oxygen-rich (Pletz et al. 1992) and carbon-rich (Jørgensen et al. 1992) cases, although the sphericity effects appear to be larger in the models by the JOLA. It is true that the JOLA tends to over-estimate the effect of molecular opacities, while completeness of the line list is crucial in the OS method. This is clearly shown for the case of H_2O, for which such a completeness has finally been achieved (Jørgensen in this volume). The completeness can easily be realized in the JOLA only if the integrated intensities are correctly given, but this makes sense only if the JOLA is valid. This must anyhow be confirmed by a complete line database.

4 An Exploratory Model Atmosphere of Brown Dwarfs

The primary role of approximate opacities is to explore new problems, as it has been in the past. One current problem to be explored is the model atmosphere of brown dwarf, which is a substellar object whose existence has been predicted by several arguments but not confirmed yet by any direct observation, in spite of extensive searches (Liebert in this volume). For further searches to be successful, it should be indispensable to understand the physical properties of the atmospheres of brown dwarfs and to predict the observable properties such as infrared spectra more carefully. As compared with the atmosphere of M dwarfs (Scholz & Wehrse in this volume), a major difference is the predominance of CH_4 in the atmosphere of brown dwarfs as noted in Sect.2.2. Thus, we examine the effect of methane opacity on the atmospheric structure and infrared spectra of brown dwarfs by the use of the approximate CH_4 opacity discussed in Sect.2.2.

For this purpose, we extended our model atmosphere code of cool dwarf stars to be used for $T_{\rm eff}$ below 2,000K (and to $T_{\rm eff} \approx$ 1,000K). This requires the opacities and chemical equilibria data down to $T \approx$ 500K. We have added the CH_4 opacity to our VAEBM72 including 9 molecules as mentioned in Sect.3. Here, however, we assume that CH_4 lines are well smeared out by the high line density due to the transitions from numerous vibrational states and/or by the pressure broadening, and thus CH_4 is included as a pseudo-continuous opacity. In fact, the JOLA may be applicable to polyatomic molecules with complicated spectra while it is not to diatomic molecules with simple spectra, as we have noted in Sect.3. Another important opacity source at high density environment of very low mass stars is the CIA of H_2, which dominates the whole infrared region by quasi-continuous absorption (Borysow in this volume). Also, convective energy transport is highly efficient up to the very surface of the atmosphere in cool dwarf stars. However, as the temperature gradient in the convective zone approaches to the adiabatic gradient at the high density of cool dwarf atmospheres, the convergence in radiative-convective equilibria is easily attained.

The predicted infrared spectra for model atmospheres of $T_{\rm eff} = 1,600$K are shown in Fig.7. The heavy solid line and the dashed line are the blanketed spectra from model atmospheres with and without CH_4, respectively. Clearly, the CH_4 opacity produces a serious effect upon the infrared spectrum. Especially, the strong ν_3 band of methane at 3.3μm appears to be comparable with the ν_1 and ν_3 bands of water vapour at 2.7μm. The CH_4 opacity also modifies the atmospheric structure as to give an appreciable change of the H_2O absorption. The thin solid line is the expected true continuum which can be observed if there are true windows in the spectrum and if they are resolved by very high resolution spectroscopy. It is interesting to observe that the expected peaking of the continuum flux at 1.6μm corresponding to the minimum of H^- absorption does not appear, but broad depression centered at about 2.2μm appears instead. This depression is caused by the CIA of the H_2 fundamental band, which is regarded as a continuous opacity rather than a line opacity in our models. Anyhow, this result clearly demonstrates the importance of the CIA at the high density and low temperature environment of brown dwarf atmospheres. Also, predicted spec-

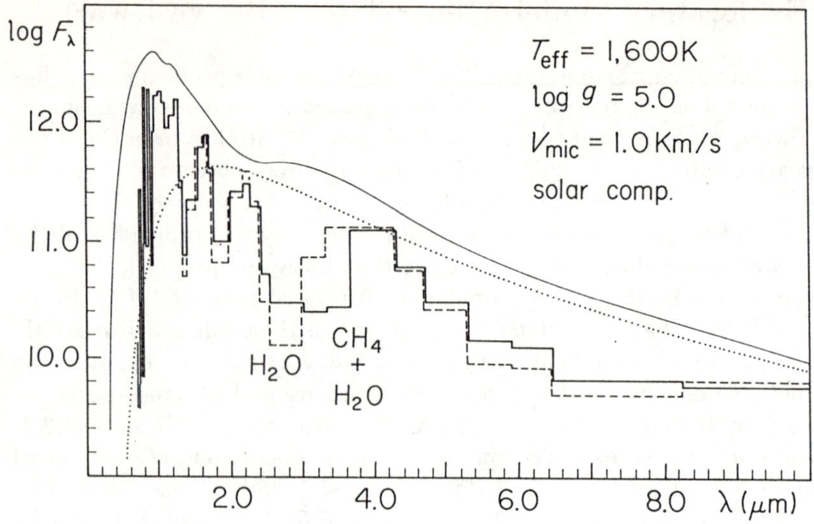

Fig. 7. Predicted infrared spectrum of brown dwarf (T_{eff}=1,600K) including CH_4 (in addition to H_2O, CO, OH, TiO etc) as opacity source is shown by the heavy solid line, and predicted spectrum based on a model without CH_4 by the dashed line for comparison. Clearly, a prominent spectroscopic signature of brown dwarf is the appearance of the CH_4 ν_3-band at 3.3μm. The continuous flux including H_2 CIA is shown by the thin solid line, indicating that the CIA is a dominant opacity source as is evidenced by a dip at 2:2μm. The black body radiation of T =1,600K is shown by the dotted line.

tra based on the model atmosphere shows large deviation from the blackbody radiation of $T = 1,600$K shown by the dotted line. Especially, there appear large excesses of flux at around 1μm. There is still a possibility that these excesses are artifacts of neglecting some opacity sources in this region. At the same time, it is true that the infrared region beyond 1.2μm is highly opaque in the brown dwarf atmospheres not only due to the combined effect of the H_2O and CH_4 opacities (see Fig.3) but also due to the quasi-continuous H_2 CIA.

The results outlined above may have some implications on the current searches for brown dwarfs. So far, many candidates for brown dwarfs have been proposed, but none was confirmed to be a true brown dwarf. Most of these candidates are bright brown dwarfs that are still warm on the cooling tracks and they are difficult to be distinguished from M dwarfs that occupy the same region on the HR diagram. Future increase of sensitivity in observations will finally make it possible to observe brown dwarfs below the main-sequence limiting luminosity in the region of $L < 6.10^{-5} L_\odot$ and $T_{eff} < 1,800$K (Burrows et al. 1993). But here, accurate determination of the luminosity (or distance) is again difficult. On the other hand, our models revealed that CH_4 can be observed only in the brown dwarfs with T_{eff} below 1,800K, and the CH_4 3.3μm band can be the most reliable spectroscopic criterion in confirming brown dwarfs. Another interesting result is that brown dwarfs are not necessarily very red but rather may show large flux excesses around 1μm. Also, brown dwarfs may not be bright at the L-band ($\lambda_0 \approx 3.5\mu$m) of IR photometry because of the strong CH_4 absorption, and brown dwarf searches by this band may not be effective.

Our exploratory model atmosphere of brown dwarfs clearly shows the importance of molecular opacities for predicting reliable spectroscopic and photometric properties of such very cool and dim objects, and many problems remain to be worked out. First, a more accurate line-by-line database for CH_4 will be required in the near future when the presence of such substellar objects will be confirmed. The latest version of the HITRAN database already includes 47,415 lines of CH_4 (Rothman et al. 1992) but, since these are mostly low excitation atmospheric lines, it would require another effort to include many highly excited lines of this complicated molecule. Second, opacity sources around 1μm are crucial for understanding the photometric properties of brown dwarfs, and more careful investigations are needed. Third, in view of the importance of CIA at high density, improved opacites such as those by Borysow & Frommhold (1990) are most welcome and will hopefully be extended to other molecules such as N_2. Fourth, grain formation and its effect on opacities may be important (Lunine et al. 1989).

5 Further Explorations on Stellar Atmospheres

Now, if the case of brown dwarf is the last exploration in model stellar atmospheres, has the role of approximate molecular opacities ended? This may to some extent be true if we confine our attention to the classical model atmospheres. In fact, the classical model atmospheres may have been well sophisticated by now. Then, one important role of the accurate classical model may be to use it as a reference by which deviations of real atmosphere from the classical model atmosphere can be examined. For example, it appeared difficult to reproduce the best empirical model of the Sun by theoretical modelling within the framework of the classical atmosphere even by the use of the accurate line opacities (Kurucz 1979). Deviations of the order of 100K appear in deeper layers as well as in the very surface layers, and this may be due to the limitation of the classical model atmosphere itself. Such a detailed test of model atmospheres by the empirical model, however, may be difficult for stars other than the Sun.

More recently, direct diagnoses of stellar atmospheres have been made possible by the high resolution spectroscopy of line profiles (e.g. Dravins 1987; Gray 1988). Further, molecular lines served to provid additional information on the solar atmosphere (Grevesse & Sauval in this volume) and to extend the similar analysis to very cool stars (Tsuji 1991). All these observations revealed the presence of somewhat ordered velocity fields in stellar atmospheres rather than the chaotic turbulent motion as assumed in the classical atmospheres. An attempt to model such a non-thermal velocity structure has been done recently by Nordlund & Dravins (1990), who have solved detailed 3D time-dependent radiation hydrodynamics of granular convection in solar type stars. For extending such a modelling of the dynamical atmospheres to very cool stars, molecular opacities should play a significant role, and approximate opacities such as by the band model method may be the best choice for such a purpose. In fact, it may be too complicating to introduce detailed line-by-line opacities based on a large database in such a computation which in itself is already highly complicated.

Also, larger deviations from the classical model atmosphere will occur in the surface layer and outer atmosphere where various kinds of non-thermal and non-equilibrium processes should take place. Because of the low temperatures, this is also the place where various molecules are formed, and molecules such as CO may play an important role in determining the atmospheric structure of cool stars because of their characteristic cooling function (Ayres 1981; Kneer 1983). The effect of CO on the dynamics of the solar atmosphere has actually been followed by time-dependent hydrodynamic codes first with a simple two-frequency opacity (Muchmore & Ulmschneider 1985), and then with more detailed ODF opacity for CO (Muchmore et al. 1988). These analyses revealed that even the simple two-frequency opacity gave the correct physics for most cases, and the ODF method may be preferred if high accuracy is required. In the atmospheres of cool stars, other molecules will play a similar role and the case of SiO in the outer atmospheres of cool giant stars has been discussed already (Muchmore et al. 1987). For extending a similar analysis to other molecules and to other stellar environments, approximate treatments for molecular opacities will again be attractive because of their simplicity and flexibility. Also, accuracy of the approximate opacities may be sufficient for these purposes, since the major contribution to the cooling function is represented by Eq.(20) which depends little on the treatments of opacities as noted in Sect.3.

Further, new observations are still revealing unexpected new phenomena. An interesting case is the IRAS carbon stars that show silicate emission: the presence of the silicate emission implies the presence of an oxygen-rich environment ($C/O < 1$) while a carbon star characterized by the carbon-rich chemistry ($C/O > 1$) must be there at the same time. One interesting possibility is to assume a presence of an accretion disk composed of the oxygen-rich material around a classical carbon star (Lloyd Evans 1990; Lambert et al. 1990). Also, the cool accretion disk will be an important ingredient in understanding the physical structures of a wide variety of objects through young stellar systems such as T Tau stars to evolved systems such as cataclysmic variables. For modelling such cool accretion disks, molecular opacities should be important and approximate methods for molecular opacities will find wide applications in such new explorations.

As may be evident from the few examples noted above, explorations on stellar atmospheres continue and approximate approach for molecular opacities will remain to be an important tool in such endeavours.

6 Concluding Remarks

Briefly summarizing, the band model such as the VAEBM provides a simple method to estimate the ODF based on a few spectroscopic constants, instead of using a large database composed of millions of lines. As a result, molecular opacities can be expressed by simple analytical formulae which are very convenient in actual applications of various kind. Of course, accuracy of such approximate methods cannot be very high as compared with that of the methods directly based on a line-by-line database. However, there are many problems that can be attacked with the accuracy to be realized by the approximate methods. Especially, approximate molecular opacities played a major role in exploring model atmospheres of almost all the types of cool stars including brown dwarfs, and will play a similar role if we are to go beyond the classical atmospheres in our explorations of stellar atmospheres.

At the same time, approximate opacities can also be used for some practical purposes, but the limitations must be well recognized. For example, we found a temperature difference as large as 100K between the model atmospheres based on the VAEBM and those based on the OS opacities. This difference is probably significant from the viewpoint of constructing self-consistent model atmospheres. However, if we remember the fact that a similar difference is also found between the best empirical model and theoretical one for the case of the Sun, it may anyhow be difficult to know the true temperatures in stellar atmospheres with an accuracy better than 100K by the present classical model atmospheres. Thus, there can be a viewpoint that the model atmospheres based on the approximate opacities can well be used for practical purposes such as abundance determinations by the quantitative analysis of high resolution stellar spectra. On the other hand, it is more difficult for approximate opacities to predict accurate emergent flux which depends directly on the details of the absorption coefficients. Especially, it is too crude to apply the predicted fluxes based on the JOLA for diatomic molecules to a direct interpretation of observed stellar fluxes. The over-simplification of the JOLA can largely be relaxed by the band model approach, but still very high accuracy cannot be expected in general. For polyatomic molecules, however, it is interesting to clarify the conditions by which the line opacity can be regarded as a pseudo-continuum due to sufficient line overlappings.

The present review necessarily focused attention to the approximate methods for molecular opacities, but this is not to suggest that the approximate molecular opacities can in any way replace the accurate opacities. Rather, we hoped to show that very simple and flexible methods may sometimes be useful, beside the accurate opacities. In fact, we should probably employ different approaches to investigate such complex systems as stellar atmospheres and stellar environments, and it may be useful to have various methods for treating molecular opacities differing in elaboration and in accuracy. Since the recent progresses in molecular science and in computer technology finally made it possible to realize the accurate opacities on a line-by-line basis, we are now in such a favourable situation as to be able to apply these high accuracy opacities to the problems that may truly require them. Under such a situation, it should still be most important to develope an accurate and complete line database for all the molecules of astrophysical interests. Such a database will be important not only as the opacity data but also as the basic data for interpreting a wide range of observations from accurate photometry to very high resolution spectroscopy. Also, only by such a database, the accuracy of approximate methods can be examined.

References

Alexander D.R., Augason G.C., Johnson H.R., 1989, *Astrophy. J.*, **345**, 1014

Auman J.R., 1967a, in *Colloquium on Late-Type Stars*, ed. M.Hack, p.313, Astron. Obs. Trieste

Auman J.R., 1967b, *Astrophys. J. Suppl.*, **14**, 171.

Ayres T.R., 1981, *Astrophys. J.*, **244**, 1064

Borysow A., Frommhold L., 1990, *Astrophys. J. Letter*, **348**, L41

Brown J.A., Johnson H.R., Alexander D.R., Cutright L., Sharp C.M., 1989, *Astrophys. J. Suppl.*, **71**, 623

Burrows A., Hubbard W.B., Saumon D., Lunine J.I., 1993, *Astrophys. J.*, **406**, 158

Carbon D.F., 1979, *Ann. Rev. Astron. Astrophys.*, **17**, 513.

Dravins D., 1987, *Astron. Astrophys.*, **172**, 200 & 211

Golden S.A., 1967, *J. Quant. Spectrosc. Rad. Transf.*, **7**, 225

Golden S.A., 1969, *J. Quant. Spectrosc. Rad. Transf.*, **9**, 1067

Goody R.M., 1964, *Atmospheric Radiation I. Theoretical Basis*, Oxford Univ. Press,

Gray D.F., 1988, *Lectures on Spectral-Line Analysis: F,G, and K Stars*, The Publisher

Grenfell T.C., 1974, *Astron. Astrophys.*, **31**, 303

Gustafsson B., Bell R.A., Eriksson K., Nordlund Å., 1975, *Astron. Astrophys.*, **42**, 407

Herzberg G., 1945, *Molecular Spectra and Molecular Structure II. Infrared and Raman Spectra of Polyatomic Molecules*, D.van Nostrand Co. Inc.

Kneer F., 1983, *Astron. Astrophys.*, **128**, 311

Kurucz R.L., 1979, *Astrophys. J. Suppl.*, **40**, 1

Johnson H.R., 1973, *Astrophys. J.*, **180**, 81

Johnson H.R., Bernat A.P., Krupp B.M., 1980, *Astrophys. J. Suppl.*, **42**, 501

Jørgensen U.G., Johnson H.R., Nordlund Å., 1992, *Astron. Astrophys.*, **261**, 263

Jørgensen U.G., 1993, *Rev. Mexicana Astron. Astrof.*, **23**, 195

Lambert D.L., Hinkle K.H., Smith V.V., 1990, *Astron. J.*, **99**, 1612

Lloyd Evans T., 1990, *Mon. Not. R. Astr. Soc.*, **243**, 349

Ludwig C.B., 1967, *Study on Exhaust Plume Radiation Predictions*, GDC-DBE67-021, General Dynamics

Lunine J.I., Hubbard W.B., Burrows A., Wang Y-P., Garlow K., 1989, *Astrophys. J.*, **338**, 314

Muchmore D.O., Kurucz R.L., Ulmschneider P., 1988, *Astron. Astrophys.*, **201**, 138

Muchmore D.O., Nuth III J.A., Stencel R.E., 1987, *Astrophys. J. Letters*, **315**, L141

Muchmore D.O., Ulmschneider P., 1985, *Astron. Astrophys.*, **142**, 393

Nordlund Å., Dravins D., 1990, *Astron. Astrophys.*, **228**, 155

Plez B., Brett J.M., Nordlund Å., 1992, *Astron. Astrophys.*, **256**, 551

Querci F., Querci M., Tsuji T., 1974, *Astron. Astrophys.*, **31**, 265

Rothman L.S. et al., 1992, *J. Quant. Spectrosc. Rad. Transf.*, **48**, 469

Saxner M., Gustafsson B., 1984, *Astron. Astrophys.*, **140**, 334

Schmid-Burgk J., Scholz M., Wehrse R., 1981, *Mon. Not. R. Astr. Soc.*, **194**, 383

Scholz M., Tsuji T., 1984, *Astron. Astrophys.*, **130**, 11

Tsuji T., 1966, *Publ. Astron. Soc. Japan*, **18**, 127

Tsuji T., 1967, in *Colloquium on Late-Type Stars*, ed. M.Hack, p.260, Astron. Obs. Trieste

Tsuji T., 1969, in *Low Luminosity Stars*, ed. S.S.Kumar, p.457, Gordon and Breach

Tsuji T., 1971, *Publ. Astron. Soc. Japan*, **23**, 553

Tsuji T., 1976, *Publ. Astron. Soc. Japan*, **28**, 543

Tsuji T., 1978, *Astron. Astrophys.*, **62**, 29

Tsuji T., 1984, *Astron. Astrophys.*, **134**, 24

Tsuji T., 1991, *Astron. Astrophys.*, **245**, 203

Watanabe T., Kodaira K., 1978, *Publ. Astron. Soc. Japan*, **30**, 21

Wehrse R., 1975, *Astron. Astrophys.*, **39**, 169

Zeidler-K.T. E.-M., Koester D., 1982, *Astron. Astrophys.*, **113**, 173

Infrared Spectroscopy and Molecules in Circumstellar Envelopes

Kenneth H. Hinkle

Kitt Peak National Observatory, National Optical Astronomy Observatories*, Tucson, Arizona, USA 85719

1 Introduction

During the last two decades there has been a tremendous renewal of interest in the circumstellar environment of evolved stars. This renewal has been mainly driven by sensitive microwave observing techniques. Microwave spectroscopy has resulted in a number of fundamental discoveries in circumstellar shells, including the discovery of maser lines and complex circumstellar chemistry. Historically, interest in circumstellar shells originates from a desire to understand the mass-loss process. It has been known for at least 50 years that circumstellar gas is the origin of interstellar gas and dust. As such the circumstellar shell becomes the pathway through which elements created in the interior of the central star enrich galactic abundances. High resolution spatial and spectral techniques and an ever improving knowledge of stellar evolution has resulted in an increased but still incomplete understanding of the mass-loss process.

There is sufficient space here only to review a small subtopic in this research area. This report is divided into four parts. First, the structure and dynamics of circumstellar shells will be briefly reviewed. Second, the same will be done for circumstellar chemistry. Third, the impact of infrared observations will be discussed in depth. Finally, the future of infrared spectroscopy will be discussed. The review will not concentrate on absorption coefficients for the molecules discussed. Jørgensen (1992) presents an excellent review of this material.

2 Circumstellar Shells

The evolved late-type stars are divided into two basic groups depending on whether the oxygen or carbon is the most abundant. The C/O ratio plays a critical role in the chemistry of the stellar atmosphere (Tsuji 1964, 1973; Scalo & Ross 1976). Stars with C<O are referred to as oxygen-rich or M-type while those with C>O are carbon-rich or C-type. Notably for chemistry the relative

* Operated by the Association of Universities for Research in Astronomy, Inc. under cooperative agreement with the National Science Foundation.

C/O abundance is of primary importance with the astrophysically important C/H and O/H abundances of secondary importance.

The simplest model of the circumstellar shell starts with a central star with a relatively thin (compared to the stellar radius), gravitationally bound photosphere. The photosphere terminates in a chromosphere or perhaps in a cool extension to the photosphere. This region is non-expanding and significant in size, extending several stellar radii above the photosphere, and terminates at the beginning of the expanding circumstellar envelope. This classical view and the evidence supporting it is discussed in reviews by Zuckerman (1980), Dupree (1986), and Goldberg (1987) among others. The great utility of simple models can be seen in their ability to model microwave line profiles and derive mass-loss rates assuming only that (1) the envelope is spherically symmetric, (2) the mass-loss rate and the terminal velocity are constant around the envelope, implying density proportional to r^{-2}, and (3) local turbulent or thermal motions are much less than the terminal velocity (see Kastner 1992 and references therein).

For circumstellar shells of cooler giant stars, which are the primary concern of this review, three principal features characterize the circumstellar shell (Glassgold & Huggins 1987). (1) The terminal wind speeds are small compared to the stellar escape velocity and are typically on the order of 10 km s^{-1}. (2) The circumstellar envelope is much larger than the stellar radius, $R_{shell}/R_{star} \sim 10^3$. The implied variation in physical conditions is enormous. (3) The mass-loss rates are large, typically in the range 10^{-5} to 10^{-7} $M_\odot$ yr^{-1}. Since the stellar masses can be as small as 1-2 $M_\odot$, the time scale for this stellar evolutionary stage is relatively short. It is generally accepted that the outflow in the expanding circumstellar envelope is driven by radiation pressure on dust (Jones et al. 1981; Dominik et al. 1990). For carbon-rich dust, infrared spatial interferometry of prototypical sources has set the dust formation temperature at around 1300 K (Danchi et al. 1990). Oxygen-rich dust has a condensation temperature of about this value.

While the simple model must be basically correct since it provides reasonable matches to many kinds of observations, there has been observational evidence for some time that the actual situation is more complex. Asymmetries in the circumstellar shell of the otherwise normal M supergiant α Ori are well documented (e.g. Bernat & Lambert 1976; Mauron et al. 1984; Hebden et al. 1987). For a sample of M giants, Bernat et al. (1979) and Bernat (1981) found multiple velocity components in the profiles of infrared CO lines seen in absorption against the stellar photosphere. The velocity and excitation temperature of these components are derivable. For some stars velocity increased as excitation temperature decreased. This can be explained as radiation pressure changes on grains as different circumstellar gases freeze onto grains (Keady et al. 1988). However, for a few stars velocity and excitation temperature were not related. This can be explained by discrete or time varying mass-loss events (Jura 1986; Olofsson et al. 1993). The extreme case of μ Cep (Bernat 1981) may require clumping in the circumstellar shell. VLBI maps of SiO masers (Moran et al. 1979) show that the SiO maser results from a large number of spots rather than the entire circumstellar shell. Alcock & Ross (1986) have shown that the masers form in 'dense cloudlets' and not in the spherically expanding circumstellar wind and suggest that the density enhanced clouds are formed by the ejection process from the photosphere. Ultimately, the circumstellar shell of a red giant becomes the expanding shell of a planetary nebula. Planetary nebulae are marked more by their deviation from spherical symmetry than their adherence to it.

While far more mass is typically in the gas than the dust (i.e. Harvey et al. 1991 estimate $M_{gas}/M_{dust} \sim 600$ in IRC+10216), the dust is an important consideration for the spectroscopist. The dust re-radiates absorbed radiation from the photosphere of the star into the infrared. A large class of objects exists where the dust has enough optical depth to obscure the photosphere. These objects are of special interest for spectroscopy of circumstellar lines. In normal stars the photosphere contributes a large number of lines to the spectrum throughout the infrared. In stars obscured by optically thick circumstellar dust only the circumstellar lines are seen. The C/O ratio determines the composition of the dust as well as the gas, and the dust composition impacts the circumstellar structure. In the oxygen-rich case, the grains are oxides of refractory elements especially silicon, while in the carbon-rich case, the grains are largely carbon combined with refractory elements (Turner 1991).

This review uses the terms column density and abundance frequently. The abundance is the number of molecules per unit volume relative to the number of hydrogen molecules per unit volume. The density decreases with radius as r^{-2}; by referencing the abundance to a stable molecule this effect is removed. The column density is the total number of molecules per square centimeter along the line of sight to the star. Many classes of molecules do not exist throughout the circumstellar shell but exist only through a portion of the circumstellar radius. Thus for an observed molecule, dividing the observed column density by the column density of molecular hydrogen does not necessarily produce a physically meaningful abundance. While this is now well understood, incorrect abundances can be found in early papers. Perhaps more meaningful for molecules created in the circumstellar shell is the peak radial abundance which is the largest value of the abundance at any point along a radius from the central star.

To determine the peak radial abundance of a molecule, line profiles must be synthesized using a model of the circumstellar shell that specifies the run of temperature, velocity, and density with radius. The determination of a peak radial abundance is possible because the physical conditions may be uniquely assigned by radius. The run of physical conditions with radius results in a distinctive line shape for each spectral line as well as a characteristic change of line strength with rotational energy. Modeling the line profile as well as the change of profile and equivalent width with rotational energy allows the modeler to determine the radial distribution of the molecule in the model. The circumstellar model can be determined semiempirically from lines of a molecule like CO, which has the stability to exist throughout the vast range of physical conditions from the cool star photosphere to the outer circumstellar radius (Keady et al. 1988). Since infrared spectroscopy typically probes the inner circumstellar shell ($10 \lesssim R \lesssim 1000$

R_*) and microwave spectroscopy probes outer regions ($R \gtrsim 1000\ R_*$), abundances derived from different techniques may be somewhat different and there is always uncertainty as to whether this reflects measurement differences or circumstellar chemistry (Keady & Hinkle 1988).

3 Circumstellar Molecules and Chemistry

The chemistry of circumstellar envelopes is a topic of considerable current interest because of both theoretical and observational challenges. Several reviews appear in the recent literature (Omont 1985; Glassgold & Huggins 1987). The circumstellar envelope challenges the theorist because the chemistry is time-dependent as the material is carried away from the star through regions with widely differing physical conditions. The simplest chemical model is the 'freeze out' model. In this model thermal equilibrium applies close to the central star. As the density of the circumstellar gas decreases the temperature drops below some critical value at which the chemical reactions stop and the abundances are 'frozen'. Freeze out models are presented by McCabe et al. (1979) and Lafont et al. (1982). In general these models work well for simple CNO molecules but fail for refractory compounds, chain molecules, and radicals. A more refined model includes chemical reactions in the circumstellar shell driven by photodissociation and photoionization . These processes are driven by interstellar ultraviolet, chromospheric ultraviolet (in unobscured circumstellar envelopes), and cosmic rays. Ions produced through photoionization, for instance $C_2H_2^+$, can lead to rapid reactions which build large molecules (Nejad et al. 1984; Glassgold et al. 1986). This model is in agreement with spatially resolved microwave images of circumstellar shells which show that complex molecules exist in the outer envelope but not near the star (e.g. Bieging & Tafalla 1993). Models are just appearing which include reactions that take place in or on grains (Brown et al. 1988). Glassgold & Huggins (1987) note that shocks in pulsating stars could also drive chemical processes.

A current list of molecules detected using microwave and infrared techniques can be found in Olofsson (1992). Much of the observational work on the chemistry in circumstellar envelopes has been done on the carbon-rich envelope around the bright, obscured carbon star IRC+10216. Interestingly IRC+10216 is such a rich source and such a bright object that a number of the molecules detected are seen only in this source. Detailed chemical models exist for IRC+10216 (e.g. Glassgold et al. 1986; Nejad & Millar 1987; Howe & Millar 1990). The models include the photodissociation and photoionization of a large number of molecules of photospheric origin. The photodissociation and photoionization chemistry of the most common molecules, H_2, CO, HCN, C_2H_2, is quite complex. Of particular interest is the photoionization of C_2H_2 which leads to the formation of complex molecules including carbon chains bearing hydrocarbons. Frenklach & Feigelson (1989) have shown that acetylene chemistry can produce benzene and ultimately polycyclic aromatic hydrocarbons.

Work on the chemistry of oxygen-rich circumstellar shells is less developed than for carbon-rich circumstellar shells. This stems from two causes. First, there is no highly obscured, bright star that would be an oxygen-rich analogy to the carbon-rich star IRC+10216. Second, in oxygen-rich circumstellar shells carbon is mainly associated into CO and the chemistry has not been considered as interesting. Recently there has been increased interest in oxygen-rich circumstellar shells as a result of the discovery of post-AGB, pre-planetary nebula objects in the form of molecule-rich bipolar nebulae (e.g. OH231.8+4.2 [Morris et al. 1987]).

Mamon et al. (1987) present a model for photochemistry and molecular ions in oxygen-rich circumstellar envelopes. This paper is restricted to photochains associated with CO, H_2O, and N_2. It was shown that the molecular ions H_3O^+, HCO^+ and H_3^+ all play an important role in the photochemistry. In particular, the polar ions H_3O^+ and HCO^+ are predicted to have a large peak abundance. Omont et al. (1993) present observations and discussion of sulfur compounds in oxygen-rich circumstellar envelopes. References to the gas phase chemistry are given by Omont et al. (1993).

4 The Infrared spectrum

Microwave spectroscopy has been shown to be tremendously successful at detecting molecules in circumstellar and interstellar environments. Infrared spectroscopy is a complementary technique. The infrared spectrum contains stretch and bending vibration transitions for molecules made up from combinations of the astrophysically important CNO group atoms with themselves and hydrogen. Figure 1 adapted from Allamandola (1984) summarizes the spectral locations of the most important transitions. In addition, the infrared has permitted rotation-vibration transitions of symmetric molecules which have no permitted pure rotation transitions in the microwave. Historically, a critical role for the infrared has been in establishing the abundances of symmetric molecules like H_2, C_2H_2, and CH_4. While these symmetric molecules also have permitted electronic transitions, the flux from cool sources is so low in the ultraviolet where most of these transitions occur that they will be mainly unobservable. (A few electronic transitions occur in the infrared, see Bernath 1992.) In addition, opacity caused by interstellar and circumstellar dust is large in the blue but at a minimum in the 2 μm infrared. In circumstellar shells, the infrared may be used to probe regions of the circumstellar shell closer to the star than can be seen in the microwave. Of the molecules previously detected in the microwave most have transitions in parts of the infrared not blocked by telluric absorption. Many of these transitions have large transition moments but have not been exploited because the relatively small column densities and/or low fluxes of the sources have exceeded the capabilities of existing infrared detectors.

4.1 Carbon-Rich Circumstellar Shells

If C<O, the CO molecule bonds all the available carbon atoms. However, if C>O, extra carbon atoms remain to combine with hydrogen, nitrogen and other

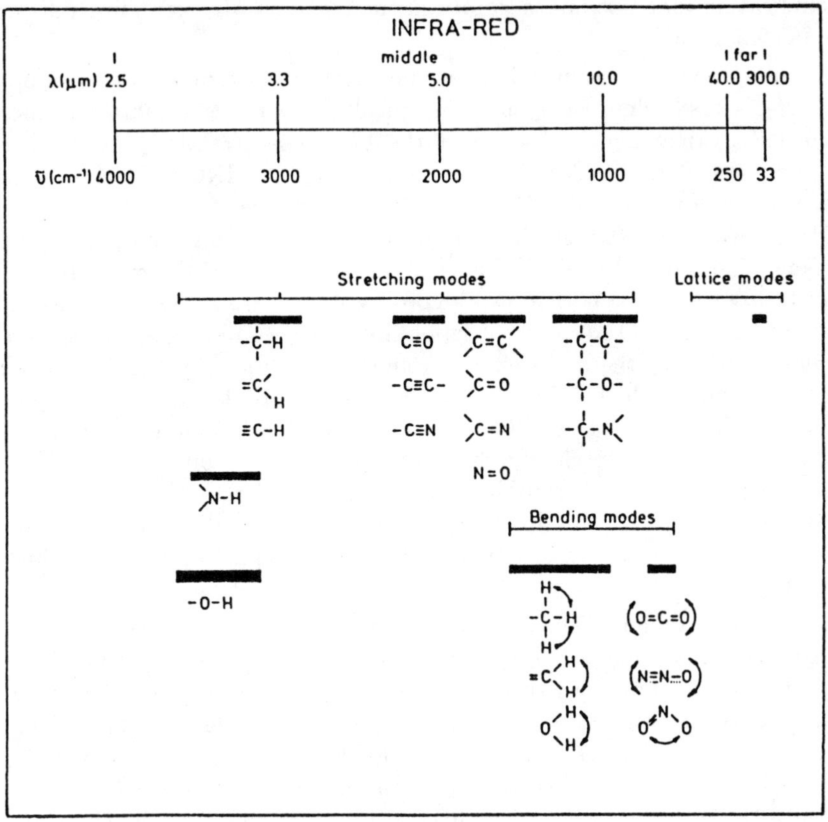

Fig. 1. Spectral locations of HCNO group molecular stretch and bend transitions in the infrared. In a gas, these transitions consist of a number of rotation-vibration lines. Figure from Allamandola 1984 as adapted by Genzel 1992

carbon atoms, as well as less abundant elements. As can be seen from Figure 1 the rotation-vibration spectra of these molecules are observable in the 3-12 μm infrared.

The most abundant circumstellar molecule is H_2, the principal constituent of circumstellar shells. Typically H_2 is 10^3 times more abundant than any other molecule. However, the homonuclear H_2 molecule has only a weak quadrupole spectrum. Oscillator strengths are 10^6 times smaller than the oscillator strengths of typical diatomic molecules. The factor of 10^6 more than compensates for the large abundance of H_2, so sources which have exceeding strong spectral lines from typical diatomic molecules need not have detectable H_2 lines. Weak circumstellar absorption lines from the 1-0 S(0) and S(1) (i.e. v"=0,J"=0 $\rightarrow$ v'=1,J'=2 and v"=0,J"=1 $\rightarrow$ v'=1,J"=3) lines have been detected in the 2 μm spectra of Mira variables (Hinkle et al. 1994a) and the 1-0 S(1) line has been detected in the spectrum of IRC+10216 (Hinkle et al. 1994b). Keady & Ridgway (1993) provide models indicating that this line is formed in the inner, slowly expanding part of the circumstellar shell.

CO is ubiquitous to circumstellar and interstellar environments and as opposed to H_2 is easily detectable spectroscopically. CO rotation-vibration lines of circumstellar origin can be observed in the 4.6 μm spectra of typical M-type and C-type stars (Bernat 1981). The obscured carbon-star IRC+10216 presents the best modeled example. The CO spectrum was explored in detail by Keady et al. (1988) who used the CO line profile to develop a semiempirical model atmosphere. The infrared provides information on sections of the circumstellar shell much closer to the star than can be observed in microwave CO. In the infrared, the individual CO rotation-vibration lines are seen in absorption against a continuum formed by thermal emission from the dust. Emission resulting from resonant scattering in the CO lines can be seen in the profiles of the CO $\Delta v=1$ low excitation lines as well as in off source spectra (Sahai & Wannier 1985). CO line profiles in IRC+10216 clearly show multiple velocity components. These components appear to result from discrete changes in the radiation pressure on grains and have been attributed to the freezing of circumstellar gases on grain mantles. The IRC+10216 low excitation CO fundamental lines are optically thick and $^{12}C^{16}O$, $^{13}C^{16}O$, $^{12}C^{17}O$, and $^{12}C^{18}O$ are all easily detected.

The next most abundant molecule in carbon-rich circumstellar shells is generally acetylene. C_2H_2 has a rich spectrum, with lines detected in IRC+10216 from 2.5 through 12 μm. With such a wide wavelength range, C_2H_2 is a useful probe of circumstellar structure, as discussed by Keady & Ridgway (1993). From the point of view of opacities the most remarkable feature of acetylene is the 3 μm band. This band results from the C_2H_2 and HCN C-H stretch (Ridgway et al. 1978). In IRC+10216 the strongest lines in this band are optically thick and the line spacing is dense. Sufficient lines are present that other much less abundant molecules could contribute to the band but would be undetectable. The $^{13}C^{12}CH_2$ isotopic lines are easily detected, in spite of the fact that IRC+10216 has a $^{12}C/^{13}C$ ratio of $\sim$40 (Wannier & Sahai 1987; Keady et al. 1988). In IRC+10216 Keady & Hinkle (1988) find evidence that the C_2H_2 abundance decreases by a factor of 5 between 100 and 1000 R_*. This suggests that the acetylene abundance is reduced through interactions with grain surfaces.

HCN has approximately the same abundance as C_2H_2 in the circumstellar shell of IRC+10216. Wiedemann et al. (1991) find an inner shell abundance of 4×10^{-5} for HCN and a column density of 7×10^{18} cm^{-2} while Keady & Hinkle (1988) measured an inner shell abundance of 5×10^{-5} for C_2H_2 with a column density of 2×10^{18} cm^{-2}. As is the case for H_2, CO and C_2H_2, the circumstellar HCN line profiles in IRC+10216 show multiple velocity components. Since these result from accelerations known from the CO to occur fairly close to the central star, they are the infrared spectral signature of a molecule of photospheric origin. The photospheric origin of HCN is not surprising, since HCN has been known for some time to be an important opacity contributor to the upper photosphere of unobscured, cool carbon stars (Jørgensen 1990).

The remaining molecules detected in the infrared all have peak radial abundances an order of magnitude, or more, smaller than the above major contributors. These will be divided into classes according to their circumstellar chemistry. Figure 2 taken from Keady & Ridgway (1993) illustrates the relative molecular abundances in the prototypical carbon rich circumstellar shell around IRC+10216. The first group are molecules of photospheric origin. These include SiO and CS. In IRC+10216, Keady and Ridgway (1993) report that CS shows evidence of decreasing column density with radius in the range observed by the infrared. In addition, the infrared CS abundance is an order of magnitude larger than the microwave CS abundance. Since the CS abundance appears to change from 10 to 1000 R_*, an interval where photolysis is not important, the depletion must be due to absorption on the grain surfaces (Keady & Ridgway 1993). On the other hand, infrared and microwave abundances of SiO in IRC+10216 are fairly well matched, implying a constant abundance out to the radius where photolysis occurs.

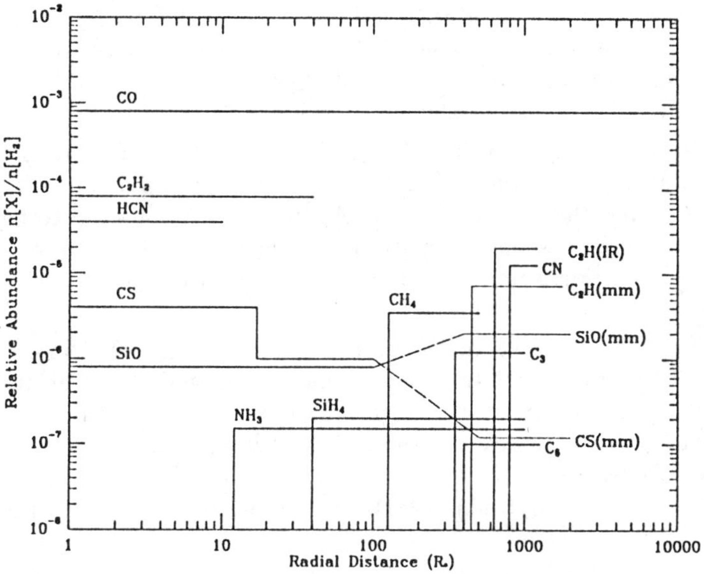

Fig. 2. Radial abundances in IRC+10216 for molecules detected in infrared spectra. The source of abundances are labeled for outer shell molecules with abundances determined from both infrared (IR) and millimeter wave (mm) observations. All abundances at R<100 R_* are infrared. From Keady and Ridgway 1993

Ammonia (NH_3), silane (SiH_4), and methane (CH_4), form a class of molecules created in the mid-circumstellar shell through grain-surface chemistry. The geometric similarity of the three molecules is notable. Infrared line profiles for these three molecules, unlike profiles for CO or C_2H_2 lines, do not have a contribution from the circumstellar gas at velocities close to the star. Detailed modeling by Keady & Ridgway (1993) shows that the lines from each molecule can be best fit by radial abundances that have a cut off at an inner radius relatively close to the star compared to the usual case for molecules detected in the microwave. In the case of NH_3 this radius is $\sim$20 R_*, for SiH_4 $\sim$40 R_*, and for CH_4 $\sim$100 R_*. Keady & Ridgway (1993) note that the production of these gases in the circumstellar shell occurs near radii where the gas is accelerated. They point to a link between grain surface chemical changes and changing photon pressure on the grains. Photolysis clearly does not play a role in forming these molecules since the circumstellar shell is well shielded from UV where they are formed.

The above evidence points to an origin of these molecules on the grain surface. Detections of these molecules and the knowledge that they were formed in the mid-circumstellar shell have been known for some time (CH_4: Clegg et al. 1982; NH_3: Betz et al. 1979; SiH_4:Goldhaber & Betz 1984). However, progress in this area has been motivated by dense interstellar clouds. To match interstellar NH_3 abundances Brown et al. (1988) and others proposed that CH_4 and NH_3 form by hydrogenation of C and N on grain mantles. Presumably a similar mechanism applies to SiH_4. Lacy et al. (1991) found observational evidence that CH_4 is made in molecular cloud grain mantles. Charnley & Miller proposed a mechanism for H_2O also to be formed on grain mantles but this hypothesis is probably not testable due to the already high water abundance in oxygen-rich environments.

A possibly related molecule is ethylene (C_2H_4). Goldhaber et al. (1987) report that ethylene has an average rotation temperature of 100 K and a column density of 4×10^{15} cm^{-2}. However, the rotational levels of ethylene did not seem to be distributed in LTE as is the case for the above three molecules. The data were not of high S/N and additional infrared observations of ethylene would be of interest.

C_2H and CN are two photolysis molecules that have been detected in the infrared spectrum of IRC+10216. Both molecules have rotational temperatures on the order of 10 K and column densities near 3×10^{15}. The CN result is in good agreement with CN microwave measurements and with formation from photolysis of HCN (Wiedemann et al. 1991). C_2H infrared and microwave results are again in agreement (Keady & Hinkle 1988). The parent molecule for C_2H is C_2H_2.

The naked carbon chain molecules (C_3 and C_5) detected in the infrared (Hinkle et al. 1988; Bernath et al. 1989) have an unknown origin. The line profiles of both of these molecules clearly indicate that they are formed in the circumstellar shell. Both C_3 and C_5 have a respectably large abundance ($\sim10^{-6}$ for C_3 and $\sim10^{-7}$ for C_5). The column density of C_5 is 1×10^{14} cm^{-2}. No parent molecule with sufficient abundance has been identified. The excitation temperatures of both molecules are $\sim$40 K. The most likely suggestion at this time is that these molecules result from photolysis of the carbonaceous grains. This is a distinct process from the grain surface chemistry discussed above.

We have searched for a number of other naked carbon chains up to C_{60}. The odd numbered chains may be less reactive than the even chains and hence more easily detectable (Kroto et al. 1987). However, at this time the detection of odd numbered chains has mainly to do with the infrared spectra. C_4 has its strongest infrared transition at 1549 cm^{-1} where it is blocked from observation by telluric water vapor. The odd numbered chains C_3, C_5, C_7, C_9 occur in atmospheric windows. An upper limit of $<2\times10^{13}$ cm^{-2} has been set for the C_7 column density (Hinkle & Bernath 1992).

The fullerenes, which are larger members of the naked carbon chain family deserve a special note. These molecules turn out to be difficult to detect spectroscopically. The C_{60} molecule has a high symmetry which results in only four vibration modes being infrared active. These are the four modes reported by Kratschmer et al. (1990): 528, 577, 1183, and 1429 cm^{-1}. Of these four only 1183 cm^{-1} is reasonably placed in the spectrum to observe astronomically. Frum et al. (1991) found that the 1183 cm^{-1} band in the solid phase was at 1169.1 cm^{-1} in the gas phase. The rotational structure of this transition is so closely spaced as to be unresolvable. In addition, there is no sharp, distinctive Q-branch which makes detection of the 1169 cm^{-1} band difficult. While we could not detect this band in IRC+10216, this could be due to hydrogenation in this source and it might be of interest to try observing a yet more carbon-rich source.

Polycyclic aromatic hydrocarbons (PAHs), the family of ring molecules of which benzene is the simplist, are the suspected carriers of a number of widely observed infrared emission features. A primary source of PAHs may be mass-losing carbons stars (Latter 1991). Detailed chemical reaction networks developed by Frenklach & Feigelson (1989) indicate that PAH can be produced from acetylene with yields in excess of 1%. The above mentioned depletion of acetylene with increasing radius in the circumstellar shell of IRC+10216 could reflect this process (Latter 1991). At a 1% yield, the PAH column density in IRC+10216 would be on the order of 10^{16}. Latter (1991) has suggested that PAHs may be taken up by grains shortly after the PAH molecule is created, only to be released later in the interstellar medium. Even if the PAHs remain in the gas phase they will be nearly impossible to detect in a circumstellar shell. As is well known, the PAHs include a large number of different molecules (see Léger et al. 1987). Microwave spectroscopy is impossible because the rotational spectra are extremely complex to the confusion limit and individual molecules in the group are not necessarily very abundant. The CH stretch frequency depends on the bonding and on the side groups attached, resulting in a PAH CH stretch at 3050 cm^{-1} while the acetylene CH stretch is at 3300 cm^{-1}. However, the prognosis for detection by high resolution techniques (i.e. line resolved spectroscopy) in the infrared is the same as in the microwave for basically the same reasons. In the interstellar medium relatively low concentrations of PAHs can be seen excited by ultraviolet radiation. In circumstellar shells this excitation mechanism is not active due to UV shielding.

In sources that become sufficiently cold before photoionization and photodissociation commence, carbon compounds form ices. CO has a sublimation temperature of 15-20 K under astrophysical conditions (Tielens et al. 1991). When CO goes into the solid state rotation is no longer allowed, thus the R and P branches disappear. However, a pure vibration transition, the Q branch which is not permitted in the gas phase, is allowed in the solid phase. The Q branch appears in the spectrum at about the location of the band origin. Non-polar compounds all have sublimation temperatures below that of polar compounds. Thus the common carbon-rich ices have sublimation temperatures considerably below that of water ice (on the order of 10-20 K rather than 100 K) (Tielens et al. 1991). Chemical studies must include the abundant molecules N_2 and O_2 both of which form ices (Tielens et al. 1991). Unfortunately these molecules are undetectable since in the solid phase neither molecule is infrared active and in the gas phase both have only very weak infrared quadrupole spectra.

4.2 Oxygen-Rich Circumstellar Shells

As indicated by circumstellar chemistry models, the more interesting objects for the infrared spectroscopist are the carbon-rich circumstellar shells. The oxygen-rich circumstellar shells are dominated by the spectrum of H_2O. Water has a rich, complex gas-phase spectrum. As early as the 1960's work by Kuiper (1963) and Johnson et al. (1968) showed that the photospheric spectrum of cool M-giants was dominated by water bands in the 1-2.5 μm region. Chemical equilibria calculations show that for $C<O$ water is a dominant constituent at temperatures less than about 3000 K (Jørgensen this volume). Only for effective temperatures below about 1800 K at relatively high gravities do other polyatomic molecules (CH_4) start to compete with H_2O (Tsuji this volume). Hinkle & Barnes (1979) noted that in oxygen-rich Mira variables water lines could be seen originating in the upper photosphere as well as in a ~1000 K region. The ~1000 K region appears to be the inner circumstellar shell (Hinkle et al. 1994). Unfortunately, water has such a complex spectrum that it is difficult to analyze on a line by line basis. In spite of significant efforts by several groups (see for example Camy-Peyret et al. 1977) literally thousands of unidentified water lines remain in the infrared spectra of astronomical sources. Mozurkewich et al. (1987) discovered a rich circumstellar water spectrum in the RV Tau star R Sct. In the case of R Sct the star is not obscured but the photosphere is too hot for water lines. While the above references are to 2 μm region spectra, water contributes lines throughout the infrared into the sub-millimeter and millimeter wavelength regions. Strong bands are obvious in M-giant photospheric spectra in the 3-4 μm region (Ridgway et al. 1984) and are the dominant feature of 10 μm sunspot spectra (Wallace & Livingston 1993). Microwave water masers originating from oxygen-rich circumstellar shells are well known.

The ubiquitous CO molecule also contributes spectral features in oxygen-rich circumstellar shells. Circumstellar CO fundamental lines are obvious in the 4.6 μm spectra of M stars and 2.3 μm first overtone lines can have a circumstellar contribution. Weak circumstellar H_2 lines may be observed in the 2 μm region in M-type giants (Hinkle et al. 1994). Circumstellar SiO is observable in the 4 and 10 μm region (Geballe et al. 1979; Tsuji et al. 1994).

Finally, absorption bands caused by ices and silicate grains are major contributors to infrared spectra of oxygen-rich objects. A detailed discussion of these topics is beyond the scope of the present review, and the reader is referred to e.g. Allamandola 1984; Allamandola & Sandford 1988; Tielens & Allamandola 1987.

See also the reviews by Alexander & Ferguson and by Sedlmayr in this volume. Linked with the above discussion of gas phase H_2O, water ice has long been known to be the origin of the 3.08 μm continuous absorption band seen toward very obscured late-type stars and in spectra of embedded sources in dense clouds (Gillett & Forrest 1973). Recent work by Smith et al. (1988) indicates that this band results from water frozen in an amorphous layer onto silicate grains. This is an interesting and very well documented case of freeze out of a gas phase constituent of the circumstellar shell onto a grain mantle.

5 The Future of Infrared Spectroscopy

Until recently infrared detectors were available only in single pixel format and were plagued with very high inherent noise levels. Recording narrow spectral elements of an astronomical source one by one with such a detector was virtually impossible. Fourier transform spectroscopy (FTS), which was used for most of the observations reported here, features a natural multiplexing strategy which encodes many spectral elements on a single detector. As long as detector noise dominates, FTS offers a gain over a single channel device on order of the number of spectral resolution elements. More or less by chance, FTS also offers a number of wonderful instrumental features: no slit; continuously adjustable and unlimited spectral resolution; well understood instrumental line profile; very large spectral coverage; very high throughput; no scattered light; accurately known, internally calibrated frequency scale without a dispersion relation; noise dominated wavelength precision; inherently photometric spectra. Unfortunately FTS, as implemented for high resolution spectroscopy, has the severe limitation that it lacks the sensitivity to reach 'faint' sources.

Infrared arrays with dimension of 256x256 pixels are now widely available in a number of materials (InSb, HgCdTe, PtSi). While infinite increase in size is not possible, arrays a factor of 2-4 larger are likely. These arrays offer a large number of pixels with each pixel having very low noise. The array detectors are integrating detectors, rather than continuous readout as were the single element detectors of the past. Assuming high quality devices of both types (noise equivalent power [NEP, see e.g. Wolfe and Zissis 1989] for the single element detector of 10^{-18} watt/$\sqrt{\text{Hz}}$ and readout noise for the integrating detector of 30 e^- with negligible dark current), then the signal to noise (S/N) and integration time in seconds (t) are related by:

$$log_{10}(\text{relative S/N [integrating/analog]}) = 0.5\log_{10}(t) + 0.5 \tag{1}$$

(Ridgway & Hinkle 1992).

From equation 1, we conclude that a grating spectrometer with an infrared array will be on the order of 100$\times$ more sensitive than an FTS. Generally, a spectrograph that takes advantage of infrared arrays is cryogenic. Infrared arrays become background limited in warm spectrographs at wavelengths longer than about 1.8 μm. Several cryogenic spectrographs have now been built to use infrared arrays (CSHELL at IRTF and CGS4 at UKIRT). These spectrographs offer two pixel resolution of about 20000, with 40000 at the limiting end of their performance. At NOAO, we have been working on a resolution 70,000-100,000 spectrograph intended for use on the 2.1 and 4 meter telescopes. This spectrograph will have obvious applications in circumstellar and interstellar spectroscopy.

References

Alcock C., Ross R.R., 1986, *Astrophys. J.*, **310**, 838

Allamandola L.J., 1984, in *Galactic and Extragalactic Infrared Spectroscopy*, ed. M.F. Kessler, J.P. Phillips (Dordrecht:Reidel), p. 5

Allamandola L.J., Sandford S.A., 1988, in *Dust in the Universe*, ed. M.E. Bailey, D.A. Williams (Cambridge: Cambridge Univ. Press), p. 229

Bernat A.P., 1981, *Astrophys. J.*, **246**, 184

Bernat A.P., Hall D.N.B., Hinkle K.H., Ridgway S.T., 1979, *Astrophys. J.*, **233**, L135

Bernat A.P., Lambert D.L., 1976, *Astrophys. J.*, **210**, 395

Bernath P.F., 1992, in *Astronomical Infrared Spectroscopy* ed. S. Kwok (ASP Conf. Series Vol. 41), p. 251

Bernath P.F, Hinkle K.H., Keady J.J., 1989, *Science*, **244**, 562

Betz A.L., McLaren R.A., Spears D.L., 1979, *Astrophys. J.*, **229**, L97

Bieging J.H., Tafalla M., 1993, *Astron. J.*, **105**, 576

Brown P.D., Charnley S.B., Miller T.J., 1988, *Mon. Not. Royal Astron. Soc.*, **231**, 409

Camy-Peyret C., Flaud J.M., Toth R.A., 1977, *J. Mol. Spec.*, **67**, 117

Clegg R.E.S., Lambert D.L., Hinkle K.H., 1982, *Mon. Not. Royal Astron. Soc.*, **201**, 95

Danchi W.C., Bester M., Degiacomi C.G., McCullough P.R. Townes C.H., 1990, *Astrophys. J.*, **359**, L59

Dominik C., Gail H.-P., Sedlmayr E., Winters J.M., 1990, *Astron. Astrophys.*, **240**, 365

Dupree A.K., 1986, *Ann. Rev. Astron. Astrophys.*, **24**, 377

Frenklach M., Feigelson E.D., 1989, *Astrophys. J.*, **341**, 372

Frum et al., 1991, *Chem. Phys. Lett.*, **176**, 505

Geballe T.R., Lacy J.H., Beck S.C., 1979, *Astrophys. J.*, **230**, L47

Genzel R., 1992, in *The Galactic Interstellar Medium*, ed. D. Pfenniger, P. Bartholdi (Springer-Verlag:Berlin)

Gillett F.C., Forrest W.J., 1973, *Astrophys. J.*, **179**, 483

Glassgold A.E., Huggins P.J., 1987, in *The M-type Stars*, ed. H. Johnson, F. Querci (NASA SP-492)

Glassgold A.E., Lucas R., Omont A., 1986, *Astron. Astrophys.*, **157**, 35

Goldberg L., 1987, in *The M-type Stars*, ed. H. Johnson, F. Querci (NASA SP-492)

Goldhaber D.M., Betz A.L., 1984, *Astrophys. J.*, **279**, L55

Goldhaber D.M., Betz A.L., Ottusch J.J., 1987, *Astrophys. J.*, **314**, 356

Harvey P.M., Lester D.F., Brock D., Joy M., 1991, *Astrophys. J.*, **368**, 558

Hebden J.C., Eckart A., Hege E.K., 1987, *Astrophys. J.*, **314**, 690

Hinkle K.H., Barnes T., 1979, *Astrophys. J.*, **227**, 923

Hinkle K.H., Bernath P., 1992, in *Astronomical Infrared Spectroscopy*, ed. S. Kwok (ASP Conf. Series Vol. 41), p. 125

Hinkle K.H., Keady J.J., Bernath P.F., 1988, *Science*, **241**, 1319

Hinkle K.H., Ridgway S.T., Martin C., 1994a, *Astrophys. J.*, submitted

Hinkle K.H., Keady J.J., Sada, P., 1994b, in preparation

Howe D.A., Millar T.J., 1990, *Mon. Not. Royal Astron. Soc.*, **244**, 444

Johnson H.L., Coleman I., Mitchell R.I, Steinmetz D.L., 1968, *Comm. Lunar Planet. Lab.*, **7**, 83 (No. 113)

Jones T.W., Ney E.P., Stein W.A., 1981, *Astrophys. J.*, **250**, 324

Jørgensen U.G., 1990, *Astron. Astrophys.*, **232**, 420

Jørgensen U.G., 1992, *Rev. Mex. Astron. Astrof.*, **23**, 49

Jura M., 1986, *Astrophys. J.*, **303**, 327

Kastner J.H., 1992, *Astrophys. J.*, **401**, 337

Keady J.J., Hall D.N.B., Ridgway S.T., 1988, *Astrophys. J.*, **326**, 832

Keady J.J., Hinkle K.H., 1988, *Astrophys. J.*, **331**, 539

Keady J.J., Ridgway S.T., 1993, *Astrophys. J.*, **406**, 199

Kratschmer W., Lamb L.D., Fostiropoulos K., Huffman D.R., 1990, *Nature*, **347**, 354

Kroto H.M., Heath J.R., O'Brien S.C., Curl R.F., Smalley R.E., 1987, *Astrophys. J.*, **314**, 352

Kuiper G.P., 1963, *Comm. Lunar Planet. Lab.*, **1**, 179 (No. 23)

Lacy J.H., Carr J.S., Evans N.J., Bass F., Achtermann J.M., Arens J.F., 1991, *Astrophys. J.*, **376**, 556

Lafont S., Lucas R., Omont A., 1982, *Astron. Astrophys.*, **106**, 201

Latter W.B., 1991, *Astrophys. J.*, **377**, 187

Léger A., d'Hendecourt L., Boccara N. ed., 1987, *Polycyclic Aromatic Hydrocarbons and Astrophysics*, (Dordrecht:Reidel)

Mamon G.A., Glassgold A.E., Omont A., 1987, *Astrophys. J.*, **323**, 306

Mauron N., Fort B., Querci F., Dreux M., Fauconnier T., Lamy P., 1984, *Astron. Astrophys.*, **130**, 341

Mozurkewich D., Gehrz R.D., Hinkle K.H., Lambert D.L., 1987, *Astrophys. J.*, **314**, 242

McCabe E.M., Connon Smith R., Clegg R.E.S., 1979, *Nature*, **281**, 263

Moran J.M., Ball J.A., Predmore C.R., Lane A.P., Huguenin G.R., Reid M.J., Hansen S.S., 1979, *Astrophys. J.*, **231**, L67

Morris M., Guilloteau S., Lucas R., Omont A., 1987, *Astrophys. J.*, **321**, 888

Nejad L.A.M., Millar T.J., 1987, *Astron. Astrophys.*, **183**, 279

Nejad L.A.M., Millar T.J., Freeman A., 1984, *Astron. Astrophys.*, **134**, 129

Olofsson H., 1992, in *Mass Loss on the AGB and Beyond*, in press

Olofsson H., Eriksson K., Gustafsson B., Carlstrom U., 1993, *Astrophys. J. Suppl.*, **87**, 267

Omont A., 1985, in *Mass-Loss from Red Giants*, ed. M. Morris, B. Zuckerman (Dordrecht:Reidel), p. 269

Omont A., Lucas R., Morris M., Guilloteau S., 1993, *Astron. Astrophys.*, **267**, 490

Ridgway S.T., Hinkle K.H., 1992, in *High Resolution Spectroscopy with the VLT*, ed. M.-H. Ulrich (ESO Conf. & Workshop Proceedings 40), p. 213

Ridgway S.T., Carbon D.F., Hall D.N.B., 1978, *Astrophys. J.*, **225**, 138

Ridgway S.T., Carbon D.F., Hall D.N.B., Jewell J., 1984, *Astrophys. J. Suppl.*, **54**, 177

Sahai R., Wannier P.G., 1985, *Astrophys. J.*, **299**, 424

Scalo J.M., Ross J.E., 1976, *Astron. Astrophys.*, **48**, 219

Smith R.G., Sellgren K., Tokunaga A.T., 1988, *Astrophys. J.*, **334**, 209

Tielens A.G.G.M., Allamandola L.J., 1987, in *Interstellar Processes*, ed. D. Hollenbach, H. Thronson (Dordrecht:Reidel), p. 397

Tielens A.G.G.M., Tokunaga A.T., Geballe T.R., Baas F., 1991, *Astrophys. J.*, **381**, 181

Tsuji T., 1964, *Ann. Tokyo Astron. Obs.*, 2nd Ser.,**9**, 1

Tsuji T., 1973, *Astron. Astrophys.*, **23**, 411

Tsuji T., Ohnaka K., Hinkle K.H., Ridgway S.T., 1994, *Astron. Astrophys.*, submitted.

Turner B.E., 1991, *Astrophys. J.*, **376**, 573

Wallace L., Livingston W., 1993, in preparation

Wannier P.G., Sahai R., 1987, *Astrophys. J.*, **319**, 367

Wiedemann G.R., Hinkle K.H., Keady J.J., Deming D., Jennings D.E., 1991, *Astrophys. J.*, **382**, 321

Wolfe W.L., Zissis G.J. eds. 1989 *The Infrared Handbook*, (Environmental Research Institute of Michigan, Ann Arbor, Michigan)

Zuckerman B., 1980, *Ann. Rev. Astron. Astrophys.*, **18**, 263

Molecular Abundances in the Envelopes Around Evolved Stars

Hans Olofsson

Onsala Space Observatory, S-43992 Onsala, Sweden

1 Introduction

Red giant stars on the asymptotic giant branch (AGB), AGB-stars, lose copious amounts of matter in a slow stellar wind (Olofsson 1993). Mass loss rates in excess of $10^{-4}\,M_\odot\,yr^{-1}$ have been measured. The primary observational consequence of this mass loss is the formation of an expanding envelope of gas and dust, a circumstellar envelope (CSE), that surrounds the star. This is a truly extended atmosphere that continues thousands of stellar radii away from the star. At the highest mass loss rates (which probably occur at the end of the AGB evolution) the CSE becomes so opaque that the photosphere is hidden and essentially all information about the object stems from the circumstellar emission. At some point on the AGB a star may change from being O–rich (i.e., the abundance of O is higher than that of C) to becoming C–rich (i.e., a carbon star where the abundance of C is higher than that of O) as a result of nuclear-processed material being dredged up to the surface. The chemical composition of the CSE will follow that of the central star, although with some time delay so that there may be some rare cases of O–rich CSEs around carbon stars. The mass loss decreases and changes its nature as the star leaves the AGB and starts its post-AGB evolution. Eventually the star becomes hot enough to ionize the inner part of the AGB-CSE and a planetary nebula (PN) is formed. The ultimate fate of the star is a long life as a slowly cooling white dwarf. The CSE will gradually disperse and its metal-enriched matter will mix with the interstellar medium, and thereby it contributes to the chemical evolution of a galaxy. The intense mass loss makes it possible for stars as massive as $8\,M_\odot$, i.e., the bulk of all stars in a galaxy, to follow this evolutionary sequence. Similar CSEs are also found around supergiants.

The AGB-CSEs form relatively well defined astrophysical laboratories. They appear to have an over-all spherical geometry. The gas expansion velocity is roughly constant throughout the envelope and it is typically of the order 10 to $15\,km\,s^{-1}$. The gas density falls as r^{-2}, where r is the radius, if the mass loss rate and expansion velocity are constant with time and radius, respectively, i.e., the number density spans quite a range within a CSE. The kinetic temperature is determined mainly by adiabatic cooling as a consequence of the expansion and heating due to gas-grain collisions as the dust grains stream through the gas, but there is also an additional cooling contributed by line radiation and there are other possible heating sources (see e.g., Truong-Bach et al. 1990). The inner region of a CSE may have temperatures as high as 1000 K, while the external teneous parts may be as cool as 10 K or less. The molecules are exposed to radiation from two sources. First, radiation from the central star which has characteristics, depending on the evolutionary stage, that fall somewhere in the range from a luminous (10^4 $L_\odot$), cool (2500 K) red giant to a much less luminous ($< 10^2$ $L_\odot$), but hot ($>$ 30000 K) star on its way to becoming a white dwarf. Second, the interstellar radiation field which is relatively isotropic, but there may be local anisotropies.

2 Circumstellar molecules detected at radio wavelengths

The first circumstellar molecule to be detected at radio wavelengths was OH in 1968, i.e., 25 years ago (Wilson & Barrett 1968). Initially the progress was slow, but in 1971 Solomon et al. (1971) detected CO from an infrared bright object, IRC+10216. This is probably the most nearby carbon star, and it happens to have a very high mass loss rate, of the order a few 10^{-5} $M_\odot$ yr^{-1} (Troung-Bach et al. 1992). It was quickly realized that a CSE around an evolved star, in particular a C–rich one, can be extremely rich in different molecular species. The reason being that the circumstellar dust effectively shields the molecules that are formed both in the stellar atmosphere and in the CSE itself, and in the C–rich case the carbon excess leads to an efficient organic chemistry. At this epoch 45 circumstellar molecules have been detected at radio wavelengths, of these 38 in the envelope of IRC+10216, Table 1. Indeed, some of them (16) are only detected in IRC+10216, and most of them (31) in fewer than five sources. The fact that so few sources completely dominate the picture means that most of what will be dicussed below is based on meagre statistics.

Among the most interesting of these circumstellar molecules are the long carbon-chain ones, the cyanopolyynes HC_nN ($n = 3, 5, 7, 9, 11$), the related hydrocarbons C_nH ($n = 2, 3, 4, 5, 6$) and H_2C_n ($n = 3, 4$), as well as HC_2N. Also nitrogen-carbon chains C_nN ($n = 1, 3$), silicon-carbon chains C_nSi ($n = 1, 4$), and sulphur-carbon chains C_nS ($n = 1, 2, 3$) are present. Simple ring molecules, like the triangular molecules SiC_2 and *cyclic*–C_3H_2, have been detected, but branched molecules are notably absent. On the more spectacular side we have the detections of the refractory species NaCl, AlCl, AlF, and KCl (Cernicharo & Guélin 1987; Lucas & Guélin 1990), and most recently the identification of three lines towards IRC+10216 as due to MgNC (Kawaguchi et al. 1993). The O–rich CSEs are apparently less rich in molecules. Here H_2O and OH dominate, but also sulphur-bearing species, in particular SO, SO_2, and H_2S, are frequently found (Sahai & Wannier 1992; Omont et al. 1993).

Table 1. Circumstellar molecules detected at radio wavelengths

Molecule	No. of objects	Molecule	No. of objects
AlCl	1	OCS	1
AlF	1	SiC_2	4
CO	≈420	SO_2	17
CN	17	ℓ-C_3H	2
CP	1	C_3N	5
CS	21	C_3S	1
KCl	1	HC_2N	1
NaCl	1	H_2CO	2
OH	>1500	NH_3	4
SiC	2	C_4H	5
SiN	1	C_4Si	1
SiO	≈50 (th), ≈200 (maser)	*c*-C_3H_2	7
SiS	6	HC_3N	7
SO	9	H_2C_3	1
C_2H	4	C_5H	1
C_2S	4	CH_3CN	3
HCN	≈110	H_2C_4	1
HCO^+	7	C_6H	1
H_2O	≈400	HC_5N	4
H_2S	20	HC_7N	2
HNC	10	HC_9N	1
MgNC	1	$HC_{11}N$	1
NH_2^+	1		

3 The nature of the circumstellar molecular radio line emission

In order to derive reliable molecular abundances from the line emission one requires "well behaved" emission, i.e., one tries to avoid emission from transitions that are anomalously excited. The extreme example of the latter is maser emission, i.e., amplified line radiation due to a transition with a population inversion, which unfortunately is not an uncommon phenomenon in the circumstellar medium (Elitzur 1992). We will not give an exhaustive discussion of this problem, but rather in brief, state our current knowledge.

There is strong maser emission from the $^2\Pi_{3/2}$, $J = 3/2$ Λ-doubling transitions of OH (Cohen 1989), from several rotational transitions and one vibrationally excited rotational line of H_2O (Menten & Melnick 1989, 1991; Cernicharo et al. 1990; Menten et al. 1990; Lewis & Engels 1991), from several vibrationally excited rotational lines of SiO and also the rare isotope variants ^{29}SiO and ^{30}SiO (Cernicharo & Bujarrabal 1992; Cernicharo et al. 1993), and at least two vibrationally excited rotational transitions of HCN (Guilloteau et al. 1987; Lucas & Cernicharo 1989). This means that in the case of OH and H_2O, where no non-masing lines have been detected, there exists no observational estimate of their circumstellar abundances. Weak maser emission from vibrationally excited rotational lines are found towards IRC+10216 for CS (Turner 1987a), SiS (Turner 1987b), and HCN (Lucas & Cernicharo 1989). Strong maser emission is usually easy to identify because of the complex line profile and/or the strong time variability.

Maser emission from ground-state transition lines is more of a problem since these are normally the ones used for estimating abundances. Fortunately, in most cases this maser emission contributes to only a minor fraction of the total line intensity. Weak maser emission has been found in the HCN(J=1–0) line towards tenuous, C–rich CSEs (Izumiura, 1990; Olofsson et al. 1993b), and the same applies to the $H^{13}CN$(J=1–0) line although here the statistics is poor (Izumiura et al. 1987; Nyman et al. 1993). The frequently observed SiO(J=2–1) line only occasionally shows evidence for weak maser features (Nyman & Olofsson 1985), while the less frequently observed SiO(J=1–0) line appears to be more affected by maser emission (Jewell et al. 1991). Surprisingly, strong maser emission has been found for ^{29}SiO and ^{30}SiO in the J=1–0, 2–1, and 5–4 lines (Alcolea & Bujarrabal 1992). The CS(J=2–1) line shows evidence for a maser feature in one case (Nyman et al. 1993). The SiS(J=1–0) line of IRC+10216 shows relatively strong maser components (Nguyen-Q-Rieu et al. 1984), and the SiS(J=5–4) line has a remarkable behaviour towards the same object that may indicate contributions from maser emission (Carlström et al. 1990).

That is, maser emission is a surprisingly common phenomenon, and even though its contribution to the total line intensity may be small, it does show that we can in general expect an excitation that is far from LTE.

4 Circumstellar molecular abundances: theory

4.1 Abundance formula

In principle, a fairly detailed modelling is required, which in some cases includes a complex non-local radiative transfer analysis (see e.g., Truong-Bach & Nguyen-Q-Rieu 1989), to obtain the abundance of a circumstellar molecule. However, for most molecules in Table 1 this is a too cumbersome approach, e.g., radiative transition rates and collisional cross sections are missing, and a simpler method is usually adopted. The model is based on the assumption that a spherical CSE of gas, expanding at a constant velocity, v_e, is formed by a constant stellar mass loss rate $\dot{M}$. The molecules are thermally excited at a temperature T_{ex} inside a shell of inner radius r_i and outer radius r_e. Molecules outside this shell do not contribute to the emission. The response of the telescope is assumed to be described by a Gaussian defined by the FWHM of the beam, B. Reasonably reliable estimates of molecular abundances are only obtained using optically thin lines, and in this limit (and in the Rayleigh-Jeans regime) the abundance of the molecule X with respect to H_2, f_X, is obtained using (see e.g., Olofsson et al. 1993b)

$$f_X = 1.7 \times 10^{-28}\, I\, \frac{v_e B D}{\dot{M}_{H_2}} \frac{Q(T_{ex})\nu_{ul}^2}{g_u A_{ul}} \frac{e^{E_l/kT_{ex}}}{\int_{x_i}^{x_e} e^{-4\ln 2\, x^2}\, dx} \tag{1}$$

where I is the line intensity integrated over velocity in units of $\mathrm{K\,km\,s^{-1}}$, v_e is given in $\mathrm{km\,s^{-1}}$, B in arc seconds, D is the distance to the source in pc, $\dot{M}_{H_2}$ is the mass loss rate of molecular hydrogen [the assumption that the major constituent of hydrogen in CSEs is molecular hydrogen is usually justified, Glassgold & Huggins (1983)] given in $\mathrm{M_\odot\,yr^{-1}}$, Q is the partition function, ν_{ul} is the frequency in GHz, g_u is the degeneracy of the upper level, A_{ul} is the Einstein A-coefficient for the transition, E_l is the energy of the lower level, and $x = r/BD$ and $x_{i,e} = r_{i,e}/BD$. Usually the emitting region is smaller than the beam and we may simplify further to get

$$f_X = 2.5 \times 10^{-15}\, I\, \frac{v_e B^2 D^2}{\dot{M}_{H_2}} \frac{Q(T_{ex})\nu_{ul}^2}{g_u A_{ul}} \frac{e^{E_l/kT_{ex}}}{r_e - r_i} \tag{2}$$

where r_i and r_e are given in cm. Essentially all reported circumstellar molecular abundances are based on Eq.(2) or variants thereof, see Sect. 4.3.

The molecular abundance estimate, obtained using Eq.(1) or Eq.(2), is critically dependent on such uncertain quantities as T_{ex}, D, $\dot{M}_{H_2}$, r_i, and r_e, and we proceed now to discuss these.

4.2 The source size

The size of the emitting region, i.e., r_i and r_e, is crucial in the calculation of the molecular abundance. Unfortunately, it has been measured only for a few sources and for a very limited number of molecular species. Additionally, the size is dependent on the particular transition in question. In principle, the only data with reasonable statistics are those on SiO (Lucas et al. 1992; Sahai & Bieging 1993), and OH (Bowers et al. 1980, 1983). For a number of species, e.g., SiS, HCN, HNC, HC_3N, C_2H, C_4H, C_3N, and SiC_2, there exist high-quality maps for only a single object, IRC+10216 (Bieging et al. 1984; Bieging & Nguyen-Q-Rieu 1988, 1989; Gensheimer et al. 1992; Takano et al. 1992; Bieging & Tafalla 1993; Dayal & Bieging 1993). Thus, in most cases the size of the emitting region has to be estimated in a different way.

The initial circumstellar abundance of a parent species is determined by the upper stellar atmosphere chemistry, $f_X(R_*)$. The abundance remains at this value until some point in the envelope where the interstellar UV photon density becomes high enough to dissociate the molecules. The location of this point is determined by the effectiveness of the dust shielding and whether the molecule is self-shielding, i.e., whether it is photodissociated in lines, or through continuous absorption. The abundance as a function of radius is determined by (Huggins & Glassgold 1982)

$$\frac{df_X}{dr} = -\frac{G_{0,X}}{v_e} \exp\left(-\frac{d(N_X, \tau_d)}{r}\right) \tag{3}$$

where $G_{0,X}$ is the unshielded photodissociation rate for a given interstellar UV field, and the shielding distance $d(N_X, \tau_d)$ is dependent on the outward dust optical depth τ_d, and , if self-shielding dominates, the outward column density N_X of the molecule in question (for some self-shielding molecules there is an additional shielding due to line absorption from other molecules, e.g., H_2 helps shielding CO). The shielding distance for pure dust absorption is given by (see e.g., Olofsson et al. 1993b)

$$d_{\rm X} = 1.4 \frac{3(Q_{\rm abs}/a)_{\rm X}}{4\rho_{\rm d}} \frac{\dot{M}_{\rm d}}{4\pi v_{\rm d}} \quad (4)$$

where $Q_{\rm abs}$ is the dust absorption efficiency, a is the size of a dust grain and $\rho_{\rm d}$ its density, $\dot{M}_{\rm d}$ is the dust mass loss rate, and $v_{\rm d}$ the dust expansion velocity (the factor 1.4 approximately takes into account the three-dimensionality of the problem). By far, the largest molecular envelopes are found for the self-shielding species CO (Mamon et al. 1988) and H_2 (Glassgold & Huggins 1983). The result for CO can be approximated by

$$R_{\rm e}({\rm CO}) = 10^{17} \left(\frac{\dot{M}}{10^{-6}\,{\rm M}_\odot\,{\rm yr}^{-1}}\right)^{0.6} \left(\frac{15\,{\rm km\,s^{-1}}}{v_{\rm e}}\right)^{0.4} \left(\frac{f_{\rm CO}}{10^{-3}}\right)^{0.5} {\rm cm.} \quad (5)$$

For comparison, a mass loss rate of $10^{-5}\,{\rm M}_\odot\,{\rm yr}^{-1}$ leads to $R_{\rm e}({\rm CO}) \approx 4\times10^{17}$ cm while $R_{\rm e}({\rm HCN}) \approx 10^{16}$ cm (Olofsson et al. 1993b). For parent species we have $r_{\rm i} \ll r_{\rm e}$ and $r_{\rm e} \leq R_{\rm e}$ ($r_{\rm e}$ is the outer radius of the brightness distribution in the simple model discussed in Sect. 4.1). The exact value of $r_{\rm e}$ depends on the molecule and the mass loss rate, i.e., whether the emitting region is photodissociation limited or brightness limited. In the former case one usually chooses $r_{\rm e}$ as the e-folding radius, i.e., such that $f_{\rm X}(r_{\rm e}) = f_{\rm X}(R_*)/e$. Typical examples of parent species are CO, SiO, HCN (in C–rich objects), and H_2S.

The abundance of a photodissociation product is initially very low, but it increases at the radius where the parent species $X_{\rm p}$ becomes photodissociated. A peak is reached since eventually the photodissociation product also becomes photodissociated, i.e., these molceules are distributed in a shell. The abundance as a function of radius is given by (Huggins & Glassgold 1982)

$$\frac{df_{\rm X}}{dr} = -\frac{G_{0,{\rm X}}}{v_{\rm e}} \exp\left(-\frac{d(N_{\rm X},\tau_{\rm d})}{r}\right) + \frac{G_{0,{\rm X_p}}}{v_{\rm e}} \exp\left(-\frac{d(N_{\rm X_p},\tau_{\rm d})}{r}\right) \quad (6)$$

In the event that the brightness distribution is photodissociation limited one usually chooses $r_{\rm i}$ and $r_{\rm e}$ as the inner and outer radii where the abundance has dropped to $f_{\rm X,peak}/e$. Typical examples of photodissociation products are CN, C_2H, and C_3N.

This model for circumstellar photodissociation has a number of weaknesses. The three dimensional geometry of the problem is often treated in a relatively crude way (Jura & Morris 1981). Only dust absorption is taken into account, and even so the absorbing properties, as well as the actual amount, of the dust particles are poorly known. The effectiveness of the self-shielding is dependent on the kinematics of the molecules in the external parts of the CSE. The interstellar UV radiation field is highly uncertain and may vary from place to place, and in certain locations it may even be anisotropic (van Dishoeck 1988, and references therein). In general, while the photodissociation coefficients are well determined for some species like H_2, CO, H_2O, and OH, they are poorly known for many molecules (see e.g., van Dishoeck 1988). We have also ignored the possibility that the photodissociation product of the parent species rapidly reacts chemically with another species. However, the major error source may be the actual density distribution of the molecules in the CSE. In a highly clumped medium, for which there is growing evidence (Olofsson et al. 1992; Bieging & Tafalla, 1993), the molecules may survive to a much bigger distance from the star than given by the above model, which is based on the assumption of a homogeneous density distribution (Bergman et al. 1993).

Finally, there are molecules whose existence depends on a circumstellar chemistry initiated by the photodissociation processes (see e.g., Millar 1988). Their radial distributions are more difficult to predict since they depend on a complicated mixture of chemical reactions. For instance, the ionization of acetylene leads to a complex ion-molecule chemistry in C–rich CSEs (Nejad & Millar 1987; Howe & Millar 1990; Glassgold et al. 1992). Neutral-neutral reactions may be responsible for the production of some carbon-chain molecules (Howe & Millar 1990). Likewise, the photodissociation of H_2O in O–rich CSEs produces OH which rapidly reacts to form other species like SO and SO_2 (Scalo & Slavsky 1980). Another interesting example concerns the production of C-bearing species in O–rich CSEs (Nejad & Millar 1988; Nercessian et al. 1989). In the latter cases the chemistry is of the neutral-neutral type and it becomes critically dependent on the existence of activation energy barriers and therefore becomes particularly uncertain. The actual choice of r_i and r_e therefore depends on relatively complex models. It is probably true that the majority of the molecules in Table 1 is due to a circumstellar chemistry. For some species we have the additional problem that grain surface reactions may be of importance. Typical examples of this kind are CH_4 (Clegg et al. 1982) and SiH_4 (Betz 1987).

In summary, the dependence of the estimated molecular abundance on r_i and r_e will lead to substantial uncertainties. In future, the situation is likely to improve considerably as more and more source sizes are actually measured observationally.

4.3 The excitation temperature

The discussion in Sect. 3 implies that a relatively complex non-LTE excitation analysis is required to obtain the excitation temperature and the correct partition function. However, at present it seems that such a detailed analysis for each molecular species is not warranted considering all other uncertainties in the circumstellar model, and in many cases it is not even possible. A similar treatment of all data based on the simplifying assumption of a uniform excitation temperature still seems appropriate. The excitation temperature T_ex can be estimated in the following way by observing two or more optically thin lines from the same species. Equation (1) may be rewritten in the form

$$\ln\left(\frac{(I/\eta)\nu_{ul}^2}{g_u A_{ul}}\right) + \ln\left(1.8\times 10^{10}\,\frac{Q(T_\mathrm{ex})}{\bar{N}_\mathrm{X}}\right) = -\frac{E_l}{kT_\mathrm{ex}} \qquad (7)$$

where $\eta = \int_{x_i}^{x_e} e^{-4\ln 2 x^2}\, dx$ and $\bar{N}_X$ is the beam-averaged column density defined by

$$\bar{N}_X = \frac{f_X \dot{M}_{H_2}}{8\pi m_H v_e B D} \tag{8}$$

and it is given in cm^{-2}, and I is given in $K\,km\,s^{-1}$, and ν_{ul} in GHz. For two arbitrary lines this results in

$$T_{ex} = \frac{E_{l2} - E_{l1}}{k} \ln\left(\frac{g_{u2} A_{ul2} I_1}{g_{u1} A_{ul1} I_2}\right) . \tag{9}$$

It is assumed that the emitting regions of the two lines are the same (this is a questionable assumption in many cases, since one expects a line from high-lying energy levels to emanate from a smaller area than a line from low-lying energy levels). Furthermore, they must be unresolved by the telescope beam. We have also used the fact that the beam size is inversely proportional to the frequency (i.e., both lines are observed with the same telescope). A statistically more reliable value for the excitation temperature is obtained by combining data from many transitions in a rotation temperature diagram where $\ln((I/\eta)\nu_{ul}^2/g_u A_{ul})$ is plotted versus E_l/k (see e.g., Cernicharo et al. 1987b). A fit of a straight line to the data in such a diagram, $y = \alpha + \beta E_l/k$, leads to an estimate of the excitation temperature

$$T_{ex} = -\frac{1}{\beta} \tag{10}$$

and also of the beam averaged column density

$$\bar{N}_X = 1.8 \times 10^{10}\, Q(T_{ex})\, e^{\alpha} \tag{11}$$

which is an alternative way of expressing the molecular abundance. In fact, this gives a statistically more reliable abundance estimate than that obtained from Eq.(1) or Eq.(2) which is based on a single line.

However, in most cases a reasonable value for T_{ex} is simply assumed, and in the worst case this is a pure guess. This is particularly unfortunate since the dependence of f_X on T_{ex} is quite strong, roughly $T_{ex} \exp(E_l/kT_{ex})$ for linear rotors and $T_{ex}^{3/2} \exp(E_l/kT_{ex})$ for symmetric and asymmetric top molecules.

4.4 The mass loss rate and the distance

The mass loss rate is often highly uncertain (van der Veen & Olofsson 1990; Olofsson 1993). It is not unrealistic to expect an uncertainty that approaches an order of magnitude in many cases. This quantity enters into the calculation of the abundance not only through $\dot{M}_{H_2}$ but also via the size of the emitting region since the photodissociation radius, the chemistry, and the excitation are all dependent on it. The combined effect is that the mass loss rate of the star is probably the major error source in the abundance estimate. In fact, it can be concluded from studies of circumstellar HCN and CS (Olofsson et al. 1993b), and SO and SO_2 (Sahai & Wannier 1992), that the mass loss rates of the stars in their samples are probably significantly higher than the accepted values.

The distances to the stars of interest here, are notoriously uncertain, but often the mass loss rate estimate has a dependence on distance which is close to D^2 and this significantly decreases the dependence of $f_{\rm X}$ on distance.

4.5 A note on the column density

Interstellar molecular abundances are frequently given in terms of column densities, i.e., the number of molecules within a column of given cross section. Occasionally, this is also done in the circumstellar case, but the relation between column density and abundance is not straightforward. If the mass loss rate, and the gas expansion velocity and the molecular abundance are assumed to be constant with time and radius, respectively, then the molecular density varies as

$$n_{\rm X}(r) = n_{\rm Xo} R_{\rm o}^2 \frac{1}{r^2} \tag{12}$$

where $n_{\rm Xo}$ is the number density of species X at the arbitrary radius $R_{\rm o}$, and $n_{\rm Xo} R_{\rm o}^2$ is related to the abundance $f_{\rm X}$ via

$$n_{\rm Xo} R_{\rm o}^2 = \frac{f_{\rm X} \dot{M}_{\rm H_2}}{8\pi m_{\rm H} v_{\rm e}} . \tag{13}$$

The column density $N_{\rm X}$ towards the star is given by

$$N_{\rm X} = n_{\rm Xo} R_{\rm o}^2 \left(\frac{1}{R_{\rm i}} - \frac{1}{R_{\rm e}} \right) \approx n_{\rm Xo} R_{\rm o}^2 \frac{1}{R_{\rm i}} \tag{14}$$

where $R_{\rm i}$ and $R_{\rm e}$ are the inner and outer radii of the molecular distribution (note that $R_{\rm i}$ and $R_{\rm e}$ are not necessarily the same as $r_{\rm i}$ and $r_{\rm e}$). The column density and the abundance are consequently related through

$$N_{\rm X} = \frac{f_{\rm X} \dot{M}_{\rm H_2}}{8\pi m_{\rm H} v_{\rm e}} \frac{1}{R_{\rm i}} \tag{15}$$

i.e., via the observationally ill-defined quantity $R_{\rm i}$. Infrared absorption line measurements naturally lead to $N_{\rm X}$, while the radio observations are dependent on the number of molecules within the beam, and hence are better defined by the abundance $f_{\rm X}$ or the beam averaged column density $\bar{N}_{\rm X}$, defined in Eq.(8). It is apparent from this discussion that a comparison between radio and infrared data is not straightforward, nor is the comparison between radio data taken with different beam sizes.

5 Circumstellar molecular abundances: observational estimates

We have in Table 2 summarised a number of molecular abundances in the CSEs of AGB-stars. Practically all data have been obtained using the method outlined in Sect. 4, but it should be borne in mind that the data set is relatively heterogeneous, and in many cases the entry applies to a single CSE, that of IRC+10216. The majority of the data comes from Cernicharo et al. (1987a,b), Lindqvist et al. (1988, 1992), Sahai & Wannier (1992), Nyman et al. (1993), Olofsson et al. (1993b), and Omont et al. (1993). The values given should be used with great care. If one is interested in the study of a number of molecules it is advisable to go back to the original data sets. It is notable that, in general, the circumstellar abundance of a molecule is much higher, often by two orders of magnitude, than its interstellar abundance (Cernicharo et al. 1987b). For essentially all species, except CO, the abundance is drastically different in C–rich and O–rich CSEs. In C–rich CSEs the linear carbon-chain molecules dominate, and the abundance decreases only weakly with increasing size of the molecule. However, only slight modifications like the substitution of CH_3 for H lower the abundances substantially, e.g., $[CH_3CN]/[HCN] \approx 4 \times 10^{-4}$ (the interstellar ratio is $\approx 10^{-1}$), and other more hydrogen-saturated species like CH_2CHCN, CH_3CH_2CN, and CH_3C_2H have not been detected. The abundances of the refractory molecules are low, possibly due to depletion on grains.

The circumstellar molecular abundances in the envelopes of post-AGB objects may differ substantially from those given in Table 2. This will be discussed further in Sect. 8.

6 Accuracy of estimated circumstellar molecular abundances

A key question in this connection is the accuracy of the reported molecular abundances. This problem is probably best illuminated by a practical example taken from a study by Olofsson et al. (1990, 1993b). Their sample of carbon stars contains a sub-sample of 30 bright carbon stars whose photospheric characteristics, in particular, the CNO abundances and the effective temperatures, have been reasonably well determined (Lambert et al., 1986). The photospheric HCN, CN, and CS abundances were estimated using the appropriate CNO abundances (a solar abundance of S was assumed) and effective temperatures in static LTE model atmospheres. The abundances at $\log \tau_{\rm Ross} = -4$ were adopted. The large spread in carbon excess [(C/O–1) ranges from <0.01 to 0.8] leads to a relatively large spread in the photospheric abundances (e.g., $6 \times 10^{-6} < f_{\rm HCN} < 5 \times 10^{-5}$ and $10^{-8} < f_{\rm CS} < 3 \times 10^{-5}$), which is an advantage in a study of this type.

These stars were observed in radio lines of CO, HCN, CN, and CS. The CO data provided the mass loss rates, i.e., the molecular hydrogen mass loss rate assuming that the photospheric and circumstellar CO abundance (relative to H_2) are equal (Olofsson et al. 1993a). The HCN, CN, and CS circumstellar

Table 2. Molecular abundances in AGB-CSEs

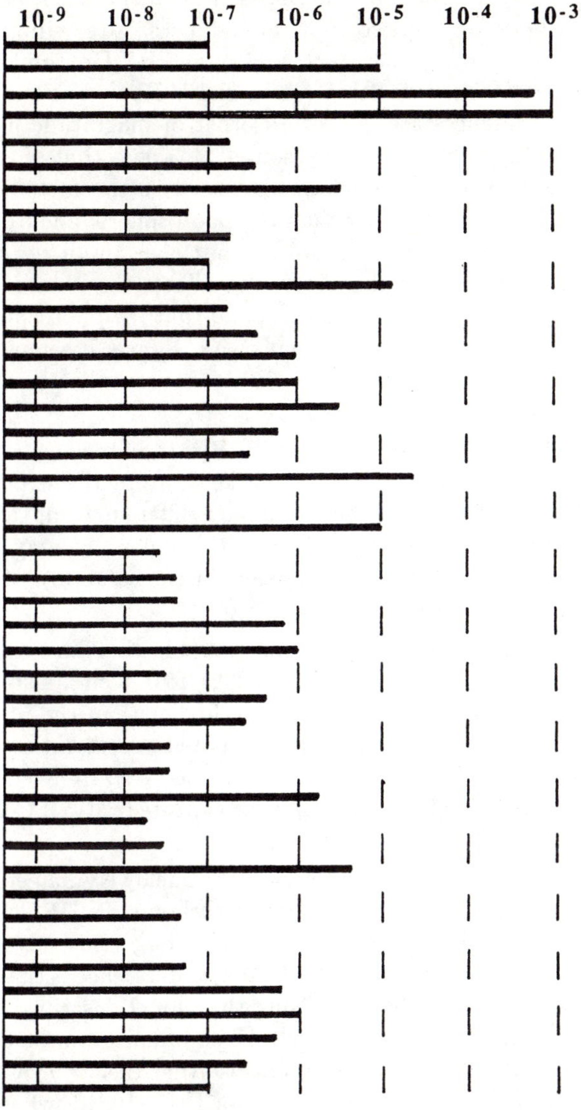

Molecule	O/C[a]	P/C[b]
CN	O	C?
	C	C
CO	O	P
	C	P
CP	C	C?
CS	O	C?
	C	P
NaCl	C	P
SiC	C	C?
SiN	C	C?
SiO	O	P
	C	P
SiS	O	C?
	C	P
SO	O	C?
C_2H	C	C
C_2S	C	C?
HCN	O	C
	C	P
HCO^+	C	C?
H_2S	O	P
	C	P
HNC	C	C
MgNC	C	C?
SiC_2	C	C
SO_2	O	C?
l-C_3H	C	C?
C_3N	C	C
C_3S	C	C?
HC_2N	C	C?
NH_3	C	C?
C_4H	C	C
C_4Si	C	C?
c-C_3H_2	C	C?
HC_3N	C	C
H_2C_3	C	C?
C_5H	C	C
CH_3CN	C	C?
H_2C_4	C	C?
C_6H	C	C?
HC_5N	C	C?
HC_7N	C	C?
HC_9N	C	C?
$HC_{11}N$	C	C?

[a] O-rich or C-rich CSE

[b] Photospheric or circumstellar origin. ? indicates that no observations exist that conclusively show that the molecule is formed in the CSE.

abundances were estimated using the model outlined in Sect. 4. It is reasonable to assume that the photospheric and circumstellar abundances of HCN should agree, since there is no known efficient mechanism for producing HCN in the envelope. The same should apply to CS. Olofsson et al. (1993b) also found a hint of a correlation between the two HCN abundances (after eliminating five somewhat peculiar stars: two ^{13}C-rich stars, Y CVn and RY Dra, and three stars whose CSEs may have detached, R Scl, U Cam, and V Hya), but the estimated circumstellar HCN abundances are almost an order of magnitude higher than the corresponding photospheric ones. Olofsson et al. argue that this is due to uncertainties in the circumstellar model, partly based on the fact that the circumstellar CS abundance is also higher than the photospheric one by about the same amount (here the statistics is much worse), but one cannot completely exclude that there is a circumstellar chemistry that affects the abundance in the relatively tenuous and hot CSEs in this sample, and that the photospheric abundances are underestimated (in particular that of N). Hence, it seems that even in this relatively well defined problem the estimated circumstellar abundances are off by at least a factor of five.

On the other hand, the results for R Scl and V Hya (and to some extent those for U Cam, Y CVn, and RY Dra) differ so markedly from the expected, that there is every reason to believe that these stars are different from the majority of the stars in the sample, in some way. However, this difference may not be one of elemental composition, but rather a difference in the structure of the CSEs and hence the excitation of the molecules.

It appears that abundance ratios are more reliable than individual abundances. In the case of CN Olofsson et al. (1993b) found circumstellar abundances that are about three orders of magnitude higher than the photospheric ones. This is likely to be much larger than the estimated uncertainties, and hence CN cannot be a parent molecule. Indeed, the estimated CN/HCN circumstellar abundance ratio is about 0.6, which is roughly the expected ratio if CN is formed from photodissociation of HCN.

Another study of interest in this connection is the one by Nyman et al. (1993). They observed the same molecular species in two carbon stars, IRC+10216 and IRAS15194−5115, with similar characteristics, including relatively high mass loss rates of the order a few 10^{-5} $M_{\odot}$ yr^{-1}. The major difference is that the latter source is about five times more distant. Using the model described in Sect. 4, they conclude that for the 13 species observed, the abundances in the two sources agree within a factor of four. A reliable result is also that the $^{12}C/^{13}C$-ratio is at least five times lower in IRAS15194−5115 than in IRC+10216, where the best data suggest a value in the range 40-50 (see Sect. 7.2). This puts some faith to the model, at least as long as the comparison is made within a homogeneous data set. Furthermore, it shows that IRC+10216 is not a unique object; it just happens to be very nearby.

Finally, an illuminating comparison is offered by two recent studies of SO and SO_2 (Sahai & Wannier, 1992; Omont et al., 1993). For some stars the estimated abundances disagree by a factor of ten between the two studies, and for some stars Sahai & Wannier obtain values that differ by a factor of ten depending on the assumptions made.

7 Circumstellar molecular line emission: a measure of elemental composition?

The elemental composition of red giants with very thin CSEs is preferrably estimated using visual and near-IR high-resolution spectra combined with detailed stellar atmosphere models (see e.g., Lambert et al. 1986; Gustafsson 1989; Smith & Lambert 1990; Lazaro et al. 1991). However, this procedure runs into difficulties as the opacity, and consequently the dust emission, of the CSE increases, and it becomes useless when the mass loss rate is so high that the star is completely obscured. Thus, the elemental composition of the most extreme objects can only be inferred from circumstellar emission, and preferably line emission. However, this is not an easy task since, according to the discussion above, it is difficult to reliably estimate circumstellar molecular abundances. Furthermore, elemental abundances are obtained from these only if the circumstellar chemistry is known. Finally, the circumstellar gas-phase elemental composition may differ appreciably from the stellar atmosphere elemental composition due to selective depletion on grains.

7.1 The C/O–ratio

The simplest, and to some extent the most basic, problem would be to distinguish between O–rich and C–rich objects. It is preferable if this can be done using simple line intensity ratios rather than abundance estimates of an individual species. It seems that SiO and HCN are the best candidates for this problem if one also takes into account the strength of the lines. In O–rich CSEs we expect SiO to be abundant while little HCN should be present since most of the carbon is tied up in CO. In C–rich CSEs we expect the opposite behaviour: HCN is abundant while little SiO is present since most of the oxygen is tied up in CO. Both species are presumably radiatively excited except at the highest mass loss rates, and so they should respond in the same manner to different excitation conditions (one potential problem here is that HCN in O–rich objects is of circumstellar origin). Thus, we expect the $I_{\rm HCN}/I_{\rm SiO}$ line intensity ratio to be a strong discriminator between O–rich and C–rich CSEs. Bujarrabal et al. (1993a) and Olofsson et al. (1993c) have recently studied samples of O–rich and C–rich red giants. Olofsson et al. found $I_{{\rm HCN,J=1-0}}/I_{{\rm SiO,v=0,J=2-1}}$-ratios in the range 0.05–0.3 and 3–10 for the 19 O–rich objects and the 6 C–rich objects, respectively, where both lines were detected. The results of Bujarrabal et al. agree with this. Thus, the discrimination between the two types of objects is clear and unambiguous.

Next in line stands the question whether a more detailed determination of the C/O–ratio can be made. This is a considerably more difficult problem. The effects of trends in the C/O–ratio are likely to be much smaller than those due to whether C is larger than O or vice versa. This is because the stability of the CO molecule, and hence its ability to store C and O, has such a profound effect on the chemistry.

In view of this we may ask whether the scatter in the I_{HCN}/I_{SiO}-ratio within the O–rich and C–rich samples discussed above carries any additional information related to the C and O abundances. This question cannot as yet be answered. Only marginal detections of SiO in two carbon stars that belong to the sample of Lambert et al. (1986) have been made (Bujarrabal et al. 1993b) and therefore no comparison between the I_{HCN}/I_{SiO} and C/O ratios can be made. For the O–rich stars Olofsson et al. (1993c) found that sources with high F_{25}/F_{12} and $F_{60}/F_{25}\,\mu$m flux ratios have only high, in a relative sense, I_{HCN}/I_{SiO}-ratios. This is interesting since the stars with $2.5\log F_{60}/F_{25} > -1.8$ fall in a region of the far-IR two-colour diagram of van der Veen & Habing (1988) that is mainly occupied by carbon stars. On the other hand, the trend of increasing I_{HCN}/I_{SiO} with increasing $F_{25}/F_{12}\,\mu$m flux ratio may indicate that it is the mass loss rate that plays the major role. In this connection it is worth noting that Bujarrabal et al. (1993a) found that the S-type stars have line intensity ratios that fall equidistant between those of O–rich and C–rich sources.

Olofsson et al. (1993b) in their study of HCN in bright carbon stars, also had in mind the possibility of actually determining the C/O–ratio from circumstellar HCN and CO data. The main problem here, apart from the previously discussed difficulty of estimating the HCN abundance, is that the CO lines are usually quite optically thick and hence should be relatively insensitive to the CO abundance. However, models suggest that the mass loss rate scales linearly with the CO line intensity and it is almost inversely proportional to the CO abundance. If such a relation is used in (2) when estimating the HCN abundance one obtains directly an estimate of the circumstellar HCN/CO abundance ratio. Of course, this estimate is likely to have an associated appreciable uncertainty. Indeed, for the moment it is much larger than acceptable if the aim is to determine the C/O–ratio. It is also important to realize that even if the uncertainty in f_{HCN}/f_{CO} can be mastered, the problem remains that the relation between f_{HCN}/f_{CO} and C/O is unique only for low values of C/O (for solar abundances of N and O this critical value is 0.12). For high values of C/O the f_{HCN}/f_{CO}-ratio becomes essentially a measure of the N-abundance, and it has in fact been used for this purpose (Jura, 1991).

Unfortunately, one has to conclude that there is still a long way to go before circumstellar data can be used to obtain detailed information about the elemental composition. However, the reward seems high enough to pursue this research with continued large efforts.

7.2 Isotope ratios

Another side of the elemental composition concerns the isotopic ratios of various elements. Here the observational data, i.e., line intensities of the same chemical species but with different isotopic substitutes (isotopomers), are usually more easily converted into astrophysically interesting quantities, the reason being that one expects the corresponding lines from the isotopomers to respond in the same way to the physical conditions. In most cases one simply lets line intensity ratios obtained for the isotopomers reflect also the isotope ratios, i.e., $^{n}X/^{m}X = I(^{n}XA)/I(^{m}XA)$. Minor corrections for the small differences in beam size and transition probability between the two isotopomers are usually introduced. This method is probably fairly reliable provided (i) that the observed lines from the two isotopomers correspond to the same transition, (ii) that great care is taken in selecting optically thin lines, and (iii) that there is no difference in the spatial distribution and in the excitation of the two isotopomers. To this we shall add that since the lines are weak there is an increased possibility of blending with other lines, or even misidentifications.

In general, the rare isotope lines from CSEs are very weak, and hence the statistics is poor, and, in fact, even worse than in the case of molecular abundances. Only in IRC+10216 does there exist an isotope-ratio setup that is worth discussing. The results for this source obtained by Wannier & Sahai (1987), Cernicharo & Guélin (1987), Kahane et al. (1988, 1992), and Wannier et al. (1992) are summarized in Table 3. The uncertainties are of the order ±25% or less. Major deviations from the terrestrial values are found for N (^{15}N is highly underabundant) and O (^{18}O is underabundant and ^{17}O is overabundant, and so $^{17}O/^{18}O$ is ≈8 times higher than the terrestrial value). Data for ^{15}N have been obtained in five more C–rich CSEs (Wannier et al. 1992), and for ^{17}O and ^{18}O in four C–rich CSEs (Kahane et al. 1992), but the majority of these sources are post-AGB objects. The results are in line with, but not identical to, those for IRC+10216. These data show the potential, but also the problem that the weakness of the lines severely limits the number of sources that can be studied.

Table 3. Isotope ratios in IRC+10216

Isotope ratio	Ratio	IRC/Solar	Species used
$^{12}C/^{13}C$	47	0.5	^{13}CS, $C^{34}S$
$^{14}N/^{15}N$	5300	20	$H^{13}CN$, $HC^{15}N$
$^{16}O/^{17}O$	840	0.3	^{13}CO, $C^{17}O$
$^{16}O/^{18}O$	1300	3	^{13}CO, $C^{18}O$
$^{28}Si/^{29}Si$	19	1	^{28}SiS, ^{29}SiS, $^{28}SiC_2$, $^{29}SiC_2$
$^{28}Si/^{30}Si$	29	1	^{28}SiS, ^{30}SiS, $^{28}SiC_2$, $^{30}SiC_2$
$^{32}S/^{33}S$	100	1	$Si^{32}S$, $Si^{33}S$
$^{32}S/^{34}S$	20	1	$Si^{32}S$, $Si^{34}S$
$^{35}Cl/^{37}Cl$	2	1	$Na^{35}Cl$, $Na^{37}Cl$, $Al^{35}Cl$, $Al^{37}Cl$

In principle, one could expect that the statistics for the $^{12}C/^{13}C$–ratio would be relatively good since there is a fair number of detections of ^{12}CO and ^{13}CO in the same objects, and there is some confidence that the modelling of the optically thick ^{12}CO emission is mastered. However, there are two additional problems. Isotope selective photodissociation and fractionation is particularly problematic in the case of CO, and the two isotopomers are expected to have different spatial distributions. Furthermore, it is expected that ^{12}CO and ^{13}CO are differently excited. Taken together this results in large uncertainties in the estimates of the $^{12}C/^{13}C$-ratio from CO data (Knapp & Chang 1985; Sopka et al. 1989).

8 Circumstellar molecular line emission: a measure of stellar evolution?

In this section we will discuss to what extent there are differences in the appearance of different molecular species as a function of evolutionary stage. Ultimately, it would perhaps be possible to use the appearance and disappearance of various molecular lines to devise a molecular line chronology of stellar evolution on and beyond the AGB. In this we include also the possibility that lines from the same species but with different excitation requirements may contain important information. Since these types of studies are still in their infancies, we will here restrict ourselves to discuss whether there are differences between the "thermal" circumstellar molecular line emission from AGB-stars, post-AGB objects, and planetary nebulae (PNe). Lewis (1989, 1990) has made a similar, but much more detailed, study for the strong maser lines.

It should not come as a surprise that the molecular abundances in the CSE around a post-AGB object may differ substantially from those in an AGB-CSE. The reason is twofold. First, the mass loss decreases drastically as the star leaves the AGB, and consequently the CSE detaches from the star. This leads to a rapidly decreasing opacity of the envelope, roughly given by

$$A_V \approx 20 \left(\frac{\dot{M}_{AGB}}{10^{-6}\,M_\odot\,yr^{-1}} \right) \left(\frac{10\,yr}{t} \right)\ \mathrm{mag} \qquad (16)$$

where A_V is the visual extinction (the relation $N_{H_2} = 10^{21} A_V$ cm^{-2} mag^{-1} which seems appropriate for interstellar clouds have been used, Bohlin et al. 1978), and t is the time elapsed since the mass loss terminated. Thus, photodissociation rapidly sets in. Second, eventually the photodissociation increases substantially due to the increasing UV flux from the now much hotter central star. The effect is that the molecular abundances change rapidly as the star leaves the AGB. In general, the abundances decrease, but also new chemical reactions are initiated and the behaviour becomes quite complex. To this we shall add changes in the excitation conditions, which will also have an effect on the molecular line pattern. There is also the possibility that some change in elemental composition may be associated with the termination of the AGB phase.

There is observational evidence that the effects discussed above do lead to changes in the molecular line emission appearance. A nice example is provided by the observations by Bujarrabal et al. (1988) of three C–rich objects in different evolutionary stages, IRC+10216 (AGB-object), CRL2688 (post AGB-object, probably less evolved than CRL618), and CRL618 (post AGB-object, a good pre-PN candidate), (see their Fig. 1). IRC+10216 has been detected in at least

38 molecular species at radio wavelengths. CRL2688 and CRL618 are still detectable in a fair number of molecular species, 18 and 14, respectively, and this despite the fact that they are at least a factor of four more distant. Bujarrabal et al. conclude that the molecular abundances in the CSEs of the two latter sources are lower than those in the CSE of IRC+10216, but not drastically so. This is somewhat surprising, but it may indicate that the circumstellar medium is clumped and that therefore the photodissociation is less effective. However, in the young C–rich PN NGC7027, in which the CO lines are still very strong, the number of detections has decreased to five, and the most prominent line, except for those of CO, is one of HCO^+ (see e.g., Deguchi et al. 1990). Another (probably C–rich) young PN is IRAS21282+5050, it shows detectable CO and HCO^+ but no HCN. This is drastically different to the results for IRC+10216 where the HCN line is very strong, while the HCO^+ line is just marginally detected (Lucas & Guélin 1990). Observations of the more evolved (probably C–rich) PNe NGC2346 (Bachiller et al. 1989) and NGC6072 and IC4406 (Cox et al. 1992) further amplify the situation. HCO^+ is remarkably strong and it appears that HCN, and in particular CN, regain strength as the object becomes more evolved. We have tried to summarize these findings in Table 4. Listed are the line intensity ratios $I_{X,J=1-0}/I_{CO,J=1-0}$, except for CN and C_2H where the N=1–0 transition has been used, of lines observed with the same telescope. This normalization with respect to the CO line is not meant to reflect changes in abundance with respect to CO (the CO line is normally too optically thick for this), but merely to show the large changes in line intensity for other species than CO, which is known to be easily excited and difficult to destroy. Information on other line intensity ratios is easily extracted from the presented data. Some of the more interesting trends are presented in Fig. 1.

Table 4. Line intensity ratios, $I_{X,1-0}/I_{CO,1-0}$, for objects in different evolutionary stages

Source	HCN	HNC	HCO^+	CN	C_2H	Evolutionary stage
IRC+10216	0.71	0.087	0.0015	0.89	0.16	AGB
CRL2688	0.66	0.054	<0.0075	0.39	0.094	early post-AGB
CRL618	0.18	0.17	0.13		0.019	post-AGB
NGC7027	0.13		0.30	0.04	<0.01	young PN
21282+5050	<0.005		0.014			young PN
NGC2346	0.08	0.08	0.064			PN
NGC6072	0.22	0.049	0.18	0.71	<0.013	PN
IC4406	0.20	0.042	0.16	0.38	<0.015	PN

The trends are much larger than the uncertainties, and it seems inescapable that most of them are due to photodissociation beyond the AGB and the initiation of a post-AGB circumstellar chemistry (Howe et al. 1992). These results indicate that the molecular line signature of the CSE can be of help in establishing the evolutionary stage of an object. One may also be impressed by the ability

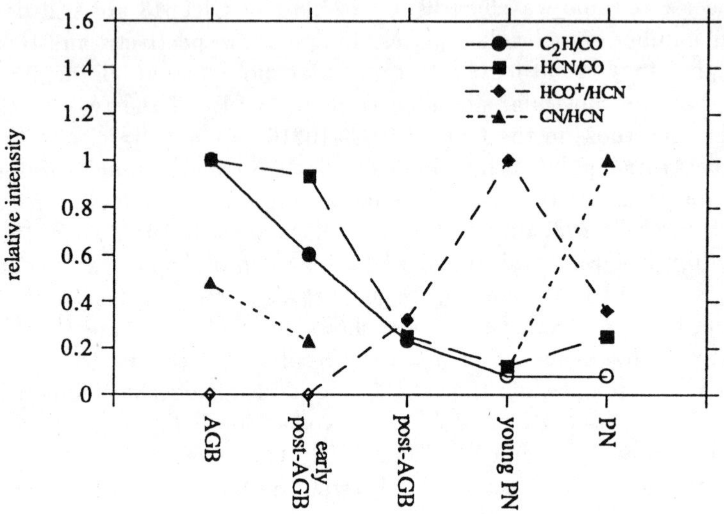

Fig. 1. Line intensity ratios as a function of the evolutionary stage. Open symbols are upper limits. See text for details

of the molecules to survive in even the harshest environments. The results for NGC6072 and IC4406 are astonishingly similar in view of this, and it is probable that there is substantial clumping in the circumstellar medium.

In many post-AGB objects there is evidence for the presence of a higher-velocity wind ($v_e \approx 50$ to 200 km s^{-1}) that may have a bipolar structure rather than a spherical outflow (Olofsson 1993). It would be of great interest to compare the abundances in this wind and the AGB-CSEs. An example of this is provided by the observations of CRL2688 by Jaminet et al. (1992). They claim that the $^{12}C/^{13}C$-ratio is four times lower in the fast wind than in the AGB-CSE. If true, this would indicate a change in this isotopic ratio on a time scale of a few hundred years or less. Also the HCN/CO abundance ratio appears lower in the fast wind.

9 Conclusions

The CSEs of stars on the AGB and beyond show a remarkably rich variety of different molecular species. However, only about a third of them have been detected in more than five sources. The relatively simple geometry and kinematics of the CSEs make it possible to use rather uncomplicated methods for estimating molecular abundances that are accurate to within a factor of about five or slightly worse in some cases. The methods have not yet reached the level of sophistication that allows the circumstellar data to be used to obtain detailed information about the the elemental composition. However, the results on isotope ratios are already of high quality. It is clear that different molecular species behave differently as the object evolves, and there is good hope that one may eventually be able to use this as a kind of molecular line chronology for stellar evolution on and beyond the AGB.

There remains much to be done in the improvement of the circumstellar models used. In particular, a detailed non-LTE, non-local radiative transfer analysis is required, but for the majority of the molecules the spectroscopic data and the collisional cross sections are lacking. Another important quantity for which the uncertainty must be considerably lowered is the mass loss rate. Also, the actual distribution of matter in a CSE, e.g., to what extent is the material clumped, may have a profound effect on the resulting abundance estimates.

Acknowledgements. I am grateful to the Knut and Alice Wallenberg foundation and the organizers of IAU Colloquium 146 for travel support. I also value the comments on the manuscript by John Black. Finally, I am grateful to Per Bergman for his support in the transformation of my manuscript into a Latex-file.

References

Alcolea J., Bujarrabal V., 1992, *Astron. Astrophys.*, **253**, 475
Bachiller R., Bujarrabal V., Martín-Pintado J., Gómez-González J., 1989, *Astron. Astrophys.*, **218**, 252
Bergman P., Carlström U., Olofsson H., 1993, *Astron. Astrophys.*, **268**, 685
Betz A., 1987, in *Astrochemistry , IAU Symp. No. 120*, eds M.S. Vardya and S.P. Tarafalar, Reidel, Dordrecht, p.327
Bieging J.H., Chapman B., Welch W.J., 1984, *Astrophys. J.*, **285**, 656
Bieging J.H., Nguyen-Q-Rieu, 1988, *Astrophys. J.*, **329**, L107
Bieging J.H., Nguyen-Q-Rieu, 1989, *Astrophys. J.*, **343**, L25
Bieging J.H., Tafalla M., 1993, *Astron. J.*, **105**, 576
Bohlin R.C., Savage B.D., Drake J.F., 1978, *Astrophys. J.*, **224**, 132
Bowers P.F., Johnston K.J., Spencer J.H., 1983, *Astrophys. J.*, **274**, 733
Bowers P.F., Reid M.J., Johnston K.J., Spencer J.H., Moran J.M., 1980, *Astrophys. J.*, **242**, 1088
Bujarrabal V., Fuente A., Omont A., 1993a, *Astron. Astrophys.*, subm.
Bujarrabal V., Fuente A., Omont A., 1993b, *Astron. Astrophys.*, subm.
Bujarrabal V., Gómez-González J., Bachiller R., Martín-Pintado J., 1988, *Astron. Astrophys.*, **204**, 242
Carlström U., Olofsson H., Johansson L.E.B., Nguyen-Q-Rieu, Sahai R., 1990, in *From Miras to Planetary Nebulae - Which Path for Stellar Evolution*, eds M.O. Mennessier and A. Omont, Editions-Frontiérs, p.170
Cernicharo J., Bujarrabal V., 1992, *Astrophys. J.*, **401**, L109
Cernicharo J., Bujarrabal V., Santarén J.L., 1993, *Astrophys. J.*, **407**, L33
Cernicharo J., Guélin M., 1987, *Astron. Astrophys.*, **183**, L10
Cernicharo J., Guélin M., Hein H., Kahane C., 1987a, *Astron. Astrophys.*, **181**, L9
Cernicharo J., Guélin M., Menten K.M., Walmsley C.M., 1987b, *Astron. Astrophys.*, **181**, L1
Cernicharo J., Thum C., Hein H., John D., Garcia P., Mattioco F., 1990, *Astron. Astrophys.*, **231**, L15

Clegg R.E.S., Hinkle K.H., Lambert D.L., 1982, *MNRAS*, **201**, 95
Cohen R.J., 1989, *Reports on Progress in Physics*, **52**, 881
Cox P., Omont A., Huggins P.J., Bachiller R., Forveille T., 1992, *Astron. Astrophys.*, **266**, 420
Dayal A., Bieging J.H., 1993, *Astrophys. J.*, **407**, L37
Deguchi S., Izumiura H., Kaifu N., Mao X., Nguyen-Q-Rieu, Ukita N., 1990, *Astrophys. J.*, **351**, 522
Elitzur M., 1992, *Astronomical Masers*, Kluwer, Dordrecht
Gensheimer P.D., Likkel L., Snyder L., 1992, *Astrophys. J.*, **388**, L31
Glassgold A.E., Huggins P.J., 1983, *MNRAS*, **203**, 517
Glassgold A.E., Omont A., Guélin M., 1992, *Astrophys. J.*, **396**, 115
Guilloteau S., Omont A., Lucas R., 1987, *Astron. Astrophys.*, **176**, L24
Gustafsson B., 1989, *Ann. Rev. Astr. Astroph.*, **27**, 701
Howe D.A., Millar T.J., 1990, *MNRAS*, **244**, 444
Howe D.A., Millar T.J., Williams D.A., 1992, *MNRAS*, **255**, 217
Huggins P.J., Glassgold A.E., 1982, *Astrophys. J.*, **252**, 201
Izumiura H.,1990, PhD thesis, Tokyo University
Izumiura H., Ukita N., Kawabe R., Kaifu N., Tsuji T., Unno W., Koyama K., 1987, *Astrophys. J.*, **323**, L81
Jaminet P.A., Danchi W.C., Sandell G., Sutton E.C., 1992, *Astrophys. J.*, **400**, 535
Jewell P.R., Snyder L.E., Walmsley C.M., Wilson T.L., 1991, *Astron. Astrophys.*, **242**, 211
Jura M., 1991, *Astrophys. J.*, **372**, 208
Jura M., Morris M., 1981, *Astrophys. J.*, **251**, 181
Kahane C., Gómez-González J., Cernicharo J., Guélin M. , 1988, *Astron. Astrophys.*, **190**, 167
Kahane C., Cernicharo J., Gómez-González J., Guélin M.:, 1992, *Astron. Astrophys.*, **256**, 235
Kawaguchi K., Kagi E., Hirano T., Takano S., Saito S. , 1993, *Astrophys. J.*, **406**, L39
Knapp G.R., Chang K.M., 1985, *Astrophys. J.*, **293**, 281
Lambert D.L, Gustafsson B., Eriksson K., Hinkle K.H., 1986, *Astrophys. J. Suppl.*, **62**, 373
Lazaro C., Lynas-Gray A.E., Clegg R.E.S., Mountain C.M., Zadrozny A., 1991, *MNRAS*, **249**, 62
Lewis B.M., 1989, *Astrophys. J.*, **338**, 234
Lewis B.M., 1990, *Astron. J.*, **99**, 710
Lewis B.M., Engels D., 1991, *MNRAS*, **251**, 391
Lindqvist M., Nyman L.-Å., Olofsson H., Winnberg A., 1988, *Astron. Astrophys.*, **205**, L1
Lindqvist M., Olofsson H., Winnberg A., Nyman L.-Å., 1992, *Astron. Astrophys.*, **263**, 183
Lucas R., Cernicharo J., 1989, *Astron. Astrophys.*, **218**, L20
Lucas R., Guélin M., 1990, in *Submillimetre Astronomy*, eds G.D. Watt and A.S. Webster, Kluwer, Dordrecht, p.97
Lucas R., et al., 1992, *Astron. Astrophys.*, **262**, 491
Mamon G.A., Glassgold A.E., Huggins P.J., 1988, *Astrophys. J.*, **328**, 797
Menten K.M., Melnick G.J., 1989, *Astrophys. J.*, **341**, L91
Menten K.M., Melnick G.J., 1991, *Astrophys. J.*, **377**, 647
Menten K.M., Melnick G.J., Phillips T.G., 1990, *Astrophys. J.*, **350**, L41

Millar T.J., 1988, in *Rate Coefficients in Astrochemistry*, eds T.J. Millar and D.A. Williams, Kluwer, Dordrecht, p. 287

Nejad L., Millar T.J., 1987, *Astron. Astrophys.*, **183**, 279

Nejad L., Millar T.J., 1988, *MNRAS*, **230**, 79

Nercessian E., Guilloteau S., Omont A., Benayoun J.J., 1989, *Astron. Astrophys.*, **210**, 225

Nguyen-Q-Rieu, Bujarrabal V., Olofsson H., Johansson L.E.B., Turner B.E., 1984, *Astrophys. J.*, **286**, 276

Nyman L.-Å., Olofsson H., 1985, *Astron. Astrophys.*, **147**, 309

Nyman L.-Å., Olofsson H., Johansson L.E.B., Booth R.S., Carlström U., Wolstencroft R., 1993, *Astron. Astrophys.*, **269**, 373

Olofsson H.: 1993, in *Mass Loss on the AGB and Beyond*, ed. H. Schwarz, in press

Olofsson H., Carlström U., Eriksson K., Gustafsson B., 1992, *Astron. Astrophys.*, **253**, L17

Olofsson H., Eriksson K., Gustafsson B., Carlström U., 1993a, *Astrophys. J. Suppl.*, **87**, 267

Olofsson H., Eriksson K., Gustafsson B., Carlström U., 1993b, *Astrophys. J. Suppl.*, **87**, 305

Olofsson H., Lindqvist M., Winnberg A., Nyman L.-Å., 1993c, in prep.

Omont A., Lucas R., Morris M., Guilloteau S., 1993, *Astron. Astrophys.*, **267**, 490

Sahai R., Bieging J.H., 1993, *Astron. J.*, **105**, 595

Sahai R., Wannier P.G., 1992, *Astrophys. J.*, **394**, 320

Scalo J.M., Slavsky D.B., 1980, *Astrophys. J.*, **239**, L73

Smith V.V., Lambert D., 1990, *Astrophys. J. Suppl.*, **72**, 387

Solomon P., Jefferts K.B., Penzias A.A., Wilson R.W., 1971, *Astrophys. J.*, **163**, L53

Sopka R.J., Olofsson H., Johansson L.E.B., Nguyen-Q-Rieu, Zuckerman B., 1989, *Astron. Astrophys.*, **210**, 78

Takano S., Saito S., Tsuji T., 1992, *PASJ*, **44**, 469

Truong-Bach, Nguyen-Q-Rieu, 1989, *Astron. Astrophys.*, **214**, 267

Truong-Bach, Morris D., Nguyen-Q-Rieu, 1992, *Astron. Astrophys.*, **249**, 435

Truong-Bach, Morris D., Nguyen-Q-Rieu, Deguchi S., 1990, *Astron. Astrophys.*, **230**, 431

Turner B.E., 1987a, *Astron. Astrophys.*, **182**, L15

Turner B.E., 1987b, *Astron. Astrophys.*, **183**, L23

van der Veen W.E.C.J., Habing H.J., 1988, *Astron. Astrophys.*, **194**, 125

van der Veen W.E.C.J., Olofsson H., 1990, in *From Miras to Planetary Nebulae - Which Path for stellar Evolution*, eds M.O. Mennessier and A. Omont, Editions-Frontiérs, p.139

van Dishoeck E.F., 1988, in *Rate Coefficients in Astrochemistry*, eds T.J. Millar and D.A. Williams, Kluwer, Dordrecht, p. 49

Wannier P., Sahai R., 1987, *Astrophys. J.*, **319**, 367

Wannier P. Andersson B.-G., Olofsson H., Ukita N., Young K., 1992, *Astrophys. J.*, **380**, 593

Wilson W.J., Barrett A.H., 1968, *Science*, **161**, 778

Polyynes and Polycyclic Aromatic Molecules in C-rich Circumstellar Envelopes

Alain Omont

Institut d'Astrophysique de Paris, CNRS, 98bis, Bld Arago, F-75014 Paris, France

1 Introduction

The aim of this review is to discuss our knowledge on molecules in the circumstellar environment of evolved stars. In particular the presence and the behaviour of various kinds of molecules with several or many carbon atoms, in relation to C-rich dust, is considered.
Such objects include mainly: (i) circumstellar shells of AGB carbon stars, either visible (such as Y CVn) or infrared (such as IRC+10216 (CW Leo)); (ii) planetary nebulae (PNe, e.g. NGC 7027); (iii) pre-planetary nebulae (PPNe, also called post-AGB stars, such as CRL 2688 or the Red Rectangle), probably in an intermediate evolution stage between the two former classes. I will not discuss more peculiar classes, such as R CrB stars and novae, for which very little is known about the presence of such molecular species.

Polyyne species are known from millimetre observations in AGB envelopes and PPNe, mostly in IRC+10216 (see Olofsson this volume). They are mainly cyanopolyynes (HC_nN) with n = 3, 4, ... 11, and C_n and C_nH linear radicals, plus a few other cases (C_4H_2, C_5, C_3N) (see Table 1). Observational results are discussed in Section 2: abundances, spatial distributions, etc.. Such information allows us to infer the basic features of their chemistry (Section 3).

In nearly all cases where their spatial distribution is known, the latter infers that they are formed mainly by photo-chemistry in the outer layers of the envelope, i.e. by chains of reactions induced by interstellar UV radiation. They thus have no direct relation with dust and their formation is surprising.

Polycyclic aromatic molecules (generally called PAHs) have been convincingly proposed as the carriers of the infrared bands observed in various interstellar and circumstellar environments. Aromatic species are presently not detected in AGB stars (see Hinkle this volume). On the other hand, PAHs are quite abundant in PNe and some PPNe (Section 4). They could thus be related to the evolution and processing of dust at the onset of the PN stage. Transition objects, PPNe, are therefore essential to elucidate this evolution. Their infrared spectroscopy, including dedicated studies, is discussed in Section 5. Together with the likely identification of PAH related species in the gas or in the dust, there remain several unidentified features. Indeed, there is no final conclusion whether PAHs are already present in the gas of AGB envelopes or they are generated at the onset of the PN stage. Because of lack of space, and also of evidence, I will not discuss the conjecture that C_{60} or its derivates could be present in such objects.

Table 1. Observed abundances in IRC+10216 (molecules denoted with a * have only been detected in the infrared). MgNC has also recently been detected.

$x(i)/H_2$	IRC+10216					
	$C_2H_2^*$					
10^{-5}	HCN					
					CN	
10^{-6}	NH_3				C_3^*	
					C_2H	
		SiS			C_4H	
			CS	HC_3N		
	$C_2H_4^*$	C_2Si			C_3N	
10^{-7}	CH_4^*	SiO		HC_5N	C_5^*	
	HNC			HC_7N	C_6H	ClAl
					C_3H_2	
			C_2S		C_3H	
10^{-8}	CP	SiN	C_3S	$HC_{11}N$	C_5H	FAl
		SiH_4^* SiC			C_4H_2	
	CH_3CN					
			H_2S	HC_2N		
10^{-9}	(HCO^+)					ClNa
		C_4Si				ClK

2 Observations of polyynes in circumstellar envelopes

2.1 Detected molecules and abundances in IRC+10216 (Table 1)

Cyanopolyynes ($HC_{2n}CN$) were the first carbon chains detected in space. Many were first detected in IRC+10216. HC_3N, HC_5N, HC_7N, HC_9N and $HC_{11}N$ are now known. Their abundances do not decrease very rapidly with n. Typically the ratio

$$HC_{2n}CN \;/\; HC_{2n+2}CN \;\sim\; 3$$

However, they are more difficult to detect for large n. Their very small rotation constant B makes many rotational levels populated with the typical rotation temperatures ~ 10–25 K and brings detectable lines into the centimetre range. The line intensities rapidly decrease despite the increase of the electric dipole momentum. They also have low vibrational levels allowing the detection of rotational lines of ν_6 and ν_7 states of HC_3N.

Pure hydrocarbon chains, $HC_{2n}H$, have not been detected. They have no permanent electric dipole momentum, and hence no radio spectrum. However, there is an infrared limit for the abundance of HC_4H (see Hinkle this volume). Note that millimetre lines of its isomer H_2C_4 have been detected.

$C_{2n}H$ radicals , with n=2,3 (C_4H, C_6H) are quite abundant. They have several electronic levels of comparable low energy. C_4H (and C_2H) have a $^2\Sigma$ ground state with a low $^2\Pi$ state. The situation is reversed for the higher members such as C_6H which have a $^2\Pi$ ground state (Pauzat et al. 1989). This made difficult the identification of C_6H in IRC+10216 (Guélin et al. 1987). These radicals have large electric dipole momenta. Rotation transitions of vibrationally excited C_4H are easily detected in IRC+10216 where it is exceptionally abundant and extended. This could be related to the likely enhancement of the line strengths of vibration transitions by vibronic coupling.

C_3H and C_5H are also detected with the expected $^2\Pi$ ground state, but with much lower abundances (Table 1).

Pure carbon chains, C_n , can be observed only by their infrared transitions (see Hinkle this volume). C_3 and C_5 have been detected with fairly large abundances in IRC+10216.

Several chain radicals with hetero-atoms have been found in IRC+10216: C_3N has a relatively large abundance, but the higher members $C_{2n+1}N$ are not detected, probably because of the smallness of their electric dipole momentum (Pauzat et al. 1991). C_3S (and C_2S) and C_4Si are also found with low abundances.

2.2 Spatial Distribution in IRC+10216

The determination of the position in the envelope of the observed molecules is essential to analyze the chemical processes. To do this clearly requires the telescope beam to be smaller than the extension of the molecular envelope. The exceptionally close proximity of IRC+10216 for a source with a large mass–loss rate (see Olofsson this volume) yields very large angular diameters (20–60"). Accordingly, its envelope is resolved with the small beams of large millimetre single dishes. For instance the beam of the IRAM 30m telescope is 12" at 1.3mm. Several studies of the extensions of various molecules have been carried out with single dishes. Indeed, the shape of the profile observed in the center of the source provides by itself a useful information on the extension of the molecule (see Olofsson). However, the large millimetre interferometers are unravelled to give complete details on the molecular distributions. The spatial distribution of most important molecules is now well studied in IRC+ 10216 (see Olofsson this volume). The most important contribution has been first provided by Bieging, Nguyen–Q–Rieu and collaborators with the Hat Creek interferometer since several years (Bieging & Nguyen–Q–Rieu 1988, Nguyen–Q–Rieu & Bieging 1990, Dayal & Bieging 1993, Bieging & Taffala 1993). The IRAM interferometer is now the most sensitive instrument and is making systematic studies of many molecular lines (see e.g. Guélin 1993).

All polyyne species studied (HC_3N, HC_5N, C_4H, C_3N) display quite characteristic structures of hollow shells (slightly clumpy, Fig. 1). It is quite likely that this is also true for all the other polyynes whose weaker lines have profiles characteristic of a large extension, but are not yet completely studied with interferometer mapping. It is thus proved that most detected polyyne molecules are synthesized in the outer shells of the envelope permeated by the interstellar UV and are not generated in the dense regions close to the star (see Section 3).

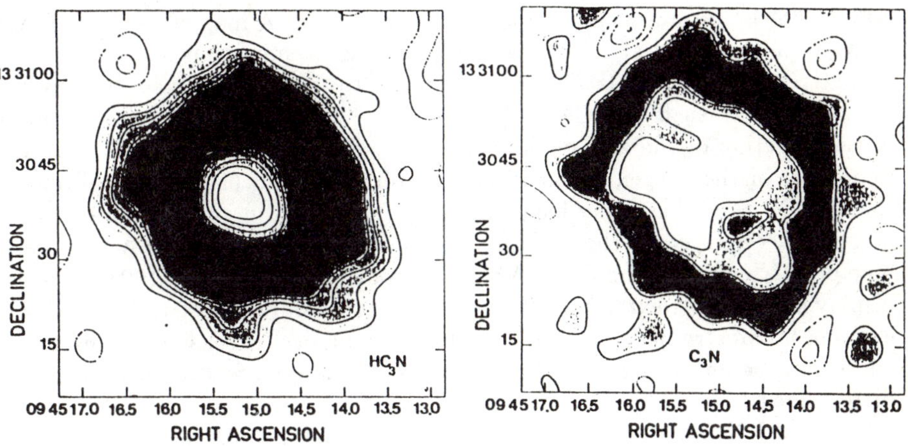

Fig. 1. Interferometer maps of 3mm lines of HC_3N and C_3N in IRC+10216 by Bieging & Taffala (1993)

2.3 Polyynes in Envelopes of Other Stars

The information is much more meagre on molecules in stars other than IRC+10216, since all millimetre molecular lines are at least 10 times weaker (see Olofsson this volume). However, the strongest lines (C_4H, C_3N, HC_5N and especially HC_3N) are easily detected in 10–20 C–rich envelopes with large mass–loss rates (a few 10^{-6} – a few $10^{-5} M_{\odot}/yr$) comparable to that of IRC+10216. The most studied sources are CIT 6 (RW LMi), IRAS (15194–5115) (Nyman et al. 1993), AFGL 3068, IRC+40540 (LP And), IRC+10401 (Cernicharo et al. 1994), etc...

There are no striking differences in the abundances compared to IRC+10216. In particular the ratio of the line intensities HC_3N/HCN is remarkably constant (Bujarrabal et al. 1994). However, in sources with large excesses of HCN, such as IRC+10401, the abundances of HC_3N, HC_5N and C_4H are also probably an order of magnitude larger than in IRC+10216. On the other hand the abundances of HC_3N and especially the ratio HC_3N/HCN are quite small in C–rich envelopes with low mass–loss rates such as Y CVn. HC_3N has never been detected in O–rich stars (Bujarrabal et al. 1994).
In C–rich pre–planetary nebulae the abundance of HC_3N (and of C_4H, HC_5N) is also generally comparable to that of IRC+10216, although somewhat smaller (Henkel et al. 1994, Lucas et al. 1994). The latter authors have performed a detailed study of many lines in AFGL2688 (CRL2688, the Egg Nebula) ; HC_3N, but not C_4H, displays modest high velocity wings. HC_3N has been mapped with the Hat Creek inteferometer (Nguyen–Q–Rieu & Bieging 1990), but not with enough sensitivity to derive detailed information about its spatial distribution. However, cyanopolyynes display important pecularities in this source as well as in the molecular envelope of the extremely young planetary nebula AFGL 618.
AFGL 2688 is the prototype of bipolar nebulae. The VLA map of HC_7N at 1.4cm by Nguyen–Q–Rieu et al. 1986 (Fig. 2) shows that its emission is relatively strong and extremely extended (r∼20–30") compared to the size of the emission of the other molecules such as HCN or NH_3 ($r \leq 10"$). This extension is at maximum in the direction of the axis of the nebula and of its bipolar flow, but it is also important in the perpendicular direction. Jura & Kroto (1990) have proposed that HC_7N, as well as other long chains, are generated in grain–grain collisions of grains accelerated by the radiation pressure of visible radiation. A millimetre map of HC_5N made using the IRAM interferometer (Lucas et al. 1994) does not display such a large extension. However, this could only reflect the difficulty of exciting millimetre lines far from the star. The modest extension of this HC_5N emission is again mostly axial, but with also a non axial feature at an intermediate velocity.
AFGL 618 (CRL 618) is indeed a nascent planetary nebula, with a very small and expanding ionized region inside a massive molecular envelope. The millimetre HC_3N lines are quite prominent and have a very large rotation temperature. Even the rotation lines of vibrationaly excited HC_3N are quite strong.
On the other hand, HC_3N has never been detected in any other, more regular, planetary nebula, even the very young one NGC 7027, despite active searches (Cox et al. 1994).

3 Chemistry of Polyynes in IRC+10216

The spatial distribution of polyynes in IRC+10216 clearly points to generation by photochemistry. Indeed, the understanding of their formation was probably the main challenge of circumstellar chemistry and the results already obtained are a major achievement. The general features of circumstellar chemistry are now well established (see e.g. reviews by Glassgold & Huggins 1988, Millar

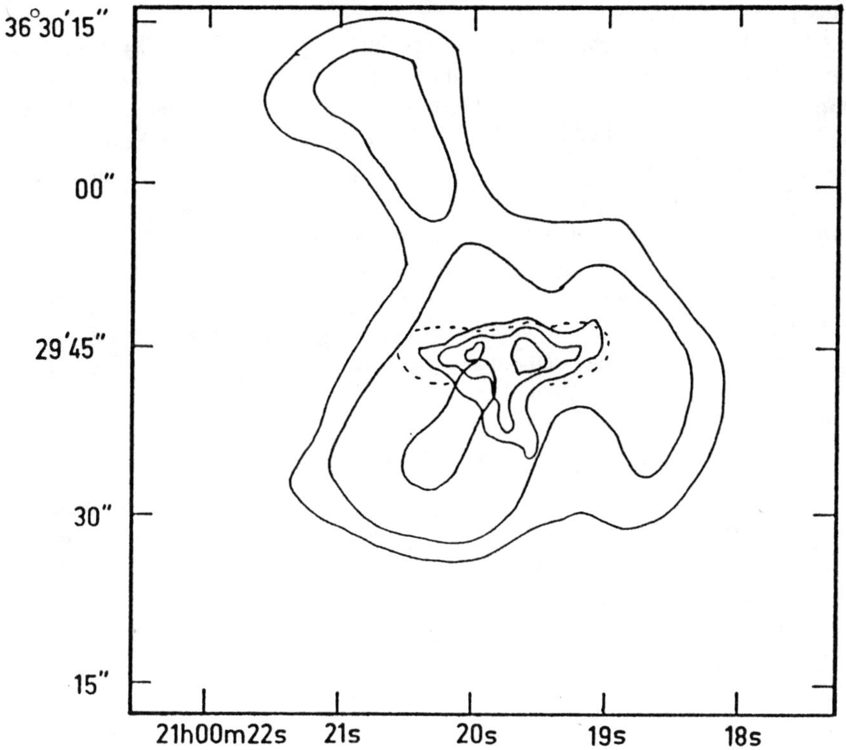

Fig. 2. VLA interferometer map of 1.4cm lines of HC_7N (outer contours) and NH_3 (inner contours) by Nguyen–Q–Rieu et al. (1986) (from Jura & Kroto 1990)

1988, Omont 1990). Thermodynamical chemical equilibrium is achieved only in the deep inner layers , dense and warm, close to the photosphere. Most stable molecules, such as H_2, CO, N_2, C_2H_2, HCN, CS, SiS, SiO, are generated there. However, calculations of equilibrium abundances (see e.g. Tejero & Cernicharo 1991) derive a relatively small abundance for HC_3N and much smaller ones for other polyynes.

The chemistry is certainly complex and poorly understood in the intermediate layers where dust and the circumstellar wind are generated. Small carbon chains appear as possible initial steps in the nucleation of C–rich dust. However, given the observed shell distribution of polyynes, their abundance must remain quite small at the end of dust formation. Photochemistry dominates the chemistry in the outer layers which are not shielded against the penetration of the interstellar UV radiation. The latter first destroys the stable molecules, such as CO, C_2H_2, HCN, etc, transported by the expansion from the inner layers. Thus, the photodestruction products, radicals such as C_2H and CN, atoms and ions such as $C_2H_2^+$ and C^+, can react with abundant species to build up more and more complex polyyne chains.

Detailed modelling of such photochemistry has been developed by Glassgold and collaborators, and Millar and collaborators. The input parameters are the envelope parameters (mass–loss rate, expansion velocity, temperature, etc), the initial abundances of stable molecules such as CO, C_2H_2, HCN coming from the inner shells, the UV intensity and the reaction and photodestruction rates. One has to model the transfer of the UV radiation coming from the interstellar medium and progressively attenuated by dust in the outer layers of the envelope. A part of the photodissociation rates can be reasonably well estimated; CO needs a special elaborate treatment (Mamon et al. 1988). However, most of the photodestruction rates of unstable species can only be roughly guessed. Ion reaction rates are reasonably well known for small species. A typical proposed reaction is (Glassgold et al. 1986, 1987)

$$C_2H_2^+ + HCN \rightarrow H_2C_3N^+ + H \tag{1}$$

and then $H_2C_3N^+$ could yield HC_3N by dissociative recombination. However, it is now recognised that such ion reactions are not efficient enough to account for the large amount of observed polyynes. Radical reactions are more effective given the large reaction rates at low temperature (Nejad & Millar 1987, 1988). For instance, possible reactions are

$$CN + C_2H_2 \rightarrow HC_3N + H \tag{2}$$

$$C_3N + C_2H_2 \rightarrow HC_5N + H \tag{3}$$

If early models still had difficulties in accounting for the long chains, the last model of Cherchneff et al. (1993), Cherchneff & Glassgold (1993) gives satisfactory results for the first cyanopolyynes, C_4H, etc, with the large reaction rates now admitted for polar molecules at very low temperature.

In conclusion, although many uncertainties remain (in particular in the detailed rates) there is very good agreement between the observed abundances and their distribution. The results of theoretical chemical modelling demonstrates that most, if not all, polyynes observed in IRC+10216 are synthesized by photochemistry in the outer layers from small molecules such as C_2H_2, HCN, etc. The situation is likely to be the same in similar AGB envelopes. It is indeed slightly paradoxal that the observed polyynes have no direct relation with C–rich dust, while they have been proposed as logical intermediates in dust nucleation, as well as natural products of grain–grain collisions (Jura & Kroto 1990).

4 Polycyclic Aromatic Molecules and Circumstellar Envelopes

Polycyclic aromatic hydrocarbons (PAHs) are also a quite natural and stable form of small carbon clusters in the hydrogen rich environment of the interstellar and circumstellar media, probably playing an important role in the nucleation of circumstellar grains (see Sedlmayr this volume). Complex mixtures of such PAHs have been convincingly proposed as the carriers of the set of characteristic infrared bands at 3.3, 6.2, 7.7, 8.6, 11.3 μm observed (Fig. 3) in various interstellar and circumstellar sources exposed to UV radiation (see e.g. Léger & Puget 1984, Allamandola et al. 1985, Puget & Léger 1989 and references therein).

The molecular envelopes of young planetary nebulae, such as NGC 7027, and of pre–planetary nebulae with early type stars, such as the Red Rectangle (HD 44179), are major sources of PAHs.
The "canonical model" of PAH IR emission induced by UV radiation (see references above), explains the very large observed vibrational temperature, practically independent of the UV intensity: free–flying PAHs, with 20–100 atoms, are small enough systems to be heated to 500–1000 K by a single UV photon. The main arguments which have lead to this interpretation are their very large stability, the high specificity of some bands (6.2 μm) characteristic of aromatic compounds and the properties of the observed vibration temperature not achievable in conventional grains. However, several problems remain concerning their stability in astrophysical conditions, as well as their degree of ionization and of de–hydrogenation and the related spectral characteristics.

In circumstellar envelopes, PAH emission is quite strong in PNe and warm PPNe, but they are not observed in AGB stars. This raises interesting questions about their generation and their excitation. Many details are known on PAH emission in the strongest sources such as NGC 7027 and the Red Rectangle.
NGC 7027 is a PN with a very hot star emitting very strong UV radiation. A dense expanding ionized central region coexists with a massive outer molecular envelope partially permeated by the UV radiation (Cox et al. 1994). The PAHs probably reside mostly at the interface between the ionized and molecular gas. PAH fluorescence is clearly induced by the strong UV radiation. It is quite possible that the PAHs are generated at the onset of the PN stage by some mechanism such as destruction and processing of dust in such a violent environment. However, any detailed evidence is still lacking, and they could as well have been present at early stages and be just revealed by the strong UV shining.

The situation has some similarities in the Red Rectangle (HD 44179) but with important pecularities. This is a PPN with a warm star (Ao), but not hot enough to ionize the circumstellar gas. There is some indication of a circumstellar disk, which could explain that the CO millimeter lines are very narrow and also very weak because of its likely photodissociation. Radiation pressure on dust in such a disk could also be at the origin of the strange photospheric abundances with a strong underabundance of iron and other refractory elements. This nebula has a very strong and peculiar red fluorescence with a molecular band structure reminiscent of unidentified visible diffuse interstellar bands, and possibly associated with PAHs.
There is presently no evidence for either the presence or absence of PAHs in ordinary AGB circumstellar envelopes. The lack of UV and even visible radiation would explain the absence of infrared bands even in the presence of PAHs. It

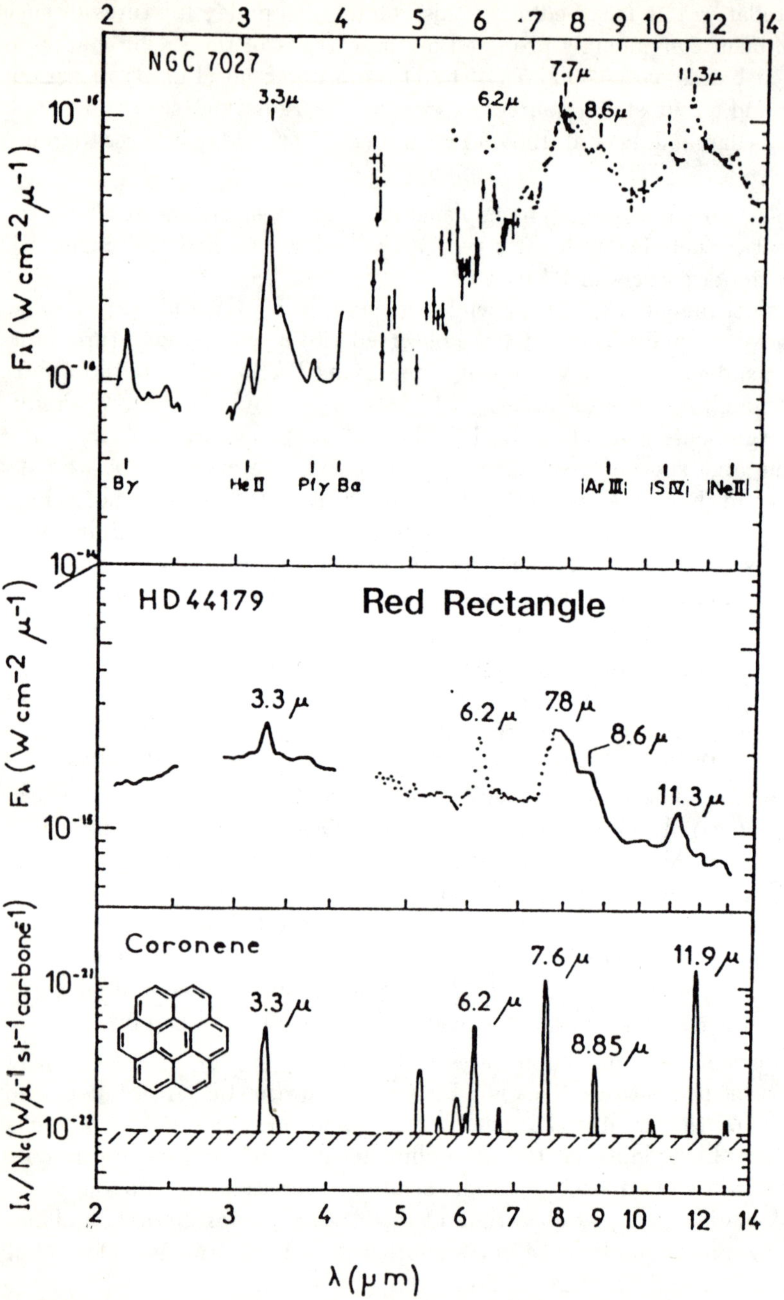

Fig. 3. Infrared features observed in NGC 7027 and HD 44179 (Red Rectangle) compared to a synthetic spectrum of coronene, from Léger & Puget (1984).

would be very difficult to see them in absorption, as well as to detect their radio rotational lines. Since they are likely intermediates in the formation of C–rich dust in the inner shells, they could well survive in sufficient amount (a few percents of the total carbon) in the whole envelope. However, a simple model of dust formation (Cherchneff et al. 1993) leads to quite small amounts of PAHs in the outer envelope.

An alternative, not unlikely possibility would be that they are formed, or just desorbed, from C–rich grains at the onset of the PPN–PN stage. The properties of dust would then rapidly evolve and a natural question is: are there spectral features related to such transitory stages of dust in PPNe ? A related question is: are there spectral features in C–rich envelopes related to PAHs either free, or incorporated in dust (i.e. any evidence of hydrogenated amorphous carbon (HAC) which is a mixture of polycyclic carbon clusters partially hydrogenated) ?

5 Infrared Spectroscopy of C–rich Circumstellar Dust

5.1 Spectra of Ordinary AGB Envelopes

They are relatively featureless because the main constituent of dust is solid carbon which has no prominent spectral feature. However, there are several strong and characteristic features:

– In the near infrared there is a strong absorption band of the gas at 3.4 μm due mainly to C_2H_2 and HCN, as well as a few other weaker features at longer wavelengths in some cases.

– The middle infrared is very well documented by "Low Resolution Spectra" (LRS) of the IRAS satellite between 7.5 and 23 μm which contain several hundred spectra of C–rich envelopes. Most of them display a strong characteristic band at 11.4 μm attributed to solid SiC.
In addition, there is quite often a less obvious feature at the lower LRS limit, which could be either an emission at 8.5 μm or absorption at 7.5 μm (gas bands ?) (see e.g. Omont et al. 1993) and references therein).
– In the far infrared, there is a strong and broad feature at about 30 μm, when the mass–loss is large enough ($\geq$ a few$10^{-6} M_\odot$/yr) (Forrest et al. 1981). Although it has been attributed to solid MgS (Goebel & Moseley 1985), its origin is still discussed. A similar feature is also often present in PNe and PPNe (Section 5.3).

5.2 Infrared Spectroscopy of cold C–rich PPNe

We have already discussed the case of "warm" PPNe such as the Red Rectangle (Section 4). In typical "cold" PPNe, such as the Egg Nebula (AFGL 2688), there is no warm dust because the expanding envelope is completely detached from the star with a nearly empty central cavity around the star. Accordingly, the IR spectrum has its maximum in the far infrared, beyond 20 μm (Fig. 4). As seen in this figure, the IRAS LRS spectra are mainly of two types: a good part of them, including AFGL 2688 and other thick envelopes, are nearly featureless.

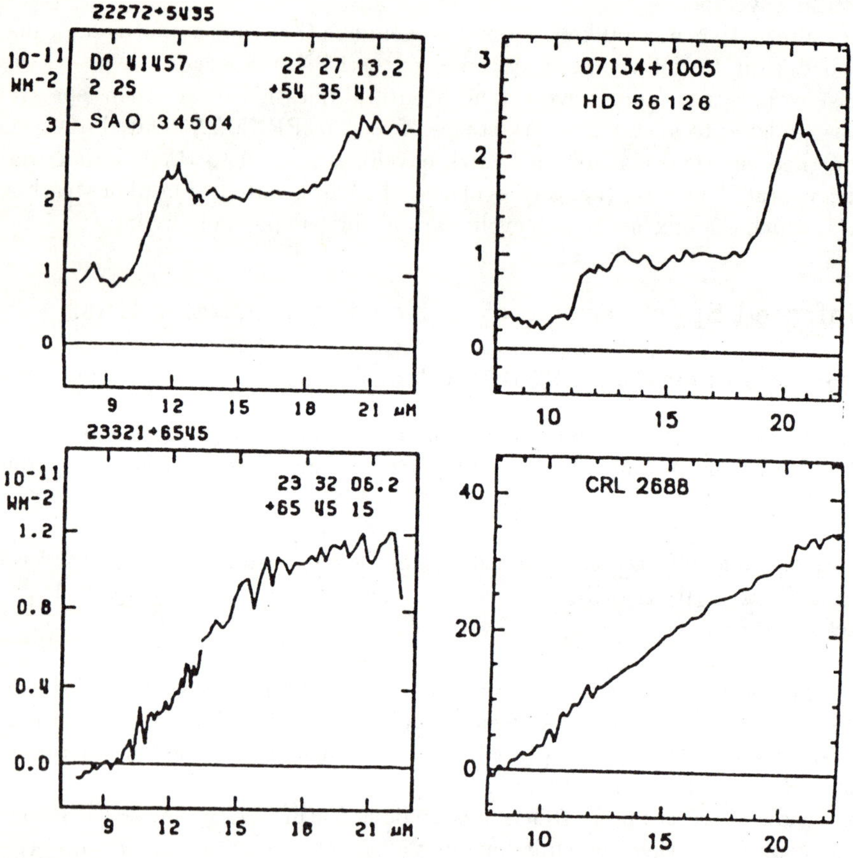

Fig. 4. IRAS Low Resolution Spectra (1986) of C–rich pre–planetary nebulae with and without an emission feature at 21 μm.

On the other hand, the LRS spectra of other sources display two strong features: a broad band at 21 μm and a plateau between 12 and 15 μm (Kwok et al. 1989). They have a relatively thin envelope so that they emit comparable amounts of energy in the visible and the far infrared. Airborne spectroscopy of the 5–8 μm range (Buss et al. 1990) shows several features reminiscent of those attributed to PAHs, but somewhat different and weaker when compared to strong classical PAH sources (Fig. 5). In addition, the 3.3 μm "PAH" feature has also been detected in some sources, as well as other features at about 3.4 μm. All these characteristics have lead Buss *et al.* to propose that they are emitted by some carbonaceous compound containing hydrogenated amorphous carbon (HAC). It should be noted that similar, although weaker, features are also present in AFGL 2688.

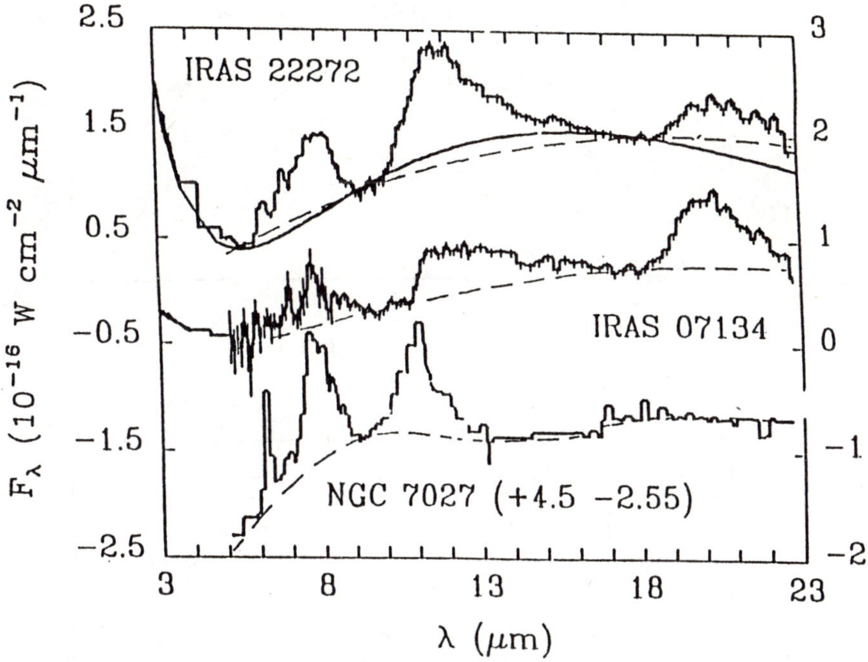

Fig. 5. 5–23 μm spectrum of two C–rich pre–planetary nebulae displaying the 21 μm emission feature, compared to the spectrum of the PAH source NGC 7027 (from Buss et al. 1990).

The origin of 21 μm band is not yet entirely clear. It seems likely, as suggested by Buss et al. (1990), that it is associated with some kind of carbonaceous compound. Such materials frequently emit at 11–15 μm, so that there is no difficulty to explain the companion feature observed in this range. However, no compound is presently known with such a strong emission at 21 μm and the other observed spectral characteristics (see e.g. Omont et al. 1994 and references therein). Many of the PAHs, such as naphtalene, have strong bands in the 19–23 μm range, but they have also other bands which are not observed. It would obviously be quite interesting to identify the carrier of the 21 μm band. It should be a transient species, characteristic of dust processing in these PPNe, since the feature is not observed in their AGB progenitors, neither in the following PNe stages. Omont et al. (1994) have recently extended the spectroscopy of these C–rich PPNe up to 50 μm by airborne KAO observations. There is no other relatively narrow feature such as the one at 21 μm.

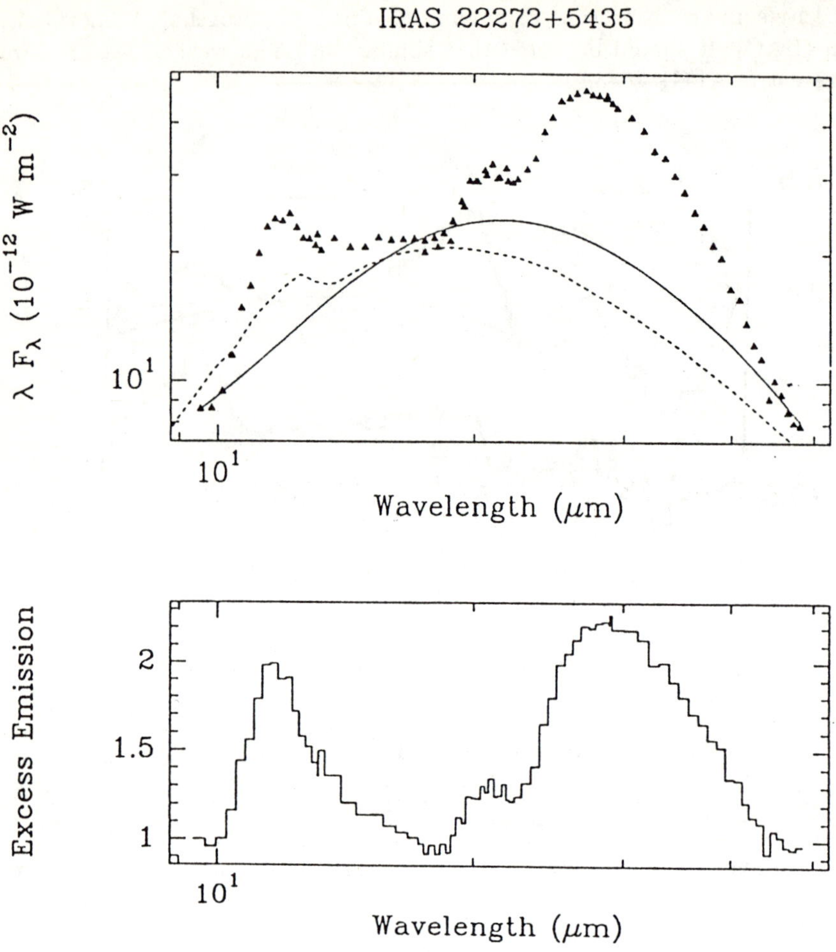

Fig. 6. Composite for infrared spectrum of IRAS 22272+5435 from Omont et al. (1994). The continuous and dashed curves represent two different estimates of the background emission.

However, most sources display a very broad excess of emission between 25 and 45 μm (Fig. 6). The intensity of this emission is quite variable following the source and apparently does not correlate with the strength of the 21 μm band. This 30 μm emission appears similar to the one observed in AGB carbon stars with large mass–loss (Forrest et al. 1981) as well as in some young planetary nebulae (Moseley & Silverberg 1986, Glaccum 1989). Both of the latter cases have been attributed to MgS by Goebel & Moseley (1985). However, the remarkable strength of the emission in sources such as IRAS 22272 and the PN IC 418 casts some doubt on this. Further laboratory data on MgS are required to check this. Other possible explanations should also be explored, such as the emission of metals (Fe, Mg...) attached to aromatic compounds (Chaudret et al.

1991, Kroto & Jura 1992).

Acknowledgments. I am glad to thank Mrs M.C. Pantalacci for the edition of this paper.

References

Allamandola L.J., Tielens A.G.G.M., Barker J.R., 1985, *Astrophys. J.* **290**, L25

Bieging J.H., Taffala M., 1993, *Astrophys. J.*, **105**, 576

Bieging J.H. and Nguyen-Q-Rieu, 1988, *Astrophys. J.*, **329**, L107

Bujarrabal V., Fuente A., Omont A. 1994, *Astron. Astrophys.*, in press

Buss Jr R.H., Cohen M., Tielens A.G.G.M., Werner M.W., Bregman J.O., Witteborn F.C., Rank D., Sandford S.A., 1990, *Astrophys. J. Lett.*, **365**, L23

Cernicharo J., Fuente A., Omont A. 1994, in preparation

Chaudret B., Lebeuze A., Rabaa H., Saillard J.Y., Serra G., 1991, *New J. Chem.*, **15**, 791

Cherchneff I., Barker J.R., 1992, *Astrophys. J.*, **394**, 703

Cherchneff I., Barker J.R., Tielens A.G.G.M., 1992, *Astrophys. J.*, **401**, 269

Cherchneff I., Glassgold A.E., Mamon G.A., 1993, *Astrophys. J.*, **410**, 188

Cherchneff I., Glassgold, A.E., 1993, *Astrophys. J. Lett.* in press

Cox P., Guilloteau S., Bachiller R., Huggins P.J., Omont A., Forveille T., 1994, in preparation

Dayal A., Bieging J.H., 1993, *Astrophys. J. Lett.*, **407**, L37

Forrest W.S., Houck J.R., Mc Carthy J.F., 1981, *Astrophys. J.*, **248**, 195

Glaccum W.J., 1989, PhD Dissertation

Glassgold A.E., Lucas R. and Omont A., 1986, Astron. Astrophys., **157**, 35

Glassgold A.E. and Huggins P.J., 1988, in "M-Type Stars", ed. by H.R. Johnson and F. Querci (NASA-CNRS), p.291

Goebel J.H., Moseley S.M., 1985, *Astrophys. J. Lett.*, **290**, L35

Guélin M., Cernicharo J., Kahane C., Gomez–Gonzalez J., Walmsley C.M., 1987, *Astron. Astrophys.*, **175**, L5

Guélin M. 1993, in "Circumstellar Media in the Late Stages of Stellar Evolution", ed. by R.E.S. Clegg

Henkel C., Cox P., Omont A. 1994, in preparation

Howe D.A. and Millar T.J., 1990, *MNRAS*, **244**, 444

IRAS Team, LRS Atlas 1986, Low Resolution Spectra Atlas, Olnon F.M. and Raymond E., *Astron. Astrophys. Supp.*, **65**, 607

Jura M., Kroto H., 1990, *Astrophys. J.*, **351**, 222

Kroto M. and Jura M., 1992, *Astr. Ap.*, **263**, 275

Kwok S., Volk K.M., Hrivnak B.J., 1989, *Astrophys. J. Lett.*, **345**, L 51

Léger A., Puget J.L., 1984, *Astron. Astrophys.*, **137**, L5

Lucas R., Kahane C., Omont A. 1994, in preparation

Mamon G.A., Glassgold A.E. and Huggins P.J., 1988, *Astrophys. J.*, **328**, 797.

Millar T.J., 1988, In Rate Coefficients in Astrochemistry, ed. by T.J. Millar and D.A. Williams (Kluwer Academic Pub.), p.287

Moseley S.H., Silverberg R.F., 1986, in: KAO Workshop, p. 233, (NASA)

Nejad L.A.M. and Millar T.J., 1987, *Astron. Astrophys.*, **183**, 279

Nejad L.A.M. and Millar T.J., 1988, *MNRAS*, **230**, 79

Nguyen Q Rieu, Winnberg A., Bujarrabal V., 1986, *Astron. Astrophys.*, **165**, 204
Nguyen–Q–Rieu, Bieging J.H., 1990, *Astrophys. J.*, **359**, 131
Nyman L.A., Olofsson M., Johansson L.E.B., Booth A.S., Carlström U., Wolstencroft R., 1993, *Astron. Astrophys.*, **269**, 377
Omont A., 1990 In "Chemistry in Space" ed. by M.Greenberg and V.Pironello (Kluwer, Dordrecht, 1990)
Omont A., Loup C., Forveille T., te Lintel Hekkert P., Habing H., Sivagnanam P., 1993, *Astron. Astrophys.*, **267**, 515
Omont A., Moseley S.H., Cox P., et al., 1994, to be submitted to *Astrophys. J. Lett.*
Pauzat F., Ellinger Y., 1989, *Astrom. Astrophys.*, **216**, 305
Pauzat F., Ellinger Y., Mc Lean A.D., 1991, *Astrophys. J. Lett.*, **369**, L13
Puget J.L., Léger A., 1989, *Ann. Rev. Astron. Astrophys.*, **27**, 161
Tejero J., Cernicharo J. 1991, Instituto Geografico Nacional, Madrid

Improvements in the Computation of Grain Opacity

David R. Alexander[1] *and Jason W. Ferguson*[2]

[1] Physics Department, Wichita State University, Wichita, Kansas 67260 USA
[2] Department of Physics & Astronomy, University of Kentucky, Lexington, Kentucky 40506 USA

1 Introduction and Background

Atoms and ions dominate the opacity of most astrophysical plasmas at temperatures above about 5000 K. In the range between 5000 K and about 1500 K, molecular processes are important contributors to the opacity. Depending upon the composition and pressure, grains may begin to condense at temperatures below about 1500 K. For the conditions of interest here, the grains which form are generally small (radius $\leq 0.25\,\mu$m) compared to the wavelength of light near the peak of the Planck curve. Because their absorption and scattering cross section is so large compared to atoms or molecules, grains dominate the mean opacity whenever they are present and must, therefore, be taken into account whenever such temperatures are encountered. Our calculations indicate that failure to do so can lead to errors as large as five orders of magnitude in the mean opacity.

Spherically extended model atmospheres with $T_{\rm eff} = 2500$ K which are either oxygen-rich (Scholz & Tsuji 1984) or carbon-rich (Jørgensen et al. 1992) reach grain condensation conditions in their outer layers. These atmospheres are also affected by a variety of temporal phenomena, such as pulsations, shock waves, and winds, which significantly affect the condensation of grains (Fix, 1969; Sedlmayr, this volume). Since these temporal processes occur more quickly than grain condensation can reach equilibrium, the size and quantity of grains present at any given moment depends upon the past history of the atmosphere. However, the computation of the time-dependent condensation of grains is not always practical. As a first approximation, there is some utility to a general tabulation of opacities which neglects these temporal constraints and assumes that the grains are in complete thermal equilibrium with their surroundings. This assumption will form the framework for this review.

Frequency-averaged absorption coefficients are often used to reduce the radiative transfer problem to a single set of wavelength-independent equations. The Rosseland mean absorption coefficient is widely used because its use reduces the radiative transfer equation to a particularly convenient form. However, the assumption used in the derivation of the Rosseland mean that the radiation field is isotropic and Planckian breaks down if the optical depth of the medium is small. Under such conditions the Planck mean absorption coefficient , defined as

$$\kappa_{\rm P} = \frac{\int \kappa_\nu B_\nu \, d\nu}{\int B_\nu \, d\nu} , \tag{1}$$

may be a more appropriate average over wavelength (see Sect. 7).

Several investigators contributed to early progress in computing grain opacity. Gaustad (1963) computed the opacity of small grains of ice, magnesium silicate, and iron, including a simple treatment of the condensation and evaporation of grains, in order to study the formation of stars. Knacke (1968) computed the absorption and scattering of $0.2\,\mu$m particles of quartz and vitreous silica in response to suggestions that silicates might be constituents of the interstellar medium. Kellman & Gaustad (1969) computed the Rosseland and Planck mean absorption coefficient for ice, vitreous silica, and graphite for use in star formation calculations.

Larger, more general tabulations of the opacity including the effects of small particles were produced by Cameron & Pines (1973) and Alexander (1975). Both of these computations employed a simple equilibrium calculation for grains, included the opacity effects of both molecules and grains, and covered a wide range of temperatures. Alexander et al. (1983) and Pollack et al. (1985) produced even more extensive results which included detailed treatment of grain equilibrium, consideration of a wider range of grain and molecular absorbers, and the effects of the grain size distribution.

In order to compute the opacity due to small particles, one must be able to answer four sets of questions:

1. Under what conditions do the grains condense? What is their chemical composition? At a given temperature and pressure, how much of each species condenses?
2. What is the distribution of the sizes of the grains that form?
3. How do the grains absorb and scatter light?
4. What are the optical properties of the grains?

Each of these topics will be considered in turn before general results are discussed.

2 Grain Equilibrium

The equilibrium abundance of molecules in the gas phase is governed by the law of mass action. For example, if one considers the reaction

$$\mathrm{A(g) + B(g)} \rightleftharpoons \mathrm{C(g)} , \tag{2}$$

then the law of mass action is a relation among the number densities of the species:

$$\frac{n_{\rm C}}{n_{\rm A} n_{\rm B}} = K(T) , \tag{3}$$

where the equilibrium constant, $K(T)$, is a function of temperature only.

When grains may be present the application of the law of mass action is modified in the following way. Consider the equilibrium of corundum, relative to the monatomic gas phase,

$$Al_2O_3(c) \rightleftharpoons 2Al(g) + 3O(g) . \tag{4}$$

The equilibrium constant for this reaction is given by

$$K_{eq} = e^{(-\Delta G_o/kT)} , \tag{5}$$

where ΔG_o is the Gibbs free energy of the reaction. Corundum is condensed if

$$K_c = n_{Al}^2 n_O^3 \geq K_{eq} . \tag{6}$$

In fact, just enough corundum condenses to produce an equality in Eq. (6).

If only a single substance condensed from the gas phase, the application of the above principles would be relatively simple. In reality, many species must be tested to determine if they are condensed and, if more than one species is condensed, then the abundance of all the condensates must be determined simultaneously.

Grossman (1972) presented the results of such calculations as curves showing the percentage of an element which is condensed as a function of temperature. As discussed by Alexander et al. (1983), these percentages can be subtracted from the elemental abundances and the material remaining in the gas phase can be treated with the normal equations of molecular equilibrium. Updated results (Sharp & Huebner 1990) which include data for several chemical compositions and several pressures were used for this study.

3 Grain Size Distribution

Particles in an astrophysical environment will be present in a variety of different sizes. This distribution of sizes should be explicitly considered when computing the grain opacity, because the extinction efficiency of a grain depends upon its size. Mathis et al. (1977) determined that grains in the interstellar medium obeyed a power law distribution of sizes given by

$$n(a) = ka^{-3.5} , \tag{7}$$

where $n(a)$ is the relative number of grains per unit volume of size a. Their analysis indicated that most interstellar particles occur in the size range from $0.01\,\mu m \leq a \leq 0.25\,\mu m$, except for iron grains, which they found to be six times smaller.

Seab & Snow (1989) found that this power law distribution was consistent with observations of ultraviolet extinction around red supergiants, although they found a minimum grain size of $0.08\,\mu m$. Le Sergeant d'Hendecourt & Lamy (1980) found a similar power law distribution for grain sizes, with an exponent of −3.85, for small interplanetary dust particles from a study of microcraters on lunar rocks.

Dominik et al. (1989) found that the size distribution in theoretical models of the dust driven winds around cool carbon stars is represented by a power law with an exponent of −5. While the grains in their model cover the same range of sizes as found by Mathis et al. (1977), the steeper size distribution in their model increases the total extinction due to grains by almost a factor of two.

4 Absorption and Scattering of Light by Small Particles

The monochromatic opacity due to absorption by grains is given by:

$$\kappa_{\text{abs}} = \frac{\pi \sum_a Q_{\text{abs}}\, n(a)\, a^2}{\rho} , \tag{8}$$

where $n(a)$ is the number of grains of size a per cubic centimeter, ρ is the total density of matter, and Q_{abs} is the efficiency factor for absorption. Note that if Q_{abs} is equal to one, then the opacity is just equal to the geometric cross section of a grain (πa^2) times the number of grains of that size, summed over the size distribution of the grains. The density appearing in Eq. (8) converts what would otherwise be a volume absorption coefficient with units of cm^{-1} into a mass absorption coefficient with units of cm^2g^{-1}. A similar definition applies to the opacity due to scattering from grains.

Mie theory provides an exact description of the interaction of small spherical particles with electromagnetic radiation (see Bohren & Huffman 1983). The exact treatment of scattering and absorption by small particles is only possible for a few other simple shapes. There is evidence, however, that particles in the interplanetary medium (Giese et al. 1978; Greenberg 1980), in the comae of comets (Greenberg & Hage 1990), and in micrometeorites collected in the atmosphere (Brownlee et al. 1976) may deviate significantly from spherical shape. It is, therefore, desirable to investigate the effects of shape on grain opacity.

Hage & Greenberg (1990) approximated non-spherical particles as aggregates of cubical subparticles in a larger cubic lattice. They found that very porous grains can produce significantly greater extinction compared to solid particles of the same total size.

Rouleau & Martin (1991) analyzed laboratory extinction measurements of amorphous carbon particles utilizing a model of aggregated particles represented by fractal clusters of subparticles. They found significant enhancement of the extinction in the infrared for these fractal clusters compared to spherical particles. They also found that a model based upon a continuous distribution of ellipsoids (CDE), to be discussed below, is a good representation of the extinction properties determined from their much more complex calculations of fractal clusters.

Huffman (1988) also emphasized the importance of shape effects in analyzing laboratory observations of extinction for a variety of different materials. He showed that the CDE approximation produces a much better fit to laboratory observations than does a model based upon isolated spheres.

Because of the general applicability of the CDE approach and its computational simplicity, it has been adopted here. In order to develop the CDE approximation, we must first present the fundamental results of Rayleigh scattering theory. Approximate results for particles of arbitrary shape which are small compared to the wavelength of light can be obtained from Rayleigh scattering theory (see Van de Hulst 1957:

$$Q_{\mathrm{abs}} = 4\,x\,\mathrm{Im}[\alpha] \quad \text{and} \quad Q_{\mathrm{sca}} = \frac{8}{3}\,x^4\,|\alpha| \,, \tag{9}$$

where $x = 2\pi\,a/\lambda$ and α is the complex polarizability per unit volume of the grain material, subject to the conditions

$$x \ll 1 \quad \text{and} \quad x\,|m| \ll 1 \,, \tag{10}$$

where $m = n + ik$ is the complex index of refraction of the grain material.

For spheres the polarizability is given by

$$\alpha = \frac{(\epsilon - 1)}{(\epsilon + 2)} \,, \tag{11}$$

where $\epsilon = m^2$ is the complex dielectric function of the grain material. For an ellipsoid (of volume V) with incident electric field parallel to a principal axis, the polarizability is

$$\alpha = V \frac{\epsilon - 1}{1 + L(\epsilon - 1)} \,, \tag{12}$$

where L is a geometrical factor dependent solely on the shape of the ellipsoid.

The CDE model assumes that the particles are ellipsoids, with all possible shapes represented with equal probability. This distribution of shapes includes long needles, thin plates, spheres, and all other possible ellipsoidal shapes. The CDE model also assumes that these ellipsoids are oriented randomly. The result of this double averaging over both shape and orientation is that the polarizability is given in the simple form (Bohren & Huffman 1983):

$$\ll\!\alpha\!\gg = \frac{1}{3}\frac{2\epsilon}{\epsilon - 1}\ \ln\epsilon - 2 \ . \tag{13}$$

Values of the polarizability obtained from Eq. (13) can then be used directly in the Rayleigh scattering formulae given above in Eq. (9) to find the scattering and absorption efficiencies for the continuous distribution of randomly oriented ellipsoids. For large x, where Rayleigh scattering approaches infinity because Eq. (10) is not satisfied, the CDE approximation is replaced by a standard Mie calculation for spheres.

5 Optical Properties of Materials of Interest

The condensation equilibrium calculations described previously show that the principal high temperature ($T > 600\,\mathrm{K}$) condensates will include a variety of magnesium and iron silicates, iron, carbon, and silicon carbide, depending upon the chemical composition. Specifically, silicates condense when the carbon to oxygen ratio (C/O) is less than one and, under conditions of extremely high density, even when $C/O > 1$. Carbon and silicon carbide only condense when $C/O > 1$, while iron condenses at any value of C/O. At lower temperatures, other materials begin to condense and the presence of composite grains needs to be considered.

A great variety of laboratory investigations of these materials have appeared in the literature in recent years. Huffman (1988, 1989) emphasized the difficulties in selecting appropriate laboratory values for use in astrophysical calculations. For example, he demonstrated that optical properties derived from bulk samples may differ significantly from those derived from observations of small particles. Shape effects of the particles produced in the laboratory may also significantly affect the results, if not accounted for in the analysis. Among the factors to be considered in selecting a particular set of optical properties are:

1. Were the grains produced in the laboratory in the same range of sizes and shapes as will be found in the astrophysical context of interest?
2. Is the chemical composition and crystal structure of the grains produced in the laboratory appropriate for the astrophysical context of interest?
3. Do the laboratory measurements include the full wavelength range of interest?
4. Does there exist a careful analysis of the laboratory data which presents results in a convenient form (e.g., index of refraction or dielectric function) for use in computing extinction efficiencies using the formulae of the previous section?

Studies of silicates include those of Day (1979), Dorschner et al. (1988), Draine & Lee (1984), Pollack et al. (1973), and Mooney & Knacke (1985). The optical properties deduced by Draine & Lee (1984) for an "astrophysical silicate" have been adopted for this study because of their completeness and relevance to the astrophysical situations of interest here. Ossenkopf et al. (1992) have also synthesized the optical properties of a dirty "astrophysical silicate" utilizing both astronomical observations and laboratory studies.

For iron, laboratory investigations by Leksina & Penkina (1967), Johnson & Christy (1974), Gorban et al. (1975), and Tanabe et al. (1983) exist. Unfortunately, none of these studies meet all the selection criteria defined above. For example, both Leksina & Penkina (1967) and Gorban et al. (1975) investigated the optical properties of bulk iron. Johnson & Christy (1974) and Tanabe et al. (1983) fail to cover the entire wavelength range of interest. The data of Gorban et al. (1975) has been adopted for this study because of the relative completeness of their spectral coverage. The dangers of extrapolating iron data to wavelengths not included in a particular data set are discussed by Ferguson (1992). Obtaining better data for the optical properties of small iron grains should be a high priority for the astronomical community.

Carbon grains present a particularly difficult problem because they can take so many different physical structures, with sometimes significantly different optical properties. For high temperature grains, the question is whether the grains form as amorphous carbon or in the crystal structure of graphite. Rouleau & Martin (1991) and Mathis (1988) summarize the existing evidence which supports the conclusion that carbon grains in the stellar environment will occur in the form of amorphous carbon.

Huffman (1988) emphasized the importance of particle clustering on interpreting laboratory measurements. Among the more complete studies of small particles are those of Koike et al. (1980), Bussoletti et al. (1987), Edoh (1983), and Tanabe et al. (1983). Draine & Lee (1984) synthesized a variety of laboratory studies of graphite over a wide range of wavelengths. Rouleau & Martin (1991) carefully analyzed the data of Bussoletti et al. (1987) to obtain optical constants for amorphous carbon in the wavelength range of interest here. The results of Rouleau & Martin (1991) for the BE1 sample of Bussoletti et al. (1987) were used in the study reported here.

The optical properties of alpha-SiC have been studied by Phillip & Taft (1960), Stephens (1980), Friedmann et al. (1981), and Borghesi et al. (1985). The thorough synthesis and analysis of laboratory data of Pégourié (1988) has been utilized in this study.

6 Results

6.1 Shape Effects

The effect of grain shape, as shown by differences between extinction computed for spherical grains and with the CDE approximation, affect grains of different compositions in radically different ways. Figure 1 shows these effects for the four grain species considered here. In each case the extinction cross section for the grain species under consideration is defined as

$$C_{\mathrm{ext}} = Q_{\mathrm{ext}}\, \pi\, a^2 \ , \tag{14}$$

where $Q_{\mathrm{ext}} = Q_{\mathrm{abs}} + Q_{\mathrm{sca}}$.

Shape effects are generally unimportant for silicate grains, compared to the other grain species, although the extinction in the CDE approximation is fifty percent greater than the Mie result for wavelengths longer than $10\,\mu$m. The

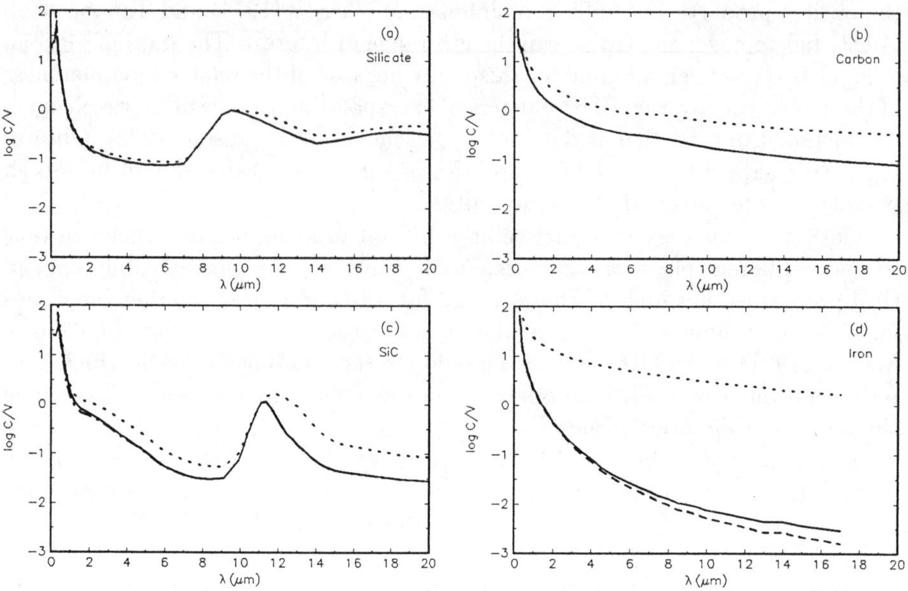

Fig. 1. Cross section for extinction per unit volume computed with Mie theory (solid curves), spherical Rayleigh particles (dashed curves), and the CDE approximation (dotted curves) for grains in the middle of the Mathis et al. (1977) size distribution.

CDE result for amorphous carbon shows a steady enhancement throughout the infrared of up to a factor of five, compared to spherical grains.

Because of the differing geometrical cross sections of grains presented to the incoming radiation field by CDE grains, sharp spectral features become more diffused in this approximation. This effect is most pronounced in the 10 μm feature of silicon carbide. This effect has been noted by Bohren & Huffman (1983) and, as they note, usually produces a better fit with laboratory data.

Shape effects are most pronounced for metallic grains. Figure 1 shows that the CDE result for iron grains is as much as three orders of magnitude greater than the Mie result. Computationally, the effect is caused by the presence of a logarithm in the computation of the polarizability in Eq. (13). When the imaginary part of the index of refraction becomes large, this term also becomes very large. It cannot be argued that the large index of refraction causes a break down in the Rayleigh approximation used in the CDE calculation, because the extinction of spheres computed with the Rayleigh approximation is actually less than the Mie result. Remember also that iron grains are smaller than the other grains considered here.

Bohren & Huffman (1983) find a similar effect for aluminum, which they show is necessary to explain laboratory extinction observations. Greenberg & Hong (1975) also find that the absorption efficiency of iron needles can be up to ten times greater than the absorption efficiency of spheres.

As a cautionary note, it should be remembered that the optical properties of iron used here are derived from observations of bulk samples. If the index of refraction of small iron particles turns out to be significantly smaller than observed for bulk iron, the extinction enhancement found for the CDE result would also be greatly reduced. This reservation once again points out the importance of obtaining more appropriate optical constants for iron.

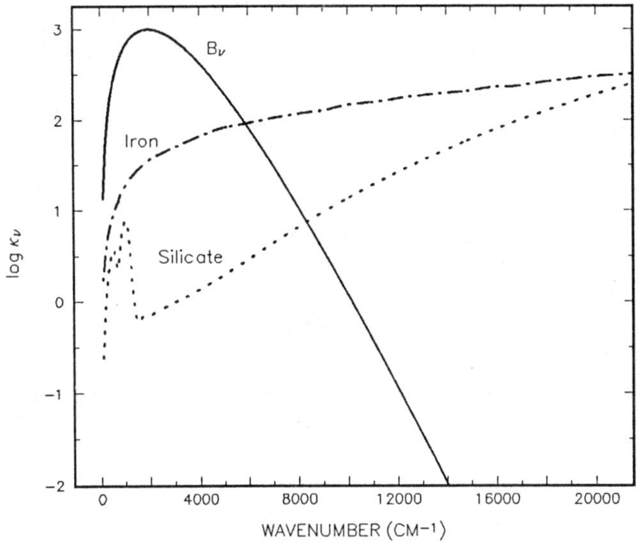

Fig. 2. Monochromatic absorption coefficients of grains for an oxygen-rich composition at $T = 1000$ K and $\rho = 10^{-10}$ gm cm^{-3}. The Planck function B_ν, which is the weighting function for the Planck mean opacity, is shown for comparison.

6.2 Results for $C/O < 1$

The total opacity of grains, as a function of wavenumber (defined as $1/\lambda$), for the solar composition (Anders & Grevesse, 1989) is shown in Fig. 2 for a representative value of temperature and density. Because of the CDE enhancement of the iron cross section, iron grains dominate the opacity at all wavelengths compared to silicate grains. Indeed, since iron condenses at slightly higher temperatures than silicates, iron dominates the opacity whenever grains are present at all temperatures and densities. This result is illustrated in Fig. 3, which shows the Planck mean opacity at three temperatures as a function of density.

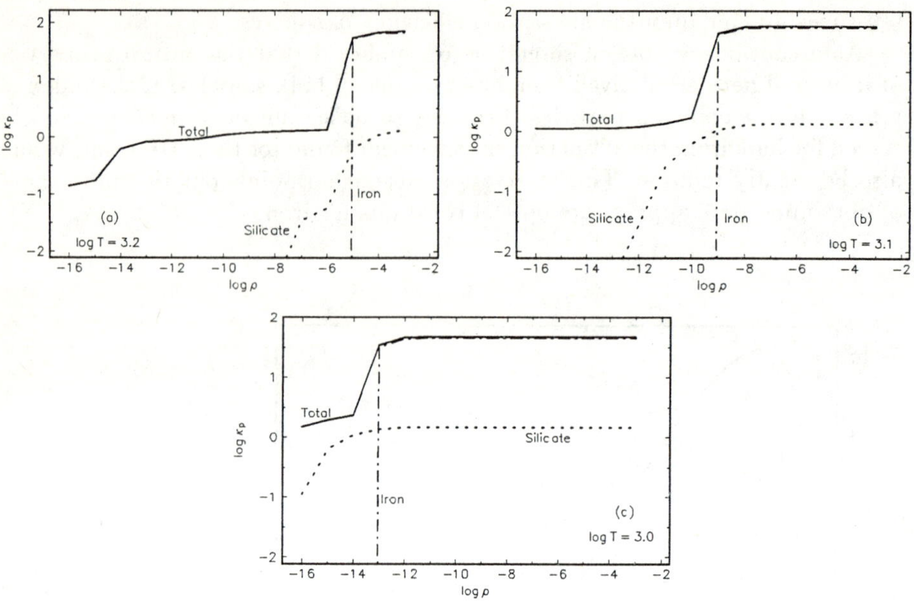

Fig. 3. The Planck mean absorption coefficient as a function of the density for three temperatures for an oxygen-rich composition. Vertical curves indicate that the species appears or disappears at that density.

6.3 Results for $C/O > 1$

Similar results are obtained for a carbon-rich mixture ($C/O = 1.05$). Figures 4 and 5 show that iron dominates the opacity for this composition as well, as would be expected because its condensation is not strongly affected by changes in the C/O ratio. The fact that silicates are present at all may seem somewhat surprising. Silicates condense under these conditions, despite the unfavourable C/O ratio, because enough carbon condenses to drive the C/O ratio of the gas phase below one.

Unlike the case for oxygen-rich compositions, iron does not dominate the Planck mean opacity under all conditions. Since carbon grains can form at lower densities than iron grains, the opacity at low densities is dominated by the opacity of carbon grains.

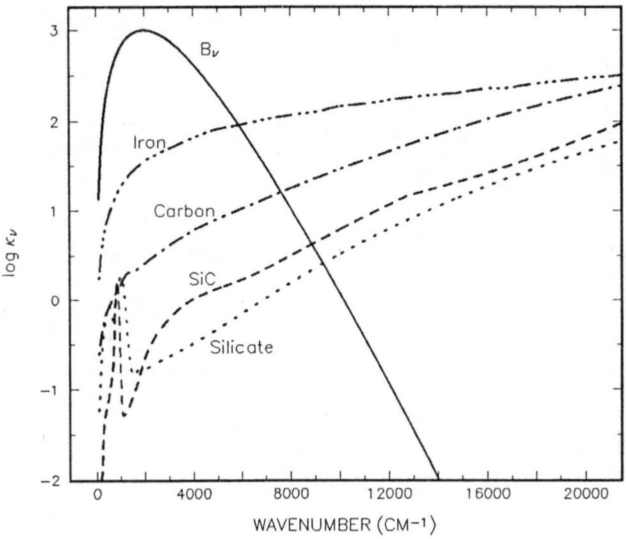

Fig. 4. Monochromatic absorption coefficient of grains for a carbon-rich composition at $T = 1000$ K and $\rho = 10^{-10}$ gm cm^{-3}. The Planck function B_ν is shown for comparison.

7 Discussion and Conclusion

The treatment of grain opacity discussed here, including the accurate treatment of condensation equilibrium, the use of a realistic distribution of grain sizes, the consideration of shape effects through the use of the CDE approximation, and the use of the most reliable optical constants available for the major condensing species, allows for the most reliable possible computation of the opacity at low temperatures. The inclusion of accurate molecular opacities, treated with the opacity sampling technique, and atomic line and continuum absorbers allows accurate opacities to be computed in the temperature range from 600 K to 12,000 K for a wide range of densities and chemical compositions. Tables of Rosseland or Planck mean opacities are available from the first author for a wide range of situations via the Internet at the address DRA @ TWSUVM.UC.TWSU.EDU.

Many factors enter into the decision of whether to use the Rosseland or the Planck mean opacity in a given situation. Clearly, model building is greatly simplified when the Rosseland mean is utilized. Nevertheless, it is important to keep in mind the nature of the average opacity one is using. Because the Rosseland mean opacity is a harmonic mean, it most strongly weights those wavelengths with small opacity. The linear Planck mean, on the other hand, more strongly weights those wavelengths with the greatest opacity. Because of the nature of the assumptions made in the derivation of the Rosseland mean, one would expect it to lead to a good approximation of the radiative transfer when the optical depth of the medium under consideration is quite large. However, if the optical depth of the medium is small, then the radiative transfer will

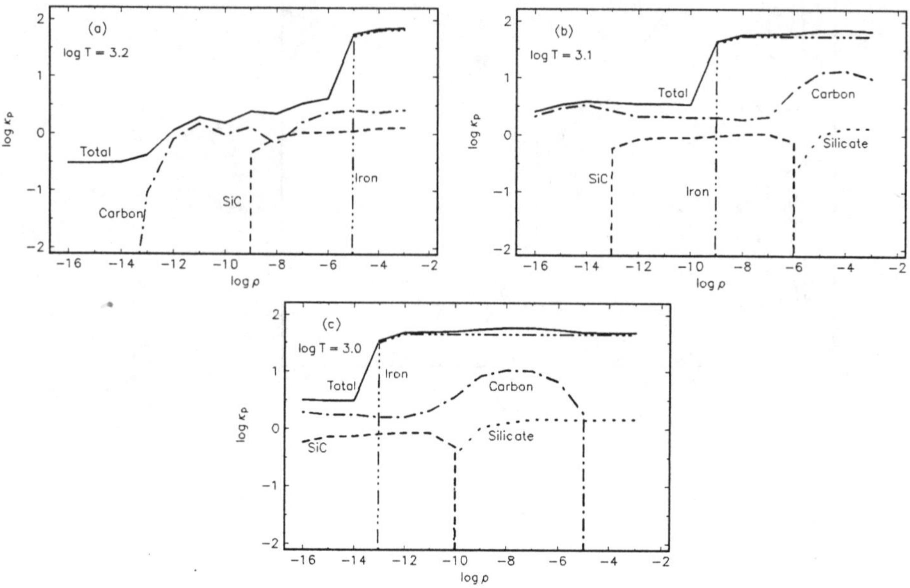

Fig. 5. The Planck mean absorption coefficient as a function of the density for three temperatures for a carbon-rich composition.

become more sensitive to the wavelengths where the opacity is highest. Hence, the Planck mean may lead to a more accurate treatment of the radiation field under those conditions than the traditional diffusion approximation (see Collins, 1989; Mihalas 1979).

Figure 6 illustrates when these considerations will be important. By comparing the ratio of the Rosseland to the Planck mean opacity, we can determine when the radiative transfer of an optically thin medium will be affected by the choice of the mean to be used. At low temperatures, where grains dominate the opacity, the two means produce virtually identical results. This result should be expected because the opacity of grains is a relatively flat and smooth function of wavelength. Similarly at high temperatures, where continuum absorption of H and H^- dominates, the two means approach each other. However, at intermediate temperatures, where molecular line sources dominate the opacity, the two means can be orders of magnitude apart. It is in this temperature regime where these considerations will become most important.

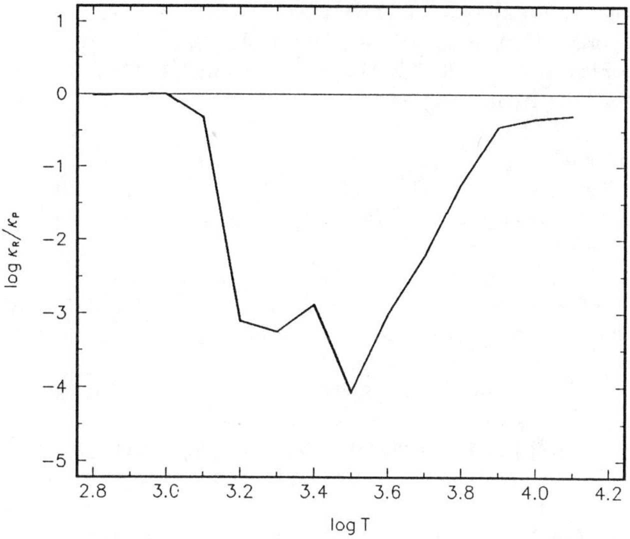

Fig. 6. The ratio of the Rosseland mean absorption coefficient to the Planck mean absorption coefficient. A reference line is shown for a ratio of one.

References

Alexander D.R., 1975, *Astrophys. J. Suppl.*, **29**, 363

Alexander D.R., Johnson H.R., Rypma R.L., 1983, *Astrophys. J.*, **272**,773

Anders E., Grevesse N., 1989, *Geochim. Cosmochim. Acta*, **53**, 197

Bohren C.F., Huffman D.R., 1983, *Absorption and Scattering of Light by Small Particles*, (New York: John Wiley & Sons), p. 82ff

Borghesi A., Bussoletti E., Colangeli L., 1985, *Astron. Astrophys.*, **142**, 225

Bussoletti E., Colangeli L., Borghesi A., Orofino V., 1985, *Astron. Astrophys. Suppl.*, **70**, 257

Brownlee D.E., Tomandl D.A., Hodge P.W., 1976, in *Interplanetary Dust and Zodiacal Light*, eds. H. Elsässer and H. Fechtig, (Berlin: Springer-Verlag), 279

Cameron A.G.W., Pine M.R., 1973, *Icarus*, **18**, 377

Collins II G.W., 1989, *The Fundamentals of Stellar Atmospheres*, (New York: W.H. Freeman & Co.)

Day K.L., 1979, *Astrophys. J.*, **234**, 158

Dominik K., Gail H.-P., Sedlmayr E., 1989, *Astron. Astrophys.*, **223**, 227

Dorschner J., Friedmann C., Gürtler J., Henning T., 1988, *Astron. Astrophys.*, **198**, 223

Draine B.T., Lee H.M., 1984, *Astrophys. J.*, **285**, 89

Edoh O., 1983, Ph. D. thesis, University of Arizona, Tucson Arizona

Ferguson J. W., 1992, Masters' thesis, Wichita State University, Wichita, Kansas

Fix J.D., 1969, *Mon. Not. R. Ast. Soc.*, **146**, 37

Friedmann C., Gürtler J., Schmidt R., Dorschner J, 1981, *Astrophys. Space Sci.*, **79**, 405

Huffman D.R., 1988, in *Experiments on Cosmic Dust Analogues*, eds. E. Bussoletti, C. Fusco, and G. Longo (Dordrecht: Kluwer Academic Publishers), 25
Huffman D.R., 1989, in *Interstellar Dust*, eds. L.J.Allamandolla and A.G.G.M. Tielens, (Dordrecht: Kluwer Academic Publishers), 329
Gaustad J.E., 1963, *Astrophys. J.*, **138**, 1050
Giese R.H., Weiss K., Zerull R.H., Ono T., 1978, *Astron. Astrophys.*, **65**, 265
Gorban N. Y., Stashcuk V.S., Cherromorets M.F., 1975, *Opt. Spectrosc.*, **38**, 568
Greenberg J.M., 1980, in *Solid Particles in the Solar System*, eds. I. Halliday and B.A. McIntosh, (Dordrecht: Reidel), 343
Greenberg J.M., Hage J.I., 1990, *Astrophys. J.*, **361**, 260
Greenberg J.M., Hong S.S., 1975, in *The Dusty Universe*, eds. G.B. Field and A.G.W. Cameron, (New York: Neale Watson Academic Publications, Inc.), 131
Hage J.I., Greenberg J.M., 1990, *Astrophys. J.*, **361**, 251
Grossman L., 1972, *Geochim. Cosmochim. Acta*, **36**, 597
Johnson P.B., Christy R.W., 1974, *Phys. Rev. B*, **9**, 5056
Jørgensen U.G., Johnson H.R., Nordlund Å, 1992, *Astron. Astrophys.*, **261**, 263
Kellman S.A., Gaustad J.E., 1969, *Astrophys. J.*, **157**, 1465
Knacke R.F., 1968, *Nature*, **217**, 44
Koike C., Hasegawa H., and Manabe A., 1980, *Astrophys. Space Sci.*, **67**, 495
Le Sergeant d'Hendecourt L.B., Lamy P.L., 1980, *Icarus*, **43**, 350
Leksina I.Y., Penkina N.V., 1967, *Phys. Metals Metallography*, **23**, No. 2, 344
Mathis J.S., 1988, *Astro. Lett. and Comm.*, **26**, 239
Mathis J.S., Rumpl W., Nordsieck K.H., 1977, *Astrophys. J.*, **217**, 425
Mihalas D., 1979, *Stellar Atmospheres*, (San Francisco: W.H. Freeman & Co.)
Mooney T., Knacke R.F., 1985, *Icarus*, **64**, 493
Ossenkopf V., Henning T., Mathis J.S., 1992, *Astron. Astrophys.*, **261**, 567
Pégourié B., 1988, *Astron. Astrophys.*, **194**, 335
Phillip H.R., Taft E.A., 1960 in *Proc. Conference on Silicon Carbide*, ed. J.S. O'Connor, (Boston: Pergamon), 366
Pollack J.B., McKay C.P., Christofferson B.M., 1985, *Icarus*, **64**, 471
Pollack J.B., Toon O.B., Khare B.N., 1973, *Icarus*, **19**, 372
Rouleau F., Martin P.G., 1991, *Astrophys. J.*, **377**, 526
Seab C.G., Snow T.P., 1989, *Astrophys. J.*, **347**, 479
Scholz M., Tsuji T., 1984, *Astron. Astrophys.*, **130**, 11
Sharp C.M., Huebner W.F., 1990, *Astrophys. J. Suppl.*, **72**, 417
Stephens J.R., 1980, *Astrophys. J.*, **237**, 450
Tanabe T., Nakada Y., Kamijo F., 1983, *Publ. Astron. Soc. Japan*, **35**, 397
Van de Hulst H.C., 1957, *Light Scattering by Small Particles*, (New York: John Wiley & Sons, Inc.), p. 63ff

From Molecules to Grains

Erwin Sedlmayr

Institut für Astronomie und Astrophysik
Technische Universität Berlin
Hardenbergstr. 36, D-10623 Berlin, Germany

1 Introduction

Cosmic dust is an ubiquiteous component of the interstellar medium, the presence of which has severe bearings both upon the observations and upon the physical description and realistic modelling of astronomical objects. In most dusty objects, the grain component is intimately interwoven with the gas component and the radiation field, and, hence strongly affects the thermodynamical, the hydrodynamical and the chemical structure of the object. Therefore any consistent description of the astrophysical dust complex has to rely upon a treatment where these aspects are properly taken into account. In general this complex comprises the following problems:

- Formation of small stable molecular clusters out of the gaseous phase
- Growth of these clusters to macroscopic specimens (primary condensates: grains, plates, etc.)
- Destruction of grains (e.g. thermal evaporation, sputtering, shattering), and
- Physical and chemical processing of already existing grains (e.g. coagulation, chemical and thermal transformation, etc....).

In this contribution, the basic concepts for treating astrophysical grain condensation are surveyed by discussing the potentials, limitations and shortcomings of the various approaches. As it is especially dedicated to the question of the *molecule-solid-transition*, particular emphasis is given to the aforementioned first and second problem, by which a physical two-step process is suggested, involving the gas phase formation of small stable clusters and their subsequent growth to macroscopic specimens, which according to their specific shape result in 3-dimensional solid grains, two-dimensional (flat or curved) plates or even in fractal like structures. We confine our scope to the investigation and physical description of *primary grain condensation*, i.e. the formation of the first condensates emerging in astrophysical environments. As in the context of grain formation, the other problems – *grain destruction* and *grain processing* – are of secondary nature, in a sense that they concern already existing dust particles, we refrain from a discussion of these important subjects (e.g. Greenberg 1989; Jenkins 1989).

2 The astronomical objects of dust formation

In order to give some frame for the dust forming objects and the range of system parameters which determine the general conditions of astrophysical dust formation (see Gehrz 1989), Fig. 1 displays the total average mass loss rate of various astronomical objects in the disk of our galaxy showing dust formation.

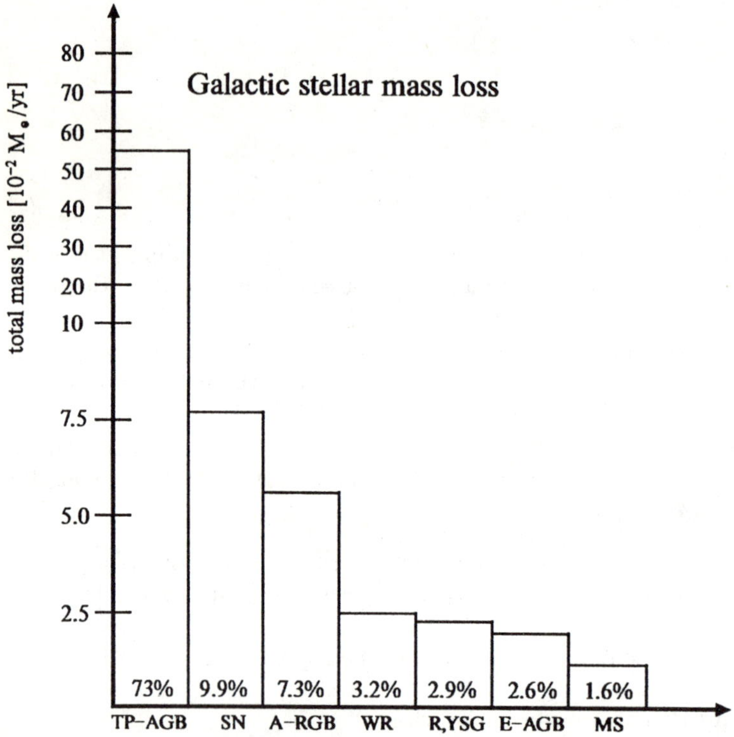

Fig. 1. Total galactic mass loss (solar masses per year) of various classes of stellar objects: TP-AGB: tip-AGB objects, SN: supernovae, A-RGB: asymptotic red giant branch, WR-stars, R, YSG: red and yellow supergiants, E-AGB: early AGB objects, MS: main sequence stars. The indicated numbers are collected from the literature and are inferred from a rather inhomogeneous material (Weinzierl 1991). Despite these inherent uncertainties, they provide a confidential order of magnitude and a clear trend which is expressed at the percentage of their relative contribution to the total mass loss.

Inspection of Fig. 1 clearly shows that nearly 90 % of the total stellar mass loss originates from cool high luminosity stars, i.e. red giants and supergiants, in particular from AGB objects, which provide by far the largest contribution to the stellar mass injection rate into the interstellar medium. The winds of TP-

AGB-stars show a large fraction of condensates, with average values of the mass loss ratio

$$\frac{\dot{M}_{dust}}{\dot{M}_{gas}} = 10^{-3} \cdot \begin{cases} 6.3 & M-stars \\ 4.5 & C-stars \\ 2.0 & S-stars \end{cases}$$

for the different spectral types (e.g. Knapp 1985; Jura 1986).

Beside these main sources of galactic dust production also supernovae and novae outflows, B[e]-stars and even some WR-stars exhibit circumstellar dust formation. As these phenomena, however, are episodic or of minor importance in our context (cf. Fig. 1) we focus the further discussion on the problem of grain formation in the expanding circumstellar shells of cool late giants and supergiants (in this context see the excellent review on the mass loss mechanisms in evolved stars of Lafon & Berruyer 1991).
This restriction is also motivated by model calculations including dust condensation applied to various dusty systems.

Table 1. The condensation timescale for different astrophysical situations. $n_{<H>}$ total density of hydrogen in the gas, $t_{\rm typ}$ typical hydrodynamical timescale in the system, $t_{\rm cond}$ condensation timescales for SiO, Fe and C.

Object	**Temperature** [K]	log $n_{<H>}$ [cm^{-3}]	log $t_{\rm typ}$ [s]	log $t_{\rm cond}$ [s] **SiO**	**Fe**	**C**
			interstellar medium			
H II regions	10000	$2-3$	≈ 12	∞	∞	∞
Intercloud medium	10000	≈ -1	?	∞	∞	∞
Diffuse clouds	100	≈ 2	?	> 18	> 18	> 18
Dark clouds	$10-20$	≈ 4	?	> 18	> 18	> 18
Molecular clouds	50	≈ 6	≈ 15	> 18	> 18	> 18
Compact H II reg.	$100-1000$	$3-4$	≈ 11	≈ 13	≈ 14	$13-14$
			explosive ejection of matter			
Novae	from 10000 down	$\leq 8-10$	$6-7$	$\approx 6-8$	$7-8$	$7-8$
Supernovae	from 10000 down	$7-9$	$7-8$	$\approx 7-9$	$8-10$	$8-10$
			massive winds of giant stars			
Cool winds	from 2000 down	$7-10$	≈ 8	$\approx 7-9$	≈ 8	≈ 8
Hot winds	from 20000 down	$6-8$	≈ 6	$\approx 8-10$	$\approx 9-10$	≈ 9

Table 1 displays the typical system parameters and timescales which for each object determine the possibility and the efficiency of grain formation. For significant condensation of a species to take place, its characteristic condensation time t_{cond} has to be shorter than the time-period t_{typ} for which favourable conditions for grain formation hold during the evolution of the considered object. From this table one concludes that only winds of cool stars, and novae and supernovae fulfil the condition $t_{cond} < t_{typ}$ necessary for effective condensation of species like {SiO, MgO, Fe, ... } and { carbon compounds }, which are expected to play a key role in primary grain formation in oxygen rich and carbon rich environments, respectively (see Sect. 4.2).

3 Astrophysical dust condensation

3.1 Embedding of the dust formation problem in physics and chemistry

Like any process of physical structure formation, also the problem of cosmic dust formation is closely related to various physical and chemical processes which either determine the physical and chemical initial and boundary conditions for the considered molecule-grain transition or are influenced by the local and global feed back of the grain formation process to the overall system. For this reason any reliable description of astrophysical grain formation basically has to refer to the couplings sketched in Fig. 2, from which the necessary methodical ingredients and the induced physical effects can be inferred.
In a general situation, many of these aspects are still unsufficiently explored or even unknown. Thus, a real improvement of this unsatisfactory situation requires extensive future work, in particular

- specific laboratory experiments,
- detailed theoretical calculations, and
- suitably designed astrophysical observations.

Regarding these aspects, at several places important work is going on which supports the hope that significant progress in the reliable description of astrophysical dust formation and the consistent quantitative modelling of dust forming objects will be on line in the near future.

3.2 The wind scenario

The basic situation of primary grain condensation can be best illustrated for a stellar outflow, where initially hot matter expands and subsequently cools and dilutes along its way outwards. Hence, both the gas temperature and the matter density show a monotonic decrease which causes the plasma to recombine and at increasing distance diatomic, polyatomic and finally complex molecules to form. At some distance, where the temperature is well below 1500 K certain atoms and molecules combine to stable clusters which by further addition of similar or different species grow to homogeneous or heterogeneous macroscopic specimens.

Fig. 3 shows a schematic sketch of this basic situation encountered in the flows of cool giants and supergiants, novae and supernovae. Though, Fig. 3 refers

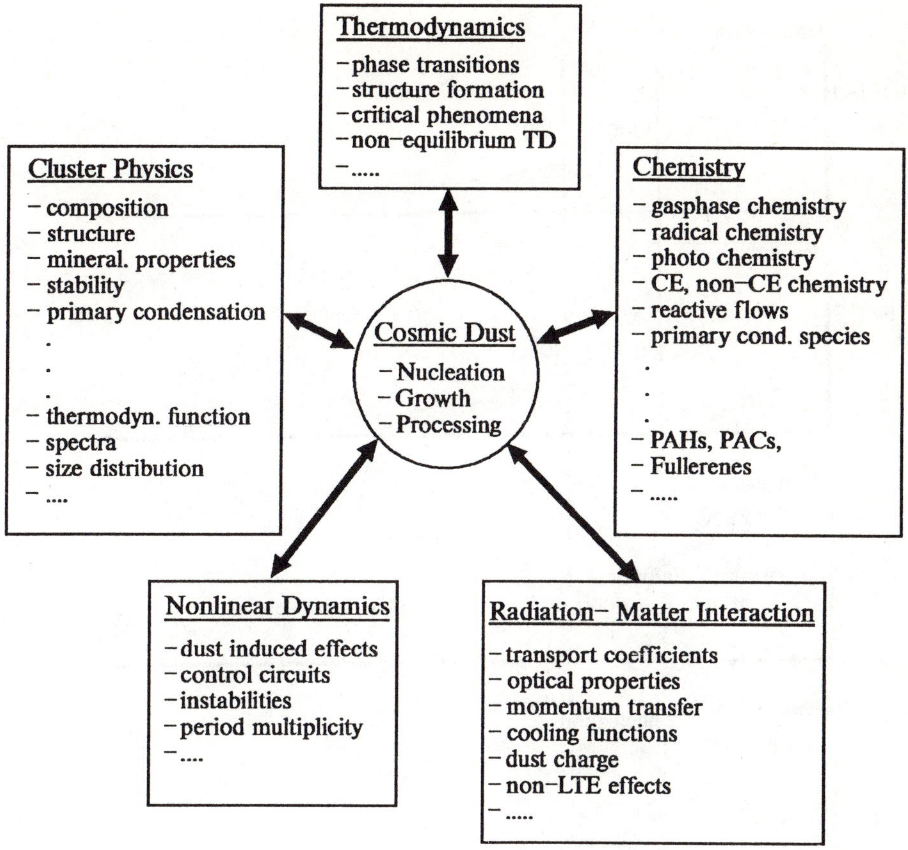

Fig. 2. Basic couplings of the cosmic dust complex to various physical and chemical disciplines which in principle have to be considered in a reliable description of dust nucleation and growth and for the consistent modelling of a dust forming object.

to the case of stationary outflows (e.g. stationary stellar winds, novae and supernovae expansions) similar scenarios hold for non-stationary expansions (e.g. circumstellar shells of Miras and LPVs). As cluster formation and grain growth are local phenomena, the efficiency of which is controlled by the local thermodynamical and chemical conditions, dust condensation in these objects depends strongly e.g. upon their specific shock structure and the related time scales (see Fleischer et al. 1992).

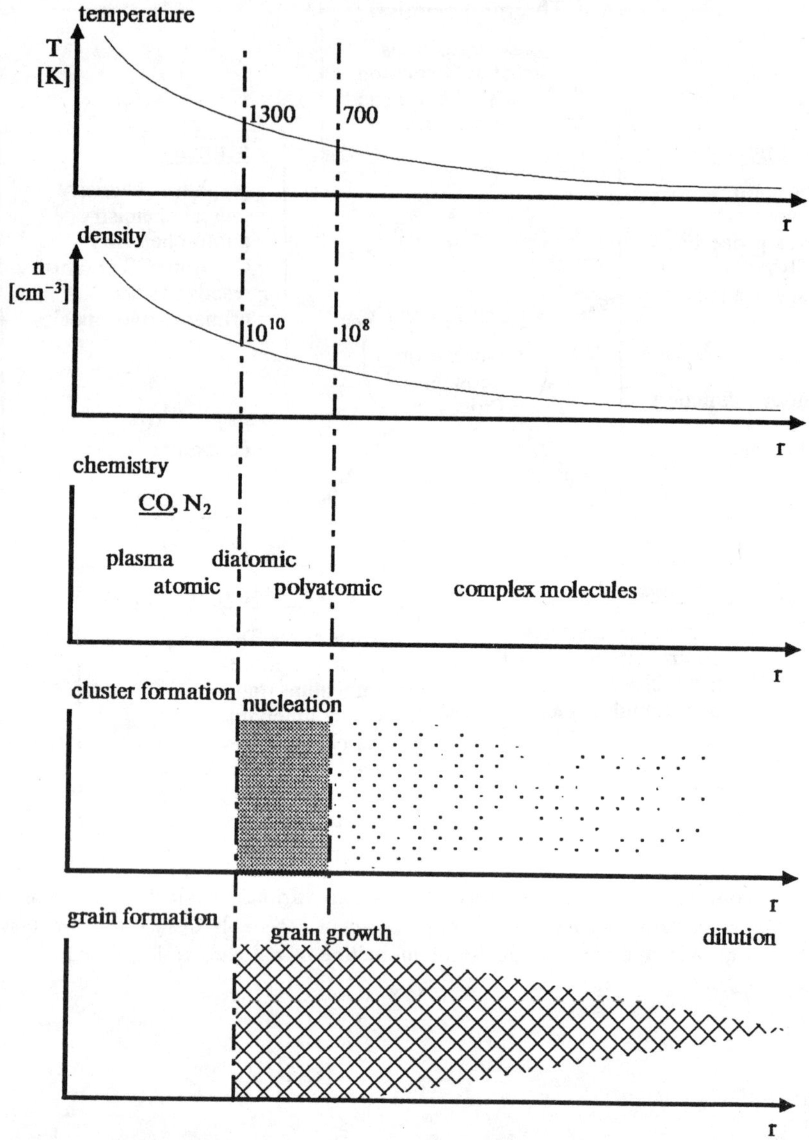

Fig. 3. Schematical radial course of temperature and density of an expanding stellar wind (cool stationary stellar winds, novae and supernovae) showing the evolution of a complex chemistry, and dust nucleation and growth. The nucleation regime is confined to a rather narrow zone with temperature between 1300 K ... 700 K having a relatively sharp inner edge. The growth regime extends from the nucleation regime outwards to a distance where either the condensable material will be totally consumed by grain growth or where it becomes too diluted for having still significant collision rates.

3.3 The selfconsistent wind model

The basic ingredients and couplings for the selfconsistent modelling of a dust forming stellar wind are depicted in Fig. 4.

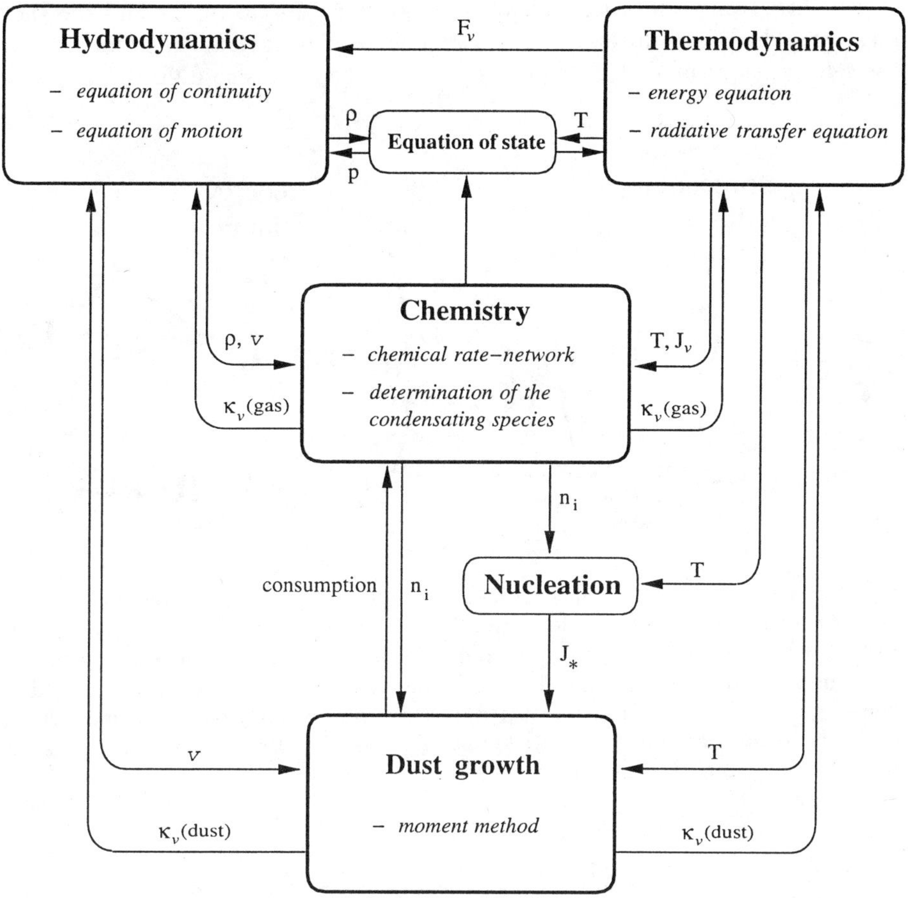

Fig. 4. Basic couplings for the selfconsistent description of dust forming stellar wind.

The upper part comprises the usual stellar atmosphere problem, whereas the lower part represents the dust formation complex, consisting of (i) a chemical subsystem which determines the dust forming species and their consumption due to condensation, and two complexes describing (ii) the formation of stable clusters out of the gas phase, and (iii) the subsequent growth of these clusters to macroscopic particles. Besides the specific physical nature of the primary dust components and their size distribution function, this box also includes the calculation of realistic transport-coefficients which provide the coupling to the hydrodynamical and the thermodynamical structure of the shell.

In this way both for momentum and energy, control circuits are induced which explain why in case of stationary winds, the dust forming shell remains stable (see Fig. 5 in Goeres & Sedlmayr 1992), in contrast to non-monotonic situations, like Miras and LPVs or RCrB stars where dust induced instabilities develop. It is just this strong interaction between the evolving dust complex and the shell structure which introduces a pronounced nonlinear behaviour and which makes the selfconsistent modelling of most dust forming systems so difficult.

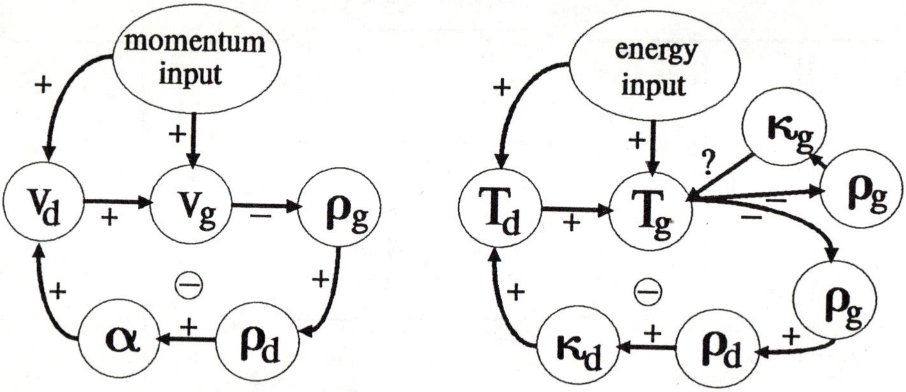

Fig. 5. Stabilising control circuits for momentum and energy induced by dust condensation in a stellar wind. Quantities with index g(d) refer to the gas (dust)–component, respectively. The negative sign of the product of the influences (symbolised by the arrows) in a closed loop guarantee dynamical and thermal stability of the system. The side chain of the energy circuit indicated by ? depends strongly on the actual form of κ_g. Thus, its influence upon T_g can be positive or negative. However, this fact does not alter the above general conclusion.

4 The role of chemistry

4.1 The primary condensates

Basically grain formation can be conceived as a chain of chemical reactions which result finally in macroscopic products. This is at least true for the first phase of the molecule-grain transition where suitable molecules combine to form small clusters which by additional chemical reactions grow to their critical size (see Sect. 5.2).

In order to find out which molecules are important for this initial step of cluster condensation, usually three heuristic criteria are applied to the system considered (e.g. Gail & Sedlmayr 1986; 1987):

i) *The clustering molecules have to be abundant species in the gas phase*
This criterion is motivated by the observational fact that circumstellar dust shells show large optical depths caused by appropriate dust components. Then the observed large mass loss ratios (cf. Sect. 2) can only be explained if a large fraction of the heavy elements take part in the condensation process.

ii) *The clustering molecules must not be blocked by high bond energies*
As a chemical process cluster formation involves reactive species. Thus all molecules having a bond energy larger than the reaction energy required for cluster formation behave inert and can be excluded. This concerns especially CO and N_2 which due to their exceptionally high bond energies are blocked for the condensation process.

iii) *The resulting clusters have to allow for high temperature condensates*
This criterion is motivated by the observations from which a temperature range from 1300 K to 700 K for the dust formation zone is deduced.

Applying these criteria to a dust forming shell having a reliable chemical composition (e.g. M-stars or C-stars) only a very small number of species remains as possible primary condensates (see Sect. 4.2).

4.2 The CO-molecule

Most important in this context is the role of the CO-molecule. Due to its exceptionally high bond energy ($E_{bond} \sim 11.09eV$) it can only be dissociated by rather energetic photons. Thus in the absence of dissociating UV-fields for an element compositions $\epsilon_O > \epsilon_C$ ($\epsilon_O < \epsilon_C$) (ϵ_X denotes the abundance of element X) nearly all *carbon* (*oxygen*) is consumed by CO formation with the consequence that in the case of M-stars (C-stars) practically no additional carbon (oxygen) compounds can be formed. This fact does not only determine the possible type of chemistry developing in such shells but also again reduces the number of possible primary condensating species, the most important of which are (e.g. Gail & Sedlmayr 1986; 1987):

- SiO, MgO, MgS, Fe — for oxygen-rich compositions (M-stars)
- C_2H_2, C_2H,... MgS, SiC, Fe — for carbon rich compositions (C-stars)

This clear picture is considerably modified if UV photons capable to dissociate CO are present. In this case in both *carbon and oxygen* rich environments always significant amounts of C atoms, C^+ ions *and oxides* are found, which give rise to a complex carbon-oxygen chemistry (see Beck et al. 1992). Such cases, however, have not been sufficiently studied in the context of circumstellar grain condensation.

4.3 The state of equilibrium

The last remark about the role of the CO molecule raises the question of the appropriate level of description for the chemistry. If the shell chemistry is dominated by kinetic reactions detailed balancing holds and consequently the assumption of chemical equilibrium (CE) is justified. Hence CE is a reasonable assumption for all situations where UV-photons can be neglected and where the expansion velocities are sufficiently low that the local expansion time scale t_{exp} is large compared to the reaction rates involved. This is true for the innermost regions of the shells of chromosphere free C- and M-stars inside the sonic point. Outside the sonic point the increasing supersonic velocity field causes a significant decrease of t_{exp} which requires expansion effects to be taken into account in the chemical reaction network, causing pronounced non-equilibrium effects, which in the extreme case of free expansion result in a frozen chemistry. If chromospheres are present in any case photoreactions dominate. In these situations the entire shell departs from chemical equilibrium and the full dynamical reaction network has to be solved to obtain reliable molecular concentrations (Beck et al. 1992; Goeres et al. 1988).

5 Cosmic grain formation

From a microscopic point of view grain formation out of the gaseous phase can be conceived as a network of chemical reactions which under favourable conditions leads to molecules of increasing complexity which finally grow to macroscopic specimens.

Such a fundamental microscopic description of grain formation is currently far from being tractable, not only due to the huge reaction network required to produce grains up to size of 1 μm, but also because of the lack of most relevant input data (see for instance the set of kinetic equations in Donn et al. 1981).

However, detailed inspection of the thermodynamic functions and of the properties of the small clusters involved shows, that under the usual conditions of astrophysical grain formation a well defined condensation barrier exists given by a corresponding *critical cluster size*. This condensation barrier allows the molecule-solid phase transition roughly to be considered as a two-step process:

1. *the formation of critical clusters*
2. *the growth of these clusters to macroscopic grains.*

5.1 The critical cluster

When molecules of a condensating species in a supersaturated situation combine to small clusters by subsequent molecule addition finally a *first* thermodynamically stable configuration is achieved (see Feder et al. 1966). This first marginally stable product is called the *critical cluster*. Along the coordinate axis N in cluster space (N is the number of monomers, i.e. the number of the basic constituents of the particle) the critical cluster is defined as that cluster of size $N_\star$ for which the *enthalpie of formation* $\Delta_f G_N^{\ominus}$ has its maximum value for given temperature and density. Hence, by definition all clusters being smaller than $N_\star$ are thermally unstable and thus preferentially will evaporate and all clusters being larger than $N_\star$ are thermally stable and thus preferentially will grow. In this way, the critical cluster naturally devides the cluster space into two disjunct regions $N < N_\star$ and $N > N_\star$, respectively, separated by a *nucleation barrier*, usually referred to as the *nucleation regime*, and the *growth regime*, respectively. The physical description of each requires different assumptions and theoretical techniques.

In astrophysical problems the critical clusters for the condensates to be expected are rather small, with N being typically 5 ... 20 at temperature where efficient nucleation takes place. Then, $N_\star$ gives also the order of magnitude for the number of chemical reactions required for a cluster to surmount the nucleation barrier.

For the conditions encountered in the innermost regions of the stellar outflows, the chemical reaction timescales usually are very small compared to the timescales governing the hydrodynamical and thermodynamical evolution of the medium. Thus, the formation of critical clusters in stellar outflows usually can be considered as a stationary process (Gail & Sedlmayr 1988; Rossi & Benevides-Soares 1988). This statement even holds for Miras and LPVs where avalanche nucleation occurs in the postshock region, where the chemical reaction timescales are small compared to the oscillation period (Gauger et al. 1990; Fleischer et al. 1992).

As by definition every critical cluster has the tendency to grow, its local growth rate $J_\star$ determines the number of grains emerging in a given environment. Thus, it is a central aim of all condensation theories to find out the critical clusters for a given thermodynamical situation and to give a quantitative expression for their growth rates.

In case of thermal equilibrium the partial pressure p_N of homogeneous clusters of size N is simply given by the law of mass action

$$p_N = p_1^N exp\{-\frac{\Delta_f G_N^{\ominus}}{kT}\}$$

with p_1 being the partial pressure of the monomers in the gaseous phase (in units of standard pressure), T the gas temperature, and k the Boltzmann constant. Under the conditions prevailing in circumstellar environments the actual value p_N is dominated by the exponential function, which at $N = N_\star$ has a pronounced minimum. For this reason, the critical cluster can be seen as a narrow *bottle neck* which the basically unstable small clusters have to surmont along their way of formation, in order to arrive at the growth regime. It is just these properties – *marginal stability* and *minimum number density* – which give the critical cluster a key role in all theories describing dust formation.

5.2 The nucleation regime

No generally applicable physical treatment of the nucleation process in astrophysical environments is yet available. Thus, depending on the actual situation and on the specific nature of the clusters to be expected, methodically different descriptions have been suggested for an approximate modelling of the nucleation regime.

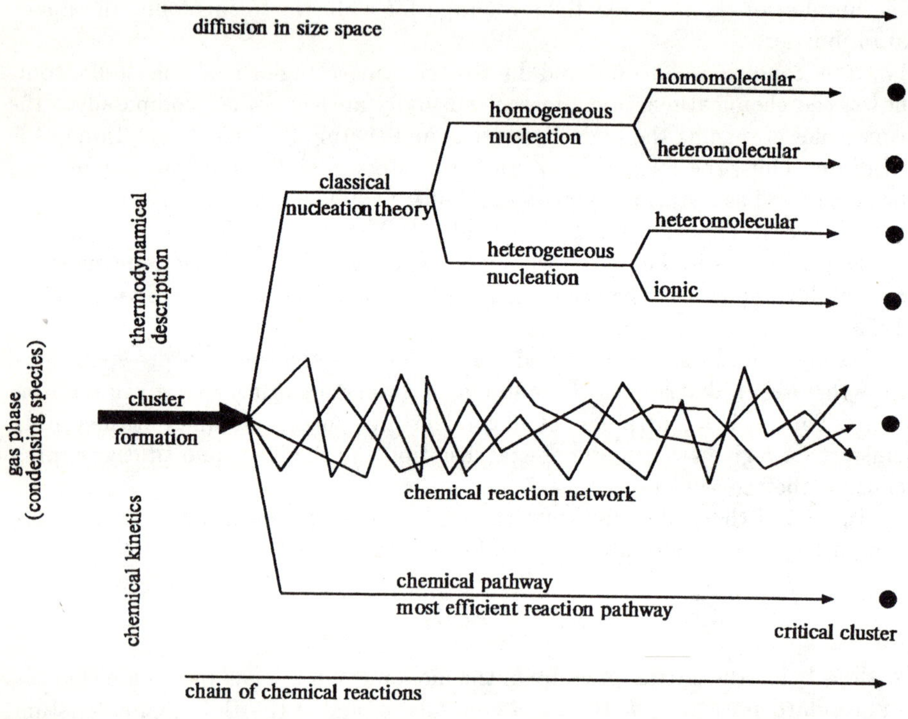

Fig. 6. Basically alternative descriptions of the nucleation regime.

Fig. 6 displays the main lines according to which the various approaches applied today in this regime can be classified with regard to the adopted physical mechanism:

5.2.1 The chemical network approach

Straightforward, one could aim to construct a *chemical reaction network* which finally results in the critical clusters. This way, however, faces the principal difficulties

- that it is not really known to which *minimum size* this network has to be extended along the evolution of specimens of increasing complexity in order to give a realistic quantitative description and how *to confine* it to such a size that it is still numerically treatable, and
- that many necessary input quantities (e.g. cross sections, reaction rates, etc..) are not available.

Therefore, at this time, the chemical network formulation is only applicable to situations where the way of cluster formation is well defined and is limited to a set of reactions, where the input quantities are known from laboratory experiments or from theory. For example, such a network has been worked out by Frenklach & Feigelson (1992), and more extended in the very important thesis of Cherchneff (see Cherchneff et al. 1991; 1992; Cherchneff & Barker 1992), who describes the first steps of the formation of *polycyclic aromatic hydrocarbon* molecules (PAHs) in the outflows of C-stars up to aromatic molecules containing five rings. But even in this narrow molecular domain such a large variety of possible reactions and products evolves, that one is finally forced to introduce *lumping*, a procedure where the reaction network is artificially reduced by fusing approximately similar reaction channels. Since immediately before effective nucleation starts the critical cluster is very large, truncated reaction networks in any case result in much too large nucleation rates. Thus the required selfregulation of the nucleation zone in its specific thermodynamic situation cannot be simulated within those rate networks.

An additional inherent difficulty of this method is, that due to the fast increase of the number of reaction channels with increasing complexity of the evolving aggregates, no obvious criterion is provided by the network of how to select the most efficient pathways leading to the critical cluster.

These basic difficulties of the chemical network approach strongly motivate alternative methods for the quantitative description of the nucleation regime, which avoid the above principle short-comings. These procedures, however, are rather schematic or highly specialized and can be classified as being to some extent asymptotic descriptions of the real problem.

5.2.2 The chemical pathway

The basic idea of this procedure is to reduce the immense variety of the different kinetically possible constructive reactions contained in the full network to the most efficient one which requires the minimum travel time in N-space for a small molecule to develop deeply into the growth regime and which thus essentially determines the grain formation rate.

Of course such a procedure requires

i) that the mechanism of cluster formation basically occurs along few chemically well defined channels, and

ii) that sufficient information is known about the reaction rates and branchings at every reaction step involved.

Then by detailed discussion of each reaction step, it is possible to select the *most efficient pathway*, i.e. that chain of cluster formation having maximum probability for bridging the molecule-cluster-region. If this pathway leads far enough into the domain of thermally stable clusters, there always exists a reaction step having minimum probability. This bottle neck, which controls the efficiency of the grain production is called the *critical reaction* with respect to the selected chemical pathway. Comparing this definition with the properties of the critical cluster introduced in Sect. 5.1 one confers that the critical reaction just defines the critical cluster $N_\star$ and its growth rate $J_\star$ along the pathway under consideration (see Gail & Sedlmayr, 1987; 1988; Goeres 1993).

In any case, the selection of a chemical pathway is a rather elaborate task which requires considerable physical and chemical insight into the actual nucleation mechanism. Therefore such a procedure is limited to specific problems, like PAH formation in C-star shells or the formation of *polyaromatic carbon* molecules (PACs) in RCrB-atmospheres (e.g. Goeres & Sedlmayr 1992). In these problems one is guided by laboratory chemistry and the wealth of data provided by pyrolysis experiments.

In this way carbon nucleation in C-star shells is described by a chemical pathway starting at acetylene (which is the most abundant condensable species) and leading via a series of polycyclic aromatic hydrocarbons of increasing complexity to flat and curved two-dimensional macroscopic structures (e.g. Keller & Sedlmayr 1983; Keller 1987; Gail & Sedlmayr 1987; Goeres 1993) which exhibit the characteristic emission features attributed to carbonaceous grains (Léger & Puget 1984; Allamandola et al. 1987)

An interesting application of this method is the formation of fullerenes in the Krätschmer-Fostiropoulos-Huffman experiment (Krätschmer et al. 1990), where according to our suggestion the nucleation pathway starts from the C_{10}-ring as monomer, leading by addition of monomers finally to closed $C_{60}-$ and $C_{70}-$ cages (Goeres & Sedlmayr 1991).

5.2.3 Classical nucleation theory

Classical nucleation theory describes the formation of critical clusters, like droplet formation in a supersaturated vapour, by means of thermodynamic quantities considering the clusters in the nucleation regime as macroscopic particles. The properties of these clusters are described either by extrapolation of the corresponding bulk phase quantities, even into the domain of very small clusters, by interpolation between the bulk phase and the corresponding molecular values, or by ab initio calculations for the actual equilibrium structure of the emerging small clusters from which their thermodynamic properties are derived (Köhler 1993). This latter procedure, however, is rather elaborate and hence is practically limited to chemically homogeneous clusters, where the internal interaction potentials are sufficiently well known (e.g. MgO-cluster formation).

In principle, classical nucleation is conceived as an evolution along the N-axis in cluster space described by a master equation for the cluster size distribution function (see Yamamoto & Nishida 1977; Gail & Sedlmayr 1988) from which by arguments similar to those presented in Sects. 5.1. and 5.2.2. (– first reaction with a negligible inverse rate –) the critical reaction step and hence the critical cluster and its growth rate can be deduced.

In real astrophysical situations several basically different mechanisms of cluster nucleation have to be distinguished with regard to the kind of the primary condensating species and to the chemical composition of the emerging clusters:

5.2.3.1 Homomolecular homogeneous nucleation

This mechanism describes the formation of chemically homogeneous clusters by direct condensation of molecules of one kind (monomers). Adopting classical nucleation theory and chemical equilibrium, this most simple approach, hitherto has been applied to the formation of

- pure SiO–, MgO, MgS– and Fe–clusters in the outflows of M-giants and supergiants, and
- amorphous C_N– clusters in the outflows of C-stars

(e.g. Gail & Sedlmayr 1986; 1987; 1988; Sedlmayr 1989)

5.2.3.2 Heteromolecular homogeneous condensation

This method describes the formation of homogeneous clusters by reactions with various kinds of molecules. For example the formation of pure amorphous carbon clusters from various contributing carbon compounds like C_1, C_2,....., C_2H, C_2H_2,...., etc. Such a process can be easily treated with slight modifications in the frame of classical nucleation theory even in case of non-equilibrium conditions, (see Gail & Sedlmayr 1988; Gauger et al. 1990). Another important application is e.g. the formation of silicates ($MgSiO_3$–, $(Mg,Fe)SiO_4$ –grains) in M-giant outflows. This formation mechanism is complicated by the fact that under circumstellar conditions only basic constituents, like MgO, SiO, Si_2O, Fe, FeO, etc., are present in the gas phase, but the corresponding nominal molecules (like $MgSiO_3$, Mg_2SiO_4, etc.) do not exist.

In the usual treatment of these situations thermal equilibrium between the various components in the gas phase and the individual constituents in the considered solid clusters is adopted, in this way calculating equilibrium abundances for the stable condensates as function of temperature and the partial pressures of the ambient condensating species. This method has been especially applied by the Japanese group in an impressive series of publications, e.g. Kozasa & Hasegawa (1987); Kozasa et al. (1989a,b).

Despite the widespread use of classical homogeneous nucleation theory in astrophysical dust formation problems (e.g. Salpeter 1977; Draine & Salpeter 1977; Draine 1979; Gail & Sedlmayr 1987; Kozasa et al. 1989 a,b), there exist some intrinsic shortcomings (see Sedlmayr 1989) which contemporary modifications aim to eliminate:

- The thermodynamical functions required in the small cluster domain, should no more be adopted as smooth functions extrapolated from the macroscopic domain, but will be provided by laboratory measurements or ab initio calculations (Köhler 1993).

- Being basically a phase transition, grain formation requires conditions far from thermal equilibrium. In the last years, generalisations of classical nucleation theory have been developed, which straightforward allow to treat situations far from thermal and chemical equilibrium (Gail & Sedlmayr 1988; Gauger et al. 1990; Dominik et al. 1993).

5.2.3.3 Heterogeneous nucleation

In the frame of classical nucleation theory heterogeneous nucleation describes the formation of clusters by chemical reactions with already existing nucleation seeds, resulting in critical clusters of a new chemical composition. In principle heterogeneous nucleation can be described by the chemical network or by the chemical pathway approach, respectively, with all difficulties and shortcomings inherent in these methods already mentioned in Sects. 5.2.1 and 5.2.2. Their application to PAH-formation in C-star shells is in fact an important example for heterogeneous nucleation.

Heterogeneous cluster formation is also expected in environments contaminated by UV-photons, where ions and charged clusters might play a major role in the condensation process. Then especially ion-neutral reactions could be important in the cluster formation process. However, until now, in the context of astrophysics, no reliable description of ionic cluster nucleation exists. In contrast to this unsatisfactory situation regarding the nucleation regime, powerful methods for treating heterogeneous *grain growth* have been developed. (see Castleman 1979; Draine 1979; and recently Dominik et al. 1993).

Heterogeneous grains are also widely discussed in the context of interstellar grain processing (see for instance the review of Tielens & Allamandola 1987 and references therein). This subject, however, is outside the scope of this contribution.

5.3 The growth regime

By definition (cf. Sect. 5.1), in a supersaturated environment every cluster larger than N_* has the tendency to grow at given temperature and density. In all cases of astrophysical interest, growth of critical and supercritical clusters is a thermodynamical process, which in principle can be described by the time evolution of the *local grain size distribution function* $f(t, N)$ according to the corresponding master equation (e.g. Gail & Sedlmayr 1988; Dominik et al. 1989; 1993). In this regime, the relevant growth timescales, at least for sufficiently large particles, are comparable or larger than the timescales governing the hydrodynamical evolution of the system. Therefore, grain growth usually has to be treated as a time dependent problem, in contrast to cluster nucleation which in most cases is a stationary process.

In many astrophysical problems, it is not necessary to have the full information provided by the local size distribution function, but it suffices to use appropriately defined moments

$$K_j = \sum_{N=N_l}^{N_u} N^{\frac{j}{d}} f(t,N) \ , \quad j = 0,1,2,\ldots$$

which directly enter into the model calculations (see Gail & Sedlmayr 1988). N_l and N_u are suitably chosen lower and upper limits of summation; d is the dimension of the considered grain with $d = 2$ for two-dimensional, $d = 3$ for three-dimensional particles, respectively.

Depending on the local chemistry and the thermodynamical conditions a large variety of resulting grain structures will evolve. Fig. 7 gives a sketch of the different types of grains which are expected to evolve in various realistic situations.

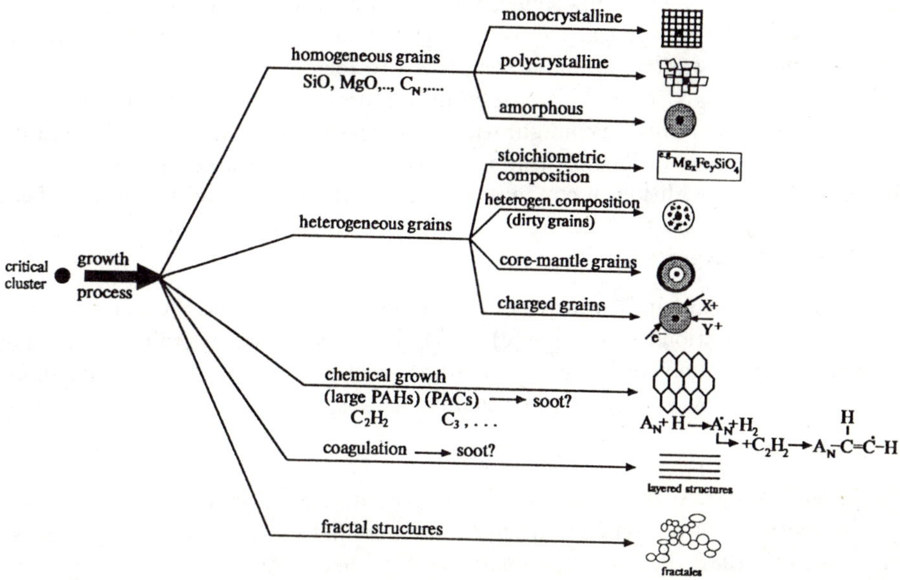

Fig. 7. Schematic sketch of different growth mechanisms and of the expected structures of the evolving grains. X^+ and Y^+ denote suitable ions contributing to the growth of charged grains.

5.3.1 Growth of chemically homogeneous grains
Homogeneous grain growth has been studied extensively both for C- and M-stars, where the formation of pure carbon grains and of SiO–, MgO–, MgS– and Fe–grains are discussed (Gail & Sedlmayr 1986; 1987; 1988).

As the applied methods only consider molecule addition controlled by a suitably defined growth timescale no answer can be given to the question concerning the real mineralogical structures of the resulting grains. To decide whether *monocrystalline*, *polycrystalline* or *amorphous* grains are produced by this condensation process four different timescales have to be compared:

- *the mean capture time* t_c for a condensating molecule to be adsorbed at the cluster surface,
- *the hopping time* t_h of this admolecule at the surface,
- *the time* t_n required for the admolecule to find a suitable lattice site to be built in, and
- *the characteristic evaporation time* t_{ev} of the admolecule.

According to the relations between these timescales different crystalline structures of the evolving grains have to be expected (Gail & Sedlmayr 1984):

a) $t_c < t_{ev}$: This is the basic necessary condition for grain growth to take place. It defines the growth regime (see Sect. 5.1).
b) $t_n = n^2 t_h < t_c$: The admolecule, after a random walk with n^2 steps, always arrives at the energetically most favourable lattice site to be built in. In this case *monocrystalline* grains are expected to form.
c) $t_n \geq t_c$ and $t_h < t_c$: In this case grain addition occurs too fast for allowing growth via equilibrium configurations. However, due to the second inequality still short range order resulting in *polycrystalline specimens* is possible.
d) $t_h > t_c$: The addition of condensating molecules to the grain occurs so fast, that even no short range order can develop. In this case *amorphous grains* have to be expected.

Along a typical wind trajectory of a cool giant, we therefore expect an evolution via the situations a) → b) → c) → d), finally resulting in grains, consisting of a monocrystalline core, a polycrystalline shell and eventually an amorphous mantle (see Gail & Sedlmayr 1984; Williams 1989)

5.3.2 Growth of heterogeneous grains
Heterogeneous grain growth can easily be treated by a straight forward extension of the methods of 5.3.1 including additional condensating species (e.g. Kozasa & Hasegawa 1987; Tielens 1989; Sharp & Huebner 1990; Dominik et al. 1993). Different situations have to be distinguished with regard to the evolving grain structures:

a) *Stoichiometric compositions*
These descriptions assume that grains of a well defined stoichiometric composition are formed in a given chemical environment. The necessary condition for this process to occur is that the gas phase partial pressures of

the individual grain constituents are larger than their corresponding vapour pressure. Thus, by definition of suitably defined *supersaturation ratios* equilibrium grain structures are calculated from a given chemistry (e.g. Kozasa & Hasegawa 1987; Sharp & Huebner 1990). A generalisation of these methods which allows to treat also non-equilibrium growth is given by Dominik et al. (1993).

b) *Heterogeneous compositions (dirty particles)*
Evolution of dirty particles simply can be treated by the method of Dominik et al. (1993) if chemically different condensating species are present having a supersaturation ratio larger than unity with regard to the considered clusters. Then, these molecules are considered to be deposited at the grain surface, which in this way (according to their individual constructive collision rates), form a chemically heterogeneous particle having a non–stoichiometric composition.

c) *Core-mantle grains*
The general condition for a species to condense is that its local partial pressure is larger than its corresponding vapour pressure in the solid. This condition is a very sensitive function of the local temperature and the chemical composition. Thus, the monotonic course of these quantities in a stellar wind defines a time order with respect to the condensation of certain species in a mass element moving outwards. As different regions in the shell usually correspond to different chemical conditions, – i.e. different abundances of the condensating species – grain growth in a stellar wind proceeds along a sequence of changing chemical environments, thus leading quite naturally to an onion-shell like grain structure (see Dominik et al. 1993).

d) *Charged grains*
In environments where ionising UV-photons are present a considerable fraction of the molecules and atoms become ionised thus producing free electrons. Calculations of the charge of clusters present in such situations show, that due to their large electron affinity macroscopic specimens usually become highly negatively charged, easily achieving values of the order $< Z > \simeq 100e$, where e denotes the electron unit (Gail & Sedlmayr 1975; Augustin 1990). This gives rise to an effective Coulomb-coupling between clusters and ions which should cause a considerable increase of the rate of constructive collisions contributing to grain growth. As this process may involve a large variety of positive ions having very different chemical nature heterogeneous grains with compositions much apart from those formed by neutral growth are expected to evolve. Until now, however, no quantitative description of this basically important growth mechanism is available, and the question about the chemical composition and the mineralogical structure of these grains is still open.

5.3.3 Chemical growth

This growth mechanism basically can be conceived as a straightforward continuation of the chemical pathway (see Sect. 5.2.2) into the growth domain. It is best illustrated by the example of PAH-formation in the expanding wind of a C-star.

In this case the growth of the critical clusters simply proceeds via a two-step process basically defined

– by formation of a radical site (marked by a dot) by hydrogen abstraction according to the reaction

$$A_N + H \rightarrow A_N^{\bullet} + H_2$$

and

– by addition of an acetylene molecule due to the reaction

$$A_N^{\bullet} + C_2H_2 \rightarrow A_N - CH = C^{\bullet} - H,$$

where A_N denotes a polyaromatic molecule consisting of N (hexagonal and pentagonal) rings. By further addition of C_2H_2 and quasi spontaneous ring closure or by a rearrangement process an aromatic molecule with size A_{N+1} results. By repetition of this process large two-dimensional flat or curved polyaromatic structures are generated (see Keller 1987; Gail & Sedlmayr 1987; Goeres 1993). For large PAHs this growth process can by described by suitably defined moment equations (see Gail & Sedlmayr 1987), the solution of which yields the total amount of the condensed material and the mean size of the resulting particles, from which by functional derivation also the corresponding grain size distribution function follows (Dominik et al. 1989, see Sect. 5.4).

5.3.4 Grain coagulation

In the context of stellar winds until now grain coagulation has not been taken into account. A first investigation of this process (Krüger 1992) applying the moment method proposed by Pruppacher & Klett (1978) suggests, that due to the large grain-grain collision timescales coagulation processes should be rather ineffective in stellar wind situations and, thus, have only a minor effect upon the emerging size distribution function. Despite this result, which has been obtained by rather severe mathematical simplifications, in particular regarding the adopted initial size distribution functions, coagulation deserves further studies. This concerns especially the formation of graphitic (layered) grains from large PAHs, where the adhesive forces seem to become relevant for specimens having more than hundred C-atoms.

5.3.5 Fractal structures

Fractal structures are likely to evolve by grain–grain collisions and thus should be of minor importance in circumstellar shells (cf. Sect. 5.3.4). However, as the question of PAH-coagulation, in particular if curved structures are involved, is still open, the formation of fractal like geometrical shapes deserves further investigation. Under the conditions prevailing in the interstellar medium, Wright (1989) has demonstrated that fractal grain structures are likely to evolve and could be important in particular for the interpretation of the far–infrared and the sub-mm emissivity ($\lambda > 100\mu$m) of the cold components of the interstellar medium, which would provide a critical test on the existence of fractal dust grains in interstellar space.

5.4 The grain size spectrum

As the growth regime generally can be described thermodynamically by means of suitably defined moment equations (Gail & Sedlmayr 1988; Gauger et al. 1990; Dominik et al. 1993) the actual grain size spectrum in a dust forming stellar wind can be directly calculated by a functional derivative of the equation for the 0-th moment of the grain size distribution function, which describes the time evolution of the grain number density (Dominik et al. 1989; 1993). From these calculations one infers that, in case of dust-driven winds, selfconsistent solutions always result in a power law like size distribution function

$$f(a) \sim a^{-\beta}$$

for the large particle wing of the final grain component, with a being the particle size and β the spectral index, which for the large variety of calculated models ranges between 4 and 5. Such a power law, however, only shows up, if dust driven winds, i.e. selfconsistent solutions of the system sketched in Fig. 4 are considered. (see Dominik et al. 1989). In situations where the stellar outflow is not primarily dust induced, but generated by some different process, like mechanical energy input, etc..., asymmetric Gaussian like grain size distribution functions are produced (Dominik et al. 1993). This should be the case for objects situated in the HR-diagram below the critical line for the possibility of dust driven winds (see Dominik et al. 1990).

In recent years considerable progress has been achieved for obtaining reliable size distribution functions for specific primary condensates. Nevertheless, much work remains to be done, in particular regarding the various processes of heterogeneous grain growth, the resulting grain sizes and grain geometries, which are most important ingredients for the calculation of reliable optical properties of the emerging grain components (see Alexander & Ferguson this volume).

6 Conclusions

Though considerable progress has been achieved in the last decade regarding cosmic dust condensation a number of important open questions still remain, e.g.:

- What is the true geometrical shape of the grains ? In all calculations adopting classical nucleation theory spherical grains are assumed to evolve.
- What is the true mineralogical structure of the emerging grains, in particular in case of heterogeneous growth ?
- What are the optical constants of these particles ?
- What is the nucleation mechanism in a partially ionised chemistry and what are the critical clusters to be formed ?
- What grain size distribution functions evolve in non-monotonic situations, like Miras and LPVs ?
- etc....

Much work addressing these questions is currently under way. Hence one can be confident, that within the near future the complex physical and chemical problem of astrophysical grain formation, which provides the key for a reliable understanding of many astrophysical objects, can be treated in a quantitatively more adequate manner for the various situations encountered.

Acknowledgments. I am very greatful to I. Birambaux, U. Bolick, B. Patzer, K. Sedlmayr, J.M. Winters and P. Woitke for assistance in preparing the manuscript.

References

Allamandola L.J., Tielens A.G.G.M., Barker J.R., 1987, In: G.E.Morfill, M.Scholer (eds.), *Physical Processes in Interstellar Clouds, Proc. NATO Advanced Study Institute*, D.Reidel Publ. Comp., Dordrecht, p.365

Augustin K., 1990, *Diplomarbeit*, Technische Universität, Berlin

Beck H.K.B., Gail H.-P., Henkel R., Sedlmayr E., 1992, *Astron. Astrophys.*, **265**, 626

Castleman A.W., 1979, *Astrophys. Space Sci.*, **65**, 337

Cherchneff I., Barker J.R., 1992, *Astrophys. J.*, **394**, 703

Cherchneff I., Barker J.R., Tielens A.G.G.M., 1991, *Astrophys. J.*, **377**, 541

Cherchneff I., Barker J.R., Tielens A.G.G.M., 1992, *Astrophys. J.*, **401**, 269

Dominik C., Gail H.-P., Sedlmayr E., 1989, *Astron. Astrophys.*, **223**, 227

Dominik C., Gail H.-P., Sedlmayr E., Winters J.M., 1990, *Astron. Astrophys.*, **240**, 365

Dominik C., Sedlmayr E., Gail H.-P., 1993, *Astron. Astrophys.*, in press

Donn B., Hecht J., Khanna R., Nuth J.A., Stranz D., Anderson A.D., 1981, *Surface Sci.*, **106**, 576

Draine B.T., 1979, *Astrophys. Space Sci.* **65**, 313

Draine B.T., Salpeter E.E., 1977, *J. Chem. Phys.*, **67**, 2230

Feder D., Russel K.C., Lothe B., Pound G.M., 1966, *Advan. in Phys.*, **15**, 111

Fleischer A.J., Gauger A., Sedlmayr E., 1992, *Astron. Astrophys.*, **266**, 321

Frenklach M., Feigelson E., 1989, *Astrophys. J.*, **341**, 372

Gail H.-P., Sedlmayr E., 1975, *Astron. Astrophys.*, **41**, 359

Gail H.-P., Sedlmayr E., 1984, *Astron. Astrophys.*, **132**, 163

Gail H.-P., Sedlmayr E., 1986, *Astron. Astrophys.*, **166**, 225

Gail H.-P., Sedlmayr E., 1987, In: G.E.Morfill, M.Scholer (eds.), *Physical Processes in Interstellar Clouds, Proc. NATO Advanced Study Institute*, D.Reidel Publ. Comp., Dordrecht, p.275

Gail H.-P., Sedlmayr E., 1988, *Astron. Astrophys.*, **206**, 153

Gauger A., Gail H.-P., Sedlmayr E., 1990, *Astron. Astrophys.*, **235**, 345

Gehrz R.D., 1989, In: L.J.Allamandola & A.G.G.M.Tielens (eds.), *Interstellar Dust, Proc. IAU Symp. No. 135*, D. Reidel Publ Comp., Dordrecht, p.445

Goeres A., 1993, In: G.Klare (ed.), *Reviews in Modern Astronomy*, **6**, Springer, p.165

Goeres A., Sedlmayr E., 1991, *Chem. Phys. Letters*, **184**, 310

Goeres A., Sedlmayr E., 1992, *Astron. Astrophys.*, **265**, 216

Goeres A., Henkel R., Sedlmayr E., Gail H.-P., 1988, In: G.Klare (ed.), *Reviews in Modern Astronomy*, **1**, Springer, p.231

Greenberg J.M., 1989, In: L.J.Allamandola & A.G.G.M.Tielens (eds.), *Interstellar Dust, Proc. IAU Symp. No. 135*, D. Reidel Publ Comp., Dordrecht, p.345

Jenkins E.B., 1989, In: L.J.Allamandola & A.G.G.M.Tielens (eds.), *Interstellar Dust, Proc. IAU Symp. No. 135*, D.Reidel Publ Comp., Dordrecht, p.23
Jura M., 1986, *Astrophys. J.*, **303**, 327
Keller R., 1987, In: A.Léger, L.d'Hendecourt, N.Boccara (eds.), *Polycyclic Aromatic Hydrocarbons and Astrophysics, Proc. NATO Advanced Study Institute*, D.Reidel Publ. Comp., Dordrecht, p.387
Keller R., Sedlmayr E., 1983, *Mitt. Astron. Ges.*, **60**, 312
Knapp G.R., 1985, *Astrophys. J.*, **293**, 273
Köhler M., 1993, *Dissertation*, Technische Universität, Berlin, in prep.
Kozasa T., Hasegawa H., 1987, *Prog. Theor. Phys.*, **77**, 1402
Kozasa T., Hasegawa H., Nomoto K., 1989a, *Astrophys. J.*, **344**, 325
Kozasa T., Hasegawa H., Nomoto K., 1989b, *Astrophys. J.*, **346**, L81
Krätschmer W., Fostiropoulos K., Huffman D.R., 1990, *Chem. Phys. Lett.* **170**, 167
Krüger M., 1992, *Diplomarbeit*, Technische Universität, Berlin
Lafon J.P.J., Berruyer N., 1991, *Astron. Astrophys. Rev.*, **2**, 291
Léger A., Puget J.L., 1984, *Astron. Astrophys.*, **137**, L5
Pruppacher H.R., Klett J.D., 1978, *Microphysics of Clouds and Precipitation*, D.Reidel Publ. Comp., Dordrecht
Rossi S.C.F., Benevides-Soares P., 1988, *Astron. Astrophys.*, **192**, 379
Salpeter E.E., 1977, *Ann. Rev. Astron. Astrophys.*, **15**, 267
Sedlmayr E., 1989, In: L.J.Allamandola & A.G.G.M.Tielens (eds.), *Interstellar Dust, Proc. IAU Symp. No. 135*, D.Reidel Publ Comp., Dordrecht, p.467
Sharp C.M., Huebner W.F., 1990, *Astrophys. J. Suppl.*, **72**, 417
Tielens A.G.G.M., 1989, In: L.J.Allamandola & A.G.G.M.Tielens (eds.), *Interstellar Dust, Proc. IAU Symp. No. 135*, D.Reidel Publ Comp., Dordrecht, p.239
Tielens A.G.G.M., Allamandola L.J., 1987, In: G.E.Morfill, M.Scholer (eds.), *Physical Processes in Interstellar Clouds, Proc. NATO Advanced Study Institute*, D.Reidel Publ. Comp., Dordrecht, p.333
Weinzierl S., 1991, *Diplomarbeit*, Technische Universität, Berlin
Williams D.A., 1989, In: L.J.Allamandola & A.G.G.M.Tielens (eds.), *Interstellar Dust, Proc. IAU Symp. No. 135*, D.Reidel Publ Comp., Dordrecht, p.367
Wright E., 1989, In: L.J.Allamandola & A.GG.M.Tielens (eds.), *Interstellar Dust, Proc. IAU Symp. No. 135*, D.Reidel Publ Comp., Dordrecht, p.337
Yamamoto T., Nishida S., 1977, *Prog. Theor. Phys.*, **57**, 1939

Opacity Problems in Protostellar Objects

Harold W. Yorke[1] *and Thomas Henning*[2] *

[1] Institut für Astronomie und Astrophysik, Am Hubland, D-97074 Würzburg, Germany
[2] Max Planck Society, Research Group: "Dust in Star-Forming Regions", Schillergäßchen 2–3, D-07745 Jena, Germany

1 Introduction

Opacity influences the transfer of radiation and is thus of decisive importance for (a) the interpretation of astronomical observations, (b) determining the thermal structure, (c) critically influencing molecular, atomic and thermonuclear processes and (d) affecting the overall dynamic evolution during the star formation process. Questions pertaining to the onset of instability of gas clouds, their subsequent fragmentation into smaller clumps, the formation of multiple systems, the initial mass function (IMF), the upper and lower mass limits, star formation efficiency and the fate of material not collected into stars cannot be considered without intimate knowledge of the opacity in this environment.

It is important to note that, when present, dust dominates the opacity by several orders of magnitude in comparison to molecular or atomic contributions (see e.g. Yorke 1985). Dust is absent in the star/protostar and its immediate vicinity, where the radiation field is too intense, i.e. where the equilibrium grain temperature exceeds the sublimation temperature or chemical sputtering takes place. This and a multitude of other processes (to be discussed in section 2) alter the mixture of grain types, sizes and shapes which can exist at different parts within the protostellar envelope.

We introduce in this context the concept of XNLTE. Whereas in Local Thermodynamic Equilibrium (LTE) the opacity can be unambiguously determined by local conditions, say the chemical composition Z_i and two state variables, density ρ and temperature T, there is a spatial interconnectivity associated with NLTE. In addition to local conditions the opacity depends on the radiation field. Usually it suffices to specify the 0th moment J_ν of the radiation intensity at all frequencies ν. In the case of XNLTE (eXtreme Non-Local Thermodynamic Equilibrium) there is also a temporal interconnectivity. The opacity depends not only on the global state of the medium but on its overall evolutionary history. This is particularly true for dust opacities in the stellar/protostellar environment.

* This work was supported by the DFG (German research Foundation). Numerical calculations were conducted on the CRAY Y-MPs at the HLRZ (Jülich) and at the LRZ (Munich) computing centers.

LTE:	$\kappa_\nu = \kappa_\nu(\rho, T, Z_i)$
NLTE:	$\kappa_\nu = \kappa_\nu(\rho, T, Z_i; J_{\nu'})$
XNLTE:	$\kappa_\nu = \kappa_\nu(\rho, T, Z_i; J_{\nu'}; \text{history})$

1.1 Basic types of dust mixtures

We shall distinguish between four basic types of interstellar dust mixtures and include for completion a fifth type, interplanetary dust (see Figure 1).

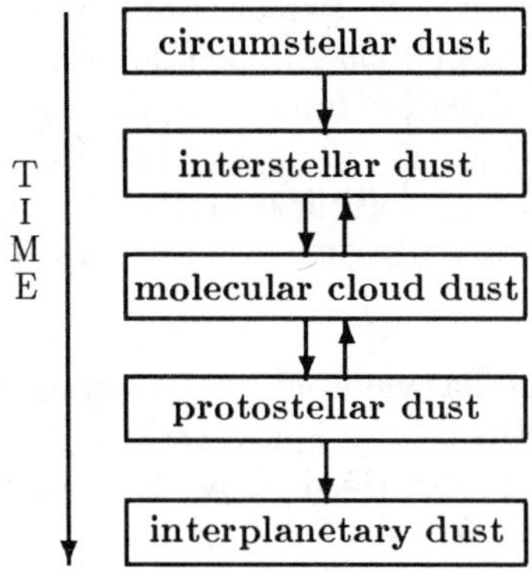

Fig. 1. Evolutionary sequence from circumstellar to interplanetary dust

Circumstellar dust. Dust forms in the outer regions of cool stars (see e.g. Sedlmayr this volume), in novae and in supernovae (see e.g. Evans 1990, Kozasa *et al.* 1991). The composition and size distribution which forms and evolves in this environment depends on the chemical abundances, the density, temperature and velocity structure and thus can be expected to vary from object to object.

Diffuse interstellar dust. In the general interstellar medium ($n_{\rm H} \lesssim 10\ {\rm cm}^{-3}$) grains from various sources have intermixed. Because of frequent reprocessing due to the passage of shocks, adsorption[1] of gaseous species, the effect of the diffuse interstellar UV field, cosmic rays and having spent several previous periods of time in the molecular cloud phase, the dust has developed into a homogeneous sample. Its overall optical properties appear to be constant within the Galaxy and similar to that observed in the interstellar medium of other galaxies. A "classical" model for the size distribution assumes a power law number density of spherical dust grains $n(a) \propto a^{-3.5}\,da$ for grain radii in the range $50\ \text{Å} \leq a \leq 0.25\mu\text{m}$ (MRN distribution, Mathis *et al.* 1977). Current grain models include a fraction of very small grains (VSG) with a radius of about 10 Å.

[1] *adsorption* = process of taking up molecules and atoms onto the surface of a solid or liquid and binding them by van der Waals forces, ionic or chemical bonds

Molecular cloud dust. Due to the comparatively high densities in molecular clouds ($n_{H_2} \gtrsim 10^3$ cm^{-3}) the adsorption of gaseous species proceeds at a faster rate, grain surface chemistry becomes more important than in the diffuse interstellar state and the grains develop dirty "icy" mantels. Theoretically, one can expect a mixture of H_2O, CH_3OH, CO and some other simple molecules (see e.g. Whittet 1993). The exact composition is very uncertain and, since it reflects the physical conditions and history of the grain's environment, should vary from cloud to cloud. Embedded in the icy coatings one can expect to find impurities such as amorphous carbon particles (see e.g. Preibisch *et al.* 1993). Effects of grain coagulation will modify the size distribution and lead to larger composite and fluffy grains (Ossenkopf 1993). Depending on particle composition van-der-Waals bonds or H-bridges determine the sticking coefficient. High relative velocities between colliding grains, say $v_{\rm rel} \gtrsim 10^4$ cm s^{-1}, may also occur and will generally be destructive, so that the overall problem of grain growth cannot be expected to be straight-forward.

Protostellar dust. In the protostellar environment densities $n_{H_2} \gtrsim 10^8$ cm^{-3} occur in the disks formed around the central hydrostatic object (which we here designate "protostar"). Processes as described above for molecular clouds are prevalent but at a much higher rate. We shall elaborate on this in the following sections.

Interplanetary dust. Interplanetary dust is the result of the extensive preprocessing taking place in the protostellar environment. Adsorption, accretion, coagulation and gravitational agglomeration, in addition to further chemical evolution and thermal annealing (repeated heating and cooling), occurs to the extent that macroscopic particles are formed. These processes, combined with the effects of destructive collisions and the destruction of the outer icy layers of comets, lead to the characteristic types of interplanetary dust observed in our Solar System. Since a detailed discussion of interplanetary dust is beyond the scope of the present work, we will not consider it further here and instead refer the interested reader to recent reviews (see e.g. Mukai 1990, Levasseur-Regourd and Hasegawa 1991).

1.2 Dust Models

When developing a new dust model the following steps are generally followed:

1. *Assume* (or attempt to calculate *ab initio*) *chemical composition, shape and size of each grain type.*
2. *Use "measured" dielectric properties.* This is not necessarily trivial, because the dielectric properties are in general not known for all materials and for all wavelengths of astrophysical interest. Also, the interpretation of laboratory measurements is often model dependent.

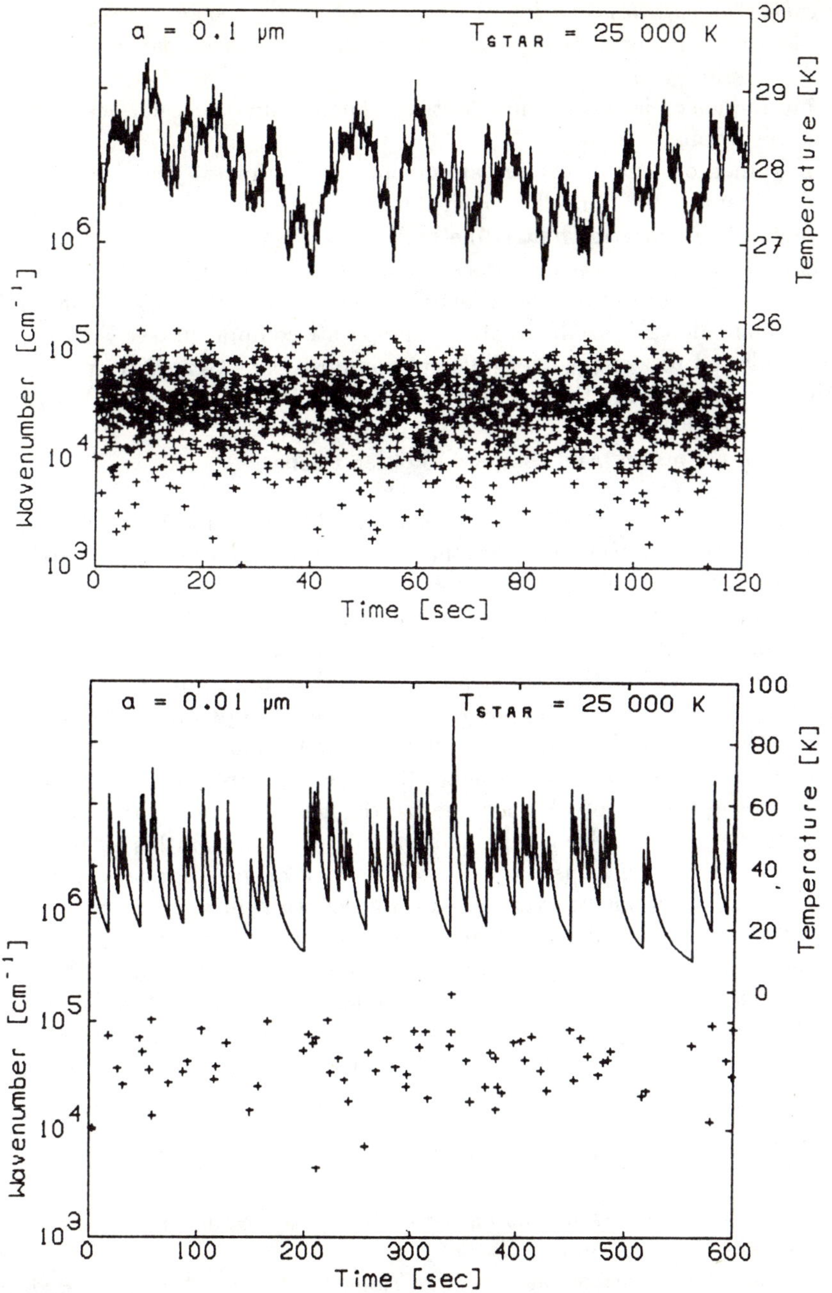

Fig. 2. Monte-Carlo simulation of the evolution of grain temperature (solid lines) for a dust particles of radius $a = 10^{-5}$ cm (top panel) and $a = 10^{-6}$ cm (bottom panel) located 0.3 pc from a B0 star. The wavenumber and arrival time for each individual photon is indicated by a '+' (see Yorke 1988 for details).

3. *Calculate the optical properties of each grain type using* e.g.
   - Mie theory (spheres), theories for rotationally symmetric particles (spheroids, finite cylinders)
   - interaction of light with small particles (lattice vibrations, surface modes, luminescence, ...)
   - new methods for irregular, anisotropic fluffy, fractal grains (discrete dipole approximation, effective medium theories combined with T-matrix method or scattering matrix method)
4. *Construct appropriate mean values.* Again, this is not always trivial. For instance, can one use the mean properties of a mixture of grains of differing sizes? Or should each size be treated as a separate component (see Figure 2)? Is one allowed to average over different compositions? Generally speaking, each averaging assumption should be carefully tested with respect to its extent of validity.
5. *Work with the model.* And compare to observations.

The common features of all current dust models (see Table 1) are (1) they contain some form of silicate and carbonaceous material and (2) they all *claim* to be able to fit the observed extinction and polarization observations. The first item can be interpreted as a general concensus that indeed carbon and silicates are important constituents of the dust, whereas the second item is indicative of the non-uniqueness of the model predictions. The basic differences in the models are the assumed size and shape distributions, the possible existence of organic mantles in the diffuse interstellar medium (ISM) and the origin of certain characteristic features, such as the near IR emission features and the 220nm ultraviolet absorption bump. We summarize the basic characteristics of some current grain models in Table 1 (updated and adapted from Mathis 1990).

Problems with current dust models are: the lack of *good* optical constants of the individual components over a broad wavelength region, our poor understanding of the influence of grain shape and porosity, our lack of detailed understanding of the observed polarization, and, of course, the non-uniqueness of the models. To a certain extent, these shortcomings can be overlooked, i.e. when one is only interested in the global effects of opacity on the spectral appearance. Often, however, the details of the dust model critically influence the evolution.

2 Evolution of Protostellar Objects

We are accustomed to viewing the formation and early evolution of stars as a four step process (see e.g. Shu *et al.* 1987):

a.) Formation of dense molecular cores and cloud fragments which presumably are supported by magnetic fields.

b.) After the onset of gravitational instability a fragment with non-zero angular momentum collapses and forms a central hydrostatic core ($M \approx 10^{-3}$ $M_\odot$, i.e. the *protostar*) surrounded by a disk. Both protostar and disk grow in mass as they accrete matter from the molecular cloud fragment.

Table 1. Characteristics of some current grain models

Brief title of theory (and authors)	Composition of grains	Size distr. of grains	Carrier of bump	Refs.
Core/mantle (Greenberg)	Silicate cores/organic refractory mantles; also small silicates, graphite	"Flat"[b]	Small graphite[c]	1,2
Core-mantle + iron (Chlewicki, Laureijs)	Similar as Greenberg, with small metallic iron particles, PAHs	"Flat"[b]	Small graphite[c]	3
Draine-Lee (or MRN)	Bare silicate, graphite	$a^{-3.5}$	Graphite	4,5
Duley/Jones/Williams	Silicate cores, hydrogenated amor. C clusters, small silicates	$a^{-3.5}$	OH-ions near Si	6
Composite grains (Mathis, Whiffen)	Silicates, amor. C, small graphite in the same grains (with free small graphite, possibly silicates)	$a^{-3.5}$	Small graphite[c]	7
Fractals (Wright)	Grains with fractal dimensions (from growth)	–	–	8
Sorrell	Porous silicates, small amor. carbon and silicates; small graphitic particles; different fractions for the diffuse ISM and molecular clouds	Discrete sizes	Small graphite[c]	9
IRAS-compatible dust (Rowan-Robinson)	Amorphous silicate and carbon grains; Small graphite and silicate grains	Discrete sizes	Small graphite[c]	10
VSG model (Désert et al.)	Silicate-core mantle grains, 3D VSG, 2D VSG	$a^{-3.5}$; VSG	3D VSG	11
PAH model (Siebenmorgen and Krügel)	Silicate and carbon grains, small graphite particles, PAHs	$a^{-3.5}$; PAHs	Small graphite[c]	12

Notes:

[a] References: 1: Hong and Greenberg (1980); 2: Chlewicki and Greenberg (1990); 3: Chlewicki and Laureijs (1988); 4: Mathis *et al.* (1977); 5: Draine and Lee (1984); 6: Duley and Williams (1988), Duley *et al.* (1989); 7: Mathis and Whiffen (1989); 8: Wright (1987); 9: Sorrell (1992); 10: Rowan-Robinson (1992); 11: Désert *et al.* (1990); 12: Siebenmorgen and Krügel (1992)

[b] "Flat" is really an exponential-type distribution.

[c] "Small" graphite means that the radii of all graphite particles are $\ll 0.22\ \mu$m, or $\lesssim 0.005\ \mu$m. Below this size, the extinction properties do not change with size.

c.) The hydrostatic core continues to grow in size primarily due to accretion from the disk as angular momentum in the disk is transported outwards. This phase is associated with molecular outflows (presumably driven by disk accretion) and sometimes with optical jets.

d.) The disk is dissipated, the outflow dies out and the result is a pre-main sequence object ($M \lesssim 3\ M_\odot$) or hydrogen-burning star ($M \gtrsim 10\ M_\odot$; e.g. Yorke 1985), possibly surrounded by a planetary system.

Phases (a), (c) and (d) are often observed, but there is no clear-cut candidate for an object in phase (b), i.e. a protostar highly embedded in a dusty envelope which is not associated with an outflow.

As material falls into the central regions dust grains can be destroyed due to a variety of effects (direct sublimation, photon or particle sputtering, passage of grains through a shock front, etc.), which causes a dramatic decrease of the opacity. This is due to the extremely large differences between the dust and gas opacity for temperatures in the range from 1000 to 2000 K. This effect can cause a number of instabilities in the accretion process. In spherically symmetric numerical calculations by Yorke and Krügel (1977) oscillations occurred in the protostellar envelope for cases $M_{\rm cloud} = 50\ M_\odot$ and $100\ M_\odot$ with periods $P \approx (r_{sub}^3/GM)^{1/2}$, where r_{sub} is the radius of destruction of a major dust component.

Yorke (1988) derived approximate formulae for the destruction radii of refractory particles ($T_{sub} \approx 1800$ K):

$$r_R \approx 2.5 \times 10^{13}\ \mathrm{cm} \left[\frac{Q_{\rm eff} L}{10^3\ L_\odot}\right]^{1/2}$$

and for the destruction of volatile (icy) coatings ($T_{sub} \approx 100$ K):

$$r_V \approx 3.4 \times 10^{16}\ \mathrm{cm} \left[\frac{Q_{\rm eff} L}{10^3\ L_\odot}\right]^{1/2}$$

where L is the luminosity of the central source and $Q_{\rm eff}$ is the effective absorption cross section of a spherical dust particle normalized by the geometric cross section. When the dust "sees" the optical photons from the central source directly then $Q_{\rm eff} \approx 1$. For the volatile component $Q_{\rm eff}$ can vary between $Q_{\rm eff} \approx 1$ (optically thin case) and $Q_{\rm eff} \approx 10^{-3}$ (optically thick). Note that these approximate formulae do not include the effect of the temperature fluctuations to be expected in very small grains (see Figure 2).

An analytical study by Wuchterl (1990) predicts the generation of local hydrodynamic instabilities for protostars of all masses if opacity gradients with temperature and/or density occur. However, it is not clear if the conditions derived by Wuchterl are not only necessary but also sufficient for large scale instabilities to develop. Recently, Balluch (1991) found in his hydrodynamical calculations of low mass protostars large scale instabilities to occur simultaneously, but with unspecified causal connections, with dust disappearance. In contrast to the Yorke and Krügel (1977) high mass protostellar models, Balluch allowed dust to reappear spontaneously with the same tabulated optical properties when permitted by local conditions. Numerically, the instability that causes the oscillations is most likely the result of the sudden and large changes of the Rosseland mean opacities used. This may be even a purely numerical effect if the opacity is part of the grid equation.

In the paper by Balluch (1991) no physical treatment of the dust destruction and production is given. Extremely large negative opacity gradients appear abruptly as soon as one exceeds the largest temperature at which the dust opacity is tabulated. The question remains how the opacity behaves in the destruction/production zone and how steep the gradient is. What is the rôle of non-equilibrium molecular chemistry and its contribution to the opacity in the dust-free zone? How do large scale oscillations affect the dust evolution? Lenzuni *et al.* (1993) made an attempt of a more realistic modelling of the dust destruction in protostellar cores and derived temperature dependent Rosseland mean opacities for this case. Dust production was not considered. Obviously, this is an XNLTE problem, which can only be solved by detailed numerical modelling.

The problem is further complicated by the coagulation and fragmentation of dusty grains in the immediate vicinity of protostars and young main sequence stars. At the relatively high densities of protostellar disks dust grains can coagulate if relative velocities are smaller than 10^2 cm s^{-1}. At much higher relative velocities the collisions would lead to a fragmentation of the grains. Both processes can change the grain distribution and therefore the dust opacities. The coagulation process leads to more open and fluffy structures which introduces a porousity parameter in addition to the size parameter for the description of the dust density distribution. The formation of fluffy "dirty-ice" grains results in increased far-infrared and submillimeter dust opacities (e.g. Preibisch *et al.* 1993). Specific estimates for the value of the opacity depend on the adopted grain model, especially on the silicate-to-carbon ratio and the abundance of grains with large refractive indices (Blum 1993).

As an example of the density, temperature and velocity structure to be expected from a collapsing 1 $M_\odot$ molecular clump during phase (b) accretion, the results of numerical calculations by Yorke *et al.* (1993) are displayed in Figure 3. Dusty material falls at almost free fall velocities until it encounters the accretion shock front (bunching of density contour lines) which surrounds the equilibrium disk. After passing through the shock front where further dust processing (adsorption, fragmentation and coagulation) can be expected, the material settles onto the disk several scale heights above the equator. One density scale height H_ρ corresponds to crossing two contour lines at a constant radial distance r from the axis. When $H_\rho \ll r$, one speaks of a "thin disk". For the disk displayed in Figure 3 $H_\rho \approx 0.1\, r$.

The temperature in the equilibrium disk depends only weakly on the height z from the equator and is thus basically a function of radial distance only. This means that such disks will not be unstable to convection in the z-direction. Only when the energy generated by viscous processes in the disk is much greater that the energy absorbed from the central source, can the disk become convectively unstable. At first dust can be expected to sediment toward the equator, its drift velocity depends on the ratio of dust surface area to mass. Coagulation will be prevalent and the opacity in the disk and the dust's scale height will be modified.

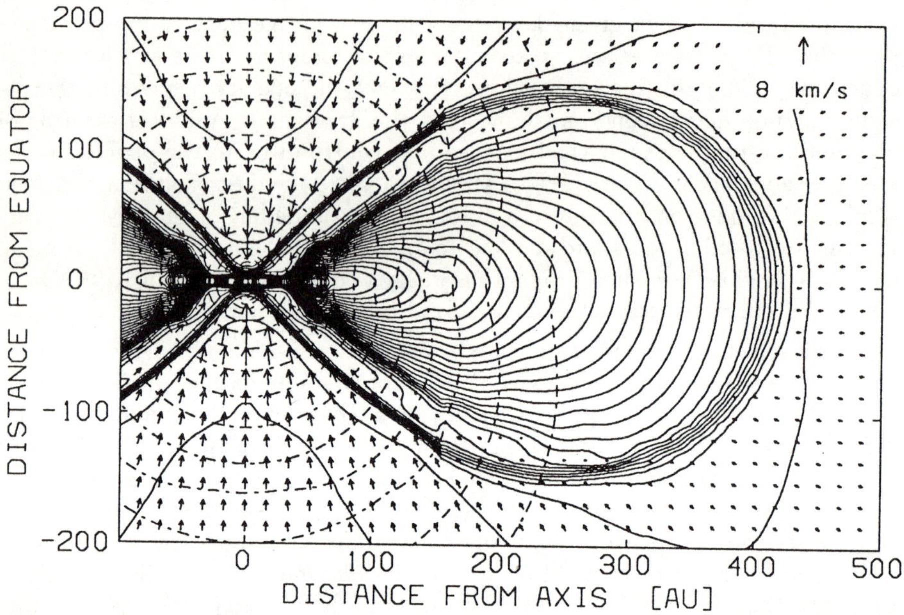

Fig. 3. Structure of temperature (dashed-dotted contour lines), density (solid contour lines) and velocity (arrows; linear scale in upper left of frame) of a protostellar disk during phase (b) at time t = 86 000 yr. The outermost density (temperature) contours (in cgs units) begin at $\log \rho = -19.4$ ($\log T = 1.3$) and adjacent contour levels are separated by $\Delta \log \rho = 0.2$ ($\Delta \log T = 0.05$).

The further evolution depends on the interplay of several poorly understood physical processes. Variations of opacity as a function of z can induce convection, which will increase the outward transport of angular momentum within the disk (and the inward mass flux). Other processes may be principally responsible for angular momentum transport (magnetic fields, turbulence, gravitational torques). The accretion of material via the disk onto the central source is accompanied by mass outflow along the axis. Random turbulent velocities will further enhance coagulation (Morfill 1985) and at the same time increase the dust's scale height. Untangling this XNLTE problem is necessary for understanding the formation of planets in general and our Solar System in particular.

References

Evans A., 1990, in *Physics of Classical Novae, IAU Coll. No.* **122**, eds. A. Cassatella, R. Viotti (Berlin: Springer), p.253

Allamandola L.J., Tielens A.G.G.M. (eds.), 1989, *Interstellar Dust, IAU Symp. No.* **135**, (Dordrecht: Kluwer).

Balluch M., 1991, *Astron. Astrophys.* **243**, 205

Blum J., Henning T., Ossenkopf V., Sablotny R., Stognienko R., Thamm E., 1993, *Proceedings of Fractal 93*, ed. M.M. Novak (Amsterdam: N. Holland Publ.), in press

Chlewicki G., Greenberg J.M., 1990, *Astrophys. J.* **365**, 230

Chlewicki G., Laureijs R.J., 1988, *Astron. Astrophys.* **207**, L11

Désert F.-X., Boulanger F., Puget J.L., 1990, *Astron. Astrophys.* **237**, 215

Draine B.T., Lee H.M., 1984, *Astrophys. J.* **285**, 89

Duley W.W., Williams D.A., 1988, *Mon. Not. Roy. Astron. Soc.* **231**, 969

Duley W.W., Jones A.P., Williams D.A., 1989, *Mon. Not. Roy. Astron. Soc.* **236**, 709

Hong S.S., Greenberg J.M., 1980, *Astron. Astrophys.* **88**, 194

Kozasa T., Hasegawa H., Nomomoto K., 1991, *Astron. Astrophys.* **249**, 474

Levasseur-Regourd A.C., Hasegawa H. (eds.), 1991, *Origin and Evolution of Interplanetary Dust, IAU Coll. No.* **126**, (Dordrecht: Kluwer)

Lenzuni P., Gail H.-P., Henning T., 1993, in preparation

Mathis J.S., Rumpl W., Nordsieck K.H., 1977, *Astrophys. J.* **217**, 425 (MRN)

Mathis J.S., Whiffen G., 1989, *Astrophys. J.* **341**, 808

Mathis J.S., 1990, private communication.

Morfill G.E., 1985, in *Birth and Infancy of Stars, Les Houches Summer School* **XLI**, eds. R. Lucas, A. Omont, R. Stora, (Amsterdam: N. Holland Publ.), p. 693

Mukai T., 1990, in *Evolution of Interstellar Dust and Related Topics*, eds. A. Bonetti, J.M. Greenberg, S. Aiello, (Amsterdam: N. Holland Publ.), p.397

Ossenkopf V., 1993, *Astron. Astrophys. Suppl.* , in press.

Preibisch T., Ossenkopf V., Yorke H.W., Henning T., 1993, *Astron. Astrophys.* , in press.

Rowan-Robinson M., 1992, *Mon. Not. Roy. Astron. Soc.* **258**, 787

Siebenmorgen R., Krügel E., 1992, *Astron. Astrophys.* **259**, 614

Sorrell W.,H., 1992, *Mon. Not. Roy. Astron. Soc.* **255**, 594

Shu F.H., Adams F.C., Lizano S., 1987, *Ann. Rev. Astron. Astrophys.* **25**, 23

Whittet D.C.B., 1993, in *Dust and chemistry in astronomy*, eds. T.J. Millar and D.A. Williams, (Bristol: Adam Hilger), p. 1

Wright E.L., 1987, *Astrophys. J.* **320**, 818

Wuchterl G., 1990, *Astron. Astrophys.* **238**, 83

Yorke H.W., 1985, in *Birth and Infancy of Stars, Les Houches Summer School* **XLI**, eds. R. Lucas, A. Omont, R. Stora, (Amsterdam: N. Holland Publ.), p. 648

Yorke H.W., 1988, in *Radiation in Moving Gaseous Media, Eighteenth Advanced Course, Swiss Society for Astrophysics and Astronomy*, (Sauverny-Versoix: Geneva Obs.), p. 195

Yorke H.W., Bodenheimer P., Laughlin G., 1993, *Astrophys. J.* **411**, 274

Yorke H.W., Krügel E. 1977, *Astron. Astrophys.* **54**, 183

Molecules in the Sun and Molecular Data

Nicolas Grevesse[1] *and A. Jacques Sauval*[2]

[1] Institut d'Astrophysique, Université de Liège 5, avenue de Cointe, B-4000 Liège, Belgium

[2] Observatoire Royal de Belgique 3, avenue Circulaire, B-1180 Bruxelles, Belgium

1 Introduction

Molecules play a unique role in solar spectroscopy in allowing to derive, with high accuracy, the temperature structure of the photospheric layers from where the solar spectrum in the near ultraviolet, visible and infrared emerges. They also allow to observe the heterogeneous structure of the outer layers as well as the subtle convective motions in the layers just above the solar convection zone. Molecules play also a unique role in defining the solar chemical composition, at least as far as the very important elements like carbon, nitrogen and oxygen are concerned. They also offer the unique opportunity to derive meaningful isotopic ratios for C and O. Although we shall not deal with sunspots in the present review, it has to be recalled that molecules are the only tracers to derive the solar abundances of Cl and F from the rotation-vibration lines of HCl and HF present in the infrared sunspot spectra (Hall & Noyes 1972; Hall & Noyes 1969).

On the other hand, the Sun itself offers a unique opportunity to produce molecular transitions, in local thermodynamic equilibrium (LTE), at rather high temperatures ($T \sim 5000$ K), higher in any case than in the laboratory, often in non LTE (see review by Johnson in this volume). Therefore, new lines and new transitions of higher excitation appear in the solar spectrum allowing to use the Sun as a "permanent laboratory source" in order to refine the molecular constants. Furthermore, the solar photosphere, where we believe the physical conditions and physical processes are better known than in any other stars, offers a unique possiblity to test basic molecular data like dissociation energies and transition probabilities.

2 Molecules and the Solar Spectrum

It is well known that the solar photospheric spectrum, from the near ultraviolet to the infrared, emerges from a very thin layer of about 1/1000 of the solar radius (700.000 km) just above the solar convection zone. In this region, the photon mean free path becomes rapidly very large whereas it was very small below it.

Solar spectra of very high resolution and high signal over noise ratio now cover the range from the ultraviolet to the infrared where the most important molecular transitions are found. Let us mention the most important of these solar atlases : Kurucz (1991) and Moore et al. (1982) for references concerning solar atlases, Delbouille et al. (1973; 3000 to 10000 Å), Delbouille et al. (1981; 1 μm to 5 μm), Kurucz et al. (1984; 296 μm to 1.3 μm), Blatherwick et al. (1983; IR), Goldman et al. (1987, 1992; IR), Farmer & Norton (1989; 2 μm to 16 μm), Livingston & Wallace (1991; 1.1 to 5.4 μm), Wallace et al. (1993; 7350 to 11230 Å), Brekke (1993; 1190 to 1730 Å). The solar atlas of Farmer & Norton (1989) is the first infrared solar spectrum obtained from above the earth atmosphere by the ATMOS FTS experiment which flew on board the Spacelab 3 Mission in May 1985. It allows to see, for the first time, the true solar spectrum from 2 to 16 μm, in a region which is heavily perturbed by telluric absorption.

Table 1. Molecular lines present in the ultraviolet and visible parts of the photospheric spectrum

Molecule	Electronic transition	Spectral region (nm)	Number of bands	Number of lines	References
H_2	$B^1\Sigma_u^+ - X^1\Sigma_g^+$	120.9 - 158.1	17	77	(1)
H_2	$C^1\Pi_u - X^1\Sigma_g^+$	120.9 - 125.4	2	2	(2)
CO	$A^1\Pi - X^1\Sigma$	136.0 - 167.0	3	70	(1)
SiO	$A^1\Pi - X^1\Sigma$	224.0 - 278.0	3	70	(3)
OH	$A^2\Sigma^+ - X^2\Pi$	281.0 - 337.8	7	457	(3) (4)
NH	$A^3\Pi - X^3\Sigma^-$	328.0 - 348.7	2	265	(4)
CH	$C^2\Sigma^+ - X^2\Pi$	309.1 - 321.5	2	92	(4)
CH	$B^2\Sigma^- - X^2\Pi$	362.7 - 411.9	3	142	(4)
CH	$A^2\Delta - X^2\Pi$	413.3 - 494.1	4	486	(4)
CN	$B^2\Sigma^+ - X^2\Sigma^+$	358.3 - 477.6	9	785	(4)
SiH^+	$A^1\Pi - X^1\Sigma^+$	399.2 - 448.9	2	50	(5)
SiH	$A^2\Delta - X^2\Pi$	410.0 - 419.1	1	148	(6)
CH^+	$A^1\Pi - X^1\Sigma^+$	422.3 - 445.7	1	14	(7)
C_2	$d^3\Pi_g - a^3\Pi_u$	466.4 - 563.5	8	555	(4)
MgH	$A^2\Pi - X^2\Sigma$	495.6 - 521.0	1	114	(4)
CN	$A^2\Pi - X^2\Sigma^+$	550.0 -2500.0	32	> 4000	(4) (8-10)

(1) Sandlin et al. (1986), (2) Bartoe et al. (1979), (3) Moore et al. (1982), (4) Moore et al. (1966), (5) Grevesse & Sauval (1970), (6) Sauval (1969), (7) Grevesse & Sauval (1971b), (8) Swensson et al. (1970), (9) Sneden & Lambert (1982), (10) Sauval et al. (1993).

In the solar spectrum, one finds transitions due to diatomic molecules only, made of the most abundant species like H, C, N and O together with fainter lines due to MgH, SiH, SiH^+, CH^+. Among the transitions, we find the electronic transitions of CO, H_2 and SiO in the UV, the electronic transitions of OH, CH, NH, C_2, and CN and in the infrared, rotation-vibration transitions of CO, CH, OH, NH and pure rotation transitions of OH and NH. We have to mention here a difficult case of molecular identification: many lines belonging to the blue and green bands of FeH were tentatively identified in the disk spectrum, essentially based on a statistical test of the wavelength coincidences (Carroll et al. 1976). Later Wing et al. (1977) and Wöhl et al. (1983) showed that there was no evidence for other FeH bands in the infrared photospheric spectrum (these IR bands are well present in sunspot spectra). Their conclusion is that the blue and green FeH bands are very probably absent in the photospheric spectrum (see Lambert, 1988, for a review of the FeH identification). Table 1 summarizes the different molecular transitions identified in the ultraviolet and visible parts of the solar photospheric spectrum (see Grevesse & Sauval, 1991, for a similar table relative to the infrared part of the solar spectrum). Detailed tables of identifications, covering atomic as well as molecular lines, may be found in Moore et al. (1966; 2935 to 8770 Å; 1982; 2095 to 3069 Å), Swensson et al. (1970; 7498 to 12016 Å), Blatherwick et al. (1983; IR), Goldman et al. (1987; IR), Geller (1992, 2 to 16 μm).

3 Computation of the Molecular Densities

Chemical equilibrium is assumed between the different atomic and molecular species. In the solar photosphere where rather few diatomic molecules are formed, the system of equations relating atomic and molecular species is rather simple. Let us write it for H, C, N and O which produce by far the most abundant diatomic molecules. We shall express it, as usual, in terms of the partial pressures of the different species :

$$\begin{aligned} P_H &= p_H + p_{H^+} + p_{H^-} + 2p_{H_2} + p_{CH} + p_{NH} + p_{OH} + \dots \\ P_C &= p_C + p_{C^+} + 2p_{C_2} + p_{CO} + p_{CN} + p_{CH} + \dots \\ P_N &= p_N + p_{N^+} + 2p_{N_2} + p_{NH} + p_{CN} + \dots \\ P_O &= p_O + p_{O^+} + 2p_{O_2} + p_{CO} + p_{OH} + \dots, \end{aligned} \tag{1}$$

where P_X is the total pressure of species X and p_X is the partial pressure of the sames species under the form indicated (neutral, ionized, molecular). The corresponding number densities, computed from P = N kT, will be represented by N_X and n_X respectively.

Other terms and other equations are added in order to include very minor species which are not or hardly seen in the solar spectrum.

The p_{X^+}/p_X are computed by the Saha law and the partial molecular pressures obey the Guldberg-Waage law :

$$p_{AX} = \frac{p_A p_X}{K_{AX}} \tag{2}$$

where K_{AX}, the so-called equilibrium constant, at temperature T, is given by

$$K_{AX} = \left(\frac{2\pi m}{h^2}\right)^{3/2} (kT)^{5/2} \frac{Q_A Q_X}{Q_{AX}} \exp(-D_0/kT) \tag{3}$$

The Q's are the partition functions for the atoms and the molecule; m is the reduced mass and D_0 is the dissociation energy of molecule AX.

If we express D_0 in eV, it is convenient to replace exp $(-D_0/kT)$ by $10^{-D_0\Theta}$ where $\Theta = 5040/T$. From (2) and (3), we derive easily the number density of molecule AX :

$$n_{AX} = n_A n_X \left(\frac{h^2}{2\pi m}\right)^{3/2} (kT)^{-3/2} \frac{Q_{AX}}{Q_A Q_X} 10^{D_0\Theta} \tag{4}$$

As in the solar photosphere $T \sim 5000K$, $\Theta \sim 1$, we see that *an uncertainty of 0.10 eV on the dissociation energy makes the molecular abundance uncertain by 0.10 dex i.e. 25%!* This problem of the accuracy needed for the dissociation energies is a very serious one. The example of the quite common CN molecule can perhaps best illustrate the point. For long years, the dissociation energy has been very uncertain with values from 7.2 to 7.9 eV. Even today, there is still a difference of about 0.12 *dex* (30 % uncertainty on the CN density) between the best experimental value, 7.77 eV, and the best theoretical result, 7.65 eV (see review by Costes & Naulin in this volume and Sect. 8.1).

Another important result is readily derived from relation (4) concerning the temperature dependence of the molecular abundances. Let us take CO as the best example, with its high dissociation energy, $D_0 = 11.09$ eV. *If we decrease the temperature by 1 % i.e. 50 K around 5000 K, the number of CO molecules will increase by 30 %!* CO is actually the most sensitive temperature indicator (see Sect. 4).

If we know all the equilibrium constants i.e. all the molecular and atomic partition functions and the dissociation energies, the system of equations (1) can easily be solved by an iterative procedure and we thus easily obtain the molecular abundances versus depth in the solar photosphere for a given set of elemental abundances and a given model atmosphere.

Computations of the molecular partition functions and equilibrium constants have been carried out by many authors. Table 2 gives an overview of the most recent investigations on this subject. Detailed discussions on the methods used to calculate exact molecular partition functions have been published, especially for CO and H_2, by Bohn & Wolf (1984), Sauval & Tatum (1984; see also Tatum 1988), Chase et al. (1985, JANAF Tables) and Irwin (1987). Most of these determinations lead to polynomial fits as a function of the temperature (or of the reciprocal temperature).

When computing the line opacity of a real molecular line, we need to use the density of molecules in a given excited level. This is computed from the Boltzmann law and this value is inversely proportional to the molecular partition function Q_{AX}. When combining this with (4), it is immediately seen, at least for minor species, that the density of a molecule in an excited level is independent of

Table 2. Partition functions and equilibrium constants for molecules of astrophysical interest

References	Q K	Total number of molecules (AB + ABC..)
(1)	K	232 (132 + 100)
(2)	Q	20 (19 + 1)
(3)	Q K	> 250 (100 + 150)
(4)	Q	66 (33 + 33)
(5)	K	53 (26 + 27)
(6)(*)	Q K	300 (300 + 0)
(7)	Q	53 (26 + 27)
(8)	Q K	> 500 (200 + 300)
(9)	Q	162 (0 + 162)

(*) Note that there is a small error in the tabulated coefficients of log K relative to 18 of the positive molecular ions, AB^+, for which I(A) < I(B). The atomic partition functions which were adopted in K (AB^+), Q (A) and Q (B^+), are incorrect and they have to be replaced by Q (A^+) and Q (B) as mentioned in the text.

(1) Tsuji (1973), (2) Scalo & Ross (1976), (3) Gurvich et al. (1978-1981), (4) Irwin (1981), (5) Rossi & Maciel (1983), (6) Sauval & Tatum (1984), (7) Rossi et al. (1985), (8) Chase et al. (1985, JANAF Tables), (9) Irwin (1988)

the molecular partition function. This is true only if the dissociation constants have been evaluated with the same molecular partition functions!

4 Molecules and the Temperature Structure

Empirical models of the solar photosphere are generally built from observations of the absolute solar continuous intensity versus wavelength, from center to limb variation of this intensity and from atomic line profiles. The best photospheric model of Holweger & Müller (1974) is built in that way; atomic spectral lines allow to extend the model to higher layers because continuum observations of high quality only cover the visible and near infrared and therefore only allow to test the temperature structure in rather deep layers. The model of Holweger & Müller (1974) is in good agreement with the most recent Harvard-Oslo solar model of Maltby et al. (1986).

We showed how rotation-vibration and pure rotation lines of OH can be used to test the temperature structure in the solar photosphere (Grevesse et al. 1984; Sauval et al. 1984). This is illustrated in Fig. 5 of Grevesse et al. (1984), where we show the resulting solar abundance of oxygen derived from rotation-vibration and pure rotation lines of different excitation energies. Using the Holweger-Müller model leads to the same abundance for all the lines whereas, using the Vernazza et al. (1976) model, an earlier version of the Harvard-Oslo model, leads to a very strong dependence of the derived abundance on the excitation energy. This can only be explained by slight temperature differences between the two models.

We showed in Sect 3. that CO is the most sensitive temperature indicator in the solar photosphere. It is definitely more sensitive than Fe I lines which have been used by different authors (in the photosphere, about 95 % of the Fe is in the form of Fe II and 5 % as Fe I).

CO lines have another advantage. With the ATMOS solar spectra from space (Farmer & Norton 1989) we entirely see all the rotation-vibration lines of the fundamental ($\Delta v = 1$) and first-overtone ($\Delta v = 2$) bands. We can follow the two sequences up to high v-values and J-values without interruption. This is quite a progress as compared with what can be seen from ground-based spectra. Thus, we have now at our disposal a very large number of CO lines, from strong to faint lines, which can be measured with very high accuracy (blends are quite rare in these spectral regions). The strongest of these CO lines are formed very high in the solar photosphere, at about the same geometrical depth as the Ca II H and K lines which indicate the chromospheric temperature rise. The faintest CO lines are formed rather deep. Therefore, *CO lines are a unique thermometer which allows to scan layers from above the so-called region of minimum temperature* between the photosphere and the chromosphere *down to rather deep layers.*

We recently reported (Grevesse & Sauval 1991) how the numerous CO lines in the infrared solar spectrum can be used to refine the photospheric temperature up to very high photospheric layers. Our new empirical model (see Fig. 3 in the hereabove mentioned paper) confirms the role of CO in the upper photospheric layers.

5 Line Shifts, Asymmetries and Convection

Observations of absolute line shifts and asymmetries in the solar line profiles are generally interpreted as the result of the presence of convective motions above the theoretical upper limit of the solar convection zone (see e.g. de Jager 1959). More recently Dravins et al. (1981, 1986) and Nadeau (1988) used a large number of Fe I and Fe II lines to make a comprehensive analysis of these shifts and asymmetries.

In a simple qualitative model of the convective motions, hot matter is moving upwards whereas cool matter moves downwards. In such a model, one sees easily when combining the line profiles shifted to the blue (hot matter, higher intensity) and to the red (cool matter, lower intensity), that the resulting line profile becomes asymmetric, each point of the line profile being shifted, relative to its laboratory position, by a different amount.

The numerous CO lines observed in the ATMOS solar spectrum have many advantages when used to study these effects. They are the most sensitive indicators of temperature differences. They are formed in a much wider range of depths than the Fe lines previously used. They allow to see what happens to the convective motions when going to extremely high photospheric layers. They do not suffer from hyperfine structure broadening. The isotopic effects are much larger than the line widths. The probability of an unknown blend slightly perturbing the line profile decreases also tremendously as one goes from the visible to the infrared. Finally, the wavenumbers of the CO lines are known with a higher accuracy than the Fe lines in the visible.

We recently showed (Grevesse & Sauval 1991; Blomme et al. 1993) how the shifts and asymmetries in the CO lines can be used to refine our knowledge of the convective motions over a very wide range of depths.

6 Solar Abundances of C, N and O

Carbon, nitrogen and oxygen have partially escaped from the meteorites. The solar spectrum is thus the only source of reliable solar system abundances of these important elements. The needs for an accurate knowledge of the solar abundances of these elements have been summarized by Lambert (1978) and Sauval et al. (1984). The most important motivation for refining our knowledge of these values is certainly the fact that they contribute to about 70 % to the metallicity (in the astronomical literature, the metallicity is the sum of the abundances, by mass, of all the elements except for hydrogen and helium). The detailed values of the C, N and O abundances are of crucial importance for the opacity calculations for stellar interiors (OPAL : Iglesias et al. 1992; OPACITY PROJECT : Seaton et al. 1992) as well as for stellar envelopes (Kurucz 1992).

C, N and O have quite a large number of indicators in the solar photospheric spectrum. Lambert (1978) was the first to show the importance of using all the indicators of the abundances of these elements, atoms as well as molecules.

In different papers (Sauval et al. 1984; Grevesse et al. 1984; Grevesse et al. 1990, 1991; Geller et al. 1991; Grevesse & Sauval 1991; Grevesse et al. 1992; Grevesse et al. 1993) we report the detailed results of new analyses. During these works, it became clear that permitted atomic lines of C I, N I and O I are probably not the best abundance indicators because of blending problems, problems with the transition probabilities (although much progress has recently been made : Biémont et al. 1991; Hibbert et al. 1991; Hibbert et al. 1993) and possible departures from LTE in O I (Kiselman 1991; Grevesse & Noels 1993). It also became clear that *the best abundance indicators are the molecular transitions*, i.e. C_2 (Swan, Phillips), CH (A-X), CH rotation-vibration, NH and OH rotation-vibration and pure rotation. Among these molecular indicators, the best ones are without any doubt the infrared transitions which are less perturbed than the transitions at shorter wavelengths and which can now be measured with high accuracy on the ATMOS solar infrared spectra from space (Farmer & Norton 1989).

Our analyses have shown that the results from the different indicators agree remarkably. The results have been slightly revised recently (Grevesse et al. 1993) in order to take into account slight changes in the model atmosphere due to a reduction in the solar abundance of iron which is a substantial electron donor. The values we recommend are $A_C = 8.55$, $A_N = 7.97$, $A_O = 8.87$, where A_X means $\log N_X/N_H$, in the usual scale where $\log N_H = 12$, with an uncertainty of about 0.05 *dex*.

It is well known that isotopic effects on atomic lines are extremely small as compared to the thermal widths of the lines in the solar spectrum. The only isotopes for which one has reliable abundance data are $^{12}C/^{13}C$ and $^{16}O/^{18}O/^{17}O$. These ratios are derived from the rotation-vibration lines of CO ($\Delta v = 1$).

In the ATMOS spectra, quite a large number of $^{13}C^{16}O$ lines are available; lines of $^{12}C^{18}O$ are easily measurable and, for the first time in the solar photospheric spectrum, we have identified and measured a few lines of $^{12}C^{17}O$. The solar isotopic ratios derived from these new measurements agree, within the error bars, with the telluric values (Anders & Grevesse 1989; 12/13 = 89, 16/18 = 499, 16/17 = 2625). They also agree with values for ^{13}C and ^{18}O obtained by Harris et al. (1987) from CO fundamental bands observed from a balloon.

7 The Sun as a "Laboratory Source" for Producing Molecular Spectra

The advantages of the Sun as a permanent source for producing molecular spectra has been recalled in the introduction. Molecules can survive in the solar atmosphere, under LTE, at rather high temperatures ($\sim$ 5000 K) whereas they are often destroyed in the laboratory, even at lower temperatures but in non LTE. The Sun thus provides a fairly complete coverage, especially in rotational excitation, reaching higher values than in the laboratory.

We have shown in different papers (Mélen et al. 1989; Farrenq et al. 1991; Grevesse & Sauval 1992; Sauval et al. 1992; Grevesse et al. 1991; Geller et al. 1991) that infrared solar spectra from space make it possible to see a large number of new high excitation lines and transitions never seen in the laboratory, and we have shown how these new data have been used to derive improved molecular constants for the ground states of CH and CO.

8 Testing Molecular Data using Solar Spectroscopy

Let us recall (Sect. 3) that the uncertainties in the line absorption coefficient of a molecular line essentially depends on the uncertainties of two molecular quantities, the line oscillator strength, f_{vJ}, and the dissociation energy, D_0, as well as on the uncertainties in the number densities of the relevant atoms, N_A and N_X, in the photospheric layers. As the atomic abundances can be derived, in most of the cases, using other indicators, it is possible to derive, from a comparison of observed and synthetic line profiles, the "solar value" of a missing molecular parameter if the other is known or a relation between the two parameters if they are both uncertain.

This however, builds on the assumption that the physical processes and physical conditions in the solar photosphere are well known. We discussed the physical conditions in Sect. 4. A very large number of observations and theoretical studies have accumulated with the years. They all converge to convince solar spectroscopists that the LTE hypothesis works pretty well in most of the cases (see e.g. Hinkle & Lambert, 1975). We are even tempted to say that the Sun is never "at fault".

We have shown (Grevesse et al. 1991), in the course of our analysis of the solar carbon abundance derived from many different indicators, how the dissociation energy of C_2 has been tested and how the transition probabilities of the C_2 Swan (0-0) band and the C_2 Phillips (0-0) and (1-0) bands have been selected among different available values. Similar tests have been made on NH (Grevesse et al. 1990; Geller et al. 1991) and on OH (Sauval et al. 1984; Grevesse et al. 1984). These solar tests of molecular data are summarized in Grevesse & Sauval (1992).

8.1 CN red system

In collaboration with R. Blomme, we have made a new analysis of a large number of lines of this system in the solar photospheric spectrum. Other recent studies of these lines have been made by Sneden & Lambert (1982); see also Lambert et al (1986), Sinha & Tripathi (1986). We have been extremely careful in the selection of the lines. This selection was based on the total half-widths of the lines. We also compared the solar spectrum with a predicted CN spectrum based on Kotlar et al (1980) for the constants and on Jørgensen & Larsson (1990) for the intensities. With these two criteria we have been able to reject most of the blended CN lines. The molecular data concerning this system are still uncertain to some extent. Recent experimental values of the dissociation energy cluster around 7.75 eV (Eres et al 1984; Wannenmacher et al. 1990; Costes et al. 1990; Huang et al. 1992) with rather small uncertainties (0.02 to 0.05 eV) whereas theoretical values are generally much lower (7.65 eV : Bauschlicher et al. 1988; 7.41 eV : Knowles et al. 1988; 7.52 eV : Larsson et al. 1983; but 8.05 eV : Rao et al. 1982). For the band oscillator strength (0-0 band), we have also a disagreement between theoretical values, which cluster around 2.35 10^{-3} (Bauschlicher et al. 1988; Knowles et al. 1988; Lavendy et al. 1984; Cartwright & Hay 1982; but 3.25 10^{-3} : Larsson et al. 1983), and recent experimental values (2.9 10^{-3} : Lu et al. 1992; 1.94 10^{-3} : Davis et al. 1986; 3.1 10^{-3} : Taherian & Slanger 1984; 3.6 10^{-3} : Jackson et al. 1982; Conley et al. 1980). We predicted the solar strengths of a large number of lines of the 0-0 band using the abundances of C and N derived by using other indicators of these abundances (Sect. 6). In order for the predicted equivalent widths to agree with the observed ones, we varied the two "free" parameters, f_{00} and D_0, which intervene as a product, $f_{00}\ 10^{D_0\Theta}$. With a high value of $f_{00} = 3.25\ 10^{-3}$, we derive $D_0 = 7.85$ eV. With a low $f_{00} =$ 2.35 10^{-3} we need a much higher $D_0 = 8.0$ eV. In the course of this analysis, we also noted that the fit between predicted and observed solar strengths becomes difficult as J" increases. This might possibly be caused by some interaction factor which was omitted in the theoretical calculations. As the Sun is so succesfull in testing molecular data for other molecules, why should it fail for CN ? We feel it is urgent to do more work on the f_{00} as well as on the D_0 of CN.

8.2 Transitions with very faint lines: SiH, SiH^+ and CH^+.

The electronic transitions from the ground states of SiH ($A^2\Delta - X^2\Pi$), SiH^+ (A $^1\Pi - X^1\sum^+$) and CH^+(A $^1\Pi - X^1\sum^+$) have been identified in the solar spectrum but the lines are very faint and heavily blended. Therefore quantitative results have rather large uncertainties.

SiH - The solar f-value, for the 0-0 band is slightly smaller, 3-4 10^{-3} (revised value based on earlier solar measurements by Sauval 1969) and 4.5 10^{-3} (Lambert & Mallia 1970), than the most recent laboratory value, $f_{00} = 5\ 10^{-3}$ (Carlson et al. 1978).

SiH^+ - The revised value based on earlier solar measurements (Grevesse & Sauval 1970, 1971a), $f_{00} = 1.3\ 10^{-3}$ agrees very well with the most recent laboratory value, $f_{00} = 1.4\ 10^{-3}$ (Hishikawa & Karawajczyk, 1993).

CH^+ - When we derived the solar f_{00}-value (Grevesse & Sauval 1971b), $f_{00} = 1.2\ 10^{-3}$, it was much smaller than the first experimental value of Smith (1971), $f_{00} = 6.8\ 10^{-2}$. This has been the origin of a pacific fight between Smith and ourselves. In the meantime, new laboratory measurements and computations have been published which converge to f_{00} between 5.45 10^{-3} (Larsson & Siegbahn 1983; theory) and 5.6 10^{-3} (Mahan & O'Keefe 1981; experiment).

Based on synthetic spectra, we have also slightly revised our solar value to $f_{00} = (2\text{-}4)\ 10^{-3}$, which is in better agreement with other determinations, showing, once more, that the Sun was right!

9 Conclusions

We have shown the very important and unique role played by diatomic molecules in solar spectroscopy. They are the best indicators of the solar abundances of carbon, nitrogen and oxygen. They are also, and especially CO, the most sensitive thermometers which make it possible to scan a wide range of layers in order to derive the temperature in these layers. They allow us to study the heterogeneous structure of the photosphere over a much wider range of optical depth, than previously possible, reaching, for the first time (through the strong CO lines of the $\Delta v = 1$ sequence), very high layers.

The Sun produces molecular transitions at temperatures higher than in the laboratory. The newly observed solar lines are used to refine the molecular constants of some ground states (CO, CH). Finally, and although it is not the ultimate goal of solar spectroscopy, solar analyses help to test basic molecular data like dissociation energies and transition probabilities.

Acknowledgements. We are much grateful to Willy Nijs (ORB) for his continuous help. N.G. thanks the Belgian Fonds National de la Recherche Scientifique for financial support.

References

Anders E., Grevesse N., 1989, *Geochim. Cosmochim. Acta* , **53**, 197

Bartoe J.-D., Brueckner G.E., Nicolas K.R., Sandlin G.D., Van Hoosier M.E., Jordan C., 1979, *MNRAS*, **187**, 463

Bauschlicher C.W., Langhoff S.R., Taylor P.R., 1988, *Astrophys. J.*, **332**, 531

Biémont E., Hibbert A., Godefroid M., Vaeck N., Fawcett B.C., 1991, *Astrophys. J.*, **375**, 818

Blatherwick R.D., Murcray F.J., Murcray F.H., Goldman A., Murcray D.G., 1983, *Atlas of South Pole IR Solar Spectra*, Dept. of Physics, Univ. of Denver

Blomme R., Sauval A.J., Grevesse N., 1993, in "*Infrared Solar Physics*", eds. D. Rabin, J.T. Jefferies and C. Lindsey, Kluwer, Dordrecht (in press)

Bohn H.U., Wolf B.E., 1984, *Astron. Astrophys.*, **130**, 202

Brekke P., 1993, *An Ultraviolet Spectral Atlas of the Sun Between 1190 and 1730 A, Astrophys. J. Suppl.* **87** ,1 (July, 1993)

Carlson T.A., Duric N., Erman P., Larsson M., 1978, *J. Phys. B.*, **11**, 3667

Carroll P.K., McCormack P., O'Connor S., 1976, *Astrophys. J.*, **208**, 903

Cartwright D.C., Hay P.J., 1982, *Astrophys. J.*, **257**, 383

Chase M.W., Davies C.A., Downey J.R., Frurip D.J., McDonald R.A., Syverud A.N., 1985, *JANAF Thermochemical Tables, 3d ed. J. Phys. Chem. Ref. Data* **14**, Sup.1

Conley C., Halpern J.B., Wood J., Vaughn C., Jackson W.M., 1980, *Chem. Phys. Lett.* **73** , 224

Costes M., Naulin C., Dorthe G., 1990, *Astron. Astrophys.*, **232**, 270

Davis S.P., Shortenhaus D., Stark G., Engleman R. Jr., Phillips J.G., Hubbard R.P., 1986, *Astrophys. J.*, **303** , 892 and **307** , 414

de Jager C., 1959, in *Handbuch der Physik*, ed. S. Flügge, Vol. LII : Astrophysics III : the Solar System, Springer Verlag, p. 80

Delbouille L., Roland G., Neven L., 1973, *Photometric Atlas of the Solar Spectrum from 3000 to 10000 Å*, Institut d'Astrophysique, Université de Liège

Delbouille L., Roland G., Brault J.W., Testerman L., 1981, *Photometric Atlas of the Solar Spectrum from 1850 to 10000* cm^{-1}, Kitt Peak National Observatory, Tucson

Dravins D., Larsson B., Nordlund A., 1986, *Astron. Astrophys.*, **158**, 83

Dravins D., Lindegren L., Nordlund A., 1981, *Astron. Astrophys.*, **96**, 345

Eres D., Gurnick M., Mc Donald J.D., 1984, *J. Chem. Phys.* **81** , 5552

Farmer C.B., Norton R.H., 1989, *A High-Resolution Atlas of the Infrared Spectrum of the Sun and the Earth Atmosphere from Space*, Vol. 1, The Sun, NASA Ref. Publ. 1224, Washington

Farrenq R., Guelachvili G., Sauval A.J., Grevesse N., Farmer C.B., 1991, *J. Mol. Spectrosc.* **149**, 375

Geller M., 1992, *A High-Resolution Atlas of the Infrared Spectrum of the Sun and the Earth Atmosphere from Space*, Vol. III, Key to Identification of Solar Features, NASA Ref. Publ. 1224, Washington

Geller M., Sauval A.J., Grevesse N., Farmer C.B., Norton R.H., 1991, *Astron. Astrophys.* **249**, 550

Goldman A., Blatherwick R.D., Murcray F.J., Van Allen J.W., Murcray F.H., Murcray D.G., 1987, *New Atlas of Stratospheric IR Absorption Spectra*, Vol. I : Line positions and and Identifications; Vol. II : The Spectra, Dept. of Physics, Univ. of Denver

Goldman A., Blatherwick R.D., Kosters J.J., Murcray F.J., Murcray F.H., Murcray D.G., 1992, *Atlas of Very High Resolution Stratospheric IR Absorption Spectra*, Dept. of Physics, Univ. of Denver

Grevesse N., Noels A., 1993, in *"Origin and Evolution of the Elements"*, eds. N. Prantzos, E. Vangioni-Flam, M. Cassé, Cambridge University Press, in press
Grevesse N., Sauval A.J., 1970, *Astron. Astrophys.* **9**, 232
Grevesse N., Sauval A.J., 1971a, *J. Quant. Spectrosc. Radiat. Transfer* **11**, 65
Grevesse N., Sauval A.J., 1971b, *Astron. Astrophys.* **14**, 477
Grevesse N., Sauval A.J., 1991, in *"The Infrared Spectral Region of Stars"*, eds. C. Jaschek and Y. Andrillat, Cambridge University Press, p. 215
Grevesse N., Sauval A.J., 1992, *Rev. Mexicana Astron. Astrof.* **23**, 71
Grevesse N., Noels A., Sauval A.J., 1992, in *"First SOHO Workshop : Coronal Streamers, Coronal Loops and Coronal and Solar Wind Composition"*, ESA SP-348, p.305
Grevesse N., Sauval A.J., Blomme R., 1993, in *"Infrared Solar Physics"*, eds. D. Rabin, J.T. Jefferies and C. Lindsey, Kluwer, Dordrecht (in press)
Grevesse N., Sauval A.J., van Dishoeck E.F., 1984, *Astron. Astrophys.*, **141**, 10
Grevesse N., Lambert D.L., Sauval A.J., van Dishoeck E.F., Farmer C.B., Norton R.H., 1990 *Astron. Astrophys.*, **232**, 225
Grevesse N., Lambert D.L., Sauval A.J., van Dishoeck E.F., Farmer C.B., Norton R.H., 1991 *Astron. Astrophys.*, **242**, 488
Gurvich L.V. et al., 1978-1981, *Termodynamicheskie Svoistva Individual' nikh Veschev, Vol. 1-3*, Eds Moscow: Soviet Acad. Sci.
Hall D.N.W., Noyes R.W., 1969, *Astrophys. J. Letters*, **4**, 143
Hall D.N.W., Noyes R.W., 1972, *Astrophys. J.* **175**, L95
Harris M.J., Lambert D.L., Goldman A., 1987, *MNRAS* **224**, 237
Hibbert A., Biémont E., Godefroid M., Vaeck N. 1991, *Astron. Astrophys. Suppl.*, **88**, 505; J. Phys. B. **24**, 3943
Hibbert A., Biémont E., Godefroid M., Vaeck N. 1993, *Astron. Astrophys. Suppl.*, **99**, 179
Hinkle K. H., Lambert D.L., 1975, *MNRAS* **170**, 447
Hishikawa A., Karawajczyk A., 1993, *J. Mol. Spectrosc.* **158**, 479
Holweger H., Müller E.A., 1974, *Solar Physics* **39**, 19
Huang Y., Barts S.A., Halpern J.B., 1992, *J. Phys. Chem.* **96**, 425
Iglesias C.A., Rogers F.J., Wilson B.G., 1992, *Astrophys. J.*, **397**, 717
Irwin A.W., 1981, *Astrophys. J. suppl.*, **45**, 621
Irwin A.W., 1987, *Astron. Astrophys.*, **182**, 348
Irwin A.W., 1988, *Astron. Astrophys. Suppl.*, **74**, 145
Jackson W.M., Payne W., Halpern J.B., Tang X., 1982, *Proc. Intern. Conf. on Lasers*, p. 72
Jørgensen U.G., Larsson M., 1990, *Astron. Astrophys.*, **238** , 424
Kiselman D., 1991, *Astron. Astrophys.*, **245**, L9
Knowles P.J., Werner H.-J., Hay P.J., Cartwright D.C., 1988, *J. Chem. Phys.* **89** , 7334
Kotlar A.J., Field R.W., Steinfeld J.I., Coxon J.A. 1980, *J. Mol. Spectrosc.* **80** , 86
Kurucz R.L., 1991, in *Solar Interior and Atmosphere*, eds A.N. Cox, W.C. Livingston, M.S.Matthews, The Univ. of Arizona Press, Tucson, p. 663
Kurucz R.L., 1992, *Rev. Mexicana Astron. Astrof.* **23**, 181
Kurucz R.L., Furenlid I., Brault J.W., Testerman L., 1984, *Solar Flux Atlas from 296 to 1300 nm*, National Solar Observatory, Tucson, Atlas Nr. 1
Lambert D.L., 1978, *MNRAS* **182**, 249
Lambert D.L., 1988, *Publ. Astron. Soc. Pacif.* **100**, 1202
Lambert D.L., Mallia E.A., 1970, *MNRAS* **148**, 313
Lambert D.L., Gustafsson B., Eriksson K., Hinkle K.H., 1986, *Astrophys. J.* **62** , 373

Larsson M., Siegbahn P.E.M., 1983, *Chem. Phys.* **76** , 175
Larsson M., Siegbahn P.E.M., Agren H., 1983, *Astrophys. J.* **272** , 369
Lavendy H., Gandara G., Robbe J.M., 1984, *J. Mol. Spectrosc.* **106** , 395
Livingston W., Wallace L., 1991, *An Atlas of the Solar Spectrum in the Infrared from 1850 to 9000 cm^{-1} (1.1 to 5.4 μm)* , Technical Report 91-001, National Solar Observatory, Tucson
Lu R., Huang Y., Halpern J.B., 1992, *Astrophys. J.*, **395**, 710
Mahan B.H., O'Keefe A., 1981, *Astrophys. J.* **248**, 1209
Maltby P., Avrett E.H., Carlsson M., Kjeldseth-Moe O., Kurucz R.L., Loeser R., 1986, *Astrophys. J.* **306**, 284
Mélen F., Grevesse N., Sauval A.J., Farmer C.B., Norton R.H., Bredohl H., Dubois I., 1989, *J. Mol. Spectrosc.* **134**, 305
Moore C.E., Minnaert M.G.J., Houtgast J., 1966, *The Solar Spectrum 2935 Å to 8770 Å* , National Bureau of Standards Monograph 61
Moore C.E., Tousey R., Brown C.M., 1982, *The Solar Spectrum 3069 - 2095 A*, Naval Research Laboratory, NRL Report 8653
Nadeau D., 1988,*Astrophys. J.*, **325**, 480
Rao V.M., Rao M.L.P. Rao, P.T., 1982, *Phys. Rev.* **26** , 1765
Rossi S.C.F., Maciel W.J., 1983, *Astrophys. Space Sci.*, **96**, 205
Rossi S.C.F., Maciel W.J., Benevides-Soares, 1985, *Astron. Astrophy.*, **148**, 93
Sandlin G.D., Bartoe J.-D.F., Brueckner G.E., Tousey R., VanHoosier M.E., 1986, *Astrophys. J. Suppl.*, **61**, 801
Sauval A.J., 1969, *Solar Physics* **10**, 319
Sauval A.J., Tatum J. B., 1984, *Astrophys. J. Suppl.*, **56**, 193
Sauval A.J., Blomme R., Grevesse N., 1993 (in preparation)
Sauval A.J., Farrenq R., Guelachvili G., Grevesse N., Farmer C.B., Norton R.H., 1992, *Astron. Astrophys.*, **265**, 355
Sauval A.J., Grevesse N., Brault J.W., Stokes G.M., Zander R., 1984, *Astrophys. J.*, **282**, 330
Scalo J.M., Ross J.E., 1976, *Astron. Astrophys.*, **48**, 219
Seaton M.J., Zeippen C.J., Tully J.A., Pradham A.K., Mendoza C., Hibbert A., Berrington K.A., 1992, *Rev. Mexicana Astron. Astrof.* **23**, 19
Sinha K., Tripathi B.M., 1986, *Bull. Astron. Soc. India* **14** , 40
Smith W.H., 1971, *J. Chem. Phys.* **54**, 1384
Sneden C., Lambert D.L., 1982, *Astrophys. J.*, **259**, 381
Swensson J.W., Benedict W.S., Delbouille L., Roland G., 1970, *The Solar Spectrum from λ 7498 to λ 12016 - A table of Measures and Identifications*, Mém. Soc. Roy. Sci. Liège 5
Taherian M.R., Slanger T.G., 1984, *J. Chem. Phys.* **81** , 3814
Tatum J.B., 1988, *Observatory* **108**, 55
Tsuji T., 1973, *Astron. Astrophys.*, **23**, 411
Vernazza J.E., Avrett E.H., Loeser R., 1976, *Astrophys. J. Suppl.*, **30**, 1
Wallace L., Hinckle K., Livingston W., 1993, *An Atlas of the Photospheric spectrum from 8900 to 13600 cm^{-1} (7350 to 11230 Å)* , Technical Report 93-001, National Solar Observatory, Tucson
Wannenmacher E.A.J., Lin H., Jackson W.M., 1990, *J. Phys. Chem.* **94** , 6608
Wing R.F., Cohen J., Brault J.W., 1977, *Astrophys. J.*, **216**, 659
Wöhl H., Engvold O., Brault J.W., 1983, *Inst. Theoret. Astrophys. Oslo* Rep. 56

Pressure-Induced Molecular Absorption in Stellar Atmospheres

Aleksandra Borysow

Physics Department, Michigan Technological University
Houghton, MI 49931, USA

1 The Nature of Collision-induced Absorption

Pressure-induced absorption arises in complexes of two or more inert atoms or molecules, due to dipole moments induced during the collisional interaction. The term "pressure-induced" still prevails in the astrophysical literature, yet "collision-induced" absorption (CIA), or "interaction-induced" absorption seems more appropriate and is commonly used elsewhere. Ordinary absorption processes in the infrared arise from individual, polar molecules interacting with electromagnetic radiation. As a consequence, the intensity of the allowed lines increases linearly with density. CIA, on the other hand, is most striking in gases composed of nonpolar, infrared-inactive molecules. Induced spectral lines are observed at rovibrational frequencies which are dipole-forbidden in single (i.e. non-interacting) molecules. Dipole transitions may, however, be *induced* in the interacting pair. The new symmetry of the electronic cloud of a collisional complex may be very different from those of the isolated molecules and thus commonly allows for a transient dipole, which then interacts with radiation. Collision-induced absorption increases quadratically in the low density limit, thus reflecting the two-body origin of the basic absorption process. At higher gas densities, ternary interactions become significant and cubic and higher-order contributions to the observable absorption are then commonly seen.

A few characteristic features of CIA will be discussed next. Interaction-induced lines are generally very broad, an immediate consequence of the fact that induced dipoles last only for a short time, roughly the mean time of a collision, t_c, which depends on temperature and reduced mass of the interacting pair. For hydrogen at room temperature, the approximate (effective) time of the duration of a collision inducing a dipole amounts to roughly $5 \cdot 10^{-14}$ s, which leads to a spectral width, $\Delta\nu \approx 1/t_c$, of approximately 600 cm^{-1}. At higher temperatures, naturally, the widths will be even larger (at 3000 K, $t_c \approx 10^{-14}$s, and $\Delta\nu \approx 3000$ cm^{-1}). In general, CIA spectral widths follow approximately a $\Delta\nu \propto T^{1/2}$ dependence (T being the temperature).

Two mechanisms are mainly responsible for the induction of a dipole in molecular encounters. Electronic overlap induction exists in atomic and molecular pairs at short distances. It is a purely quantum effect caused by the Pauli principle: due to the overlap of the electronic clouds, some states become inaccessible to electrons. As a result, the redistribution of the electronic charge occurs and a transient dipole is being induced in a pair at close distances (roughly smaller than 3 Å). The other mechanism is a multipolar induction which, to a large extent, is a classical process characteristic of the molecular gases. The electric fields surrounding any molecule may be represented by a superposition of multipolar fields. For example, the lowest, non-vanishing multipole moment of H_2 is the quadrupole, or that of CH_4 is the octopole. A collisional partner in the electric multipolar field of one molecule is polarized and thus interacts with radiation. Multipolar induction is the dominant induction mechanism for most common molecules, except the ones having small polarizabilities. Classical multipole induction provides a rather accurate approximation of the actual induced dipole function at large intermolecular distances (those larger than 3 Å or so).

Based on these simple facts, one is able to explain the structure of the CIA spectra of molecular pairs. Figure 1 shows a typical spectrum of H_2 pairs at room temperature. The spectral intensities marked on the $y-$axis are those corresponding to one amagat[1] of H_2. For a given number density of a gas, one needs to multiply the plotted absorption spectrum by the number of amagats squared. The frequencies are given in cm^{-1} units. Broad induced absorption

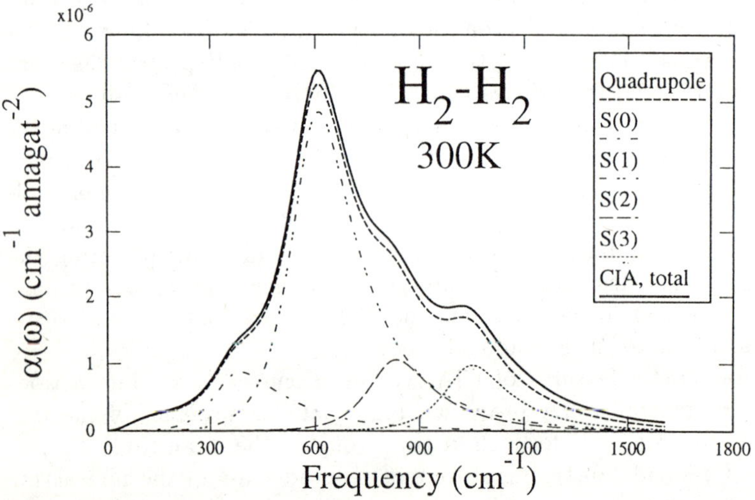

Fig. 1. Rotranslational collision-induced absorption spectrum of H_2–H_2 at 300 K (solid line). Dashed line denotes the spectral contribution due to the quadrupolar induction. $S(J)$ denote lines due to separate rotational transitions, $J \rightarrow J+2$ for populated rotational states J = 0, 1, 2, 3.

[1] 1 amagat corresponds very nearly to the number density of gas under normal conditions, and is equal to $2.687 \cdot 10^{19}$ cm^{-3}.

lines appear at the rotational transition frequencies of the hydrogen molecule, mainly with $J \to J+2$ transitions (S lines).

To illustrate this, let us first imagine that a dipole is induced during a collision between two (different) *atoms*. At large separations, the dipole will be initially zero. With decreasing interatomic distance, the dipole will increase as the electronic clouds begin to overlap. The dipole strength will reach a maximum near the point of the closest approach. After that, it will fall off to zero again as the separation increases. On the whole, the strength of the induced dipole will be significant only during the "collision". The Fourier transform of the autocorrelation function of this dipole function $\mu(t)$ will give rise to a broad spectral density profile, which is centered at zero frequency, commonly called a purely "translational" spectrum.

Next, let us briefly consider the case of two interacting *molecules*. The dipole now largely arises from multipole induction. If for the moment we focus on H_2–X collisions (X may be a helium atom or another hydrogen molecule, or indeed any other nonpolar species), the quadrupolar field of a rotating hydrogen molecule will introduce an additional modulation of the time-dependent induced dipole function, $\mu(t)$, which will result in the appearance of a similar, broad "translational" spectral profile, only shifted from zero frequency to the rotational transition frequencies of the H_2 molecule ($S(J)$ lines, Fig. 1). This simplified model does, in fact, describe the rotranslational (RT) CIA spectra quite well. What must be understood, is that it is a hydrogen molecule which undergoes rotational transitions corresponding to the $S(J)$ lines. The presence of the collisional (interacting) partner is essential for the appearance of these rotational lines, which otherwise are dipole-forbidden in a single H_2 molecule. The simultaneous change of the orbital and rotational angular momenta taking place during the collision, make the rotational transitions temporarily "allowed". Each rotational transition in hydrogen molecule, is accompanied by an absorption (or emission) between the translational states of a collisional pair. It results in the the "translational" broadening of the $S(J)$ profiles.

If we also consider the possibility of vibration of a hydrogen molecule, the vibrational transition frequency introduces additional oscillations of the induced dipole, causing a diffuse line to appear also at the rovibrational transition frequency of the hydrogen molecule (rovibrational CIA spectra). In this case, all three degrees of freedom participate in the absorption process, translation, rotation and vibration. After the collision, the hydrogen molecule is found in a higher rotation-vibration state. Summarizing, we will expect CIA spectra to cover the entire infrared region, from the "purely translational" (centered at zero frequency), to the rotranslational (Fig. 1) and the rovibrational and even the overtone bands. The latter ones will appear around $v \to v'$ transitions, with $v, v' = 0,1,2,3 \ldots$ At low temperatures only the ground vibrational level is populated, so $v = 0$, and the transitions to $v'=1$ (the fundamental band) and $v' = 2, 3, \ldots$ (the first and higher overtones) are observed. In principle, at elevated temperatures, a non-zero population of higher v states will give rise to the absorption "hot bands" which, however, have not yet been observed in the laboratory.

We are concerned with hydrogen and helium, the major components of stellar atmospheres. To complete the picture, in Fig. 2 we show the RT spectrum of H_2–He pairs at the temperature of 300 K. Even though quadrupolar induction clearly takes place (just as in pure H_2 gas), the spectral profile looks quite different. The dashed line marks the quadrupolar term. However, another spectral component, absent in Fig. 1, is shown here (dotted line): the so-called "isotropic overlap". This term, which is independent of the molecular orientations, is zero for the H_2–H_2 RT spectra, due to symmetries inconsistent with a dipole moment. This essential difference between the H_2–H_2 and H_2–He spectra persists also more or less in the rovibrational spectra. We will refer to this difference when we consider the various CIA models proposed in the past.

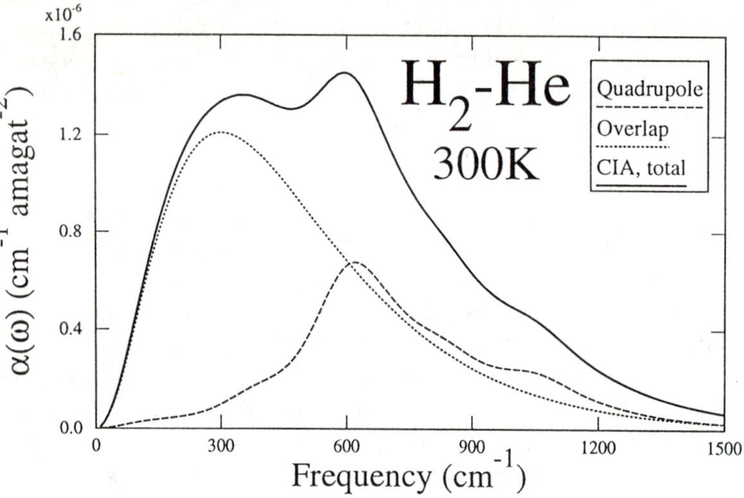

Fig. 2. Rotranslational collision-induced absorption spectrum of H_2–He at 300 K (solid line). The dashed line marks the spectral contribution due to quadrupolar induction and the dotted line marks the contribution due to isotropic overlap induction.

Collision-induced spectra are an intrinsic property of dense matter. In fluids that are not too dense, the induced spectra may be represented in the form of a series in powers of the gas density, ϱ^n. For nonpolar gases, the *binary* spectra, i.e., the coefficient of the ϱ^2 term, will be the first non vanishing term. At low densities this absorption will hardly be measurable, yet with increasing density it may become quite striking and the three-, four-, etc., body contributions may also be discernible. For hydrogen and helium, predominantly binary spectra are observed at densities below 10 amagats, except at a few frequencies where the intercollisional process affects the profiles. The characteristic "intercollisional dip" has been observed in the rovibrational CIA spectra, and has been explained by van Kranendonk (1968) as a many-body process, which is due to the destructive interference of the radiation process occurring during collisions. This interference effect is due to the negative correlations existing between the dipole moments induced in consecutive collisions. The main "dip" appears at the $Q(J)$ branch ($\Delta(J) = 0$, $\Delta(v) \neq 0$), but weaker intercollisional dips have also been measured at the $S(J)$ ($\Delta(J) = 2$) lines. The dip, unlike the CIA spectra, broadens linearly with the increasing density of the gas and its intensity does not follow the ρ^2 dependence. An excellent review paper, covering all the details of this phenomenon, which is not part of the *binary* CIA discussed here, has been written by Lewis (1985).

As mentioned above, the basic CIA induction process is "weak". Induced dipole strengths are roughly two orders of magnitude smaller than those of typical molecules like H_2O or HCl. Since the intensity is proportional to the dipole moment squared, CIA may typically be rather negligible in the presence of allowed dipole absorption unless the gas density is high and/or the metallicity (i.e. abundance of elements heavier than helium) is low. In cool, dense stellar atmospheres of low-metallicity stars, composed basically of hydrogen and helium, CIA may dominate the opacity at low (i.e. infrared) frequencies, owing to the absence of dipole moments which are not interaction-induced.

Finally, we mention another important property of CIA spectra. Unlike the well known pressure- broadened line profiles, where the spectral width increases with increasing density, CIA profiles are generally independent of density unless gas densities are very high. That means that a variation of density amounts to a simple rescaling of the profiles, by a factor proportional to density squared. The shape depends on the temperature of the gas and the reduced mass of the collisional partners, i.e. on factors which determine the time scale of molecular encounters. Of course, it also depends on the intermolecular potential and the induced dipole function.

For more details concerning collision-induced absorption, the reader is referred to the review papers by Borysow et al. (1987, 1991), Borysow & Frommhold (1985), and various other contributions to the same conference proceedings (i.e., Birnbaum et al. 1984) or related proceeedings edited by Szudy (1989) and Frommhold & Keto (1990). The first monograph on the subject (Frommhold 1993) is due to appear this year.

2 Stellar Applications

It is well-known that collision-induced absorption plays a significant role in the atmospheres of cool stars, such as white dwarfs, brown dwarfs, M dwarfs and the hypothetical Population III stars. In general, CIA will be important in all cool, zero-metallicity stellar environments. The term "cool" is meant to imply low densities of free electrons, i.e. largely neutral environments where molecules exist. Below, we will give a brief historical review of how CIA became recognized as a significant source of opacity in the various stellar atmospheres.

The importance of pressure-induced absorption for the opacity of the cool stars has been first studied by Linsky (1969), following Trafton's (1964) pioneering study on planetary atmospheres. He pointed out that sources of continuous opacity may influence the radiation field more strongly than the molecular bands, because most radiative energy ("flux") escapes between the individual lines, i.e. where the absorption is weakest. Linsky's focus has been on late-type stars, with temperatures lower than 4000 K and rather high densities, between 10^{-3} and 10 amagats (pressures from 0.01 to 100 atmospheres). He demonstrated that at temperatures below 2500 K practically all hydrogen exists in the molecular form, H_2. Therefore, molecular hydrogen plays an important role in the atmospheric (infrared) opacity. He provided the first models of CIA spectra of H_2–H_2 and H_2–He at temperatures ranging from 600 to 4000 K. At that time, no accurate computations of CIA existed; good experimental results at such temperatures are even now unavailable for H_2–H_2 and H_2–He systems. The models have been based on Trafton's (1966) early quantum mechanical computations for H_2–H_2 pairs at temperatures below 600 K. Linsky's model was based on semiempirical extrapolation procedures. The associated uncertainty of these results could only recently be determined in quantitative terms, when exact quantum calculations became available. Similar work for H_2–He has been even more uncertain: the spectral profiles were assumed to be identical to those of pure H_2 pairs, and the intensity was rescaled by some empirical factor. Figure 3 compares the quantum mechanical models of the RT CIA computed for the H_2–H_2 and H_2–He systems at the temperature of 3000 K. The figure demonstrates the significant difference between the *shapes* of the two spectra at 3000 K, yet the difference is significant at all temperatures (compare also Figs. 1 and 2). Despite its shortcomings,

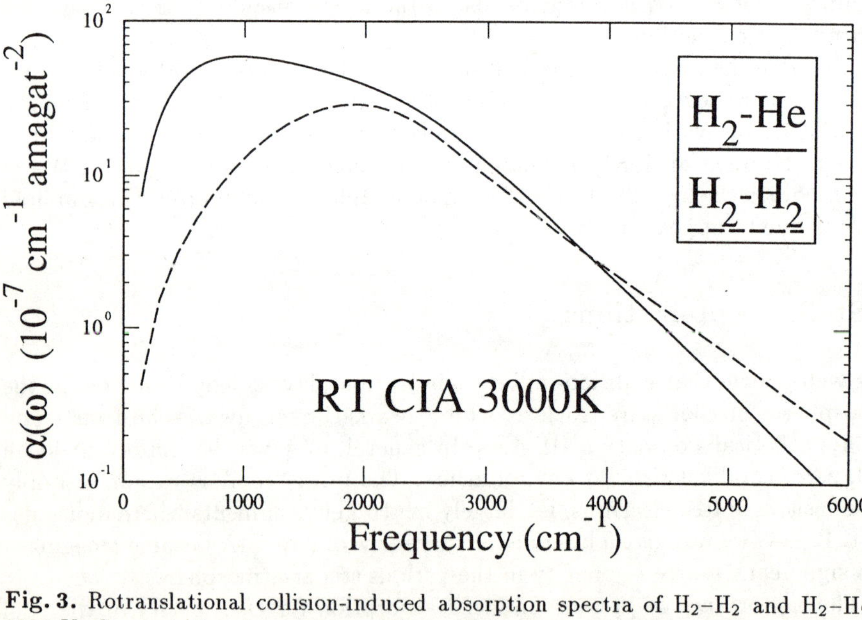

Fig. 3. Rotranslational collision-induced absorption spectra of H_2–H_2 and H_2–He at 3000 K. Computations based on models by Borysow et al. (1988) and Borysow (1993).

Linsky's work unmistakably pointed out the significance of collision-induced absorption for stellar atmospheres. Linsky concluded that at temperatures as low as 2500 K, CIA is the primary source of the continuous opacity. Linsky's model profiles of high temperature CIA have been used for two decades by scientists modelling the cool, hydrogen-rich atmospheres. They enabled astronomers to get an independent confirmation of the importance of CIA in various stellar environments and presented a real motivation toward the development of more realistic models. Advanced, new computations, and their comparison with Linsky's data will be discussed in Sec. 4.

At about the same time, Tsuji (1969) considered molecular opacities in cool stellar atmospheres, including various sources of continuous opacity. He estimated the rovibrational CIA spectrum of hydrogen pairs at 3000 K in the fundamental band of H_2 using a simple, semiempirical procedure. Nonetheless, he has reached independently the same conclusion that CIA is an important source of opacity at high pressures, and that model atmospheres developed without accounting for it are likely to result in unreliable effective temperatures.

In the following decade collision-induced absorption in stellar atmospheres attracted more attention. Shipman (1977), working on the atmospheres of cool white dwarfs, has computed the first model atmosphere which, except for absorption sources accounted for before, also included Linsky's models. He considered pure hydrogen atmospheres at temperatures between 4000 and 8000 K. His primary goal was to determine the role H_2 molecules play in these atmospheres. He found that at a temperature of 4000 K CIA contributes essentially all the opacity at such wavelengths where the flux is emitted! At 5000 K CIA provided only 10% of the total opacity, on account of the larger abundance of electrons and ions, giving rise to the relatively much stronger opacities. These findings have significant implications: depending on whether CIA is included or not, different conclusions may be reached regarding the temperatures, radii and masses of cool stars. Mould & Liebert (1978) in their paper on the atmospheric composition of cool white dwarfs have included CIA due to H_2–H_2 and H_2–He pairs, using again Linsky's data. They computed new atmospheric models at temperatures between 4000 and 7000 K. The overall importance of CIA is not explicitly stated, but the question regarding the reliability of Linsky's predictions was raised. The fact that no experimental data existed to test these estimates has been a serious drawback of any analysis making use of these results.

Population III stars are hypothetical first-generation, zero-metallicity stars, assumed to have formed in the early stages of the Galaxy formation. The stars are assumed to have condensed out of primordial matter and are therefore mainly composed of hydrogen and helium. At temperatures from 1000 to 7000 K, CIA may constitute the major source of opacity in such stars. Palla (1985) pointed out, for the first time, that the more conventional opacity sources, e.g., Rayleigh scattering by H, He and H_2, e^- scattering, bound-free and free-free absorption by H, H^-, and H_2^+, are insufficient. When CIA was finally included for the computations of the Rosseland mean opacities, the magnitude of the error of omitting CIA of H_2–H_2, H_2–He was striking (Palla 1985). Whereas at high temperatures, typically above 3500 K, CIA plays a minor role and other opacity sources dominate, CIA accounts for *all* the opacity at the lower temperatures. To be more specific, at the lowest temperature considered (1500 K) the difference between the Rosseland mean, computed with and without accounting for CIA, amounted to more than one order of magnitude at a density of 10^{-6} amagats, more than three orders of magnitude at $\rho = 10^{-3}$ amagats, and up to five orders of magnitude at the highest densities considered ($\rho = 0.1$ amagats). In Palla's work, atmospheres composed of hydrogen (71%) and helium (28%) were considered and Linsky's (1969) data were again input. One year later, Stahler et al. (1986) confirmed Palla's findings. The conclusion was now made even stronger: at *all* densities between 10^{-8} and 0.1 amagats, and temperatures below 2000 K, CIA has been found to be the dominant process of all continuous opacity sources accounted for, such as electron scattering, Rayleigh molecular scattering, H^- absorption and CIA. At higher temperatures, up to 3000 K, CIA remains the dominant process at the higher densities. In other words, the source of opacity due to collision induced absorption cannot be neglected. Surprisingly enough, it has been demonstrated that, in the absence of other sources of opacity, CIA may still dominate the atmospheric absorption at densities as low as 10^{-8} amagats (it needs to be recalled that $\alpha(\omega) \propto \rho^2!$).

A few years later, Lenzuni et al. (1991) presented a detailed analysis of zero-metallicity stars. They considered the temperatures between 1000 and 7000 K and densities from 10^{-8} to 10^3 amagats. The Rosseland mean, and the monochromatic opacities have been computed, which included collision-induced absorption of H_2–H_2 and H_2–He pairs. For the first time models based on quantum mechanical computations were used in the analysis (for H_2–H_2 in the fundamental band: Borysow & Frommhold 1990, and for all rovibrational bands of H_2–He: Borysow et al. 1989, Borysow & Frommhold 1989). The remainder of the hydrogen bands has been taken either from Linsky, or modeled independently by the authors. We mention that in the case of the RT hydrogen band an incorrect extrapolation of a low temperature model has been used for H_2–H_2 spectra. Despite these shortcomings, the authors demonstrated in an impressive way how CIA depends upon *both* temperature and density. For a fixed temperature at high enough density, CIA may become the major opacity source. At fixed density and low enough temperatures, more or less the same effect is achieved. CIA manifests itself mainly at frequencies below 15,000 cm^{-1}. Additionally, it has been pointed out that accurate data concerning the first and possibly the second overtone may be of the utmost importance in the Rosseland mean opacity because, when CIA is dominant, the monochromatic opacity goes through a pronounced minimum at the frequencies corresponding to these overtone bands. Since the Rosseland opacity depends on the *inverse* of the total opacity, the frequencies at which the minima occur will enter the calculations with the greatest weights. The importance of reliable CIA data in the studies of the protostars seems therefore crucial, especially at the higher frequencies ($v = 0 \rightarrow v' = 2, 3$) for which model calculations still do not exist. As suggested by Lenzuni & Saumon (1992) if the inclusion of more accurate CIA predictions of the higher overtones will result in a larger opacity, it may profoundly affect the currently established minimum mass of H burning in the Population III stars.

Finally, Lenzuni & Saumon (1992) pointed out the likely importance of the many-body effects which are significant at high enough densities. Rough models of these effects have been included and they seem to be important at densities larger than 200 amagats. It needs to be pointed out that whereas spectral moments (integrated intensities) can be estimated for three-body spectra (see, for example, Moraldi & Frommhold 1989), there is no method yet developed to model spectral densities accounting for the many-body effects, not even for the lowest, 3-body effects in CIA. First attempts in modelling the three body spectral contributions were, however, quite promising (see, for example, Chap. 3, in Frommhold 1993), and efforts in this direction continue.

In general, CIA models should be important also in the modelling of the atmospheres of the low mass zero-metallicity stars, M dwarfs (Allard 1994) and the brown dwarfs (Burrows et al. 1993) with H_2–rich atmospheres. Ongoing research (see, for example, Saumon et al. 1994) indicates an increasing interest in collision-induced opacities in the studies of such stellar atmospheres. With more evidence continuously delivered, most likely a greater interest in new CIA spectral contributions will emerge. With it, the need for more calculations, presently not available, is certain to ensue. Even at present, it looks unlikely that the importance of CIA of H–H_2, H^+–H_2, He^+–H_2, He^{++}–H_2 is necessarily negligible for stars with weakly ionized atmospheres (Yorke 1993).

3 Quantum Mechanical Computations

Hydrogen and helium are very light. Even at elevated temperatures, the thermal de Broglie wavelength is still fairly large so that a decidedly quantum mechanical behavior results. For example, at temperatures as high as 7000 K a wavelength is comparable to that of argon at room temperature, i.e. a few tenths of an Ångstrom. As long as the de Broglie wavelength is comparable to the typical intermolecular distances, quantum mechanics, rather than classical theory, has to be used. Moreover, characteristically for all "quantum" systems, only a relatively small number of partial waves (translational orbital angular momentum) needs to be accounted for in the quantum computations involving H_2–X pairs. According to the correspondence principle, only the presence of the large quantum numbers indicate a rather "classical" behavior of the system.

As mentioned above, the "binary" spectra, i.e. those due to pairs of molecules (atoms) constitute the major contribution to CIA, at least at low densities ($\rho \leq$ 10 amagats). The theory for computing quantum mechanical CIA for pairs of molecules is well developed (Birnbaum et al. 1984, Frommhold & Meyer 1987). The theory is based on the assumption of an isotropic potential. It assumes that for the treatment of molecular dynamics, one can neglect the molecular orientations. Such an approximation simplifies the computations in a fundamental way: rotational and translational motions are decoupled. In that case, the spectra amount to a simple convolution of a well known spectral density function of free rotors with the translational profile. It is the latter one where all the computational effort goes: it involves evaluating the induced dipole transition matrix elements between various translational states. For isotropic intermolecular potentials, this requires solving the radial Schrödinger equation, which is by far less complicated than solving a close-coupled set of equations which results when the anisotropy of the system is fully accounted for. The isotropic interaction assumption is quite satisfactory for systems including H_2, certainly at low temperatures and if highly vibrational excitation is not involved.

Additionally, induced dipole moments are now available for H_2-H_2 and H_2-He pairs. Whereas for heavy, highly polarizable molecules the multipolar induction approximation gives generally quite agreeable results, it is not sufficient when one deals with H_2 and He. State of the art, ab initio quantum computations (Meyer et al. 1989a and Borysow et al. 1990) are fortunately available, which make the computational task reliable.

These highly sophisticated induced dipole surfaces, when combined with the best models of the interaction potential, have been employed in the computations of the rotranslational and rovibrational ($v = 0 \to v' = 0, 1, 2$) CIA spectra of pure hydrogen and hydrogen-helium mixtures. In each case, remarkable agreement with the measured data (which exist at temperatures below room temperature) has been found, confirming the soundness of both input data and the applied theory. The reader is referred to Meyer & Frommhold (1989), Meyer et al. (1989b), Borysow et al. (1990) or Meyer et al. (1989a) where such comparisons are presented in full detail. In fact, the success of the method has initiated a long lasting project, which has provided numerous computer models for the planetary scientists involved in atmospheric studies (Borysow & Frommhold 1991). The case of the high temperatures is somewhat more challenging. Whereas at low temperatures the data can be confirmed by existing measurements, the same cannot be said about the computations for temperatures of 1000 K and higher. At high temperatures, the input data have to be carefully selected since induced dipole functions at shorter range are significant which are not yet thoroughly tested by measurements. The same caution must be used when an intermolecular potential is used at temperatures much higher than those of experiments. It is in fact the largest source of uncertainty for the theoretical estimates of the high-temperature behavior of the CIA spectra. The *numerical* accuracy of quantum line shape computations can be kept within the 1% uncertainty range or so. Under the circumstances, however, it would appear that such computations offer unsurpassed reliability relative to the earlier models if cautiously extended to stellar temperatures. For example, with a given set of input data, the theory provides a temperature dependence of the spectra, which cannot be obtained from any one of the semi-empirical extrapolations proposed heretofore.

4 Spectral Models for Stellar Atmospheres

It is desirable that quantum mechanical results are available to the community of interested scientists, over a wide range of temperatures and frequencies. For the modelling of atmospheres, one would like to have a quick access to accurate predictions. For that purpose, we have presented our numerical quantum profiles in the form of spectral "model lineshapes". The exact computational results have been fitted to carefully selected analytical model profiles as functions of temperature and frequency, making sure that the accuracy of the original data is not lost (see, for example, Borysow & Frommhold 1991, Moraldi et al. 1988 or Borysow et al. 1987).

These analytical profiles are extensively tested and match the quantum mechanical profiles within a few percent over the entire molecular band of concern (see, for example, Borysow et al. 1989). FORTRAN programs exist (see BF^{2-4}, Table 1) which utilize these simple lineshapes to generate reliable CIA data over a huge range of temperatures and frequencies. These are readily available from the author by e-mail (aborysow@phy.mtu.edu).

Table 1 presents a list of all available high temperature CIA data, new and old. Linsky's (1969) work is denoted JL; RP stands for the *ab initio* model by Patch (1971). BF denotes models based on quantum mechanical computations: BF^1: Borysow et al. (1988), BF^2: Borysow et al. (1989), BF^3: Borysow & Frommhold (1989) and BF^4: Borysow & Frommhold (1990). Bullets • indicate hydrogen bands at which reliable input data exist, making the quantum computations of the future CIA models possible at temperatures up to 7000 K.

Table 1. Existing CIA models for stellar applications.

Band	H_2–H_2	H_2–He
$0 \rightarrow 0$ RT CIA	JL(600–3000 K) • ↗ 7000 K	BF^1 (40–3000 K) • ↗ 7000 K
$0 \rightarrow 1$ RV CIA	JL(600–3000 K) BF^4 (600–5000 K) RP (300–7000 K)	BF^2 (20–7000 K)
$0 \rightarrow 2$ RV CIA	JL(600–3000 K) • ↗ 7000 K	BF^3 (20–7000 K)
"hot bands"		BF^3 (20–7000 K) various bands $v \rightarrow v'$ $v, v' = \{0,1,2,3\}$

The new ab initio computations permit for the first time the most reliable comparison with Linsky's often used semi-empirical models. As mentioned above, most crucial for the accuracy of the quantum computations is the induced dipole and the intermolecular potential functions input. For H_2 the assumption of the isotropic interaction is suitable, at least as far as our experience with CIA at low temperatures shows. These essential dipole and potential functions have been examined in a great detail. The theoretical computations agree with all experimental data which exist at temperatures up to 300 K. Up to date, no laboratory measurements exist at higher temperatures and greater uncertainty results, for example, due to the choice of the potential model. Nonetheless, it seems that the most recent models must be more reliable than the empirical extrapolation used heretofore.

To present just a few examples, let us focus on selected hydrogen bands. Starting with H_2-H_2 pairs and the rotranslational band: a preliminary study by Borysow (1993) shows that whereas Linsky's predictions agree quite well with the quantum computations at 600 K, they overestimate the value of the absorption coefficient compared to the quantum results at 1000 K by approx. 10%, and underestimate them at 3000 K (~40%). The shape of these spectra agrees reasonably well with state-of-the-art quantum calculations.

In the fundamental band the disagreement is even larger. Figure 4 shows the RV CIA spectra of H_2 pairs at 3000 K. The peak intensity is now off by a factor of two at 1000 K and a factor of four at 3000 K. Moreover, the spectral shape differs somewhat. For comparison, the results of ab initio computations made by Patch (1971) are also plotted. They do not agree with the current model either. A cautionary note is due at this point. An advanced model has been proposed

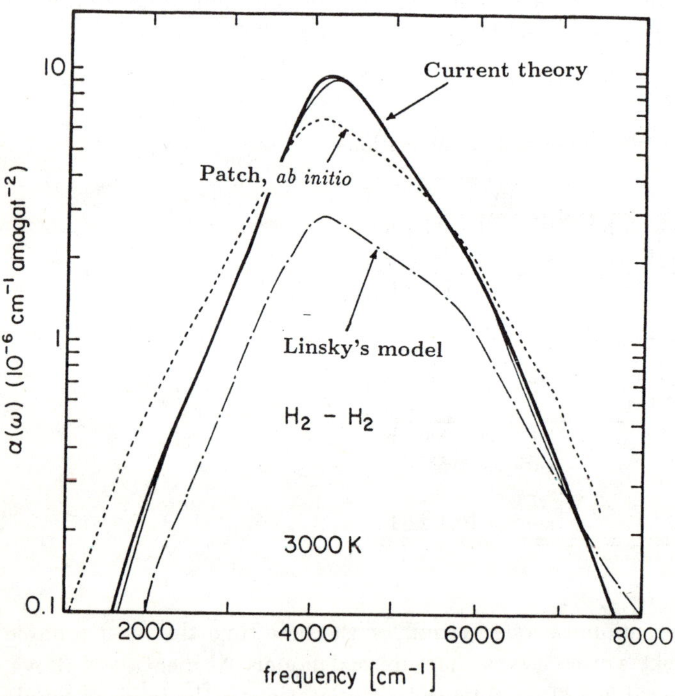

Fig. 4. Rovibrational collision-induced absorption spectrum of H_2-H_2 at 3000 K in the fundamental hydrogen band. The solid line represents the current model, reproduced from Borysow & Frommhold (1990); the dash-dotted line represents for comparison Linsky's (1969) results; the dashed line corresponds to the early calculations by Patch (1971).

(Borysow et al. 1985), which enables one to reproduce the RT CIA spectra of hydrogen pairs at temperatures *below* 300 K. This model was never intended to be used at higher temperatures. If one does extend it to stellar temperatures (as, for example, has been done by Lenzuni et al. 1991), the resulting errors amount to as much as a factor of three at temperature of 3000 K (Borysow 1993), with the current estimates being smaller.

Let us turn next to the discussion of H_2–He models. For the rotranslational band, Linsky assumed that the spectral shapes of H_2–H_2 and H_2–He are the same. He has rescaled the intensity of the pure hydrogen spectra accounting for the smaller polarizability of helium. This assumption is too crude as the comparisons in Figs. 1 and 2, and in also Fig. 3 show.

The RV CIA spectra of H_2–He have been computed by Linsky for the fundamental band only. They can be compared with the quantum model (Borysow et al. 1989). We will not show the spectra here. It is enough to mention that the maximum intensity of the two models differs by roughly a factor of four at 3000 K and a factor of two at 1000 K, with the earlier model predicting too small intensity. Additionally, Linsky's spectra appear as if they were shifted towards the lower frequencies.

5 Summary

In the last two decades it has been demonstrated that collision induced absorption may contribute significantly to the opacity of cool stellar atmospheres, predominantly composed of hydrogen and helium. It is now certain that CIA may actually be a dominant source of opacity at low enough temperatures and/or high enough densities. Additionally, it has been pointed out that in such cases the accuracy of CIA intensities used in the atmospheric models may be critical in determining various stellar parameters like, for example, the effective temperature. It is clear that all future model atmospheres of the cool stars like white dwarfs, M dwarfs, brown dwarfs, the protostars, and in general, the zero-metallicity, low mass stars, must account for CIA infrared bands of H_2–H_2 and H_2–He. Fortunately, reliable and readily available quantum mechanical models exist which can be utilized in the current analyses. Future work should concentrate on modelling of the higher overtones of H_2–H_2 absorption, and of the three body spectral contribution to the CIA at high temperatures.

Acknowledgments. The author would like to thank Dr. Lothar Frommhold for careful reading of the manuscript, and his critical comments regarding the presentation of the material.

References

Allard F., 1994, In: P. Thejll & U. G. Jørgensen (eds.), *Poster session proceedings of IAU Coll. 146*, Copenhagen University, p. 1

Birnbaum G., Chu S.-I., Dalgarno A., Frommhold L., Wright, E. L., 1984, *Phys. Rev.*, **29A**, 595
Birnbaum G. (ed.), 1985, *Phenomena Induced by Intermolecular Interactions, NATO ASI Series, B: Physics*, (New York: Plenum Press), vol. 127
Borysow A., 1993, *unpublished*
Borysow A., Frommhold L., 1989, *Astrophys. J.*, **341**, 549
Borysow A., Frommhold L., 1990, *Astrophys. J. Lett.*, **348**, L41
Borysow A., Frommhold L., 1991, *J. Geophys. Res.: Planets*, **96**, 17,501
Borysow A., Frommhold L., Dore P., 1987, *Int. J. of IR and mm Waves*, **8**, 381
Borysow A., Frommhold L., Moraldi M., 1989, *Astrophys. J.*, **336**, 495
Borysow A., Frommhold L., Meyer W., 1990, *Phys. Rev.*, A, **41**, 264
Borysow A., Moraldi M., Frommhold L., 1991, In: *Trends in Chemical Physics*, (India: Council of Scientific Research Integration), **1**, p. 83
Borysow J., Frommhold L., 1985, In: G. Birnbaum (ed.), *Phenomena Induced by Intermolecular Interactions*, (New York: Plenum Press), p. 67
Borysow J., Trafton L., Frommhold L., Birnbaum G., 1985, *Astrophys. J.*, **296**, 644
Borysow J., Frommhold L., Birnbaum G., 1988, *Astrophys. J.*, **326**, 509
Burrows A., Hubbard W., Lunine J., 1993, *Astrophys. J.*, **345**, 939
Frommhold L., 1993, *Collision-Induced Absorption in Gases*, Cambridge Monographs on Atomic, Molecular and Chemical Physics 2, (New York: Cambridge University Press), 1 edition
Frommhold L., Meyer W., 1987, *Phys. Rev.*, A **35**, 632
Frommhold L., Keto J. (eds.), 1990, *Spectral Line Shapes*, (New York: American Institute of Physics, AIP), volume 6
Lewis J. C., 1985, In: G. Birnbaum (ed.), *Phenomena Induced by Intermolecular Interactions*, (New York: Plenum Press), p. 215
Lenzuni P., Chernoff D. F., Salpeter E. E., 1991, *Astrophys. J. Suppl.*, **76**, 759
Lenzuni P., Saumon D., 1992, *Rev. Mexicana Astron. Astrof.*, **23**, 223
Linsky J. L., 1969, *Astrophys. J.*, **156**, 989
Meyer W., Frommhold L., 1989, *Phys. Rev.*, A **34**, 2771
Meyer W., Borysow A., Frommhold L., 1989, *Phys. Rev.*, A, **40**, 6931
Meyer W., Frommhold L., Birnbaum G., 1989, *Phys. Rev. A*, **39**, 2434
Moraldi M., Borysow A., Frommhold L., 1988, *Phys. Rev.*, A, **38**, 1839
Moraldi M., Frommhold L., 1989, *Phys. Rev.*, A, **40**, 6260
Mould J., Liebert J., 1978, *Astrophys. J.*, **226**, L29
Palla F., 1985, In: G. H. F. Diercksen et al. (eds.), *Molecular Astrophysics*, D. Reidl Publ. Co., p. 687
Patch R. W., 1971, *J. Quant. Spectr. and Rad. Transfer*, **11**, 1331
Saumon D., Bergeron P., Lunine J. I., 1994, In: P. Thejll & U. G. Jørgensen (eds.), *Poster session proceedings of IAU Coll. 146*, Copenhagen University, p. 98
Shipman H. L., 1977, *Astrophys. J.*, **213**, 138
Stahler S. W., Palla F., Salpeter E. E., 1986, *Astrophys. J.*, **302**, 590
Szudy J. (ed.), 1989, *Spectral Line Shapes*, volume 5. Ossolineum Publishing House
Trafton L. M., 1964, *Astrophys. J.*, **140**, 1340
Trafton L. M., 1966, *Astrophys. J.*, **146**, 558
Tsuji T., 1969, In: S.S.Kumar (ed.), *Low Luminosity Stars*, (New York: Gordon and Breach Science Publ.), p. 457
van Kranendonk J., 1968, *Can. J. Phys.*, **46**, 1173
Yorke H., 1993, Private communication

Continuous Molecular Opacities and Photodissociation

Ewine F. van Dishoeck

Leiden Observatory, P.O. Box 9513, 2300 RA Leiden, The Netherlands

1 Introduction

It is well known that molecules have much more complex electronic spectra than atoms due to their rotational and vibrational motions. Most molecular spectra observed in the laboratory and included in model atmospheres involve transitions between two bound electronic states. However, in contrast to atoms, molecules can also have bound–free transitions into states that are repulsive in at least one nuclear coordinate, even below the first ionization potential. These transitions not only lead to dissociation of the molecule, but also provide a source of continuous opacity which is usually not taken into account in model atmospheres. Moreover, photodissociation plays an important role in the non-LTE chemistry of stellar atmospheres and circumstellar envelopes (see Olofsson this volume). In this chapter, a brief overview of our current knowledge of cross sections for continuous absorption and photodissociation is given.

2 Absorption Processes

The processes of bound–bound and bound–free absorption are illustrated in Fig. 1 for a diatomic molecule. In the first case, a discrete spectrum consisting of sharp lines occurs corresponding to transitions into the various vibration–rotation levels of the excited electronic state. In the second case, a broad continuous spectrum arises. The peak of this cross section occurs close to the vertical excitation energy and its width is determined by the steepness of the repulsive potential: the steeper the potential, the broader the cross section. The shape reflects that of the ground–state vibrational wave function; the case illustrated in Fig. 1 refers to a molecule initially in its lowest $v'' = 0$ vibrational level. Cross sections for absorption out of higher vibrational levels $v'' > 0$ would show a nodal structure.

Continuous absorption starts at energies just above the lowest dissociation energy of the molecule, but usually peaks at much higher energies. An extreme case is formed by the H_3^+ ion: its dissociation energy is ~5 eV, but its first continuous absorption channel only occurs around ~18 eV. For simple hydrides such as

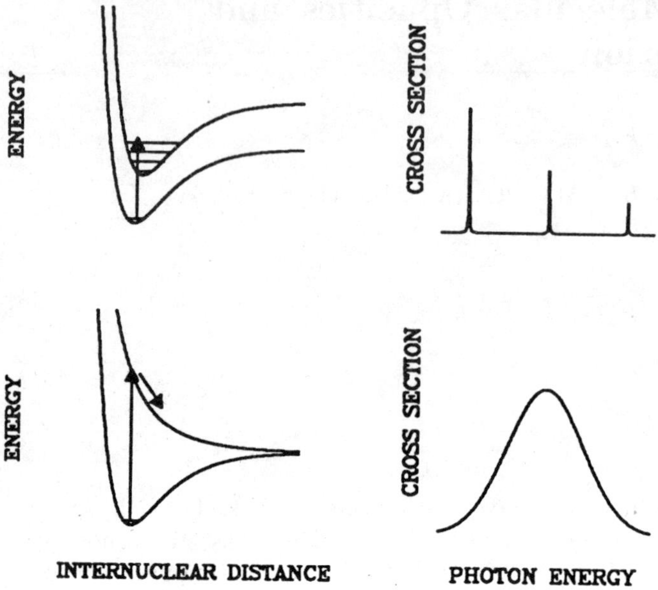

Fig. 1. Potential energy curves illustrating the processes of bound–bound and bound–free absorption for a diatomic molecule.

CH, OH, NH, CH_2 and H_2O, the dissociation energy to ground–state products is typically 4–5 eV, so that continuous absorption can occur at λ <3000 Å. For heavy diatomic and polyatomic molecules containing only first–row atoms (e.g. C_2, CO, CN), the dissociation energy is usually much larger, ranging from 6.2 eV (C_2) to 11.1 eV (CO). Continuous absorption for these molecules starts at wavelengths <2000 Å (C_2) to <1100 Å (CO) and is much less important. For heavy molecules involving second row atoms such as SiC, however, the dissociation energy is only ~4 eV, so that continuous absorption for these species is also likely to be significant.

3 Photodissociation Processes

Virtually all of the continuous absorptions shown in Fig. 1 lead to dissociation of the molecule, since spontaneous emission back to the electronic ground state is a comparatively slow process. This process of *direct photodissociation* is the most efficient way for molecular dissociation to occur, but indirect processes can play a significant role as well. These processes are summarized in Fig. 2, and are initiated by discrete absorptions into bound excited states. More detailed reviews of the photodissociation of astrophysical molecules have been given by van Dishoeck (1988) and Kirby & van Dishoeck (1988).

In the process of *predissociation*, the excited bound state interacts non–radiatively with a nearby repulsive electronic state, resulting in eventual dis-

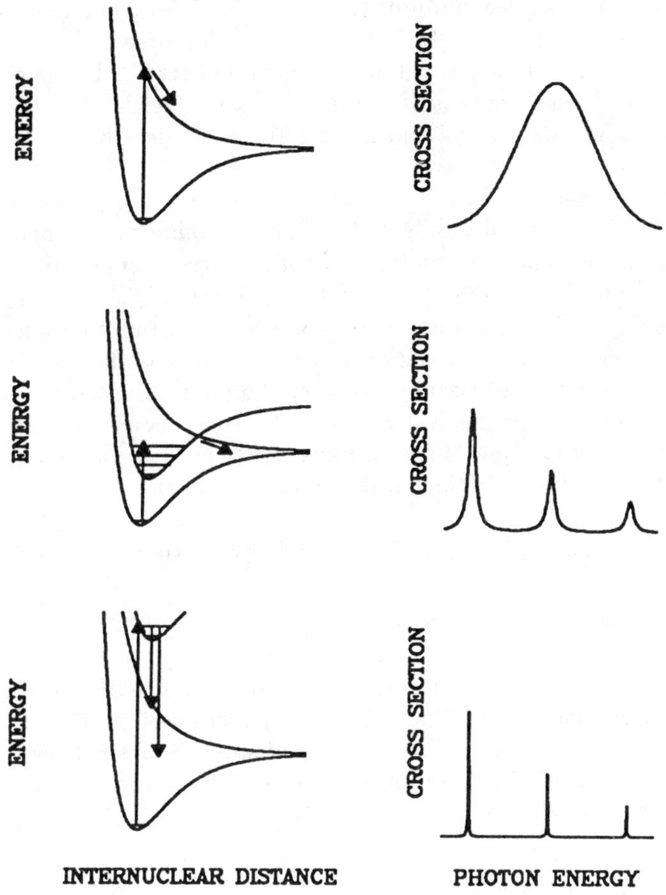

Fig. 2. Potential energy curves illustrating the processes of photodissociation for a diatomic molecule. Top: direct photodissociation; middle: predissociation; bottom: spontaneous radiative dissociation. The corresponding photodissociation cross sections as functions of energy are shown as well.

sociation of the molecule. The strength of the interaction depends on the type of coupling (e.g. spin–orbit) and on the energy level involved, but the corresponding predissociation rates are typically comparable to or larger than the rates for spontaneous emission. The effective photodissociation cross section consists in this case of a series of discrete peaks. The strength of each peak reflects the product of the oscillator strength of the initial absorption and the dissociation efficiency of the level involved. The width is controlled by the sum of the radiative and predissociation rates and is typically of order 1 cm^{-1} if the predissociation is rapid.

If the excited bound states are not predissociated, *spontaneous radiative dissociation* can still be effective through emission of photons into the continuum of a lower–lying repulsive state or the vibrational continuum of the ground electronic state. The efficiency of the process is in this case determined by the competition with spontaneous emission into lower–lying bound states. The photodissociation cross section again consists of a series of discrete peaks, but the peaks are not broadened and have small widths (<0.01 cm^{-1}) determined by the total radiative lifetime.

Because a molecule has many excited electronic states, in general all of these processes will occur. However, usually only one of them dominates the photodissociation of a molecule under astronomical circumstances. For example, the photodissociation of simple hydrides like CH, CH^+, NH, OH, H_2O and CH_2 proceeds mostly through the direct process. On the other hand, the photodissociation of CO is controlled by predissociation processes, whereas that of H_2 occurs by spontaneous radiative dissociation. Whether the photodissociation is dominated by continuous or line processes has important consequences for the radiative transfer through, for example, a circumstellar shell, because in the latter case the lines can become optically thick if the abundance of the molecule is sufficiently high.

The photodissociation rate k_{pd} (in s^{-1}) of a molecule by continuous absorption is given by

$$k_{pd}^{cont} = \int \sigma(\lambda) I(\lambda) d\lambda \ \ \mathrm{s}^{-1},$$

where σ is the cross section for photodissociation in cm^2 and I is the mean intensity of the radiation in photons cm^{-2} s^{-1} Å^{-1} as a function of wavelength λ in Å. For the indirect processes of predissociation and spontaneous radiative dissociation, the rate of dissociation by absorption into a specific level of a bound upper state u from lower level ℓ is

$$k_{pd}^{line} = \frac{\pi e^2}{mc^2} \lambda_{u\ell}^2 f_{u\ell} \eta_u x_\ell I(\lambda_{ul}) \ \ \mathrm{s}^{-1},$$

where $f_{u\ell}$ is the oscillator strength, η_u is the dissociation efficiency of the upper level which lies between 0 and 1, and x_ℓ is the fractional population in level ℓ. The total rate is obtained by summing all the continuous and discrete processes.

In circumstellar envelopes, the radiation responsible for dissociating the molecules is provided mostly by the interstellar radiation field, which extends from the Lyman limit at extreme ultraviolet wavelengths (912 Å) to infrared wavelengths. In the atmospheres of late–type stars, many fewer ultraviolet photons are found. A recent discussion of the interstellar radiation field has been given by van Dishoeck (1993). Reviews of interstellar photodissociation rates have been presented by van Dishoeck (1988) and Roberge et al. (1991).

4 Experimental Data

Laboratory measurements of absorption cross sections have been obtained over a broad wavelength range for most chemically stable species, including H_2O, NH_3, CH_4 and C_2H_2 (see Lee 1984 for a summary). Most of these experiments have been performed at rather low spectral resolution in which individual lines of transitions to bound excited states are not resolved. Although the continuous cross sections due to absorption into repulsive excited states are quite accurate in such measurements, the cross sections for discrete absorptions are very uncertain. Moreover, most experiments do not measure the fluorescence efficiency for such discrete transitions, so that it is often difficult to obtain information on the probability η_u for predissociation processes. Very high spectral resolution experiments at low pressures are essential to obtain reliable data for these discrete absorptions. Note that even for chemically stable molecules, little information is available on cross sections at extreme ultraviolet wavelengths below 1200 Å, CO being the major exception (see below). Although some beautiful experiments have been performed using far–ultraviolet lasers, these are usually limited to only a few specific wavelengths. Experimental data on the photodissociation of reactive molecules such as the radicals CH, OH, CH_2, and C_2H, are still virtually non–existent. Reviews on cross section measurements have been given e.g. by Hudson (1971), Okabe (1978), Ashfold et al. (1979), Lee (1984) and Huebner et al. (1992).

One important circumstellar molecule for which significant experimental progress has been made in the last 6 years is CO. CO is an enormously stable molecule with a dissociation energy of 11.09 eV corresponding to a threshold of 1118 Å. The detailed laboratory spectra of Letzelter et al. (1987) established that no continuous absorption occurs longward of 912 Å, but that the CO photodissociation is dominated by line absorptions in about 30 bands, most of which are strongly predissociated. Subsequent experiments by Stark et al. (1991, 1993), Eidelsberg & Rostas (1990), Eidelsberg et al. (1991), Smith et al. (1991), Levelt et al. (1992), Eikema et al. (1993) and Drabbels et al. (1993) have extended this work to other isotopes and much higher spectral resolution. Although some uncertainties still persist in the oscillator strengths and predissociation rates of individual lines (Kirby & Cooper 1989; Stark et al. 1992; Chan et al. 1993), these are not expected to change much the overall photodissociation rate. The depth dependence of the CO photodissociation rate in circumstellar envelopes is affected not only by self–shielding, but also by shielding by H and H_2 because they absorb in the same wavelength region. ^{12}CO, in turn, can shield the less abundant ^{13}CO and $C^{18}O$ species. Thus, a complete numerical simulation of the entire spectrum of ^{12}CO, ^{13}CO, $C^{18}O$, H_2 and H is required to compute correctly the attenuation at each depth into the envelope (van Dishoeck & Black 1988; Viala et al. 1988; Mamon et al. 1988).

5 Theoretical Calculations

The procedure for obtaining absorption and photodissociation cross sections from theory has been outlined by Kirby & van Dishoeck (1988) for diatomic species and by Schinke (1993) for polyatomic molecules. The first step is to obtain potential energy curves or surfaces of the ground and excited electronic states using ab initio quantum chemical methods (see Langhoff & Bauschlicher and Peyerimhoff, this volume). Such calculations can also provide the transition dipole moments connecting the excited states with the ground state. The second step is to solve the dynamical equations for the nuclear motions on these potential surfaces. This problem can be formulated either in a time-independent or a time-dependent manner. The former method is easy to apply to one-dimensional dissociation involving diatomic molecules, but becomes more cumbersome for multi-dimensional systems. Therefore, the time-dependent method is usually adopted to study the photodissociation of polyatomic molecules.

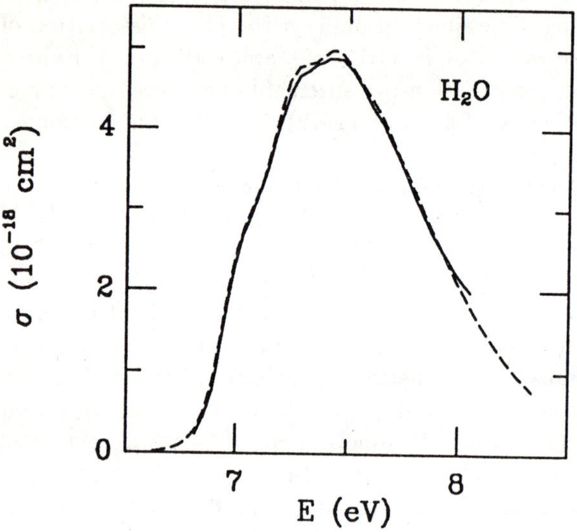

Fig. 3. Comparison between experimental and theoretical cross sections of H_2O in the $\tilde{A}\ ^1B_1 - \tilde{X}\ ^1A_1$ absorption band. The experimental data (full line) are from van Hemert (1980, unpublished results); the theoretical results (dashed line) from Kroes (1992, unpublished results) (see also Engel et al. 1992).

Theory can provide accurate results for small molecules (including radicals) in which the number of electrons is of order 30 or less. In practice, reliable results are limited to the lowest five electronic states per molecular symmetry. For such states, the uncertainty in the transition energies is typically 0.1–0.2 eV and the uncertainty in the transition dipole moments is 10% or less. The subsequent dynamics are in principle exact and are usually converged to better than 1 %, although strong interactions between states can complicate the calculations. Overall, the cross sections have typical uncertainties less than 30% in absolute value and less than 0.2 eV in position.

As an example of the accuracy of the calculations, Fig. 3 shows a comparison between experiment and a fully 3-dimensional theoretical study of the absorption of H_2O into its first excited electronic state. The theoretical cross sections were obtained using ab initio potential energy surfaces and a computed transition dipole moment, and contained no adjustable parameters except for a shift of −0.06 eV in position. The agreement with experiment is remarkable, both in magnitude and shape of the cross section.

6 Examples

6.1 OH and CH

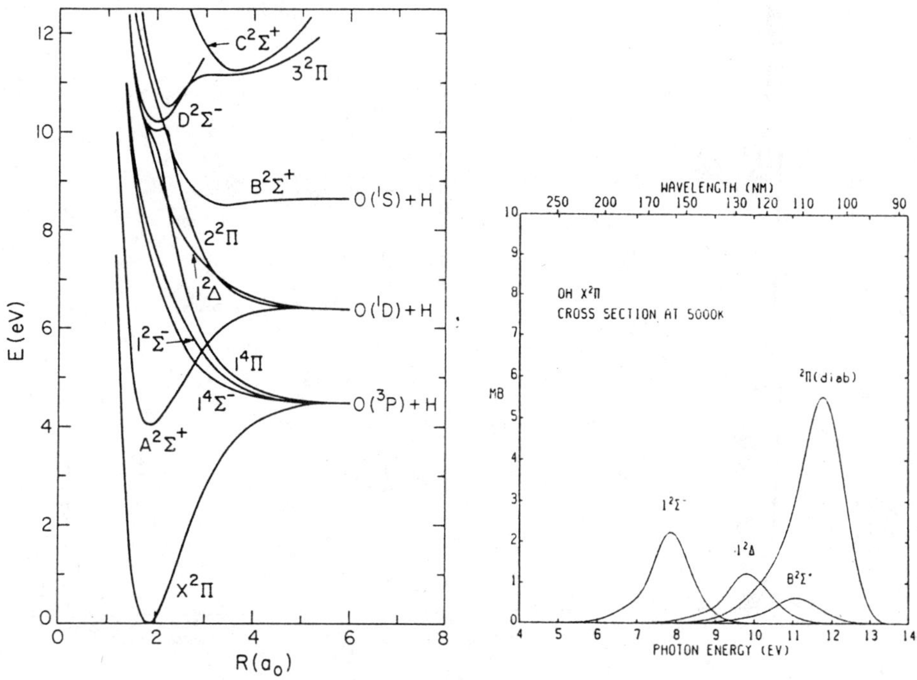

Fig. 4. Potential energy curves (left) and photodissociation cross sections at 5000 K (right) for the OH molecule (from: van Dishoeck & Dalgarno 1983; Kurucz et al. 1987).

The absorption and photodissociation processes of the OH radical have been studied in detail by van Dishoeck & Dalgarno (1983, 1984). In addition to the well–known discrete absorption in the $A^2\Sigma^+$–$X^2\Pi$ transition at ~3000 Å, several continuous absorption channels have been found in the 1000–1900 Å range. The cross sections presented in Fig. 4 were obtained assuming that the population distribution over the lower vibrational levels is characterized by a temperature of 5000 K (Kurucz et al. 1987). Over the wavelength range 1000–3000 Å, the integrated cross section for discrete transitions, $\int \sigma_{\rm line}(\lambda) d\lambda \approx 1 \times 10^{-15}$ cm^2 Å, is smaller than that for continuous absorption, $\int \sigma_{\rm cont}(\lambda) d\lambda \approx 2 \times 10^{-15}$ cm^2 Å.

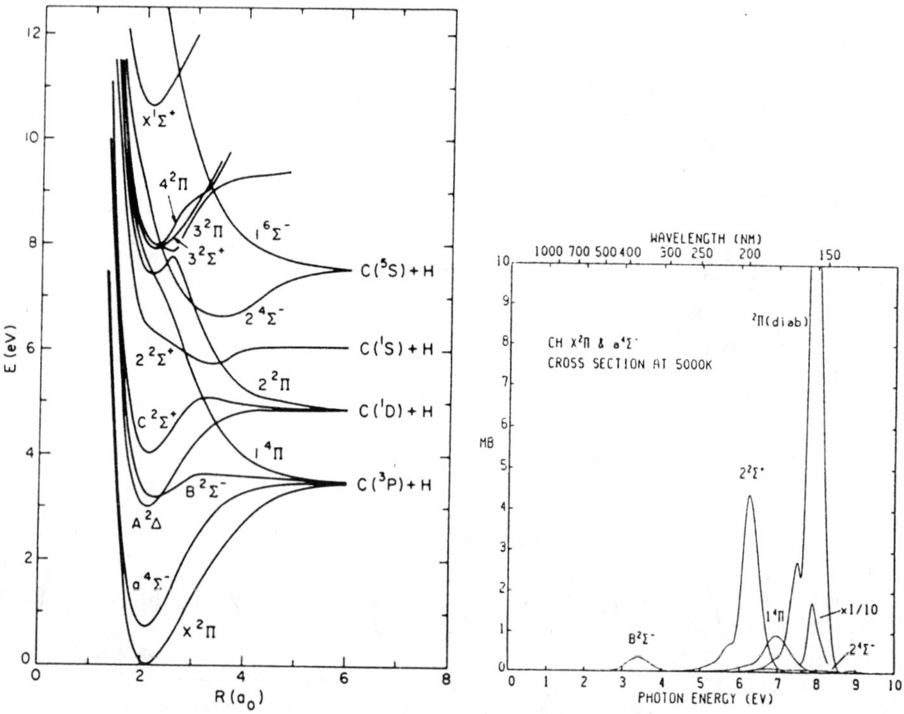

Fig. 5. Potential energy curves (left) and photodissociation cross sections at 5000 K (right) for the CH molecule (from: van Dishoeck 1987; Kurucz et al. 1987).

A similar study has been performed for CH by van Dishoeck (1987), which is illustrated in Fig. 5. In this case, the integrated absorption in discrete lines, $\int \sigma_{\rm line}(\lambda) d\lambda \approx 3.6 \times 10^{-15}$ cm^2 Å, is comparable to the integrated continuous absorption, $\int \sigma_{\text{cont}}(\lambda)d\lambda \approx 3.5 \times 10^{-15}$ cm^2 Å, over the wavelength range 1000–5000 Å.

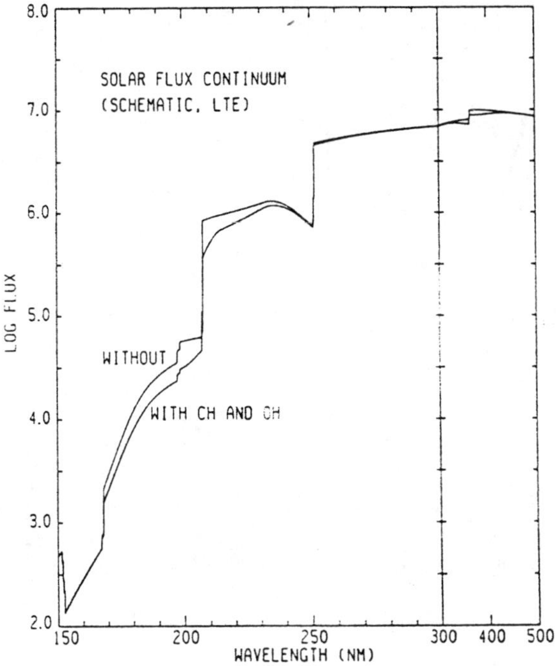

Fig. 6. Solar ultraviolet continuum flux (in erg cm^{-2} s^{-1} sr^{-1} nm^{-1}) computed with and without inclusion of OH and CH continuous absorption.

The effect of the CH and OH opacities on the solar continuum has been investigated by Kurucz et al. (1987) using Allen's (1978) LTE solar model in Kurucz's spectrum synthesis program (Kurucz & Avrett 1981) without including line absorptions. The result is presented in Fig. 6. Below 1500 Å, the continuum of the Sun is formed mostly in the chromosphere where the contribution of OH and CH is negligible. However, between 1700 and 2500 Å, OH and CH affect the continuum ultraviolet flux by tens of percent. When line opacity is included in the models, the relative effect will be smaller, but it remains relatively important in regions between the lines.

6.2 NH and CH_2

The potential energy curves and cross sections for the NH molecule have been computed by Kirby & Goldfield (1991). As for OH and CH, several continuous dissociation channels are found, although they start at quite high energies, λ <1700 Å.

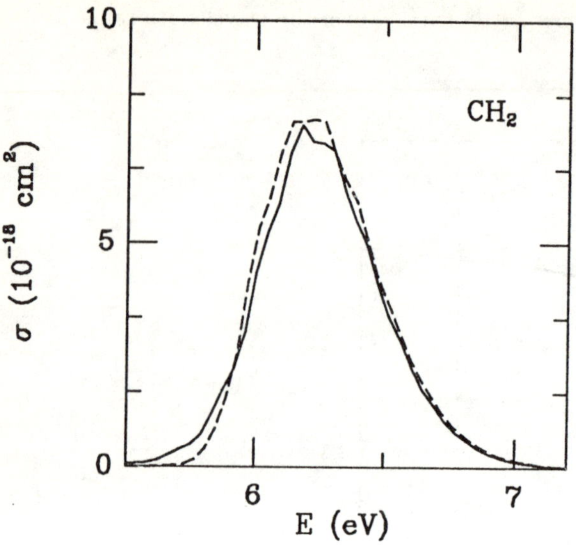

Fig. 7. Photodissociation cross sections for absorption into the first excited triplet state of CH_2 obtained with a fully 3-dimensional dynamical calculation (full line) and with a 2-dimensional calculation keeping the angle fixed at α_{eq}=134° (dashed line) (from: Kroes et al. 1993).

The photodissociation processes of the CH_2 molecule have recently been investigated in a series of papers by Beärda et al. (1992) and Kroes et al. (1993). Fully 3-dimensional potential energy surfaces and transition dipole moment functions have been obtained for the lowest ten states of triplet symmetry of the molecule. The subsequent dynamics for dissociation in the first absorption band have been solved in 2- and 3-dimensions, with good agreement between the two cases. If this result would apply more generally, substantial computational savings could be obtained. The cross sections are presented in Fig. 7.

7 Conclusions

It has been shown that continuous molecular opacity due to bound-free absorptions above the dissociation energy may have a non-negligible effect on stellar atmosphere continua. For many simple molecules of astrophysical relevance, the integrated cross sections for continuous absorption are comparable to those for discrete absorptions. The continuous absorptions also lead to photodissociation of the molecules, a process which plays an important role in the non-LTE chemistry in stellar atmospheres and circumstellar envelopes.

References

Allen R.G., 1978, Ph. D. Thesis, Univ. of Arizona

Ashfold M.N.R., Macpherson M.T., Simons J.P., 1979, *Topics Curr. Chem.*, **86**, 1

Beärda R.A., van Hemert M.C., van Dishoeck E.F., 1992, *J. Chem. Phys.*, **97**, 8240

Chan W.F., Cooper G., Brion C.E., 1993, *Chem. Phys.*, **170**, 123

Drabbels M., Heinze J., ter Meulen J.J., Meerts W.L., 1993, *J. Chem. Phys.*, in press

Eidelsberg M., Rostas F., 1990, *Astron. Astrophys.*, **235**, 472

Eidelsberg M., Benayoun J.J., Viala Y., Rostas F., 1991, *Astron. Astrophys. Suppl.*, **90**, 231

Eikema K.S.E., Hogervorst W., Ubachs W., 1993, *J. Chem. Phys.*, submitted

Engel V. et al., 1992, *J. Phys. Chem.*, **96**, 3201

Hudson R.D., 1971, *Rev. of Geophys. and Space Physics*, **9**, 305

Huebner W.F., Keady J.J., Lyon S.P., 1992, *Astron. Ap. Space Sci.*, **195**, 1

Kirby K., Cooper D.L., 1989, *J. Chem. Phys.*, **90**, 4895

Kirby K., Goldfield E.M., 1991, *J. Chem. Phys.*, **94**, 1271

Kirby K., van Dishoeck E.F., 1988, *Adv. At. Mol. Phys.*, **25**, 437

Kroes G.J., van Dishoeck E.F., Beärda R.A., van Hemert M.C., 1993, *J. Chem. Phys.*, **99**, 228

Kurucz R.L., Avrett E.H., 1981, *Smithsonian Astrophysical Observatory Special Report*, **391**, 145 pp

Kurucz R.L., van Dishoeck E.F., Tarafdar S.P., 1987, *Astrophys. J.*, **322**, 992

Lee L.C., 1984, *Astrophys. J.*, **282**, 172

Letzelter C., Eidelsberg M., Rostas F., Breton J., Thieblemont B., 1987, *Chem. Phys.*, **114**, 274

Levelt P.F., Ubachs W., Hogervorst W., 1992, *J. Chem. Phys.*, **97**, 7160

Mamon G.A., Glassgold A.E., Huggins P.J., 1988, *Astrophys. J.*, **328**, 797

Okabe H., 1978, *Photochemistry of small molecules,* (Wiley, New York)

Roberge W.G., Jones D., Lepp S., Dalgarno A., 1991, *Astrophys. J. Suppl.*, **77**, 287

Schinke R., 1993, *Photodissociation Dynamics* Cambridge: Cambridge University

Smith P.L., Stark G., Yoshino K., Ito K., Stevens M.H., 1991, *Astron. Astrophys.*, **252**, L13

Stark G., Yoshino K., Smith P.L., Ito K., Parkinson W.H., 1991, *Astrophys. J.*, **369**, 574

Stark G., Smith P.L., Ito K., Yoshino K., 1992, *Astrophys. J.*, **395**, 705

Stark G., Yoshino K., Smith P.L., Esmond J.R., Ito K., Stevens M.H., 1993, *Astrophys. J.*, **410**, 837

van Dishoeck E.F., 1987, *J. Chem. Phys.*, **86**, 196

van Dishoeck E.F., 1988, in *Rate Coefficients in Astrochemistry*, T.J.Millar & D.A.Williams (eds.), Dordrecht: Kluwer, 49

van Dishoeck E.F., 1993, to appear in *Infrared Cirrus and Diffuse Interstellar Clouds*, W.L. Latter&R. Cutri (eds.), Astr. Soc. Pacific

van Dishoeck E.F., Black J.H., 1988, *Astrophys. J.*, **334**, 771

van Dishoeck E.F., Dalgarno A., 1983, *J. Chem. Phys.*, **79**, 873

van Dishoeck E.F., Dalgarno A., 1984, *Astrophys. J.*, **277**, 576

Viala Y.P., Letzelter C., Eidelsberg M., Rostas F., 1988, *Astron. Astrophys.*, **193**, 265

Effects of Non-Local Thermodynamic Equilibrium (NLTE) on Molecular Opacities

Hollis R. Johnson

Astronomy Department SW 319, Indiana University, Bloomington, IN 47405, U.S.A.

1 Introduction and Basic Concepts

Molecules increase in importance in stellar atmospheres as the temperature falls along the spectral sequence for all luminosity classes. Among their many roles molecular features provide the basis for spectral classification (temperature and chemical composition) in M, S, and C stars; they constitute important opacity sources across the spectrum from the ultraviolet to the infrared; they influence the energy budget and hence the kinetic temperature in the outer photospheric layers; they may contribute significantly to the radiative levitation of the atmosphere; they form masers; they may cause or facilitate mass loss; and they are indicative of physical conditions and processes from the photosphere to the outer circumstellar envelope.

Most of the known molecular opacity is produced by only a few molecules – H_2^+, H_2^-, pressure-induced H_2 opacity, and opacities of CH, NH, MgH, SiH, CO, SiO, CN, CS, and SiS, of which CN is usually the most important. In normal oxygen-rich stars there are also TiO, VO, H_2O, and OH, and in carbon-rich stars there are also C_2, HCN, C_2H_2, and C_3. Finally, there are such hitherto neglected molecules as ZrO, FeH, C_2H, AlOH, and perhaps several more. Despite the importance of molecular opacities, little attention has been given to the effects of NLTE, as is clear from recent reviews (see Tsuji 1986).

Competition between the processes of molecular formation and destruction determines the number density of each molecular species and each energy level. If these processes occur at any but their equilibrium rates, they will lead to overpopulation or underpopulation of either molecular states or the density of the molecules themselves. In chromospheres, shocks, winds, molecular-cluster and dust-formation regions, and maser-formation regions, we expect major departures from LTE. However, our focus here is on the photosphere.

In strict thermodynamic equilibrium (TE) each process is exactly balanced by its inverse process at every velocity and frequency, and the distribution functions of equilibrium statistical mechanics hold. Such an ideal regime is approached deep inside a star. However, TE is not expected to hold exactly in the presence of temperature gradients or anisotropies in the radiation field, both of which must occur in stellar atmospheres. Although strict TE cannot be expected to hold, astrophysicists have long employed a slightly relaxed form of TE – local thermodynamic equilibrium (LTE). In this, one still expects (a) particle velocities are Maxwellian because of the short mean free path and high elastic collision frequency; (b) electrons are distributed among atomic and molecular states according to the Boltzmann law; (c) atoms are distributed among stages of ionization according to the Saha law; and (d) molecular dissociative equilibrium (sometimes called chemical equilibrium) holds, but (e) the radiation field might not be planckian because of the long mean free path of the photons and the necessary anisotropy of the radiation field near the stellar surface.

In fact, the specific intensity of radiation, I_ν, is described by

$$\mu \frac{dI_\nu}{dr} = \xi_\nu - \chi_\nu I_\nu, \tag{1}$$

where I_ν is the specific intensity [erg cm^{-2} s^{-1} Hz^{-1} sr^{-1}, for example], r is an outwardly directed distance, χ_ν is the linear extinction coefficient [cm^{-1}], ξ_ν is the volume emission coefficient [erg cm^{-3}s^{-1}sr^{-1}Hz^{-1}], and $\mu = \cos\theta$, where θ is the angle between the normal to the infinitesmal surface and the direction of the pencil of radiation. If we define the optical depth, $d\tau_\nu = -\chi_\nu\, dr$, and the source function, $S_\nu = \xi_\nu/\chi_\nu$, we obtain the familiar form (see also Scholz & Wehrse this volume) of the equation of radiative transfer

$$\mu \frac{dI_\nu}{d\tau_\nu} = I_\nu - S_\nu. \tag{2}$$

In TE, $I_\nu = B_\nu(T)$, the Planck function. However, in a stellar atmosphere, this is not generally true. The assumption that all the other distribution functions (a) –(d) above are nevertheless valid leads to the condition of "local" thermodynamic equilibrium (LTE), and this is a "standard" assumption in classical stellar atmospheres. The reader will surely notice the inconsistency of the LTE idea that the radiation field might depart from its LTE value without influencing the distributions of electrons, molecules or atoms. This logical inconsistency is a small price to pay, however, for the convenience of using the Boltzmann and Saha equations to calculate number densities. The question is not: Is LTE valid? Of course it is not strictly valid in stellar atmospheres because of the inconsistency noted. The question is: How bad are the LTE results?

A broad view of NLTE, including all departures from the idealized situation described above, is illustrated in Fig. 1. which shows a stylized representation of the formation and destruction of a diatomic molecule AB from its constituent neutral atoms A and B, the populations of each of which might be influenced by interactions with other molecules or other stages of ionization. Molecules might also be formed by any of several other processes, such as exchange reactions.

NLTE effects can enter in many places, as shown in Fig. 2. Populations of levels, and hence source functions and optical depths, depend on the competing rates of production and destruction. If there is a steady state, often called statistical equilibrium, then, for the number density of the particles corresponding to any state j (which could be an atomic or molecular energy level), $dn_j/dt =$ (production rate of n_j) − (destruction rate of n_j) = 0. That is, the sum of all processes leading to the creation of a particle in state j is equal to the sum of all processes taking particles out of state j (see Mihalas 1978; Anderson 1989).

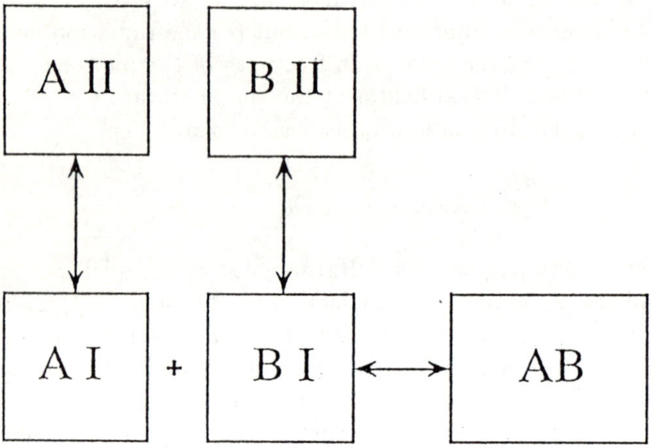

Fig. 1. Schematic representation of the formation of a diatomic molecule AB from neutral atoms A and B.

Each process involves a forward and a reverse rate or process. Considering only states i and j, if P stands for the rate of any process, we can describe the steady-state equilibrium between i and j by

$$n_i P_{ij} = n_j P_{ji}, \tag{3}$$

where n_i is the number density of particles of species i, and P_{ij} is the rate per particle of any process transferring particles from one state i to another state j. Normally, many states will be connected to the j'th state, and there will be many terms in the equation. The ratio of the number densities of j and i in the simple case where no other interactions occur is given by

$$\frac{n_j}{n_i} = \frac{P_{ij}}{P_{ji}}. \tag{4}$$

In TE the rates P_{ij} and P_{ji} will balance in such a way that their ratio is just the relevant equilibrium distribution function. More generally, for an energy level j in a system of N levels,

$$n_j \Sigma_i P_{ji} + n_j P_{jK} = \Sigma_i n_i P_{ij} + n_K P_{Kj}; \quad j = 1, 2 ... N, \tag{5}$$

where K represents the continuum or the dissociated state.

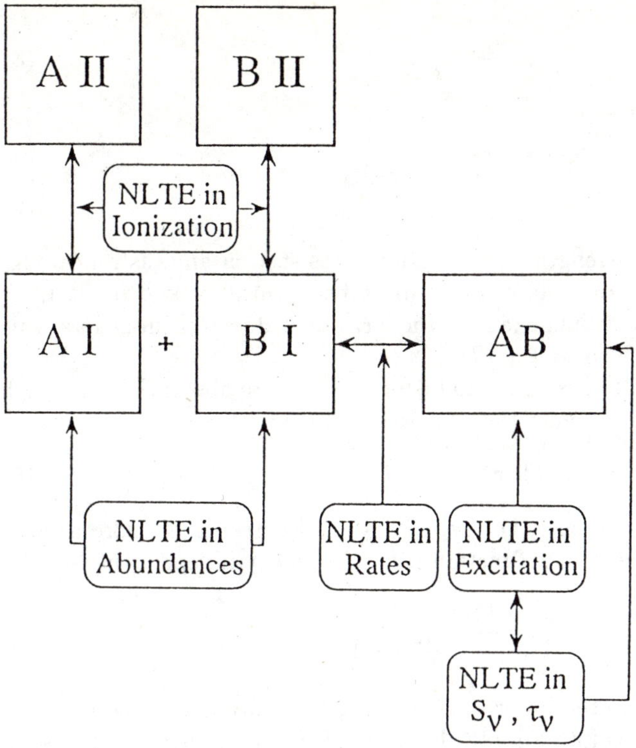

Fig. 2. Schematic representation of the points (formation and destruction rates, excitation mechanism, spectral lines, and the pool of constituent atoms) at which departures from LTE (NLTE) in the rates might occur for the simple situation shown in Fig. 1. Any NLTE effect might influence the populations of the molecular energy levels and the opacity.

In general, each transition can be made radiatively or collisionally, so that

$$P_{ij} = R_{ij} + C_{ij}, \tag{6}$$

where R and C are the radiative and collisional rates. Since the colliding particles have a maxwellian velocity distribution, we can compute the collisional rate, C, as an integral over velocity and cross section

$$C_{ij} = C_{ij}(Q, n, T), \tag{7}$$

where Q represents a cross section. In steady-state equilibrium, all the relevant cross sections must be known, and astrophysicists depend on quantum chemists and laboratory spectroscopists for these values. Note that the density and temperature of the colliding species (usually electrons, H or He atoms, or H_2 molecules) are local quantities which furthermore are in equilibrium, so that if one computes

$$\frac{n_j}{n_i} = \frac{C_{ij}}{C_{ji}}, \tag{8}$$

the result will be the TE value.

For the radiative rates one has

$$R_{ij} = R_{ij}(f, I_\nu), \tag{9}$$

where f is an oscillator strength or absorption cross section and, as repeatedly stressed at throughtout this book, these must be accurately known. If $I_\nu = B_\nu(T)$, the ratio of n_j/n_i would again be the TE value. However, in general this is not true, for I_ν (τ_ν) given by Eq. (2).

Molecular lines are often assumed to be formed by pure absorption, in which case the source function becomes the Planck function

$$S^l_\nu(\tau_\nu) = B_\nu(T). \tag{10}$$

Even under these circumstances, however, I_ν is not Planckian! A more general approximation to the line source function is the frequency-independent form

$$S^l_\nu = \frac{2h\nu^3}{c^2} \frac{1}{\frac{n_i g_j}{n_j g_i} - 1}, \tag{11}$$

where j is the upper and i the lower state and g is the statistical weight.

Substitution of Eq. (5) into Eq. (12) leads to a form which is a linear combination of pure absorption and complete redistribution (pure non-coherent scattering):

$$S^l_\nu = \epsilon B_\nu(T) + (1 - \epsilon)\bar{J}, \tag{12}$$

where

$$\bar{J} = \int_0^\infty J_\nu \phi_\nu d\nu, \tag{13}$$

ϕ_ν is the normalized absorption coefficient, and ϵ is related to the fraction of photons re-emitted thermally.

To obtain a physical feel for the radiation field, one might imagine that I_ν or its angle average, J_ν, can be parameterized as

$$J_\nu = W B_\nu(T_r), \tag{14}$$

where W is a dilution factor (the fractional solid angle filled by incoming radiation at the point of interest) and T_r is the radiation temperature. Thus J_ν might differ from $B_\nu(T)$ either because of the dilution of the radiation field near the surface ($W < 1$) or because the (wavelength-dependent!) "radiation temperature" (T_r) departs from the local electron temperature (T).

The complexity of the true (NLTE) situation is now apparent. The number densities of the relevant atomic and molecular states and species depend upon the radiative and collisional rates, and the radiative rates depend on the radiation field (Eq. 9). The radiation field (I_ν) depends upon the source functions and optical depths (Eq. 2), which depend upon the number densities of relevant states (Eqs. 1 and 11). Often the coupling between the radiation field and the energy levels is very strong, and special mathematical techniques are needed to solve the system of equations.

As is clear from the previous (Eqs. 5, 9, and 14), departures from LTE in level populations will occur if (a) radiative rates exceed collisional rates and (b) the radiation field is non-planckian ($I_\nu \neq B_\nu$ (T)). Stated differently, level populations will be driven to their LTE values if collisional rates (which occur at their equilibrium values) dominate radiative rates or if radiative rates proceed at their equilibrium value (the intensity is the Planck function). Deep in the photosphere particle densities are high ($\sim 10^{15}-10^{18}$ cm^{-3}), and collisional rates are high. Furthermore, optical depths are large, and the radiation field approaches a planckian radiation field. Under these conditions, we expect LTE to be an excellent approximation. In the outer photosphere particle densities are lower (10^{11}-10^{14} cm^{-3}), and electron densities are far lower, and departures from a plankian radiation field arise both from the existence of the surface ($\tau =$ 0) with its consequent anisotropy in the radiation field and from the frequency dependence of the absorption coefficients and therefore of the surface flux. Effects of NLTE are to be expected. In chromospheres, coronae, winds, and CSE's, NLTE effects dominate, and the full coupled radiative-transfer and steady-state equations (or even non-static equations) must be employed.

Two diametrically opposite philosophical approaches to NLTE in stellar photospheres have therefore arisen. One position, the one almost always taken in calculations of molecular opacities, is that the assumption of LTE with its attendant simplicity should be made until clearly discrepant results force us to the tedious NLTE calculations. The other position is to begin NLTE calculations at the outset. There is hardly any middle ground.

For which molecular levels and transitions are NLTE effects likely to be important? Molecular energy states are classified as rotational, vibrational, and electronic. Electronic levels and transitions involve energies of a few (2–6) eV, whereas the average kinetic energy per particle is of the order of 0.3 eV. Furthermore, the agents of electronic collisions in molecules, as in atoms, are electrons, the density of which decreases rapidly toward lower temperatures as abundant elements (H, C, N, O) become neutral. These transitions are therefore quite susceptible to NLTE effects (Thompson 1973; Mount et al. 1975; Hinkle & Lambert 1975; Anderson 1989; Ayres & Wiedemann 1989). Fortunately, many electronic molecular transitions fall in the violet and ultraviolet spectral regions where, due to the low stellar flux, their effects on the overall opacity and therefore on the atmospheric structure is minor. By contrast, rovibrational transitions have energies of the order of 0.5 eV, and rotational transitions have energies of the order of 0.05 eV. Since these are close to the average energy per particle in cool stellar atmospheres, many particles are available for collisions. Furthermore, these latter collisions are more geometric than coulombic and are mediated by neutral particles, especially neutral hydrogen atoms, which are relatively abundant even in the low-density atmospheres of red-giant stars.

Direct measurements from spacecraft show that both the absolute stellar ultraviolet flux and the fractional ultraviolet flux in molecule-rich giants are weaker than in warmer stars but still quite significant (Steiman-Cameron et al. 1985; Schrijver 1987; Judge 1989). In fact, radiation temperatures (T_r) at violet and ultraviolet wavelengths may easily exceed the electron temperature in the outer photosphere and temperature-minimum region, especially if chromospheric photons are also included (Luttermoser 1993). For example, the color (radiation) temperature in α Ori never falls below 2950 K at any wavelength, and it rises sharply at wavelengths shortward of 300 nm to values of 3,000-4,000 K (Carpenter et al. 1993). By contrast, the temperature minimum must have an electron temperature near 2400 K. The ultraviolet radiation field is relatively much stronger in cool dwarfs. It is this relatively strong ultraviolet radiation field that causes the overionization and overexcitation in both molecules and atoms found in the outer photosphere in several studies (see individual molecules).

We now summerize research on a few specific molecules.

2 H_2

Because it is the most abundant element, the equilibrium of hydrogen is fundamental to all hydrogen-containing molecules and their opacities. Although it has no bands of significance except through pressure-induced opacity (Borysow this volume), the molecule H_2 has an important indirect influence on the opacity through its influence on the abundance of H^-, whose opacity is important or dominant in all molecule-producing atmospheres except H-deficient objects.

A pioneering study of the actual (NLTE) equilibrium for both H_2 and H^- was reported in a classic paper by Lambert & Pagel (1968), who set up and solved a small network of equations governing the equilibrium of H^-. From considerations of several mechanisms of formation and destruction, they found the most important destruction mechanisms for H^- in stars of spectral type F0-K0 to be photodetachment ($H^- + h\nu \rightarrow H + e$) and associative detachment by neutral hydrogen atoms ($H^- + H \rightarrow H_2 + e$). Rates then available led to their conclusion that "NLTE effects are probably absent." Because of its indirect influence on the H^- abundance, the formation of H_2, for which 3-body recombinations ($3\ H \rightarrow H_2 + H$) seemed to dominate, was also examined. Relative to the reservoir of neutral hydrogen the number density of H_2 was also found to be in TE. A larger network of equations governing the equilibrium of H^-, H, H^+, H_2, and $H_2{}^+$ in a form suitable for calculation by computers was set up, but not solved, by Lites & Mihalas (1984). Most modern calculations of H^- and H_2 have cited this early research and have either assumed H^- to be in LTE relative to H (Vernazza et al. 1981) or have calculated it to be so from a simplified set of equations (Anderson 1989). In both these latter calculations H_2 was assumed to be in TE relative to H I and it was analytically eliminated from the H^- network of equations.

As a check on the abundance of H_2, and therefore on dissociative equilibrium calculations, observations of the infrared quadrupole transitions of H_2 are of paramount importance, and a fair sample of these are newly available (Hinkle et al. 1993), including observations of about 30 M, S, and C stars of all variability classes and of the mira variable star χCyg throughout its cycle. A few H_2 lines, especially the 1-0 S(0) and S(1) lines, were seen in most stars (the strongest lines being in S-type miras), and their strengths were roughly in agreement with the expectations of dissociative (chemical) equilibrium calculations. The great strengths of the stronger observed lines requires their formation over a large range of the stellar atmosphere, and, in extreme cases, up to 0.1 $M_{\odot}$ is contained in the atmosphere. Part of the line was found to originate in near-photospheric circumstellar material slightly shifted in velocity relative to the photosphere. It had earlier been suggested that if H_2 is produced by three-body recombination, the recombination would be so slow in the low densities in mira atmospheres that H would remain neutral (not associated) between shocks (Bowen 1988). However, these observations disclose that, after the passage of a pulsationally driven shock through the atmosphere, the H_2 lines reappear almost as quickly as do CO and TiO lines, forcing the conclusion that recombination proceeds more rapidly than expected, which in turn demands higher densities of H I or consideration of processes other than three-body recombination. An increase in the rate of association of H into H_2 would tend to drive the number density of H_2 closer to its LTE value. However, in such a complex system as shocks in the pulsating atmosphere of a mira variable star, a completely self-consistent NLTE solution is required, and that is still in the future.

3 CH

An early (perhaps the earliest) calculation of NLTE in CH (Hinkle & Lambert 1975), noted that the three observed electronic transitions of CH: $A^2\Delta-X^2\Pi$, $B^2\Sigma^- -X^2\Pi$, and $C^2\Sigma^+-X^2\Pi$ have very low electronic oscillator strengths, so that their contribution to the opacity in the Sun is small, while the stronger transitions lie so far to the ultraviolet (137.0 nm) that their contribution to the opacity is negligible. That conclusion is strengthened as one goes toward lower temperatures.

A detailed calculation of the CH (A-X), CH (B-X), and CH (C-X) systems in the solar atmosphere by the methods outlined in Section 8 was made by Anderson (1989). Interesting departures from LTE were found in all three systems in the outer solar photosphere in the sense that higher-lying levels were overpopulated relative to lower-lying levels. The departures begin sharply above log m = 2.0 in this radiative-equilibrium model, where m is mass column density (g cm^{-2}), just outside the temperature-minimum region in the real solar atmosphere. The physical reasons for the over excitation are related to the relatively strong ultraviolet radiation field as sketched in the last part of the Introduction.

A recent comparison of values for the carbon abundances in the solar atmosphere (Grevesse et al. 1991) shows that an LTE abundance analysis of the CH (A-X) electronic system yields the same value for the carbon abundance as do the rotation-vibration bands in the ground state, and that value agrees closely with the mean value of other abundance indicators. This agreement is important, for it demonstrates that any NLTE effects are confined to the outermost layers where they may slightly affect the line cores but probably do not affect the opacity (which for CH in the sun is of minor importance anyway).

4 CN

Bands of the red system of CN ($A^2\Pi - X^2\Sigma^+$) are conspicuous in the red/near-IR spectral region (0.7−1.5 μm) and provide important opacity in cool stars of all chemical composition, but we know of no evidence from either observation or theory for departures from the predictions of LTE in these lines.

Calculations of NLTE for the violet system of CN ($B^2\Sigma^+ - X^2\Sigma^+$) have been carried out for the Sun (Mount & Linsky 1975) and for Arcturus (Mount et al. 1975). Because of the obvious computational difficulties of such a complex system, the analysis was carried out only for rotational transitions from vibrational states $v''=0$ to $v'=0$, which form the bandhead at 388.3 nm. This amounted to 32 overlapping doublets corresponding to the P1 and P2 divisions of the P branch; the weaker Q branch was ignored. All rotational transitions in both the upper and lower electronic states were assumed to be in LTE. The equivalent-two-level-atom formulation of the source function was adopted, the line absorption coefficient was taken to be doppler, and complete redistribution was assumed in the photon redistribution function. For the Sun, significant NLTE effects, as much as a factor of 2.0 in the ratio $S^l/B_\nu(T)$, were found for all four atmospheric models tested, including some with inhomogeneities. It was concluded that this factor could cause an increase of 0.2 dex in the abundance of carbon. However, these workers obtained an abundance of C lower by 0.2−0.3 dex than the value of 8.60 commonly accepted (Grevesse et al. 1991), and this unexplained difference has cast an uncertainty over their results.

Because of the low density in the atmosphere of the red giant Arcturus, values of ϵ were so low that most of the band head was formed by (pure non-coherent) scattering. Possible errors in the abundance of C and N resulting from the neglect of NLTE are again approximately 0.2 dex, which is of the same order as errors introduced by uncertainties in the absolute calibration of the spectra, the molecular oscillator strengths, and the turbulence. The authors warn against an extrapolation of these results to other molecules and other stars (Mount et al. 1975).

5 CO

Exactly twenty years ago there appeared two calculations that set the stage for much of the later interesting work on CO: the tentative suggestion that CO might be an important coolant in stellar atmospheres of intermediate and low temperatures (Johnson 1973) and the pioneering calculations of NLTE in CO (Thompson 1973). Both of these intertwined effects have been treated in much more detail since then. Early calculations showed a very large cooling effect; subsequent calculations show a smaller but still important effect (Ayres & Weidemann 1989; Mauas et al. 1990; Wiedemann & Ayres 1991; Plez et al. 1992). CO is important to the energy budget in the outer layers where its strong lines are formed because most other lines have become optically thin. However, we concentrate on the NLTE aspects here, leaving the tantalizing questions of CO cooling and possible consequent bifurcation to another occasion.

Research on CO has been partly driven by observations of the strikingly deep cores of the strongest lines, which are still not fully explained (see Tsuji 1988, 1991; Judge 1989). Simply stated, certain CO lines in the fundamental band are so strong their cores must be formed in the same part of the atmosphere as the Ca II H and K lines or even the Mg II h and k lines. Both of these latter sets of lines show emission features reflecting the chromospheric temperature rise. Theory suggests these CO lines are formed by pure absorption (LTE), and, if so, they must also develop emission cores (Carbon et al. 1976; Ayres & Wiedemann 1989; Wiedemann & Ayres 1991). However, no emission features are observed in spectra of any star (Heasley et al. 1978; Tsuji 1988), including such diverse objects as Sun, Arcturus, and TX Psc (a carbon star). How to understand this conflict? Either CO must be subject to departures from LTE so that the radiation temperatures in its line cores do not correspond to electron temperatures in the stars, or the chromospheres must be inhomogeneous, with cool regions (where CO is formed) and warmer regions (where the resonance lines of Ca II and Mg II are formed). In fact, this latter point of view has been expressed by several authors, but the matter appears to be far from settled.

Pioneering work (Thompson 1973) based upon available cross sections showed that (a) rotational equilibrium must hold throughout the atmospheres of almost all stars, (b) vibrational equilibrium might break down in the the low-density atmospheres of the coolest giants, and (c) ionization and chemical equilibria ought to be in LTE. A later comprehensive overview of line formation in several molecules (Hinkle & Lambert 1975) again considered rotational, vibrational, and electronic transitions and attempted to distinguish observationally between lines formed by pure absorption and scattering (favoring the latter), disclosed the uncertainties of the cross sections used earlier, and emphasized the importance of collisions with neutral hydrogen for ground-state rovibrational transitions.

A calculation of NLTE in CO in Arcturus through the solution of the equations of radiative transfer and statistical equilibrium (Carbon et al. 1976) illustrated how sensitively the results depended upon the poorly known cross sections for H-CO collisions. They delineated the regimes of temperature and pressure in which departures from LTE might be expected in rovibrational transitions, and found that LTE would hold (for the "best" cross sections) throughout the atmosphere except perhaps the outermost layers. In addition, these authors pointed out that the CO results could not be reconciled with a chromospheric model and pointed toward an inhomogeneous outer atmosphere.

To address the problem of NLTE in CO directly, Ayres & Wiedemann (1989) undertook a careful NLTE calculation of the rotation-vibration lines in the Sun. A NLTE calculation for a molecule is a formidable problem. Consider a molecule of 10 vibrational levels, each with 121 rotational levels. Imagine well over 1,000 radiative transitions to be accounted for. That is the molecular model adopted. Fortunately, because the collisions are dominated by neutral hydrogen (for which reliable cross sections were then known), the value of the fractional absorption parameter, ϵ, essentially the rate of collisional to radiative de-excitation, is fairly large, of the order of 0.01 to 0.10. The solution of the radiative transfer equation can therefore be found by straightforward lambda iteration, and the solution thermalizes at relatively shallow depths. They found that the deep cores seen in strong infrared CO lines in Sun and Arcturus could not be explained by NLTE. This interesting result, that LTE holds for infrared CO lines, must not be confused with the fact of fluorescence, a selective NLTE effect, seen in some ultraviolet CO lines (Ayres 1986).

The NLTE research on the rotation-vibration transitions and levels within the ground electronic state ($X^1\Sigma^+$) in CO was later extended to F-K stars (Wiedemann & Ayres 1991) based on standard radiative- equilibrium, LTE models (Gustafsson et al. 1975; Bell et al. 1976), where again only slight departures from LTE were found in atmospheres for stars with log (surface gravity) ≥ 1.5, which includes all dwarfs and most giants. Model makers had previously employed the LTE assumption, and it has been used in all later work on CO opacities (Plez et al. 1992), line formation (Jørgensen & Johnson 1992), and cooling calculations (Mauas et al. 1990).

As part of a novel calculation of a self-consistent model for the solar photosphere under the constraints of hydrostatic and radiative equilibrium but without the LTE constraint (Anderson 1989), calculations were carried out for several molecules, including the CO fourth positive system (A-X). Interesting departures from LTE were found in the sense that in the outer solar atmosphere (outside the temperature minimum in a real red-giant atmosphere) all levels were overexcited; that is, all levels were overpopulated relative to LTE. The cooling effects of CO were confirmed (Anderson 1989), but the effects are much more complex than previously treated because of the heating and cooling effects of strong atomic lines. Questions of bifurcation are outside our considerations here.

6 TiO

Following H_2O, TiO is the dominant molecular opacity in oxygen-rich atmospheres, and the red and infrared spectral regions are blanketed by millions of lines of several systems: α ($C^3\Delta - X^3\Delta$), β ($c^1\Phi - a^1\Delta$), γ' ($B^3\Pi - X^3\Delta$), γ ($A^3\Phi - X^3\Delta$), δ ($b^1\Pi - a^1\Delta$), ϕ ($b^1\Pi - d^1\Sigma$), ϵ ($E^3\Pi - X^3\Delta$). Some TiO opacities have been employed by all constructors of cool-star atmospheres (Brown et al. 1989; Bessell et al. 1989; Plez et al. 1992; Jørgensen 1993). As far as we know, no NLTE analysis has been carried out for TiO in any star. Comments regarding possible NLTE effects on TiO are made in Section 8.

7 Polyatomic Molecules

Even though polyatomic molecules provide very important opacities – H_2O in oxygen rich stars and HCN, C_2H_2, and C_3 in carbon stars– we know of no research on NLTE in these molecules. We note that more complex molecules have more pathways for any given transition i $\leftrightarrow$ j, which increases opportunities for thermalization (LTE). Furthermore, all the important spectral bands of the molecules (except perhaps the A-X transition of C_3) are due to transitions within the electronic ground state (infrared rovibrational transitions), and are therefore less affected by NLTE as described in Section 1. For the near future, we expect all calculations of dissociative equilibrium among polyatomic molecules to continue to use LTE. Formation and destruction of PAHs (polycyclic aromatic hydrocarbons), polyynes, fullerenes, and other complex molecules are discussed by Omont (this volume).

8 NLTE in the Pool of Atoms

While scientists have worried about possible departures from LTE in stellar atmospheres, because almost all calculations to date have used this marvelously simplifying assumption, effects of NLTE in photospheres where molecules exist have not been easy to find. In this section we are concerned whether ionization equilibrium exists among molecule-forming atoms since such departures would change the pool of neutral atoms from which molecules might form. A few examples must suffice.

Most diatomic molecules are composed of H, C, N, and O, which share the trait of sufficiently high ionization energy (11.3–14.5 eV) that they are almost entirely neutral in all stars cooler than the Sun. Even though NLTE effects (overionization) are present (Ayres & Weidemann 1989; Anderson 1989; Luttermoser & Johnson 1992; Luttermoser 1993) and the number of ions is therefore greatly increased over LTE calculations, the enormously greater number of neutral atoms is not significantly diminished. Thus, the number density of molecules composed of these abundant atoms is not affected by NLTE.

A pioneering calculation of ionization equilibria in cool giant stars was carried out by Auman & Woodrow (1975) for the elements Na, Al, K, and Ca in several model photospheres for red-giant stars ranging from 2,000 to 4,000 K in effective temperature. Each element was represented by a model atom with 9 energy levels. Atmospheric models based on LTE were used as starting values, ultraviolet radiation fields were calculated from the models, and the radiative-transfer and steady-state equations were solved self-consistently. These results were used to obtain new models, and the process was iterated to convergence to create NLTE model atmospheres. In the outer atmospheric layers overionization was found for all elements considered, and this overionization increased with decreasing temperature and gravity up to a factor of about 10 for $T_{\rm eff} = 3500$ K, for example.

A dependence of the (LTE) abundance of Fe I and Ti I on the excitation potential as well as the equivalent width of spectral lines in the star Pollux (K0

III) has been put forward as evidence of departures from LTE in the photosphere (Ruland et al. 1980). A similar effect in Ti I and Zr I was noticed in a study of 33 G and K giants by Brown et al. (1983), who found that the Zr abundance was 0.4 dex lower, relative to Ti, than in the Sun when weak Zr I and Ti I lines were used for the abundance determination. However, these "abundance" effects disappeared when lines of Ti II and Zr II were employed, indicating that NLTE effects were responsible. These authors warn other researchers to use only the most abundant species of an element for abundance determinations.

A landmark study of complex atoms in stars was the pioneering research on line formation of Fe in the solar atmosphere (Athay & Lites 1972), who carried out a solution of the coupled steady-state and radiative transfer equations for a model atom of 15 terms and 13 radiative bound-bound transitions. This work was later extended to similar stars (Lites & Cowley 1974) with somewhat simpler (Fe I + Fe II atoms (12 atomic levels and 5 radiative bound-bound transitions).

Theoretical calculations for more complex model atoms (69 spectroscopic terms and 67 transitions) of Fe I and Fe II (Steenbok 1985) in the Sun and in Pollux have largely explained theoretically the results obtained by Ruland et al. (1980), thus putting the research on NLTE in complex atoms on a rather more solid footing. This same work was later elaborated further with an even more complex atom (Solanki & Steenbok 1988), but even this model atom was a rather crude approximation to the real iron atoms.

A novel calculation of a theoretical solar atmospheric model based upon radiative and hydrostatic equilibrium but without the LTE assumption has been made (Anderson 1989). In this calculation, the equations of statistical equilibrium (similar to Eq. 5) and radiative transfer (Eq. 2) were solved self-consistently for those atoms and molecules which account for most of the line blanketing: model neutral atoms of H, He, C, Na, Mg, Al, Si, Ca, and Fe and model molecules for CH, CN, CO, and SiO.

This is a numerically difficult and computationally expensive undertaking, and several simplifications were necessary. In particular, the "multifrequency/-multigray" approximation was employed for the opacity, which effectively reduces a complex atom or molecule to a set of representative multiplets for which more-or-less exact solutions can be obtained, and from which extrapolations to the complete solution can be made. Molecular dissociative equilibrium was treated in LTE. Although the abundance of their ions was increased relative to LTE, molecule-forming atoms remained the dominant stage of ionization, and molecular formation was therefore treated self-consistently in LTE.

In the outer photospheric layers the temperature in a NLTE model falls far below the temperature of an LTE model atmosphere. Molecular lines, treated by the complete-redistribution (CRD) formulation (Eqs. 11 and 12), departed from the predictions of pure absorption, and this resulted in interesting effects on (low-frequency) cooling and (high-frequency) heating. Again it was found that molecular energy levels are overexcited, sometimes by a large factor, in the outer atmosphere (above $\log m = 2.0$).

More recently, Takeda (1991) has noted the deficiencies in the atomic models of iron adopted by previous investigators because of the limited number of levels and transitions considered and has solved self-consistently the radiative-transfer and steady-state equations for a model Fe I atom of sufficiently many levels and lines to include the effects of huge numbers (hundreds) of highly excited states and weak lines. These were, of course, treated by a statistical procedure, in which an actual multiplet or term is represented by a single fictitious level with an appropriate transition probability/statistical weight in a manner similar to that of Anderson (1989). The model atmosphere was chosen to represent Arcturus (K2 III). Because the photon re-emission mechanism could not be known for all lines, calculations on opacities in the ultraviolet were made for both pure absorption and for pure coherent scattering. Although the NLTE factors differ by a small factor ($\sim 1-3$) depending on the inclusion or exclusion of scattering, the results differ by an order of magnitude if line-blanketing is included (as it should be) in the ultraviolet. In other words, the extent of Fe I overionization is mostly fixed by the treatment of ultraviolet fluxes and therefore line opacities. The results show a considerable overionization of Fe I in the outer atmosphere, above $\log \tau(500\ \mathrm{nm}) = -3.0$ (log column mass density ~ 1.0). Cores of strong and intermediate lines are formed above this region, and these results suggest a possibly significant overionization of Fe I.

If this result for Fe can be extrapolated to Ti, it might prove very interesting for the formation of TiO lines. Similarly, a NLTE calculation for Mg would test NLTE effects on MgH.

9 Conclusions and Discussion

Although NLTE effects in molecular opacities might arise in several contexts, the problems of calculating these are complex and poorly studied. We make the following comments.

1. Rotational levels, which have energy differences of 0.00–0.05 eV and whose lines fall in the red and infrared region, are in LTE because of the high collisional rate (due to neutral hydrogen) relative to the modest radiative rates.

2. Vibrational levels, with energy differences of 0.1-0.5 eV, are also likely to be in LTE for the reasons given above although departures may occur in some molecules in the outermost layers of cool supergiants or miras. These transitions are responsible for much of the opacity in cool stars.

3. Molecular electronic transitions, like atomic transitions, require energies of several eV. Collisions in these transitions depend on electrons, which are relatively rare in these cool objects, where abundant elements are generally neutral. Significant departures in populations of energy states may result, although no NLTE effects have been found observationally. Bands from electronic transitions in many molecules fall in the violet and ultraviolet spectral regions where their opacities are not significant (but note such exceptions as CN) for the thermal structure (though they are significant for violet and ultraviolet spectral synthesis) even in the Sun. They rapidly become less important in cooler stars.

4. We speculate that the most significant effect of NLTE on molecular opacities will arise from overionization of neutral atoms, which might reduce the pool from which molecules can be made. This effect will not occur for molecules composed of the abundant elements H, C, N, and O because their large (11-15 eV) ionization energies assure they will be predominantly neutral throughout cool-star atmospheres. We suggest NLTE effects might arise in such an important opacity as TiO if Ti I (χ_{ion} = 6.8 eV) is actually overionized when it would have been neutral in LTE. Although not a significant opacity source, MgH, which is valuable for pressure and gravity determinations, might be affected in the same way if Mg I (χ_{ion} = 7.6 eV) is overionized.

5. LTE will continue to be employed for the foreseeable future in most calculations of molecular opacities, model atmospheres, and elemental and isotopic abundances. Research in the outer photosphere, chromosphere, and CSE will require NLTE. The exact region of overlap is a subject of current research.

Acknowledgement. The author expresses appreciation to the Organizing Committee of Colloquium 146 and to Indiana University for travel grants. Helpful discussions with Lawrence Anderson and Tom Ayres are acknowledged.

References

Anderson L.S., 1989, *Astrophys. J.*,**339**, 558
Athay R.G., Lites B.W., 1972, *Astrophys. J.*, **176**, 809
Auman J.R., Woodrow J.E.J., 1975, *Astrophys. J.*, **197**, 163
Ayres T.R., 1986, *Astrophys. J.*, **308**, 246
Ayres T.R., Wiedemann G.R. 1989, *Astrophys. J.*, **338**, 1033
Bell R.A., Eriksson K., Gustafsson B., Nordlund Å, 1976, *Astron. Astrophys. Suppl.*, **23**, 37
Bessell M.S., Brett J., Scholz M., Wood P.R., 1989, *Astron. Astrophys. Suppl.*, **77**, 1
Bowen G.H., 1988, *Astrophys. J.*, **329**, 299
Brown J.A., Johnson H.R., Alexander D.R., Cutright L.C., Sharp C., 1989, *Astrophys. J. Suppl.*, **71**, 623
Brown J.A., Tomkin J., Lambert D.L., 1983, *Astrophys. J.*, **265**, L93
Carbon D.F., Milkey R.W., Heasley J.N., 1976, *Astrophys. J.*, **207**, 253
Carpenter K.G., Robinson R.B., Wahlgren G.M., Linsky J.L., Brown A., 1993, *Astrophys. J.*, submitted
Grevesse N., Lambert D.L., Sauval A.J., van Dishoeck E.F., Farmer C.B., Norton R.H., 1991, *Astron. Astrophys.*, **242**, 488
Gustafsson B., Bell R.A., Eriksson K., Nordlund Å, 1975, *Astron. Astrophys.*, **42**, 407
Heasley J.N., Ridgway S.T., Carbon D.F., Milkey R.W., Hall D.N.B., 1978, *Astrophys. J.*, **219**, 790
Hinkle K.H., Lambert D.L., 1975,it MNRAS, **170**, 447
Hinkle K.H., Martin C.L., Ridgway S.T., 1993, *Astrophys. J.*, in press
Johnson H.R., 1973, *Astrophys. J.*, **180**, 81
Jørgensen U.G., 1993, *Astron. Astrophys.*, in press
Jørgensen U., Johnson H.R., 1991, *Astron. Astrophys.* **244**, 462
Jørgensen U.G., Johnson H.R., Nordlund Å., 1992, *Astron. Astrophys.*,**261**, 263

Judge P.G. 1989, *Evolution of Peculiar Red Giant Stars*, ed. H.R. Johnson & B. Zuckerman, Cambridge Press, Cambridge, **303**
Lambert D.L., Pagel B.E.J., 1968,*MNRAS*, **141**, 299
Lites B.W., Cowley C.R., 1974, *Astron. Astrophys.*,**31**, 361
Lites B.W., Mihalas D., 1984, *Solar Phys.*, **93**, 23
Luttermoser D.G. 1993, *Astrophys. J.*, in preparation
Luttermoser D.G., Johnson H.R. 1992, *Astrophys. J.*,**388**, 579
Mauas P.J., Avrett E.H., Loeser R. 1990, *Astrophys. J.*, **357**, 279
Mihalas D., 1978, *Stellar Atmospheres (2nd ed.), Freeman, San Francisco*, chap.12
Mount G.H., Ayres T.R., Linsky J.L., 1975, *Astrophys. J.*, **200**, 383
Mount G.H., Linsky J.L., 1975,*Solar Physics*, **41**, 17
Plez B., Brett J.M., Nordlund Å., 1992, *Astron. Astrophys.*, **256**, 551
Ruland F., Holweger H., Griffin R., Griffin R., Biehl D., 1980, *Astron. Astrophys.*, **92**, 70
Schrijver C.J., 1987, *Astron. Astrophys.*, **172**, 111
Solanki S.K., Steenbok W., 1988, *Astron. Astrophys.*, **189**, 243
Steenbok W., 1985,*in Cool Stars with Excesses of Heavy Elements*, ed. M. Jaschek, P.C. Keenan, Reidel, Dordrecht, **231**
Steiman-Cameron T.Y., Johnson H.R., Honeycutt R.K., 1985, *Astrophys. J.*, **291**, L51
Takeda Y., 1991, *Astron. Astrophys.*, **242**, 455
Thompson R.I., 1973, *Astrophys. J.*, **181**, 1039
Tsuji T., 1986, *Ann. Rev. Astron. Astrophys.*, **24**, 89
Tsuji T., 1988, *Astron. Astrophys.*,**197**, 185
Tsuji T., 1991, *Astron. Astrophys.*, **245**, 203
Vernazza J.E., Avrett E.H., Loeser R., 1981, *Astrophys. J. Suppl.*, **45**, 635
Wiedemann G., Ayres T.R. 1991, *Astrophys. J.*, **366**, 277

Dissociation Energies and Partition Functions of Small Molecules

Michel Costes and Christian Naulin

URA 348 CNRS, Photophysique et Photochimie Moléculaire,
Université Bordeaux I, 33405 Talence cedex, France

1 Introduction

For a simple dissociation equilibrium

$$AB \rightleftharpoons A + B \tag{1}$$

the equilibrium constant K_p at temperature T can be written in terms of partial pressures p or densities n in the form given by Equation 6 of Tatum (1966)

$$K_p(T) = \frac{p_A p_B}{p_{AB}} = kT \frac{n_A n_B}{n_{AB}} = \left(\frac{2\pi m kT}{h^2}\right)^{\frac{3}{2}} kT \frac{Q_A Q_B}{Q_{AB}} \, e^{-D_0/kT} \tag{2}$$

where m is the reduced mass of AB, k the Boltzmann constant, h the Planck constant, Q_A, Q_B, Q_{AB} the internal partition functions of the species, and D_0 [1] the dissociation energy of AB. This equation derives from a more general expression of $K_p(T)$ for a chemical reaction that is demonstrated in all physical-chemistry textbooks treating statistical thermodynamics (see for example Atkins 1990). Partition functions are weighted Boltzmann factors

$$Q = \sum_n g_n \, e^{-\epsilon_n/kT} \tag{3}$$

[1] The dissociation energy D_0 invoked throughout the paper refers to the energy difference between the zero point energies (ZPE) of the products of dissociation and the molecule. This is the microscopic (molecular) definition of the dissociation energy which is measured spectroscopically. The macroscopic (molar) definition uses the notations D_0^0 or D_0°, with the subscript referring to the temperature $T = 0$ K and the superscript to the standard state. Indeed, dissociation energies determined thermochemically are calculated from standard enthalpies of formation at 0 K (see Equation 6 and others in the text, where we have dropped the superscript for clarity). The two definitions give identical values. They differ from the dissociation energy labelled D_e as often found in theoretical work which, for a diatomic molecule, is the energy difference between the separated ground-state atoms and the minimum of its potential energy curve.

with ϵ_n corresponding to the energy of the eigenstates of degeneracy g_n. Provided all the eigenvalues are known, Q values can be calculated exactly. Derivations and calculations of the partition functions can be found in textbooks or in reviews relating to astrophysics. The monographs of Tatum (1966) and Sauval & Tatum (1984) which include calculations of partition functions and equilibrium constants for 300 diatomic molecules of astrophysical interest are particularly relevant. Approximations such as the so-called high temperature approximation for the rotational partition function and the Morse potential for the vibrational partition function have been used to yield simple analytical expressions that only depend on the temperature. Nevertheless, comparison with the extensive calculations obtained by direct summation over calculated eigenstates reveals no striking difference. For example, the CN partition function computed between 1000 and 8500 K in that manner by Jørgensen & Larsson (1990) differs at most by 4 % from the values of Sauval & Tatum (1984).

While the knowledge of accurate molecular partition functions is essential to compute equilibrium constants, this requirement is not strictly necessary for abundance analyses of atomic species derived from molecular opacities. Indeed, the absorption coefficient, κ_{vJ}, of a molecular line in a medium in thermodynamic equilibrium at temperature T relative to a particular quantum state vJ of energy ϵ_{vJ} is

$$\kappa_{vJ} \propto n_{\mathrm{AB}(v,J)}\, f_{vJ} = \frac{n_{\mathrm{AB}}\,(2J+1)\,\mathrm{e}^{-\epsilon_{vJ}/kT}}{Q_{\mathrm{AB}}}\, f_{vJ} \tag{4}$$

where f_{vJ} is the line oscillator strength. It follows that the molecular partition function cancels out when combining Equation (2) and (4):

$$\kappa_{vJ} \propto \frac{n_{\mathrm{A}} n_{\mathrm{B}}}{Q_{\mathrm{A}} Q_{\mathrm{B}}} \left(\frac{2\pi mkT}{h^2}\right)^{-\frac{3}{2}} (2J+1)\,\mathrm{e}^{-\epsilon_{vJ}/kT}\,\mathrm{e}^{+D_0/kT}\, f_{vJ} \;. \tag{5}$$

However, when the product $n_{\mathrm{A}} n_{\mathrm{B}}$ is extracted from a calculated equilibrium constant, care should be taken to use the same partition functions that were entered in the calculation of the equilibrium constant. Sauval & Tatum (1984) have stressed the importance of giving a set of both equilibrium constants and partition functions to avoid inconsistent comparisons.

The dissociation energy always remains a critical factor in the calculation of equilibrium constants or in the evaluation of abundances from molecular opacities. Consider again the example of CN: an uncertainty of 0.3 eV in the dissociation energy results in a variation of the equilibrium constant by a factor of 2 for the solar photosphere, and even more for cooler objects, when turning from the lower to the upper bound ($D_0 = 7.6$ to 7.9 eV, as found in many references).

2 Current Status of the Knowledge of Dissociation Energies

It is not self evident that any value of a dissociation energy quoted with low error limits should be taken for granted, even when it is found in a review or a database considered as being authoritative on the subject. Compilations, once they have been recognized, have the tendency to last too many years before they are finally considered as obsolete. Quoted values can disagree, and when establishing comparisons, it is not at all clear that the best value should be in the more recently published review.

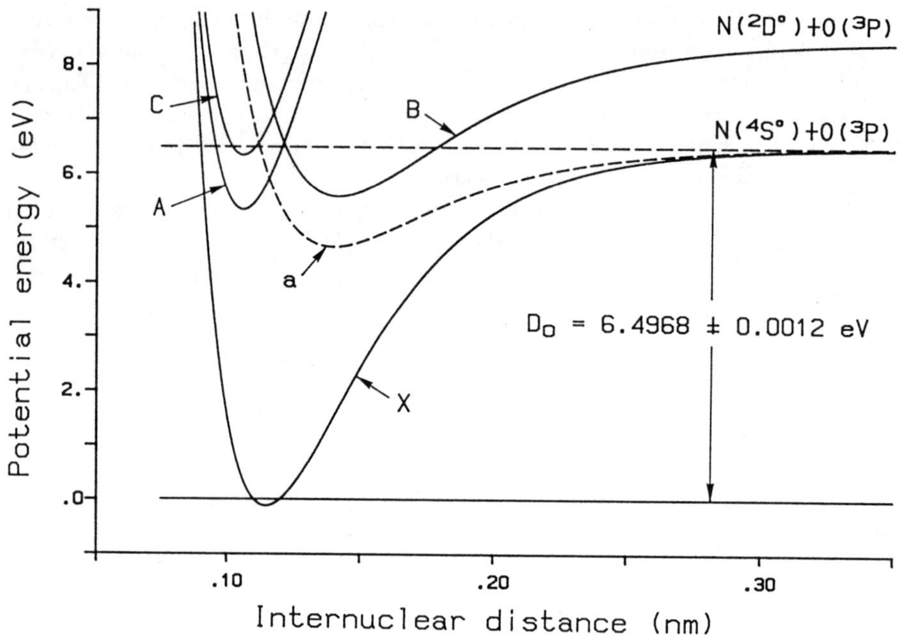

Fig. 1. Potential energy curves of some low-lying states of NO: $X^2\Pi_r$, $a^4\Pi_i$, $A^2\Sigma^+$, $B^2\Pi_r$ and $C^2\Pi_r$.

The example of the NO molecule can be taken to illustrate briefly the latter point. Following the last complete revision of the JANAF tables (Chase et al. 1985), the quoted standard enthalpies of formation of NO, N and O at 0 K are respectively $\Delta_f H_0\,(\mathrm{NO}) = 0.9305 \pm 0.0018$ eV per molecule (89.775 ± 0.17 kJ mole^{-1}), $\Delta_f H_0\,(\mathrm{N}) = 4.8797 \pm 0.0010$ eV (470.82 ± 0.10 kJ mole^{-1}) and $\Delta_f H_0\,(\mathrm{O}) = 2.5578 \pm 0.0010$ eV (246.79 ± 0.10 kJ mole^{-1}). Since

$$D_0\,(\mathrm{NO}) = \Delta_f H_0\,(\mathrm{N}) + \Delta_f H_0\,(\mathrm{O}) - \Delta_f H_0\,(\mathrm{NO}) \qquad (6)$$

the $D_0\,(\mathrm{NO})$ value can be deduced to be 6.5071 ± 0.0038 eV, which is not consistent with the value given in the compilation of constants of diatomic molecules provided by Huber & Herzberg (1979): $D_0\,(\mathrm{NO}) = 6.4968 \pm 0.0012$ eV. Yet, the value selected by Huber & Herzberg is based on papers by Callear & Pilling (1970) and Dingle et al. (1975) that could have been considered in the 1985 JANAF tables. Indeed, these authors established that only a very few rotational levels of the $C^2\Pi_r$ Rydberg state, predissociated by the interaction with the $a^4\Pi_i$ state, lie below the dissociation limit (Fig. 1). Direct spectroscopic evidence had been found by Dingle et al. by inverse predissociation in the two body recombination reaction of N and O atoms. This process populates $C^2\Pi_r$ rovibrational levels situated above the dissociation limit. Consequently, observation of the $C^2\Pi_r \rightarrow A^2\Sigma^+$ infrared bands reveals a break-off in the rotational structure (sudden drop in intensity) for levels below the dissociation limit, which yields precisely the dissociation energy, confirmed in subsequent papers.

Another current source of difficulty is to establish the complete path of the determination, as a value of a dissociation energy given in a paper is in general dependent on previous determinations of dissociation energies or standard enthalpies of formation at 0 K of other molecules. For example $D_0\,(C_2\text{–}H)$ can be deduced as follows:

$$D_0\,(C_2H\text{–}H) = \Delta_f H_0\,(C_2H) + \Delta_f H_0(H) - \Delta_f H_0\,(C_2H_2) \tag{7}$$

$$D_0\,(C_2\text{–}H) = \Delta_f H_0\,(C_2) + \Delta_f H_0(H) - \Delta_f H_0\,(C_2H) \tag{8}$$

$$D_0\,(C_2) = 2\Delta_f H_0\,(C) - \Delta_f H_0\,(C_2)\ . \tag{9}$$

Hence, eliminating $\Delta_f H_0\,(C_2H)$ and $\Delta_f H_0\,(C_2)$ yields:

$$D_0\,(C_2\text{–}H) = 2\Delta_f H_0\,(C) - D_0\,(C_2) + 2\Delta_f H_0(H) - \Delta_f H_0\,(C_2H_2) - D_0\,(C_2H\text{–}H)\ . \tag{10}$$

The result is that the certainty of $D_0\,(C_2\text{–}H)$ calculated concomitently from the measurement of $D_0\,(C_2H\text{–}H) = 5.69 \pm 0.03$ eV (131.3 ± 0.7 kcal mole^{-1}) by Ervin et al. (1991), also depends on the certainty of four other values. Ervin et al. have retained three JANAF values: $\Delta_f H_0\,(C) = 7.371 \pm 0.005$ eV (711.19 ± 0.46 kJ mole^{-1}), $\Delta_f H_0(H) = 2.23906 \pm 0.00006$ eV(216.035 ± 0.006 kJ mole^{-1}), $D_0\,(C_2) = 6.11 \pm 0.11$ eV (140.95 ± 2.5 kcal mole^{-1}), with enlarged uncertainty for the latter value, and discarded the fourth: $\Delta_f H_0\,(C_2H_2) = 2.443 \pm 0.008$ eV(235.755 ± 0.79 kJ mole^{-1}), replaced by 2.371 ± 0.007 eV (54.68 ± 0.17 kcal mole^{-1}) from Pedley et al. (1986), which results in $D_0\,(C_2\text{–}H) = 5.04 \pm 0.11$ eV (116.3 ± 2.6 kcal mole^{-1}). Following the same scheme but using their own measurement of $D_0\,(C_2)$: 6.30 ± 0.02 eV (145.2 ± 0.5 kcal mole^{-1}), Urdahl et al. (1991) have taken the four other values from Ervin et al. to deduce $D_0\,(C_2\text{–}H) = 4.86 \pm 0.04$ eV (112.0 ± 0.8 kcal mole^{-1}).

Dissociation energy values can be classified according to the methods of their determination. The best known values are those obtained directly by spectroscopy for molecules showing a predissociative behaviour, such as NO as discussed above, or a convergence limit in a progression of bands in the absorption spectrum, such as the one found in the Schumann-Runge ($B^3\Sigma_u^- \leftarrow X^3\Sigma_g^-$) transition of O_2: this has set a value of $D_0(O_2) = 5.1156 \pm 0.0019$ eV (41260 ± 15 cm^{-1}) (Brix & Herzberg 1954). A high level of accuracy is found for these experiments and the absolute uncertainty can be as low as 0.1 meV. These dissociation energies can be considered as reference values, which constitute the frame of the complex body of enthalpies of formation at 0 K. Fortunately, a reasonable number of dissociation energies enter in this category, which includes the dissociation energies of the diatomic permanent gases and also a few diatomic radicals such as OH. Such methods have not been successful with strongly bonded radicals because they require tunable vacuum-ultraviolet (vuv, below 200 nm) light so far only provided by conventional light sources. For such light sources, the density of species required is too high to be provided by the methods available for generation of radicals in the laboratory. Nonetheless, the availability of new pulsed laser sources, truly tunable over wide wavelength ranges in the vuv, will certainly set up for more determinations of this type in the near future.

Beside the values determined by the spectroscopic methods mentioned above, the accuracy of the others should always appear questionable. In the following it is not our intention to review all the field but rather to indicate trends of what we believe the best and most promising methods could be.

3 Determinations of Dissociation Energies by Ab Initio Methods

Ab initio methods have considerably improved in recent years, although they have not yet reached the accuracy of the best experimental determinations. The results of the Gaussian–2 theory deserve to our opinion special attention. In the first paper of the series devoted to this theory, Curtiss et al. (1991) have given an impressive list of 133 calculated total atomisation energies, i.e. the rupture of all chemical bonds (ΣD_0), including 67 diatomics and 33 triatomics and the rest being molecules with 4 to 8 atoms. The average value of the absolute deviation compared to the experimental values taken for a very large majority from JANAF (1985) or from Huber & Herzberg (1979) is found not to exceed 0.05 eV for a selected set of 53 molecules. The disagreement exceeds 0.2 eV for only 15 molecules, but most of the experimental values concerned can only be considered as roughly known dissociation energies or preliminary estimations not supported by convincing determinations. In fact, the major failure of the theory pointed out by the authors themselves, has been encountered for SO_2 with $(\Sigma D_0)_{\text{theory}} = 10.79\,\text{eV}$ ($248.9\,\text{kcal mole}^{-1}$) compared to $(\Sigma D_0)_{\text{exp.}} = 11.013 \pm 0.007\,\text{eV}$ ($1062.61 \pm 0.66\,\text{kJ mole}^{-1}$) as derived from JANAF tables.

Variations in the Gaussian–2 theory have also been presented with calculations performed at higher or lower levels of treatment of the correlation energy (Curtiss et al. 1992, 1993). No great changes in accuracy have been pointed out in these studies which reveal the validity of the assumptions made in the initial treatment and the versatility of the method to treat small and large systems as well.

The Gaussian–2 method thus appears of good reliability. It can furnish a useful basis when a dissociation energy is unknown, and in case of gross disagreement with the experimental value, a situation which holds for MgO: $(D_0)_{\text{exp.}} = 3.53\,\text{eV}$ (cited as "quite uncertain" by Huber & Herzberg) and $(D_0)_{\text{theory}} = 2.57\,\text{eV}$ ($59.2\,\text{kcal mole}^{-1}$), the theoretical value should be preferred.

4 Determinations of Dissociation Energies from Product Internal Energy Release of Chemical Reactions

4.1 Method

The method is based on the observation of nascent products from gas phase reactions produced under well defined conditions. It uses an atom-exchange reaction of type

$$A + BC \rightarrow AB + C \ . \qquad (11)$$

The energetics of each independent reactive event are determined by applying the energy conservation law (Fig. 2)

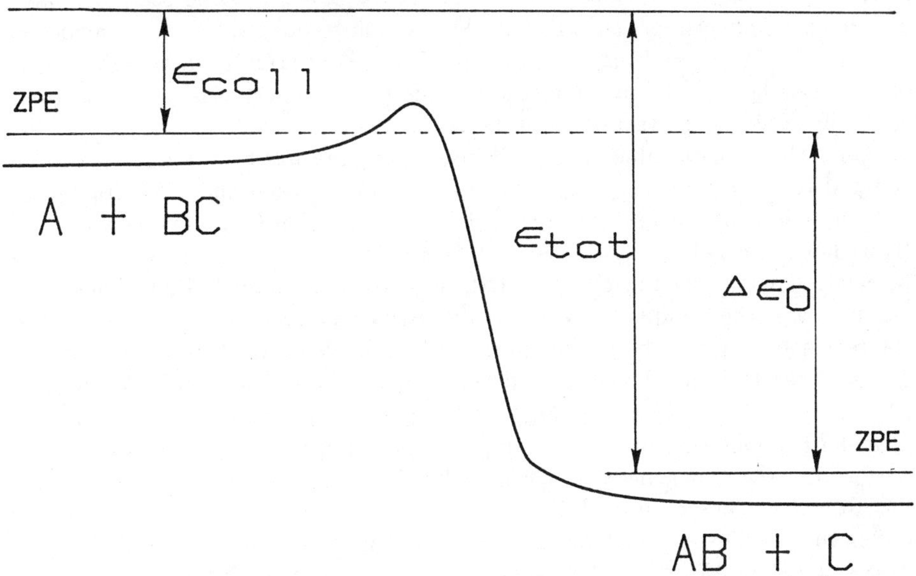

Fig. 2. Schematic representation of the potential energy along the reaction path, for an exoergic process.

$$\epsilon_{tot} = \epsilon_{tr} + \epsilon_i(A) + \epsilon_i(BC) - \Delta\epsilon_0 = \epsilon'_{tr} + \epsilon'_i(AB) + \epsilon'_i(C) \qquad (12)$$

where ϵ_{tot} is the total energy available to the reaction products. The left side of Equation (12) refers to reactants: ϵ_{tr} is their relative translational energy (or kinetic energy of relative motion in the centre-of-mass frame) and the ϵ_i s stand for their internal energies. The right side of Equation (12) refers to the products of the reaction with ϵ'_{tr} defining their relative translational energy (also known as recoil energy) and the ϵ'_i their respective internal energies. The sum

$$\epsilon_{coll} = \epsilon_{tr} + \epsilon_i(A) + \epsilon_i(BC) \qquad (13)$$

is also defined as the collision energy. The reaction energy

$$\Delta\epsilon_0 = D_0\,(\mathrm{BC}) - D_0\,(\mathrm{AB}) \tag{14}$$

is the difference between ZPE of products and reactants and obviously corresponds to the difference of the dissociation energies of BC and AB bonds.

In a typical experiment, one observes the energy release into the AB product to determine the highest energy level of AB populated by the reaction. This determination can be achieved by detecting chemiluminescence if the reaction directly produces AB in an excited radiative state. The spontaneous emission is then dispersed through a monochromator, giving information on the population densities of the upper electronic state of the transition observed. However, more detailed information can be obtained using laser-induced fluorescence (LIF): in that case, the absorption spectrum of AB is explored while scanning laser wavelength; furthermore, an extremely high signal-to-noise ratio can be achieved by collecting all (i.e. not dispersed) fluorescence photons emitted, at right angles of the laser beam. The latter method yields population densities in the lower electronic state of the transition probed.

Once the highest value of the AB internal energy has been found, it can be equated to the total energy ϵ_{tot} with the assumptions that these AB fragments of highest internal energy have recoiled from companion C fragments produced (i) without internal energy and (ii) with a relative translational energy that can be neglected (i.e. setting $\epsilon_{\mathrm{i}}'(\mathrm{C})$ and ϵ_{tr}' to zero in Equation (12)). These two assumptions, the validity of which is discussed and illustrated later on, allow the reaction energy to be evaluated from the knowledge of the collision energy. $D_0\,(\mathrm{AB})$ can thus also be calculated provided $D_0\,(\mathrm{BC})$ is accurately known.

The collision energy spread $\delta\epsilon_{\mathrm{coll}}$ plays an essential role in the determination of a sharp population limit. The simplest operating conditions, achieved when performing the experiment in a cell at low pressure, are those that yield the largest collision energy spread. Better definition of the collision energy is achieved either in a beam-gas experiment where a beam of A atoms, generally with a Maxwell distribution of velocities, is scattered by AB molecules in a low pressure cell, or in a crossed molecular beam experiment. Finally, the best that can be done is to use crossed, supersonic molecular beams because these beams have high velocity resolution and give highly efficient cooling of the internal degrees of freedom ($\epsilon_{\mathrm{i}} = 0$). Thus, the collision energy is well defined and almost equal to the reactant relative translational energy calculated from measured reactant velocities v_{A} and v_{BC}:

$$\epsilon_{\mathrm{tr}} = \frac{1}{2}\,\mu\left(v_{\mathrm{A}}^2 + v_{\mathrm{BC}}^2 - 2v_{\mathrm{A}}v_{\mathrm{BC}}\cos\alpha\right) \ . \tag{15}$$

Here, μ stands for the reactant reduced mass, and α for crossing angle of the two molecular beam axes. It is important to operate with collimated beams in order to keep the dispersion in translational energy arising from collisions with velocity vectors at angles around the mean value α as low as possible.

4.2 Application of the Method in a Crossed Beam Experiment

The instrument used for these experiments combines techniques of pulsed, crossed, supersonic molecular beams, LIF detection and atom generation by laser vaporization. The experimental conditions obtained with this instrument (Costes et al. 1987 and 1989) are particularly relevant for an accurate determination of dissociation energies by the method examined in the following.

Determination of the excitation limit. The excitation limit is determined by analysing the experimental spectra. This is achieved by generating synthetic spectra including the spectroscopy of the transition of interest (for instance, $B^2\Sigma^+ \leftrightarrow X^2\Sigma^+$ for AlO produced by the reaction Al + O_2), transition probabilities, apparatus function and rovibrational population distributions. The following distribution function, derived from the information-theoretic approach to the analysis of state-to-state reaction dynamics (Bernstein 1982) has been introduced to model the rotational populations:

$$P(v,J) = P^{\circ}(v,J)\, e^{-\beta_{\mathrm{r}}\,\phi(J)} \;. \tag{16}$$

The first term represents the prior expectation and the exponential term the deviance to the prior defining the reduced surprisal parameter, β_{r}; the $\phi(J)$ function reads

$$\phi(J) = \frac{F(J)}{\epsilon_{\mathrm{tot}} - \epsilon_{\mathrm{i}}'(\mathrm{C}) - G(v)} \tag{17}$$

where $G(v)$ and $F(J)$ stand for vibrational and rotational energies of the diatomic fragment, AB, for corresponding vibrational and rotational quantum numbers v and J.

Taking a statistical distribution as the prior expectation yields:

$$P^{\circ}(v,J) \propto (2J+1)\,\{\epsilon_{\mathrm{tot}} - \epsilon_{\mathrm{i}}'(\mathrm{C}) - G(v) - F(J)\}^{1/2} \;. \tag{18}$$

In the fitting procedure, the surprisal parameter, β_{r}, the energy available to the products, ϵ_{tot}, and the respective weights of the vibrational levels are adjusted to give a close match to the experimental spectra.

When the distribution is close to statistical (i.e. $\beta_{\mathrm{r}} \rightarrow 0$), the rotational population sharply falls off in the vicinity of the excitation limit, which thus allows for a clear observation of this limit. In this very favourable case, the rotational distribution function is only slightly affected by the β_{r} value. The spectrum can thus be fitted with a single parameter (apart from the vibrational populations): the energy available to products, ϵ_{tot}.

This case is exemplified by the LIF spectra of AlO (from Al + O_2 reaction), as shown in Fig. 3. Only the part of the spectra in the region of the excitation limit is displayed. Synthetic spectra computed with extreme values of ϵ_{tot} = 0.213 to 0.243 eV are also displayed. Among the different features, the (1-0)P(53) line appears to be clearly underestimated at ϵ_{tot} = 0.213 eV (Fig. 3-b), and overestimated at ϵ_{tot} = 0.243 eV (Fig. 3-c). The best fit is obtained when a ϵ_{tot} value of 0.228 eV is introduced into the calculation.

In this example, it is noteworthy that a variation of ± 15 meV significantly alters the spectrum appearance, thus emphasizing the sensitivity of the method.

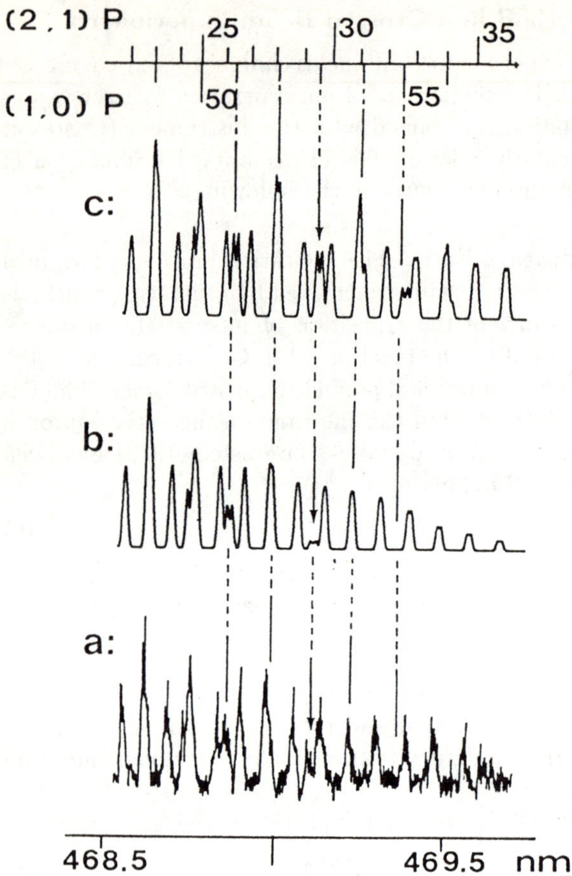

Fig. 3. Part of the LIF spectrum of AlO($B^2\Sigma^+ \leftrightarrow X^2\Sigma^+, \Delta v = 1$) from the Al + $O_2 \rightarrow$ AlO + O reaction at $\epsilon_{tr} = 0.083$ eV: (a) experimental, (b) calculated for $\epsilon_{tot} = 0.213$ eV and (c) calculated for $\epsilon_{tot} = 0.243$ eV; rotational line labelling refers to the N quantum number (total angular momentum apart from spin), with $J = N \pm \frac{1}{2}$ for a doublet state.

Determination of the reactant translational energy. The reactant relative translational energy is determined, using Equation (15), by measuring the reactant beam velocities. The dispersion results from several uncertainties: the error on the velocity measurement of both beams, which does not exceed 5 % in the worst case, the velocity spread around the mean values, which ranges between 10 and 20 % full-width at half-maximum, and the effect of collisions at angles different from the mean value $\alpha = 90°$.

Determination of the reactant and companion-fragment internal energies. Reactants are cooled down within the supersonic expansion. Consequently, only the lowest levels of the ground-state manifold of the BC molecule are significantly populated, with residual internal energy not exceeding a few meV. The situation for the atom beam is less favourable, because the cooling processes have to start from the very high energy imparted to the atoms in the laser-generated plasma. In particular, if the ground state is a multiplet, all spin-orbit components can be significantly populated: in the case of aluminium atoms, for instance, both spin-orbit levels of the ground state, $^2P_{1/2}$ and $^2P_{3/2}(\epsilon = 14\,\mathrm{meV})$, are almost equally populated. This is taken into account in the fitting procedure. In the case of carbon atoms, where the respective weights of individual spin-orbit components cannot be measured by LIF, because the absorption lines lie too far in the vuv, the spin-orbit energy is introduced by broadening the uncertainty domain by the value of the energy of the highest excited component ($\epsilon\,(^3P_2) = 5.4\,\mathrm{meV}$).

The companion fragment can be produced in any internal state. Unfortunately, it is out of reach to determine its internal energy distributions for each internal state of the first outcome. The following cases have been considered:

(i) 3-atom case: the companion fragment (second outcome) is an atom; a statistical distribution over spin-orbit terms is assumed; the rotational distribution function of the diatomic product (first outcome) is then computed as a linear combination of individual distribution functions calculated with an internal energy and a statistical weight corresponding to each spin-orbit energy of the outcoming atom; the procedure is repeated for each spin-orbit component of the incoming atom;

(ii) 4-atom case: the second outcome now is a diatom as in the example reported here ($\mathrm{Si} + \mathrm{N_2O} \rightarrow \mathrm{SiN} + \mathrm{NO}$). Without measurement of the recoil energy, it becomes generally not possible to fit the experimental spectra with a single distribution function. However, this restriction no longer holds when the data can be fitted to a completely statistical distribution, as found for the endoergic $\mathrm{Si} + \mathrm{N_2O}$ reaction at collision energies close to the reaction threshold. Such a statistical distribution is computed by direct summation over the number of states of the second diatom outcome that can be populated by the reaction at a given total energy, for each state of the first outcome, assuming all states equiprobable.

4.3 Results: Determination of Dissociation Energies of AlO, CN and SiN

***D_0*(AlO).** The dissociation energy of AlO has been derived from the crossed-beam study (Costes et al. 1987) of the reaction:

$$\mathrm{Al}(^2P_J) + \mathrm{O_2}(X^3\Sigma_g^-) \rightarrow \mathrm{AlO}(X^2\Sigma^+) + \mathrm{O}(^3P_J)\ . \tag{19}$$

As illustrated above, the excitation limit, and hence the reaction energy $\Delta\epsilon_0 = 0.145 \pm 0.030\,\mathrm{eV}$ is derived from the analysis of LIF spectra of the AlO product. This combined with the $\mathrm{O_2}$ dissociation energy $D_0(\mathrm{O_2}) = 5.1156 \pm 0.002\,\mathrm{eV}$ gives the AlO dissociation energy $D_0(\mathrm{AlO}) = 5.26 \pm 0.03\,\mathrm{eV}$.

This reaction was first studied by Dagdigian et al. (1975) in a beam-gas arrangement using an effusive Al source. They deduced a value of the AlO dissociation energy $D_0(\mathrm{AlO}) = 5.27 \pm 0.04\,\mathrm{eV}$. This result was confirmed by Pasternack

& Dagdigian (1977), again in a beam-gas study but with a velocity selected, effusive Al source: $D_0(\mathrm{AlO}) = 5.27 \pm 0.02\,\mathrm{eV}$.

Our result is in excellent agreement since the value of the AlO dissociation energy is found within the uncertainty range of these two previous studies (note that the JANAF value is based – among others – on the determination of Pasternack & Dagdigian 1977).

D_0(CN). The dissociation energy of CN has been derived from the crossed-beam study (Costes et al. 1990) of the reaction:

$$\mathrm{C}(^3P_J) + \mathrm{NO}(X^2\Pi_\mathrm{r}) \rightarrow \mathrm{CN}(X^2\Sigma^+) + \mathrm{O}(^3P_J) \ . \qquad (20)$$

The same method as for AlO was used. The exoergicity of the reaction, $\Delta\epsilon_0 = -1.27 \pm 0.05\,\mathrm{eV}$ is deduced from the determination of the excitation limit of the CN product. Introducing the value of the NO dissociation energy (6.4968 eV) recommended in the compilation of Huber & Herzberg (1979) yields $D_0(\mathrm{CN}) = 7.77 \pm 0.05\,\mathrm{eV}$.

A first group of experimental determinations of D_0(CN) in the last twenty years has used spectroscopic measurements on $\mathrm{CN}(B^2\Sigma^+ \leftrightarrow X^2\Sigma^+)$ violet transitions for laboratory sources: Arnold & Nicholls (1973): $7.89 \pm 0.13\,\mathrm{eV}$, Engleman and Rouse (1975): $7.66 \pm 0.05\,\mathrm{eV}$, Colket (1984): $7.92 \pm 0.07\,\mathrm{eV}$, or $(A^2\Pi_\mathrm{i} \leftrightarrow X^2\Sigma^+)$ red transitions in the solar spectrum: Sinha & Tripathi (1986): $7.71 \pm 0.05\,\mathrm{eV}$. All these determinations which are based on the use of Equation (5) with three parameters to work with (the source temperature, the band oscillator strength of the transition analysed and the CN dissociation energy) cannot be considered as very accurate. Only the shock-tube experiments of Arnold and Nicholls, performed at three very different temperatures, could yield both the band oscillator strength and the dissociation energy. Furthermore, the uncertainties quoted reflect more a deviation from an assumed fit, except the one given by Arnold and Nicholls, than a value including possible systematic errors.

A second group of determinations has used the laser photofragmentation of the cyanogen molecule, (NC–CN) $\rightarrow$ CN + CN. This type of experiment directly yields D_0(NC–CN): $5.83 \pm 0.04\,\mathrm{eV}$ ($47000 \pm 300\,\mathrm{cm}^{-1}$, Eres et al. 1984), $5.84 \pm 0.04\,\mathrm{eV}$ ($47100 \pm 300\,\mathrm{cm}^{-1}$, Wannenmacher et al. 1990), $5.84 \pm 0.02\,\mathrm{eV}$ (Huang et al. 1992). It also yields indirectly $\Delta_\mathrm{f} H_0$ (CN)

$$2\Delta_\mathrm{f} H_0\,(\mathrm{CN}) = D_0(\mathrm{NC-CN}) + \Delta_\mathrm{f} H_0(\mathrm{NC-CN}) \qquad (21)$$

and D_0 (CN)

$$D_0\,(\mathrm{CN}) = \Delta_\mathrm{f} H_0\,(\mathrm{C}) + \Delta_\mathrm{f} H_0\,(\mathrm{N}) - \Delta_\mathrm{f} H_0\,(\mathrm{CN}) \ . \qquad (22)$$

$\Delta_\mathrm{f} H_0\,(\mathrm{N}) = \frac{1}{2} D_0(\mathrm{N_2})$ is a reference value, already cited, and $\Delta_\mathrm{f} H_0\,(\mathrm{C}) = 711.19 \pm 0.46\ \mathrm{kJ\,mole^{-1}}$ can reasonably be considered likewise. Thus, the confidence on the value of D_0 (CN) obtained: $7.75 \pm 0.04\,\mathrm{eV}$, $7.74 \pm 0.04\,\mathrm{eV}$ and $7.74 \pm 0.03\,\mathrm{eV}$ depends mostly on the confidence on the third standard enthalpy of formation at 0 K: $\Delta_\mathrm{f} H_0(\mathrm{NC-CN}) = 307.219 \pm 1.8\,\mathrm{kJ\,mole^{-1}}$ following JANAF.

There has been also a number of theoretical determinations. Only the more recent studies will be considered here. Bauschlicher et al. (1988) used the multireference configuration interaction (MRCI) method, with an extended basis set of atomic orbitals: the pure calculation yielded $D_0(\mathrm{CN}) = 7.53\,\mathrm{eV}$, which was rescaled to $D_0(\mathrm{CN}) = 7.65\,\mathrm{eV}$, by comparison with the experimental $D_0(\mathrm{N_2})$ value and the computed one using the same basis set. Bauschicher et al. estimated the accuracy of their determination to be within 0.06 eV. The result of the Gaussian-2 theory (Curtiss et al. (1991): $D_0(\mathrm{CN}) = 7.63\,\mathrm{eV}$ (176.0 kcal mole^{-1})) falls into this interval. We have taken great care in our own work to check for all possible systematic errors. Our D_0 value of $7.77 \pm 0.05\,\mathrm{eV}$ could only suffer from two of these: (i) an error on $D_0(\mathrm{NO})$ and (ii) the failure of the C + NO reaction to populate the $\mathrm{CN}(X^2\Sigma^+)$ manifold up to the total energy available to reaction products, which could only be checked by measuring the recoil energy distribution. The former appears very unlikely and if the latter was occuring, it would lead to a lower limit of $D_0(\mathrm{CN})$, increasing the difference with the ab initio results. Finally, it should be noted that the D_0 values derived from the photodissociation of (NC – CN) can also be influenced by these two types of possible systematic errors.

D_0(SiN). The dissociation energy of SiN has been derived from the determination of the excitation limit of SiN produced in a crossed-beam experiment (Naulin et al. 1993) by the reaction:

$$\mathrm{Si}(^3P_J) + \mathrm{N_2O}(X^1\Sigma^+) \rightarrow \mathrm{CN}(X^2\Sigma^+) + \mathrm{NO}(X^2\Pi_r)\ . \tag{23}$$

An experimental value of $0.25 \pm 0.13\,\mathrm{eV}$ has been found for the reaction energy, which can be expressed as:

$$\Delta\epsilon_0 = D_0(\mathrm{N{-}NO}) - D_0(\mathrm{SiN})\ . \tag{24}$$

The N–NO dissociation energy can be derived from dissociation energies of the diatomics N_2, O_2 and NO, already cited as reference values, and standard formation enthalpy of N_2O at 0 K (85.48 ± 0.4 kJ mole^{-1} following JANAF):

$$D_0(\mathrm{N{-}NO}) = D_0(\mathrm{N_2}) + \frac{1}{2}D_0(\mathrm{O_2}) - D_0(\mathrm{NO}) - \Delta_\mathrm{f}H_0(\mathrm{N_2O}) = 4.93 \pm 0.01\,\mathrm{eV} \tag{25}$$

Introducing the obtained values for $D_0(\mathrm{N{-}NO})$ and $\Delta\epsilon_0$ in Equation (24) yields: $D_0(\mathrm{SiN}) = 4.68 \pm 0.14\,\mathrm{eV}$. This experimental value is substantially different from the imprecise JANAF value $D_0(\mathrm{SiN}) = 5.64 \pm 0.65\,\mathrm{eV}$ (130 ± 15 kcal mole^{-1}) but is in good agreement with the theoretical determination of Curtiss et al. (1991): $D_0(\mathrm{SiN}) = 4.58\,\mathrm{eV}$ (105.6 kcal mole^{-1}). It is again important to note that if the reaction fails to populate the highest energetically accessible levels of SiN, the present determination would be an underestimate. Since the theoretical $D_0(\mathrm{SiN})$ value already lies close to the lower limit of the experimental determination, a higher experimental value would significantly depart from the theoretical prediction. Given the accuracy and consistency of the Gaussian-2 method for a huge number of species, such a discrepancy for SiN seems unlikely.

5 Determinations of Dissociation Energies from Product Energy Release of Laser-Photofragmentation Reactions

For a photodissociation reaction induced by the absorption of n laser-light photons of energy $h\nu$:

$$ABC + nh\nu \rightarrow AB + C \ . \tag{26}$$

The total energy shared by the photofragments is written, similarly to Equation (12):

$$\epsilon_{\mathrm{tot}} = nh\nu + \epsilon_{\mathrm{i}}(ABC) - D_0(AB - C) = \epsilon'_{\mathrm{tr}} + \epsilon'_{\mathrm{i}}(AB) + \epsilon'_{\mathrm{i}}(C) \ . \tag{27}$$

Different experiments can be used. The main difference between these and the method described in section 4 is that even a rather simple experiment performed with two pulsed lasers, by photolyzing the molecule in a cell at low pressure and probing the nascent AB photofragments, can furnish fairly accurate results. This is because the spread in relative translational energy, which is always difficult to fight in reactive scattering experiments, is now replaced by the bandwidth of the photolyzing laser. In the worst case, an excimer laser delivers the high flux of necessary photons at most useful lines of 193 nm, 248 nm and 308 nm with a bandwidth of 150 cm^{-1}. Double dye laser outputs which furnish the opportunity to scan through the absorption spectrum can be emitted in less than 1 cm^{-1} at wavelengths approaching 200 nm. Furthermore, if cooling of the internal energy of the sample is achieved by using a molecular beam, the total energy can be incomparably well defined.

As in section 4, the validity of the measurement is dependent on the same kind of assumptions if only $\epsilon'_{\mathrm{i}}(AB)$, but not ϵ'_{tr}, is probed. However, a high flux of photoproducts along with a perfect timing of the experiment allow for using more powerful techniques like photofragment translational spectroscopy (PTS). The photofragment, after leaving the photolysing zone in the molecular beam and travelling without suffering collisions through a time-of-flight (TOF) tube, is then detected by mass spectrometry or by laser techniques such as resonance-enhanced-multiphoton ionisation (REMPI). The vibrational and even the rotational structure can be apparent in the TOF spectrum, and the threshold in the translational energy spectrum reveals the value of ϵ_{tot} and thus D_0(AB–C).

These PTS techniques have been for example used in the photodissociation of SO_2 by Effenhauser et al. (1990) who have detected the SO and S photofragments from reactions

$$SO_2 + (2h\nu \text{ at } 248 \text{ or } 308\,\mathrm{nm}) \rightarrow SO + O \tag{28}$$

and

$$SO_2 + (2h\nu \text{ at } 248 \text{ or } 308\,\mathrm{nm}) \rightarrow S + O_2 \ . \tag{29}$$

All observed kinetic thresholds of the various pathways leading to the different electronic states energetically accessible were compatible with D_0(SO–O) = 5.628 eV (543 kJ $mole^{-1}$) and $\Delta\epsilon_0$ = 5.897 eV (569 kJ $mole^{-1}$) for the $SO_2 \rightarrow S + O_2$ reaction. Using $D_0(O_2) = 5.1156 \pm 0.0019$ eV it can also be deduced that the total atomisation energy is $\Sigma D_0 = 11.013\,\mathrm{eV}$ and that $D_0(\mathrm{SO}) = 5.385\,\mathrm{eV}$. Using the JANAF values for $\Delta_f H_0(\mathrm{S}) = 274.73 \pm 0.25\,\mathrm{kJ\,mole^{-1}}$, $\Delta_f H_0(\mathrm{SO_2}) = -294.299 \pm 0.21\,\mathrm{kJ\,mole^{-1}}$, $\Delta_f H_0(\mathrm{SO}) = 5.03 \pm 1.3\,\mathrm{kJ\,mole^{-1}}$ and $\Delta_f H_0(\mathrm{O}) = \frac{1}{2} D_0(\mathrm{O_2})$ (mentioned above) yields the following values: $D_0(\mathrm{SO{-}O}) = 5.660 \pm 0.017\,\mathrm{eV}$ ($546.12 \pm 1.61\,\mathrm{kJ\,mole^{-1}}$), slightly overestimated, $\Sigma D_0 = 11.013 \pm 0.007\,\mathrm{eV}$ ($1062.61 \pm 0.66\,\mathrm{kJ\,mole^{-1}}$), in perfect agreement, and $D_0(\mathrm{SO}) = 5.353 \pm 0.015\,\mathrm{eV}$ ($516.49 \pm 1.65\,\mathrm{kJ\,mole^{-1}}$), subsequently slightly underestimated by the same amount as found in $D_0(\mathrm{SO{-}O})$.

Sophisticated laser techniques including production of vuv radiation at the Lyman-alpha wavelength, 121.6 nm, and coupled with H atom PTS experiments have been recently used in the laser photodissociation of HCN by Morley et al. (1992). The HCN dissociation energy obtained, $D_0(\mathrm{H{-}CN}) = 5.423 \pm 0.019\,\mathrm{eV}$ ($43740 \pm 150\,\mathrm{cm^{-1}}$) reduces the uncertainty compared to the JANAF value $5.36 \pm 0.19\,\mathrm{eV}$ ($517.3 \pm 18.4\,\mathrm{kJ\,mole^{-1}}$) based on $\Delta_f H_0(\mathrm{H}) = 216.035 \pm 0.006\,\mathrm{kJ\,mole^{-1}}$, $\Delta_f H_0(\mathrm{CN}) = 436.8 \pm 10\,\mathrm{kJ\,mole^{-1}}$ and $\Delta_f H_0(\mathrm{HCN}) = 135.53 \pm 8.4\,\mathrm{kJ\,mole^{-1}}$.

6 Conclusion

Abundance analyses of atomic species derived from molecular opacities are more sensitive to the values of dissociation energies than to partition functions that are introduced into the calculations. Some of the worse cases, such as the dissociation energy of CN, now appear to be resolved. Most dissociation energies will be determined with high accuracy in the near future, by the extensive use of laser-based methods employing tunable vuv light.

References

Arnold J.O., Nicholls R.W., 1973, *J. Quant. Spectrosc. Radiat. Transfer*, **13**, 115

Atkins P.W., *Physical Chemistry*, 4th Ed., 1990, (Oxford University Press, Oxford)

Bauschlicher C.W., Langhoff S.R., Taylor P.R., 1988, *Ap. J.*, **332**, 531

Bernstein R.B., 1982, *Chemical Dynamics via Molecular Beam and Laser Techniques*, (Oxford University Press, New York)

Brix P., Herzberg G., 1954, *Can. J. Phys.*, **32**, 110

Callear A.B, Pilling M.J., 1970, *Trans. Faraday Soc.*, **66**, 1618

Chase Jr. M.W., Davies C.A., Downey Jr. J.R., Frurip D.J., McDonald R.A., Syverud A.N., JANAF thermochemical tables, 3rd Ed., 1985, *J. Phys. Chem. Ref. Data*, **14**, (Suppl. 1)

Colket M.B. III, 1984, *J. Quant. Spectrosc. Rad. Transfer*, **31**, 7

Costes M., Naulin C., Dorthe G., Vaucamps C., Nouchi G., 1987, *Faraday Discuss. Chem. Soc.*, **84**, 75

Costes M., Naulin C., Dorthe G., Daleau G., Joussot-Dubien J., Lalaude C., Vinckert M., Destor A., Vaucamps C., Nouchi G., 1989, *J. Phys. E: Sci. Instrum.*, **22**, 1017

Costes M., Naulin C., Dorthe G., 1990, *Astron. Astrophys.*, **232**, 270

Curtiss L.A., Raghavachari K., Trucks G.W., Popple J.A., 1991, *J. Chem. Phys.*, **94**, 7221

Curtiss L.A., Carpenter J.E., Raghavachari K., Popple J.A., 1992, *J. Chem. Phys.*, **96**, 9030
Curtiss L.A., Raghavachari K., Popple J.A., 1993, *J. Chem. Phys.*, **98**, 1293
Dagdigian P. J., Cruse H. W., Zare R. N., 1975, *J. Chem. Phys.*, **62**, 1824
Dingle T.W., Freedman P.A., Gelernt B., Jones W.J., Smith I.W.M., 1975, *Chem. Phys.*, **8**, 171
Effenhauser C.S., Felder P., Huber J.R., 1990, *Chem. Phys.*, **142**, 311
Engleman Jr. R., Rouse P.E., 1975, *J. Quant. Spectrosc. Radiat. Transfer*, **15**, 831
Eres D., Gurnick M., McDonald J.D., 1984, *J. Chem. Phys.*, **81**, 5552
Ervin K.M., Gronert S., Barlow S.E., Gilles M.K., Arrison A.G., Bierbaum V.M., DePuy C.H., Lineberger W.C., Barney Ellison G., 1990, *J. Am. Chem. Soc.*, **112**, 5750
Huang Y., Barts S.A., Halpern J.B., 1992, *J. Phys. Chem.*, **96**, 425
Huber K.P., Herzberg G., 1979, *Molecular Spectra and Molecular Structure IV. Constants of Diatomic Molecules.*, (Van Nostrand, New York)
Jørgensen U.G., Larsson M., 1990, *Astron. Astrophys.*, **238**, 424
Morley G.P., Lambert I.A., Ashfold M.N.R., Rosser K.N., Western C.M., 1992, *J. Chem. Phys.*, **97**, 3157
Naulin C., Costes M., Ghanem N., Dorthe G., 1993, *Chem. Phys. Letters*, **202**, 452
Pasternack L., Dagdigian P. J., 1977, *J. Chem. Phys.*, **67**, 3854
Pedley J.B., Naylor R.D., Kirby S.P., 1986, *Thermochemical Data of Organic Compounds*, 2nd ed., Chapman and Hall, London
Sauval A.J., Tatum J.B., 1984, *Ap. J. Suppl. Ser.*, **56**, 193
Sinha K., Tripathi B.M., 1986, *Bull. Astron. Soc. India*, **14**, 40
Tatum J.B., 1966, *Pub. Dom. Ap. Obs. Victoria*, **13**, 1
Urdahl R.S., Bao Y., Jackson W.M., 1991, *Chem. Phys. Letters*, **178**, 425
Wannenmacher E.A.J., Lin H., Jackson W. M., 1990, *J. Phys. Chem.*, **94**, 6608.

Atomic Opacities

M. J. Seaton

Department of Physics and Astronomy, University College London,
Gower St., London WC1E 6BT, UK

1 Introduction

Let $I_\nu(r, \theta)$ be the intensity of radiation of frequency ν, at a distance r from the centre of a star and in a direction making an angle θ to the outward normal. In a plane-parallel approximation the equation of radiative transfer is

$$\cos(\theta)\frac{\mathrm{d}I_\nu}{\mathrm{d}r} = -\kappa_\nu I_\nu + j_\nu \tag{1}$$

where κ_ν is the opacity and j_ν the emissivity. The net outward flux of radiant energy is

$$F_\nu(r) = \int_{-1}^{+1} I_\nu(r, \theta)\cos(\theta)\,\mathrm{d}\cos(\theta) \tag{2}$$

and the flux integrated over all frequencies is

$$F(r) = \int_0^\infty F_\nu(r)\,\mathrm{d}\nu. \tag{3}$$

The source function is $S_\nu = j_\nu/\kappa_\nu$ and the optical depth is

$$\tau_\nu(r) = \int_r^\infty \kappa_\nu\,\mathrm{d}r. \tag{4}$$

In thermodynamic equilibrium at temperature T, $S_\nu = B_\nu(T)$ where

$$B_\nu(T) = (2h\nu 3/c^2)/\left[\exp(h\nu/kT) - 1\right]. \tag{5}$$

The relation $S_\nu = B_\nu(T)$ can often be used even when one does not have full thermodynamic equilibrium; such situations are referred to as LTE (local thermodynamic equilibrium — see Johnson, this volume).

The expression for the opacity is

$$\kappa_\nu = \sum_i N(i)\sigma_\nu(i) \tag{6}$$

where $N(i)$ is the number-density of a species of type i and $\sigma_\nu(i)$ is a cross section for absorption or scattering of radiation. In all opacity work there are two main physical problems to be tackled: obtain the number-densities $N(i)$ and the cross-sections $\sigma_\nu(i)$ used in Eq. (6).

For the astronomy work there are three cases to be considered:

1. **NLTE**, non-LTE. For the case of non-LTE one must take acount of all relevant radiative and collisional processes in order to calculate $N(i)$. The source-function S_ν depends on I_ν, and hence on the solution of the equation of radiative transfer. NLTE effects are important for many problems concerned with the atmospheres of hot stars.
2. **LTE**. This is an adequate approximation for many stars. The level populations $N(i)$ are calculated using thermodynamic formulae. One may require large amounts of atomic and molecular data for radiative processes, but similar data for collisional processes are not required. The opacity κ_ν depends only on chemical composition and on the local variables of temperature T and mass-density ρ.
3. **Large optical depths**, $\tau_\nu \gg 1$. For these cases LTE can be assumed. The transfer equation can be solved in the *diffusion approximation* (see, for example, Mihalas 1978) giving for the monochromatic radiative flux

$$F_\nu = -\frac{4\pi}{3}\frac{1}{\kappa_\nu}\frac{\mathrm{d}B_\nu}{\mathrm{d}r}. \tag{7}$$

The net flux is

$$F = -\frac{4\pi}{3}\frac{1}{\kappa_R}\frac{\mathrm{d}B}{dT}\frac{\mathrm{d}T}{dr} \tag{8}$$

where $B(T) = \int B_\nu(T)\,\mathrm{d}\nu$ and where κ_R is the *Rosseland mean opacity* defined by

$$\frac{1}{\kappa_R}\frac{\mathrm{d}B}{\mathrm{d}T} = \int_0^\infty \frac{1}{\kappa_\nu}\frac{\mathrm{d}B_\nu}{\mathrm{d}T}\,\mathrm{d}\nu. \tag{9}$$

In the present review I will be mainly concerned with the calculation of Rosseland mean opacities. They are of crucial importance for all work on the interiors of the stars. They are also important for work on stellar atmospheres, since the lower boundary conditions for an atmosphere are usually taken to be such as to match results obtained using the diffusion approximation.

2 Equations of State — EOS

Leaving aside the NLTE case, the calculation of level populations $N(i)$ is a problem of thermodynamics. For the molecular work that may be a matter of considerable intricacy but does not (so far as I am aware) present any great difficulties of principle. Such difficulties do arise when one considers the higher temperatures of stellar interiors. Let $N(j)$ be the number density for a given chemical element in ionisation stage j and let $N(i,j)$ be the number in level i, $N(j) = \sum_i N(i,j)$. The familiar Boltzmann equation is

$$N(i,j) = N(j)g(i,j)\exp[-E(i,j)/kT]/Q(j) \tag{10}$$

where $g(i,j)$ is the statistical weight, $E(i,j)$ the excitation energy of the level and $Q(j)$ the partition function,

$$Q(j) = \sum_i g(i,j)\exp[-E(i,j)/kT]. \tag{11}$$

The sum in Eq. (11) is divergent! Consider the case of the hydrogen atom and put $i = n$, the principal quantum number. As n increases the energies $E(n)$ tend to a constant value (the ionisation limit) but $g(n) = 2n^2$ so that the integral diverges like n^2. The origin of this difficulty is, of course, that the highly excited states have very large radii, eventually becoming larger than the mean inter-particle separations in the plasma. In order to overcome the difficulty it is necessary to consider how the atomic states are modified by interactions with the plasma environment. It is not my purpose here to enter into further details about how that can be done: two references to recent work are Rogers (1991) and Hummer & Mihalas (1988).

In principle, of course, the same difficulties do arise in the molecular work but they are of little practical importance since at lower temperatures the exponential factor in Eq. (11) becomes very small long before the divergence of the partition function starts to be a problem.

3 Recent atomic opacity calculations

For the calculation of atomic opacities one considers transitions of the following types: free-free (inverse bremsstrahlung); bound-free (photo-ionisation and photo-detachment); bound-bound (spectrum lines); and scattering (mainly by free electrons, also Rayleigh scattering at lower temperatures).

For many years astronomers have used Rosseland mean opacities from the Los Alamos Opacity Library, which I will refer to as LAOL (for references see Hübner (1985) and Weiss et al. (1990)). More recent calculations have been made by two independent teams: OPAL (Rogers & Iglesias (1992), and Iglesias et al. (1992) — OPAL is the name of their computer code): and OP (Seaton et al. (1993) — the Opacity Project).

OPAL and OP use very different approaches to the EOS problems. OP uses the "chemical picture" in which atoms (and atomic ions) are assumed to exist but to be perturbed by the plasma environment. The OP work is restricted to stellar envelopes, that is to say to regions in which one need not take account of plasma perturbations in the calculation of the cross sections $\sigma_i(\nu)$. OPAL uses the "physical picture" and includes allowance for plasma perturbations in calculating the cross sections. OPAL go to deeper layers than those considered by OP.

The two projects also have different approaches to the calculation of atomic radiative data. OPAL uses parametric model potentials with parameters adjusted to give good agreement for energy levels of free atoms. All cross-section data are calculated on-line as required and are not archived (that allows for the possibility of the cross sections being dependent on plasma conditions). The OP work, on the other hand, is based on very extensive *ab initio* calculations for free atoms. Quite a large team has been involved (see Berrington et al. (1987) and Seaton et al. (1992)). Most of the work is done using elaborate R-matrix methods. Some supplementary calculations for large number of iron lines are also made using configuration-interaction codes. OP atomic data are archived in a database system TOPbase (see Cunto & Mendoza (1992), and Cunto et al. (1993)).

Agreement between OPAL and OP results is generally satisfactory. For temperatures in the region of $T \simeq 2 \times 10^5$ the new opacities are larger than those from LAOL by factors as large as 3 or more. That has important consequences for theories of stellar pulsations. The enhancements are due to much more detailed treatments of atomic physics, in OPAL and OP compared with LAOL. Such detailed treatments are important for all of the "metals" (all elements other than H and He), and are particularly important for ions of iron. For $T \simeq 2 \times 10^5$ the main ionisation stages for iron are Fe XV to Fe IX with ground-configurations $3s^2$ to $3s^2 3p^6$. Those ions have large number of spectral terms belonging to configurations of the type $3s^x 3p^y 3d^z$. Transitions between such terms, and excitations from them, give very large numbers of spectrum lines — even more when fine-structure splittings are taken into account. Molecular spectroscopists who think that their spectra are complicated because there are many lines should take a look at the complexity of the iron spectra in the vicinity of 2×10^5 K, as shown on Figure 22a of Rogers & Iglesias (1992) or Figure 10 of Seaton et al. (1993).

4 Fitting and smoothing

The Rosseland mean opacity κ_R depends on temperature T, density ρ and chemical composition. Work on stellar structures requires, for each specified composition, a knowledge of $\kappa_R(T, \rho)$ and of the first two derivative,

$$\left.\frac{\partial \kappa_R}{\partial T}\right|_\rho \quad \text{and} \quad \left.\frac{\partial \kappa_R}{\partial \rho}\right|_T .$$

Higher derivatives may also be required. The calculations of $\kappa_R(T, \rho)$ are made with intervals in T and ρ such that quadratic or cubic interpolations can be used. There is, however, one practical difficulty which arises.

In order to resolve all structures in the monochromatic opacities κ_ν it would be necessary to use 10^6 or even 10^7 frequency points in evaluating the Rosseland integral, which would be prohibitively expensive. In practice the calculations are made with some 10^4 frequency points. The technique is that of *opacity sampling*, which is also used in much of the molecular work. It has been checked that the use of sampling does not lead to errors larger than two or three per cent in κ_R which is not in itself of much importance, since for other reasons κ_R will not be accurate to better than a few per cent. The practical difficulty is that the errors can change in a more-or-less random fashion as one passes from one (T, ρ) grid-point to the next, and can therefore lead to quite large errors in calculated derivatives, which take on an unpleasant "wiggly" appearance.

I have recently shown that the problem can be overcome using a modest amount of smoothing (Seaton (1993)). I use indices i and j for grid-points in T and ρ and in the vicinity of a grid-point (i, j) I consider the 25 neighbouring points $(i + n, j + m)$, $n = -2$ to $+2$ and $m = -2$ to $+2$; and make a least-squares fit to a bi-cubic involving 16 coefficients. The value of κ_R at the point (i, j) is then replaced by the value obtained from the fit. It is found that the differences between the original and the smoothed data are no larger than the errors which can be expected to arise from the use of opacity sampling; and using the smoothed data the calulated derivatives have a much more reasonable and pleasing appearance.

5 Access to atomic opacity data

5.1 OPAL opacities

OPAL opacity tables are available from the Lawrence Livermore National Laboratory *via* `ftp dplasma.llnl.gov`, user anonymous and `cd /pub/opacity`.

5.2 TOPbase

The OP atomic data database, TOPbase, is installed at the Centre des Donées Astronomiques de Strasbourg (CDS). Using `ftp 130.79.128.5` with `anonymous` login and `cd /pub/topbase` one can obtain the TOPbase User Manual `userman.tex` as a LaTeX file. To access the database itself one uses `telnet 130.79.128.5`, `-account:topbase`, and `-password:Seaton+`

5.3 OP opacities

OP opacity data for any required chemical mixture, and the fitting-and-smoothing codes (OPALFIT and OPFIT), can be obtained on sending e-MAIL messages to one of the following:-

- Dr. A.E. Lynas-Gray, Astrophysics, Keble Road, Oxford, OX1 3RH, UK.
  Telephone: +44–865–73363.
  Electronic mail (INTERNET): aelg@oxds02.astro.ox.ac.uk
- Dr. Anil K. Pradhan, Department of Astronomy, 174 W. 18th St., The Ohio State University, Columbus, Ohio 43210-1106, USA.
  Telephone: +614-292-5850.
  Electronic mail (INTERNET): pradhan@seaton.mps.ohio-state.edu

6 The marriage of atomic opacities with molecular opacities

Throughout most of a stellar interior the radiative opacity is determined almost entirely by atomic processes, the temperatures being too high for molecules to exist with appreciable abundances. That is no longer true for the outer layers of cooler stars. Those who calculate Rosseland mean opacities are constantly faced with the complaint that their tables do not go to sufficiently low temperatures. That position can be rectified only when molecular contributions are included.

Those who calculate molecular opacities have given particular attention to the outer layers of cool stars, that is to say regions in which the opacity is largely dominated by molecular contributions. For the calculations of Rosseland mean opacities there is an urgent need to bridge the gap between regions in which molecular contributions are dominant and those in which the atomic contributions are dominant. It is required to calculate monochromatic opacities, κ_ν, with inclusion of all contributions, before evaluation of the Rosseland integral Eq. (9). I hope that the present book will provide a stimulus for such work to be done.

Acknowledgements. I thank Carlos Iglesias and Forrest Rogers for permission to give details concerning `ftp` access to their opacity data.

References

Berrington K.A., Burke P.G., Butler K., Seaton M.J., Storey P.J., Taylor K.T., Yu Yan, 1987, *J. Phys. B.*, **20**, 6363

Cunto W., Mendoza C., 1992, *Rev. Mex. Astron. Astrofis.*, **23**, 107

Cunto W., Mendoza C., Ochsenbein F., Zeippen C.J., 1993, *Astron. Astrophys.*, in press

Hummer D.G., Mihalas D., 1988, *Astrophys. J.*, **331**, 794

Hübner W.F., 1985, In: P.Sturrock, T.Holzer, D.Mihalas (eds.), *Physics of the Sun*, (Dordrecht: Reidel), p.33

Iglesias C.A., Rogers F.J., Wilson B.G., 1992, *Astrophys. J.*, **397**, 717

Mihalas D., 1978, *Stellar Atmospheres, 2nd ed.*, (San Francisco: Freeman)

Rogers F.J., 1992, In: S. Eliezer, R.A. Ricci (eds.), *High Pressure Equations of State: Theory and Applications*, (New York: North Holland), p.77

Rogers F.J., Iglesias C.A., 1992, *Astrophys. J. Supp.*, **79**, 507

Seaton M.J., Zeippen C.J., Tully J.A., Pradhan A.K., Mendoza C., Hibbert A., Berrington C.A., 1992, *Rev. Mex. Astron. Astrofis.*, **23**, 19

Seaton M.J., Yu Yan, Mihalas D., Pradhan A.K., 1993, *Mon. Not. R. Astr. Soc.*, in press

Seaton M.J., 1993, *Mon. Not. R. Astr. Soc.*, submitted

Weiss A., Keady J.J., Magee N.H., 1990, *Atomic Data Nucl. Data Tables*, **45**, 209

Computation of the Absorption Coefficient for Diatomic Molecules

Mats Larsson

Physics Department I, The Royal Institute of Technology, S-100 44 Stockholm, Sweden

1 Introduction

Molecules in stellar atmospheres play key roles, not only as promoters of our understanding of stars, but also as actors affecting the structures of the atmospheres. In particular cool star atmospheres are affected by molecular opacities, and during recent years it has become clear that also weak molecular lines are of crucial structural importance. Extensive molecular data are thus needed in order to correctly model the structure of a stellar atmosphere and the transport of radiation through it. The basis for how such molecular data are acquired constitutes the theme of the present article. As will follow below, results from different subfields of molecular physics are needed in order to obtain a complete picture of the absorption of radiation by molecules. This article will only consider allowed molecular electronic transitions, i.e. dipole transitions between *different* electronic states; however, the presentation and the formulae may easily be modified to be valid for transitions between levels within a single electronic state.

2 Basic Concepts and Formulae

Molecules are more complicated entities than atoms, and it could be useful to start this section with an introduction of the concept of a single rotational line in a molecular spectrum (Whiting et al. 1980). A rotational line is produced by the sum of all allowed transitions between degenerate magnetic substates, as illustrated in Fig. 1.

For some molecular electronic states there is a small splitting called Λ-doubling, originating from the interaction between the rotation of the nuclei and the electrons angular momentum, which can be difficult to resolve even by high resolution spectroscopy. Thus in some cases a rotational line has been defined as including both components of a Λ-doublet (Tatum 1967 and Schadee 1971). If one adopts the definition illustrated in Fig. 1, however, it follows that a Λ-doublet is composed of two rotational lines (Whiting & Nicholls 1974). Figure

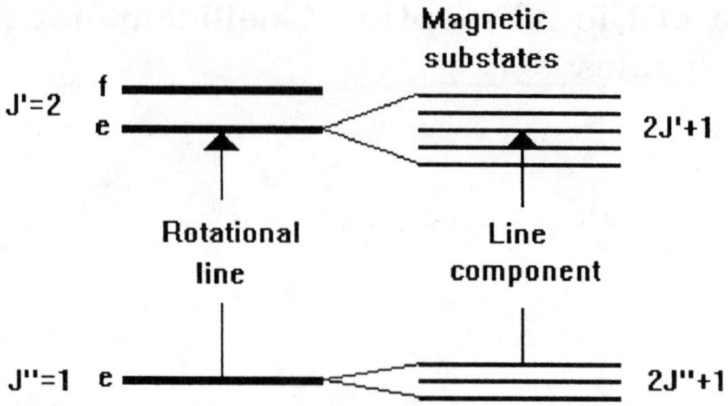

Fig. 1. A single molecular rotational line. The energy levels J'=2 and J''=1 belong to two different electronic states. The upper electronic state is affected by Λ-doubling, which means that for each value of J there is one level of e-type symmetry and one of f-type symmetry (see text). A single molecular rotational absorption line originates from a transition that terminats in one of the Λ-doublets ($J(e) = 2 \leftarrow J(e) = 1$). A rotational level of the upper or lower electronic state, $J(e/f)$, is degenerate in the magnetic substates only. This is the origin of the $2J+1$ degeneracy factor for a rotational level. The various line components of which a rotational line is composed can only be observed in the presence of a magnetic field (Zeeman effect).

2 further illustrates the point. It shows the energy level structures for $^2\Pi$ and $^2\Sigma^+$ states and the first rotational line in each of the twelve different branches of a $^2\Pi$ - $^2\Sigma^+$ transition. The Λ-doubling occurs in the $^2\Pi$ state and the levels are labelled according to their total parity, $\pm$, and their e/f symmetry. The $\pm$ and e/f labels are two different bookkeeping devices for the same physical property, the e/f labels being more convenient to use in connection with optical spectroscopy since they label a set of levels that usually can be described with a single energy level expression.

The absorption coefficient K_ν describes the attenuation of a beam of photons of frequency ν due to molecular absorption according to the expression

$$I_\nu = I_\nu^0 e^{-K_\nu \Delta x} \tag{1}$$

where I_ν^0 and I_ν are the intensities before and after transmission through a column of length Δx of the molecular gas. For a transition $n \leftarrow m$ the integrated absorption coefficient can be expressed as

$$\int K_\nu d\nu = \frac{\pi e^2}{m_e c} N_m f_{nm} \tag{2}$$

where N_m is the number density in the lower state m of the molecule and f_{nm} is the absorption oscillator strength (dimensionless). The labels n and m need to be specified if more useful formulas are to be developed. Let the upper state be specified by n=$|\Lambda', S', \Sigma', v', J', (e/f)'\rangle$ and the lower state by

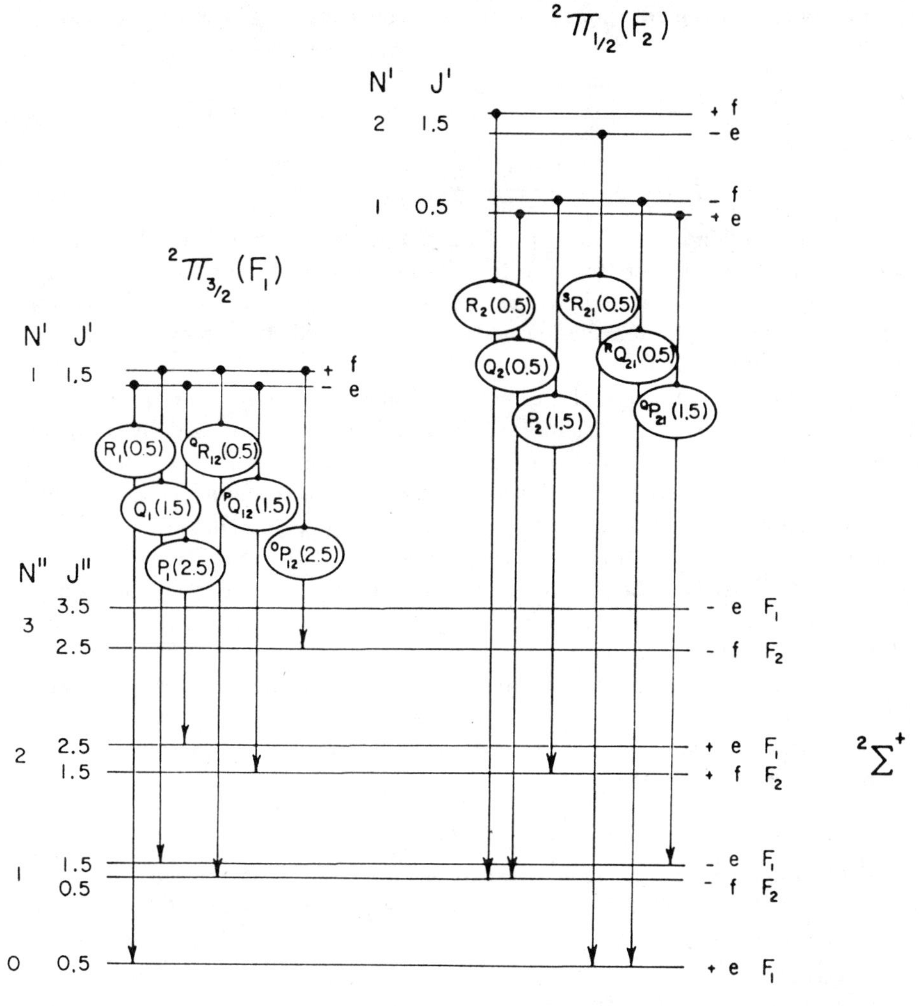

Fig. 2. Energy level diagram for a $^2\Pi - ^2\Sigma^+$ transition. The lowest rotational line of each of the twelve different branches is shown. The N quantum number is the total angular momentum of the molecule apart from the electron spin. Each filled circle in the upper state levels represents a transition from a single rotational level. (from Jørgensen & Larsson 1990).

$m=|\Lambda'', S'', \Sigma'', v'', J'', (e/f)''\rangle$, where Λ is the electronic state labelling (when the projection of the electrons angular momentum on the internuclear axis is Λ=0,1 or 2 this implies Σ, Π or Δ states, respectively), S is the total electron spin, Σ is the specification of the spin multiplet[1], v is the vibrational quantum number, J is the rotational quantum number and (e/f) is the symmetry labelling.

The line absorption oscillator strength $f_{(\Lambda' S' \Sigma' v' J' (e/f)' \leftarrow \Lambda'' S'' \Sigma'' v'' J'' (e/f)'')}$, or, for short, $f_{v'J',v''J''}$, is related to the line strength $S_{J'J''}$ (in atomic units of $e^2 a_0^2$) through

$$f_{v'J',v''J''} = \frac{8\pi^2 m_e c a_0}{3h} \nu_{v'J',v''J''} \frac{S_{J'J''}}{2J''+1} \tag{3}$$

where $\nu_{v'J',v''J''}$ denotes the energy difference between the upper and lower state expressed in wave numbers (cm^{-1}). In keeping with the tradition in molecular spectroscopy, the upper state is written first in the expression for the absorption oscillator strength although the lower state is the initial state in the absorption process. In order to avoid confusion an arrow $\leftarrow$ has been included in the full expression for f to indicate the direction of the transition. The definition of the line strength, $S_{J'J''}$, for a molecular transition was proposed by Whiting et al. (1980) and corresponds to Condon and Shortley's definition of the line strength for an atomic line (Condon & Shortely 1935). It should be noted that $S_{J'J''}$ is a shortened notation for $S_{(\Lambda' S' \Sigma' v' J' (e/f)' \leftrightarrow \Lambda'' S'' \Sigma'' v'' J'' (e/f)'')}$. The line strength is symmetric with respect to the direction of the transition, hence the two-directional arrow. The twelve rotational lines shown in Fig. 2, the first in each of the different branches in a $^2\Pi - ^2\Sigma^+$ transition, each has a separate value of $S_{J'J''}$ and $f_{v'J',v''J''}$.

It is in general a good approximation to separate the total wavefunction of a molecule into an electronic, a vibrational and a rotational part (the Born-Oppenheimer approximation). If this separation is carried through the line strength can be expressed as

$$S_{J'J''} = |\langle v'_{J'}|R_e(r)|v''_{J''}\rangle|^2 \mathcal{S}_{J'J''} \tag{4}$$

Here $R_e(r)$ denotes the electronic transition moment (in atomic units of ea_0), $\mathcal{S}_{J'J''}$ is the Hönl-London factor (dimensionless), and $|v'_{J'}\rangle$ and $|v''_{J''}\rangle$ are the vibrational wavefunctions for the upper and lower states; the index J is used to indicate that the vibrational wavefunctions are obtained from the effective potential energy curve of the rotating system. Hence vibration-rotation interaction is taken into account. The concepts introduced in Eq. (4) will be discussed in more detail in the next section.

The transition rate per particle (transition probability) $A_{v'J',v''J''}$ (s^{-1}) for a spontaneous radiative transition $(\Lambda' S' \Sigma' v' J' (e/f)') \rightarrow (\Lambda'' S'' \Sigma'' v'' J'' (e/f)'')$ is related to the electronic transition moment and Hönl-London factor through

$$A_{v'J',v''J''} = \frac{64\pi^4 e^2 a_0^2}{3h} \nu^3_{v'J',v''J''} \frac{|\langle v'_{J'}|R_e(r)|v''_{J''}\rangle|^2 \mathcal{S}_{J'J''}}{(2J'+1)} \tag{5}$$

[1] This should not be confused with a Σ state, which implies that Λ=0; in Fig. 2 the $^2\Pi_{3/2}$ state is obtained for Σ=1/2 and $^2\Pi_{1/2}$ for $\Sigma = -1/2$.

Relations between various variables describing the radiative properties of diatomic molecules can be found in the paper by (Larsson 1983).

It should finally be noted that it is very common to use the concept of *band* absorption oscillator strength and to discuss transition probabilities in terms of an electronic-vibrational transition $(\Lambda'v') \rightarrow (\Lambda''v'')$. In these cases $f_{v'v''}$ and $A_{v'v''}$ are obtained as an averaged sum (Larsson 1983).

3 Calculations of absorption oscillator strength

From an astrophysicists point of view the oscillator strength is often the desired radiative property to be used for describing absorption of radiation by molecules. In fact, it is even more appropriate to say that it is the gf-value, i.e. the absorption oscillator strength multiplied by the degeneracy factor for the lower state. For the line absorption oscillator strength this factor is always $(2J''+1)$, while if the band absorption oscillator strength is used, the degenary factor depends on the electronic transition (Larsson 1983). In the line data tape for the CN molecule, which will be described below, the gf-value is always $(2J''+1)f_{v'J',v''J''}$.

The separation of the molecular wavefunction into three parts, as expressed in Eq. (4), makes it natural to divide the discussion of how to calculate gf-values into three parts. In the first part we will discuss how to obtain the electronic transition moment, in the second part how to calculate vibrational wavefunctions, and in the third part how to obtain Hönl-London factors.

The definition of the electronic transition moment for a molecular transition is intimately connected to the exact way the electronic and rotational parts of the total wavefunction are separated. We need not go into the details here; it suffices to note that, according to the recommendations by Whiting et al. (1980), the electronic transition moment for a parallel transition $(\Lambda' = \Lambda'')$ should be expressed as

$$R_e = \langle \Lambda' S' \Sigma' | \mu_z | \Lambda'' S'' \Sigma'' \rangle \tag{6}$$

and for a perpendicular transition $(\Lambda' = \Lambda'' \pm 1)$

$$R_e = \langle \Lambda' S' \Sigma' | \frac{1}{\sqrt{2}} (\mu_x \pm i\mu_y) | \Lambda'' S'' \Sigma'' \rangle \tag{7}$$

where μ_z and $\frac{1}{\sqrt{2}}(\mu_x \pm i\mu_y)$ are different components of the electronic transition moment operator. These expressions are valid for spin-allowed transitions, which implies that $S' = S''$.

Prior to the development of modern *ab initio* quantum chemistry methods, the electronic transition moment could only be obtained by means of rather hazardous inversions of experimental data obtained from intensity measurements or measurements of fluorescence lifetimes. Nowadays, with the level of accuracy that can be obtained with computer based state-of-the-art quantum chemistry methods (see Langhoff & Bauschlicher, Malmqvist, and Peyerimhoff, this volume), the fastest and most accurate way to obtain $R_e(r)$ is to calculate it from *ab initio* molecular electronic wavefunctions. The procedure is to calculate the electronic wavefunctions for the upper ($|\Lambda' S' \Sigma'\rangle$) and lower ($|\Lambda'' S'' \Sigma''\rangle$) electronic states for a number of different values of the internuclear separation. For a diatomic molecule it is in general sufficient to carry out these calculations at 10-20 different internuclear distances. The energy values obtained in such calculations constitute a set of data which, by interpolation, provide approximate potential curves for the two states. The electronic wavefunctions can then be used to evaluate the matrix elements given in Eqs. (6) and (7), and the electronic transition moment function, $R_e(r)$ is obtained by interpolation between the 10-20 internuclear distances used in the calculation. It should finally be pointed out that if *ab initio* calculated electronic transition moments are used one should make sure that they are defined in accordance with Eqs. (6) and (7) (for a brief discussion of this issue, see Whiting et al. (1980) and Larsson (1983)).

Once the electronic transition moment function has been obtained, the question is how to assess its accuracy. The method of choice is to compute the radiative lifetime of the upper state and compare these results with available experimental results. This is readily done by summation of $A_{v'J',v''J''}$ in Eq. (5) over all lower states $v''J''$. Ideally, experimental results obtained with different techniques agree, and the comparison of the theoretical results with experiment provides a direct assessment of the theoretical method. For some molecules this is the case, as for example CH (Larsson & Siegbahn 1983). In other cases, if carefully designed, *ab initio* calculations can even be used to assess different experimental techniques (Bauschlicher & Langhoff 1987). Unfortunately there are also molecules for which there seems to be a never ending confusion both on the experimental and theoretical side. As we shall see below, CN ($A^2\Pi - X^2\Sigma^+$) is one of these molecules.

When the electronic transition moment function, $R_e(r)$, has been obtained, the next step is to calculate the matrix elements $\langle v'_{J'}|R_e(r)|v''_{J''}\rangle$, as follows from Eqs. (3) and (4). The vibrational wavefunctions $|v_J\rangle$ are obtained as solutions to the Schrödinger equation describing the oscillatory motion of the two nuclei in a potential $V_0(r)$ for the non-rotating molecule (if one should be a purist, it is the wavefunctions for $|v'_{J'=\Lambda'}\rangle$ and $|v''_{J''=\Lambda''}\rangle$ which are obtained). Provided that potential curves are available for the two electronic states, the calculation of wavefunctions $|v_J\rangle$ is straightforward and demands orders of magnitude less computer time as compared to *ab inito* calculations of the electronic wavefunctions. There are several methods that can be used to obtain potential curves. We have already mentioned that *ab initio* calculation is one such method, and in fact the only available route if there are no experimental data to resort to. However, if results from high resolution molecular spectroscopy are available, it is better to use these data together with an inversion procedure to obtain the potential curve. The Rydberg-Klein-Rees (RKR) procedure is the most widely used method among molecular spectroscopists for deriving $V_0(r)$. The method uses energy levels and molecular constants derived from molecular spectroscopy and, via a semiclassical quantization condition, generates pairs of turning points which define a potential curve (Lefebvre-Brion & Field 1986). There are also slightly modified versions of the RKR procedure, for example the Klein-Dunham method (Jarmain 1971), identical to RKR except for very light molecules. It is rare that experimental data allows the construction of an RKR potential all the way to the dissociation limit. In practise one has to make an extrapolation of the potential outside the range of internuclear distances covered by the RKR turning points. One way to do this is to extend the curve with a Morse potential, another way is to make an extension with an *ab initio* potential.

When the potential energy curves have been constructed, the vibrational wavefunctions $|v'_{J'}\rangle$ and $|v''_{J''}\rangle$ are calculated by numerical integration of the Schrödinger equation. Normally the computer programme used to obtain the potential $V_0(r)$ (RKR or Klein-Dunham methods) also provides the necessary routines to obtain $|v_J\rangle$. The potential $V_0(r)$ does not include the rotational motion of the nuclei (hence the index 0 in $V_0(r)$). When a molecule rotates, an interaction between the vibrational and rotational motion is induced. This interaction can be accounted for in calculations of line absorption oscillator strengths by adding a term $(h/8\pi^2\mu r^2)J(J+1)$ to $V_0(r)$[2]. In the calculation of the line data tape for the CN molecule (Jørgensen & Larsson 1990), the $J(J+1)$-term was always added to $V_0(r)$ and the calculation of vibrational wavefunctions was carried out separately for each value of J. The computer programme TRAPRB (Jarmain 1971, Jarmain & McCallum 1971) was used to obtain the $|v_J\rangle$ wavefunctions.

The calculation of the matrix elements $\langle v'_{J'}|R_e(r)|v''_{J''}\rangle$ is performed using the molecule-fixed components of the electronic transition moment operator (see Eqs. (6) and (7)), i.e. a coordinate system attached to the rotating molecule. Observations of molecular line intensities, however, are performed in a laboratory-fixed coordinate system. Hence a transformation of coordinate system is needed, and it is this transformation which introduces the Hönl-London factors (Hougen 1970), or rotational line intensity factors as they are also called. The Hönl-London factors enter the expression for the line strength according to Eq.(4). Originally the Hönl-London factors were calculated analytically (Kovacs 1969) and involved approximations to make the calculations tractable. More recently computer codes have been developed which calculate the Hönl-London factors numerically by means of suitable transformations (Whiting 1972, Whiting 1973, Whiting & Nicholls 1974). The theory for Hönl-London factors has also been formulated in a comprehensive way by Veseth (1986) using group theory.

If the Hönl-London factors are summed over all transitions from the group of $(2\text{-}\delta_{0,\Lambda''})(2S''+1)$ levels[3] with a specific J'' value to the $(2\text{-}\delta_{0,\Lambda'})(2S'+1)$ levels with $J' = J''$ and $J' = J'' \pm 1$, one obtains $(2\text{-}\delta_{0,\Lambda'}\delta_{0,\Lambda''})(2S''+1)(2J''+1)$ (Whiting & Nicholls 1974). If, for example, the Hönl-London factor for each branch in the $^2\Pi$ - $^2\Sigma^+$ transition in Fig.2 is summed for a specific value of J'' one obtains

$$\Sigma\mathcal{S} = \mathcal{S}(R_1) + \mathcal{S}(Q_1) + \mathcal{S}(P_1) + \mathcal{S}(^QR_{12}) + \mathcal{S}(^PQ_{12}) + \mathcal{S}(^OP_{12})$$

---

[2] μ is the reduced mass of the molecule and should not be confused with the notation for the transition moment operator in Eqs. (6) and (7).

[3] $\delta_{0,\Lambda''}$ is a Kronecker delta, i.e. $\delta_{0,\Lambda''}=1$ for $\Lambda''=0$ and $\delta_{0,\Lambda''}=0$ for $\Lambda'' \neq 0$.

$$+\mathcal{S}(R_2)+\mathcal{S}(Q_2)+\mathcal{S}(P_2)+\mathcal{S}(^S R_{21})+\mathcal{S}(^R Q_{21})+\mathcal{S}(^Q P_{21})$$
$$=(2-\delta_{0,\Lambda'}\delta_{0,\Lambda''})(2S''+1)(2J''+1)=4(2J''+1) \tag{8}$$

The sum rule is symmetrical in J' and J'' which means that the $''$ notation on J is redundant. This completes the calculation of $f_{v'J',v''J''}$ according to Eqs. (3) and (4).

4 Molecular absorption coefficients

It follows from Eq. (2) that, in order to calculate the absorption coefficient for a molecular transition, one needs, in addition to the oscillator strength, also the number density in the lower state (N_m in Eq. (2)). In a molecule the population in an electronic state is distributed over a large number of rotational and vibrational levels ($m = |\Lambda'', S'', \Sigma'', v'', J'', (e/f)''\rangle$). The number of particles, N_m, in a specified level $E_m = E_{\Lambda'',S'',\Sigma'',v'',J'',(e/f)''}$ is related to the total number of particles in all levels, N, according to the relation

$$\frac{N_m}{N}=\frac{(2J''+1)e^{-E_m/kT}}{Q_{int}} \tag{9}$$

where k is the Bolzmann constant, T is the temperature, $(2J''+1)$ is the degeneracy factor of the initial level, $e^{-E_m/kT}$ is the Boltzmann factor of the initial level, and Q_{int} is the internal partition function. Q_{int} is defined as the sum of the weighted Boltzmann factors of all discrete energy levels and can be expressed as (Tatum 1967)

$$Q_{int}=Q_{el}Q_{vib}Q_{rot} \tag{10}$$

Here the partition function has been separated, just as the line strength in Eq. (4), into one electronic partition function, Q_{el}, one vibrational partition function, Q_{vib}, and one rotational partition function, Q_{rot}. The reader is referred to (Tatum 1967) for a detailed treatment of how to develop partition functions.

In order to calculate Q_{int} one should sum over all states in a molecule (electronic, vibrational and rotational). This is impracticable and in practice one has to make some approximations. In particular it is not necessary, or even desirable, to extend the summation over all electronic states. Recipes for calculating partition functions have been published by Tatum (1966) and Sauval & Tatum (1984). In order to assess how complete a calculation of an absorption coefficient over an extended wavelength region is, it is valuable to compare how large part of the partition function that is recovered. This can be done by summation over all energy levels included in the calculation of the absorption coefficient and by comparison with the partition function calculated according to the method described by Sauval & Tatum (1984) (see also Jørgensen & Larsson 1990).

5 Molecular opacity data

In order to be useful for astrophysical applications, molecular absorption coefficients must be calculated for a large number of lines including very weak ones. Such calculations are performed on the basis of line lists containing gf-values for molecular lines. Molecular line lists have been developed by several groups based on essentially the methods outlined in the earlier sections. However, the applicability of a line list depends of course critically on the accuracy and completeness of the the underlying molecular data, and to what extent various approximations have been used to simplify the calculation of gf-values. It is not the intention of the present article to review and assess the various line data lists that are available for molecular astrophysics. The extensive work by Kurucz has been reviewed recently (Kurucz 1991) and is also described in another chapter of this book. Line lists for vibrational transitions within the ground electronic states of SiO, CO, CS and SiS have been developed by Tipping & Chackerian (1981), Chackerian & Tipping (1983) and Pineiro et al. (1987a, 1987b). Electronic transitions in CN, C_2 and CO have been treated by Querci et al. (1974). The reader is referred to (Jørgensen 1992) for a more complete review of the subject.

The line data list developed (named SCAN-CN) for the CN "red" system by Jørgensen and the present author (Jørgensen & Larsson 1990) has already been mentioned in the earlier sections. The SCAN-CN data base was developed exactly as outlined in Section 3. It contains 1.1 x 10^6 lines for each isotope combination $^{12}C^{14}N$, $^{12}C^{15}N$, $^{13}C^{14}N$ and $^{13}C^{15}N$. It carries, for each rotational line of the $A^2\Pi - X^2\Sigma^+$ transition, the vibrational quantum number of the $A^2\Pi$ and $X^2\Sigma^+$ states, the rotational quantum number of the $X^2\Sigma^+$ state (more specifically $J''+0.5$), the type of branch (see Fig. 2), the direction of the transition[4], the gf-value ($=(2J''+1)f_{v'J',v''J''}$), the vacuum line wavenumber (in cm^{-1}) and the excitation energy (in cm^{-1}) above $X^2\Sigma^+(v''=0, J=0.5, F_1)$. Some comments should be made regarding the line wavenumbers. The wavenumber of a particular rotational line was obtained by subtracting the energy of the lower level, $E_{\Lambda'',S''\Sigma'',v'',J'',(e/f)''}$, from the energy of the upper level, $E_{\Lambda',S'\Sigma',v',J',(e/f)'}$. The energy levels were calculated by applying analytical expressions describing the energy level structure of the $A^2\Pi$ and $X^2\Sigma^+$ states. This is nowadays not the optimal way to calculate energy levels in molecules and not the optimal use of molecular parameters derived from spectroscopy (see Lefebvre-Brion & Field 1986), but it served the practical purpose of generating a large number of lines in a computationally simple way. When SCAN-CN is used for opacity calculations, the accuracy of individual line positions is of limited concern. Owing to the identification system in the data base it can also conveniently be used to generate synthetic spectra. The deviation between calculated and measured line position was checked for the (6,0) and (0,0) bands (Jørgensen & Larsson 1990) and was found to be typically 0.2 cm^{-1} and 0.07 cm^{-1}, respectively. No systematic check

[4] For a low lying state like $A^2\Pi$ there is always the possibility for a transition $X^2\Sigma^+ \leftarrow A^2\Pi$, where the arrow $\leftarrow$ is used to indicate that the $A^2\Pi$ state is the lower state in the absorption process.

against all measured lines were performed and there may be lines which deviate more than 0.2 cm^{-1}. If SCAN-CN is used for the purpose of spectral analysis one should be aware of these limitations and exercise the necessary caution.

We will end this section with a brief discussion of the radiative properties of the $A^2\Pi - X^2\Sigma^+$ transition of CN. This discussion will serve to illustrate the difficulties one sometimes have to face in order to assess the accuracy of molecular data, be it experimental or theoretical. When the first planning of the SCAN-CN line list began, it seemed like the radiative properties of the CN "red" system was fully understood. Lifetime measurements of the six lowest vibrational levels of the $A^2\Pi$ state (Taherian & Slanger 1984) were in very good agreement with results from *ab initio* calculations (Larsson et al. 1983). However, more recent *ab initio* calculations using a slightly different approach gave lifetimes about 30% longer (Bauschlicher et al. 1988, Knowles et al. 1988). In terms of computational efforts these calculations are the most extensive so far for the CN "red" system. Furthermore, measurements of oscillator strengths for the $A^2\Pi - X^2\Sigma^+$ transition (Davis et al. 1986) supported the more recent *ab initio* work. At this point it seemed like there was sufficient evidence for a modification of the gf-values in the SCAN-CN tape, and in (Jørgensen & Larsson 1990) it is in fact suggested that the gf-values should be multiplied by a factor 0.734 in order to agree with the electronic transition moment function calculated by Bauschlicher et al. (1988) and Knowles et al. (1988). But the situation has once again changed. Lu et al. (1992) have measured radiative lifetimes by means of the laser induced fluorescence technique and found lifetimes even shorter than those predicted by Larsson et al. (1983). Thus it seems like the the CN "red" system is bound to remain an enigma for yet some time to come; after the experimental efforts by Lu et al. (1992) it is once again time for theoretical work in order to shed some new light on the problem.

6 Final remarks

It has been the intention of this review to provide a guidance for astrophysicists in need of data for modelling the absorption of radiation by molecules in the stellar environment. Much of the material can be found elsewhere, however, one would need to turn to several different sources in order to find it. At a first glance it would seem like a diatomic molecule, at least compared to a complex stellar environment, would constitute a rather simple system. However, diatomic molecules present many chances for errors, and they are very demanding systems for both experimentalists and theorists. This, among other things, is what makes them fascinating systems for studies in their own right.

Acknowledgements. The author would like to thank J.R. Mowat and L. Veseth for valuable comments on the manuscript.

References

Bauschlicher C.W., Langhoff S.R., 1987, *J. Chem. Phys.*, **87**, 4665

Bauschlicher C.W., Langhoff S.R., Taylor P.R., 1988, *Astrophys. J.*, **332**, 531

Chackerian Jr. C., Tipping R.H., 1983, *J. Mol. Spectrosc.*, **99**, 431

Condon U., Shortley G.H., 1935, *The Theory of Atomic Spectra*, Cambridge Univ. Press, London

Davis S.P., Shortenhaus D., Stark G., Engleman Jr. R., Phillips J.G., Hubbard R.P., 1986, *Astrophys. J.*, **303**, 892, ibid. **307**, 414

Hougen J.T., 1970, *Natl. Bur. Std. Monogr.*, **115**

Jarmain W.R., 1971, *J. Quant. Spectrosc. Radiat. Tranfer*, **11**, 421

Jarmain W.R., McCallum J.C., 1971, Program TRAPRB Handbook CRESS York University, Toronto, Ontario

Jørgensen U.G., 1992, *Revista Mexicana de Astronomia y Astrofisica*, **23**, 49

Jørgensen U.G., Larsson M., 1990, *Astron. Astrophys.*, **238**, 424

Kurucz R.L., 1991, in *Stellar Atmospheres, Beyond Classical Models*, eds Crivellari et al., NATO ASI, Kluwer, Dordrecht

Kovacs I., 1969, *Rotational Structure in the Spectra of Diatomic Molecules*, Hilger, London

Knowles P.J., Werner H.-J., Hay P.J., Cartwright D.C., 1988, *J. Chem. Phys.*, **89**, 7334

Larsson M., 1983, *Astron. Astrophys.*, **128**, 291

Larsson M., Siegbahn P.E.M., 1983, *J. Chem. Phys.*, **79**, 2270

Larsson M., Siegbahn P.E.M., Ågren, H.A., 1983, Astrophys. J., **272**, 369

Lefebvre-Brion H., Field R.W., 1986, *Perturbations in the Spectra of Diatomic Molecules*, Academic, Florida

Lu R., Huang Y., Halpern J.B., 1992, *Astrophys. J.*, **395**, 710

Pineiro A.L., Tipping R.H., Chackerian Jr. C., 1987a, *J. Mol. Spectrosc.*, **125**, 91

Pineiro A.L., Tipping R.H., Chackerian Jr. C., 1987b, *J. Mol. Spectrosc.*, **125**, 184

Sauval A.J., Tatum J.B., 1984, *Astrophys. J. Suppl.*, **56**, 193

Schadee A., 1971, *Astron. Astrophys.*, **14**, 401

Taherian M.R., Slanger T.G., 1984, *J. Chem. Phys.*, **81**, 3814

Tatum J.B., 1966, *Publ. Dominion Astrophys. Obs.*, **13**, No. 1

Tatum J.B., 1967, *Astrophys. J. Suppl.*, **14**, 21

Tipping R.H., Chackerian C.Jr., 1981, *J. Mol. Spectrosc.*, **88**, 352

Veseth L., 1986, *Symmetries and Rotational Line Intensities in Diatomic Molecules*, Report 86-05, Institute of Physics, University of Oslo, Oslo

Whiting E.E., 1972, Ph.D. Thesis, York University, Toronto, Ontario

Whiting E.E., 1973, Computer Program for Determining Rotational Line Intensity Factors for Diatomic Molecules, NASA TN D-7268

Whiting E.E., Nicholls R.W., 1974, *Astrophys. J. Supp.*, **27**, 1

Whiting E.E., Schadee A., Tatum J.B., Hougen J.T., Nicholls R.W., 1980, *J. Mol. Spec.*, **80**, 249

Computation of Opacities for Diatomic Molecules

Robert L. Kurucz

Harvard-Smithsonian Center for Astrophysics, 60 Garden St., Cambridge, MA 02138, USA

1 Atomic and Molecular Data

In this section I briefly describe my efforts to improve the atomic and molecular line data. This work is described in more detail in Kurucz (1992a). In subsequent sections I briefly describe three methods for computing opacity and the models and spectra that result from using them.

My model calculations in the 1970s used the distribution-function line opacity computed by Kurucz (1979a,b) from the line data of Kurucz & Peytremann (1975). We had computed gf values for 1.7 million atomic lines for sequences up through nickel using scaled-Thomas-Fermi-Dirac wavefunctions and eigenvectors determined from least squares Slater parameter fits to the observed energy levels. We also collected all published data on gf values and included them in the line list whenever they appeared to be more reliable than the computed data (that work is ongoing, but I am running behind).

After the Kurucz-Peytremann calculations were published, I started work on line lists for diatomic molecules beginning with H_2, CO, and SiO. Next, Lucio Rossi of the Istituto Astrofisica Spaziale in Frascati, John Dragon of Los Alamos, and I computed line lists for electronic transitions of CH, NH, OH, MgH, SiH, CN, C_2, and TiO. In addition to lines between known levels, these lists include lines whose wavelengths are predicted and are not good enough for detailed spectrum comparisons but are quite adequate for statistical opacities. All these data are listed in Table 1.

In 1983 I recomputed the opacities using the additional atomic and molecular data described above which totalled 17,000,000 lines. These opacities were used to produce improved empirical solar models (Avrett et al. 1984), but were found to still not have enough lines. In detailed ultraviolet spectrum calculations, half the intermediate strength and weak lines were missing. This discrepancy was caused by missing iron group atomic lines that go to excited configurations that had not yet been observed in the laboratory and so had not been included in the Kurucz-Peytremann calculations.

I was granted a large amount of computer time at the San Diego Supercomputer Center to carry out new calculations. I included as many configurations for

Table 1. Molecular Line Lists

FILE	NUMBER OF LINES	WAVELENGTH (NM) FIRST	LAST	COMMENT
H2	28486	84.4941	184.4573	Lyman and Werner
HYDRIDES	331268	203.6264	3245.1715	CH,NH,OH,MgH,SiH all isotopes
COAX	396004	111.3365	460.6524	CO 4th Pos all isotopes
COIR	118920	963.5078	9998.8853	CO vib-rot all isotopes
SIOAX	760378	177.3128	546.2833	SiO A-X all isotopes
SIOEX	947015	143.0468	462.4214	SiO E-X all isotopes
CNAX12	484709	292.5406	99912.800	$^{12}C^{14}N$ Red
CNAX13	503631	295.1871	99924.784	$^{13}C^{14}N$ Red
CNAX15	289887	341.5994	99776.188	$^{12}C^{15}N$ Red
CNBX	323818	201.9947	715.7552	CN Violet all isotopes
C2AX	406236	271.8071	9999.6289	C_2 Phillips all isotopes
C2BA12	462803	389.6061	99957.754	$^{12}C^{12}C$ Ballik-Ramsay
C2BA13	387809	511.6587	99980.746	$^{12}C^{13}C$ Ballik-Ramsay
C2BA33	313319	527.1495	80319.395	$^{13}C^{13}C$ Ballik-Ramsay
C2DA12	329771	344.9621	2541.6874	$^{12}C^{12}C$ Swan
C2DA13	253014	374.0201	991.5625	$^{12}C^{13}C$ Swan
C2DA33	217545	376.2757	959.4157	$^{13}C^{13}C$ Swan
C2EA12	487232	176.0052	862.4585	$^{12}C^{12}C$ Fox-Herzberg
C2EA13	332214	177.3769	560.9010	$^{12}C^{13}C$ Fox-Herzberg
C2EA33	260883	178.6938	543.2282	$^{13}C^{13}C$ Fox-Herzberg
TIOAX6	325116	528.5640	1875.4212	$^{46}Ti^{16}O\ \gamma$
TIOAX7	327496	528.7059	1771.6149	$^{47}Ti^{16}O\ \gamma$
TIOAX8	647203	490.4263	2146.0045	$^{48}Ti^{16}O\ \gamma$
TIOAX9	332113	528.9757	1864.4024	$^{49}Ti^{16}O\ \gamma$
TIOAX0	334222	529.1046	1860.9887	$^{50}Ti^{16}O\ \gamma$
TIOBX6	343598	246.6118	1232.9218	$^{46}Ti^{16}O\ \gamma'$
TIOBX7	345990	246.3570	1209.4093	$^{47}Ti^{16}O\ \gamma'$
TIOBX8	861488	235.9828	1362.8784	$^{48}Ti^{16}O\ \gamma'$
TIOBX9	350530	246.7795	1187.2650	$^{49}Ti^{16}O\ \gamma'$
TIOBX0	352802	247.3491	1186.9649	$^{50}Ti^{16}O\ \gamma'$
TIOCX6	232718	350.0947	1562.0368	$^{46}Ti^{16}O\ \alpha$
TIOCX7	234318	350.0471	1554.5985	$^{47}Ti^{16}O\ \alpha$
TIOCX8	503950	257.0800	1316.1735	$^{48}Ti^{16}O\ \alpha$
TIOCX9	237430	350.0113	1540.7715	$^{49}Ti^{16}O\ \alpha$
TIOCX0	238827	349.9744	1534.3348	$^{50}Ti^{16}O\ \alpha$
TIOEX6	241398	630.6768	2002.6215	$^{46}Ti^{16}O\ \epsilon$
TIOEX7	242954	631.0335	1997.7022	$^{47}Ti^{16}O\ \epsilon$
TIOEX8	510074	549.4411	2273.1493	$^{48}Ti^{16}O\ \epsilon$
TIOEX9	245978	631.7067	1988.4477	$^{49}Ti^{16}O\ \epsilon$
TIOEX0	247466	632.0252	1984.0965	$^{50}Ti^{16}O\ \epsilon$
TIOBA	310376	541.4458	2250.2486	TiO δ all isotopes
TIOCA	223850	446.7588	1669.0456	TiO β all isotopes
TIOFA	153496	432.0083	816.6252	TiO f-a all isotopes
TIOBD	158000	593.4372	2662.9648	TiO ϕ all isotopes
TIOED	102107	332.9690	562.5112	TiO e-d all isotopes

each ion as I could fit into a Cray. Hamilitonian parameters were determined by combining least squares fits for levels that have been observed with computed Hartree-Fock integrals (scaled) for higher configurations. All configuration interactions were included. My computer programs have evolved from Cowan's (1968) programs. Transition integrals were computed with scaled-Thomas-Fermi-Dirac wavefunctions and the whole transition array was produced for each ion. Radiative, Stark, and van der Waals damping constants and Landé g values are automatically produced for each line. The first nine ions of Ca through Ni produced 42,000,000 lines. Most of those lines have uncertain wavelengths because they go to predicted rather than measured levels. I have produced a single tape edition of these data for distribution that has all the lines with reliable wavelengths between laboratory determined energy levels.

I hope to spend the next year improving the line data. I will extend the atomic calculation to elements lighter and heavier than the iron group which I have already computed. I will recompute the energy levels and line lists whenever new laboratory analyses become available and I will make the predictions available to laboratory spectroscopists. Several of the iron group calculations have already been revised. Because computers are now more powerful, more configurations can be treated. This should account for more of the missing infrared lines because they are usually transitions between highly excited levels. I will also try to include as much hyperfine and isotopic splitting data as possible (Kurucz 1993b).

Returning to the diatomic molecules, the line lists in Table 1 are in need of revision. All the transitions are electronic except for the CO vibration-rotation bands. Most of the isotopic positions are unreliable. Most of the calculations were done more than 15 years ago. I plan to include all the improvements in the laboratory analyses since that time and I plan to add all the significant vibration-rotation bands and many ultraviolet electronic bands. The newer analyses are based on FTS spectra and produce dramatic improvements in energy levels and line positions. However, they still do not go to high enough V and J. Farrenq et al. (1990) have actually been able to use the solar spectrum itself to analyse CO to high J (see also Grevesse & Sauval this volume).

The procedure I use for generating line lists is straightforward and produces all the lines up to a specified cut-off lower energy level. I start with all the known energy levels. I set up a model rotational Hamiltonian and then I do a least squares fit to determine the rotational constants. Once the fit has converged, I use the Hamiltonian to generate all possible eigenvalues and eigenvectors. The eigenvalues are replaced by the observed energies where they are known. I compute the RKR potential and then all the vibrational wavefunctions. I then integrate over measured or computed transition moments taken from the literature to get the transition integrals. The transition integrals are divided into transition arrays in the adopted basis and are transformed to observed coupling using the eigenvectors. Given enough computer time, I can readily generate thousands of energy levels and millions of lines. Since the known energy levels are used when available, the line wavelengths are correct for lines between known energy levels, regardless of perturbations. Lines to predicted levels are as accurate as the least squares fitting procedure. Radiative damping constants are automatically produced for each line.

Line lists for the triatomic molecules are needed for work on cool M stars. I hope to obtain them from Jørgensen & Jensen (1993) and other groups.

All the atomic and molecular data will be published on CD-ROMs. These will include both a replacement for Kurucz-Peytremann (1975) and all lines with good wavelengths for molecules listed in Table 1 for use in spectrum synthesis calculations. Highly compressed versions of the atomic and diatomic data have already been published for use in opacity calculations as indicated in Table 2. This allows anyone to compute opacities or opacity-sampled model atmospheres for arbitrary abundances. For each major atomic or molecular calculation I expect to publish a CD-ROM on which I will provide tables of all laboratory measurements, together with my computed energy levels, eigenvector compositions, damping constants, Landé g values, lifetimes, branching ratios, and gf values. These should be of value to both laboratory and astronomical spectroscopists.

2 The Distribution Function Opacity Program

The opacities are calculated with a version of my spectrum synthesis computer program that has been under development since 1965 and has been described by Kurucz & Avrett (1981). The algorithms for computing the total line opacity are extremely fast because maximum use is made of temperature and wavelength factorization and pretabulation. Up to a million point spectrum can be computed in one run. There is no limit to the number of spectrum lines that can be treated.

The spectrum calculations require a pre-existing model atmosphere. For computing opacities it is just a table of temperatures and pressures. Quantities that need be computed only once for the model atmosphere are pretabulated, i.e., number densities, partition functions, and doppler widths. The equation of state and continuum opacities are computed with a version of the model atmosphere program ATLAS (Kurucz 1970). Solar abundances are taken from Anders & Grevesse (1989). Photoionization continua are put in at their exact positions, each with its own cross-section and with the series of lines that merge into each continuum included so that there are no discontinuities in the spectrum.

Hydrogen line profiles are computed using a routine from Peterson (1979) that approximates the Vidal et al. (1973) profiles, works to high n, and includes Doppler broadening, resonance broadening, van der Waals broadening, and fine-structure splitting. Autoionization lines have Shore-parameter Fano profiles. Other lines have Voigt profiles that are computed accurately for any value of the damping parameter a which includes radiative, Stark, and van der Waals broadening. For small a, table lookup is used for speed. Line wings are truncated when they are weaker than 0.001 times the continuum. Lines weaker than 0.001 times the continuum are not considered.

The opacity spectrum is computed only for zero microturbulent velocity. Other velocities are obtained by convolution. The spectrum is computed at 3,500,000 points uniformly spaced in $\Delta\lambda/\lambda$ (proportional to Doppler velocity) from 8.97666 nm to 10000 nm. The spectral resolution of 500,000, or 0.6 km/s, is enough to resolve line cores. The distribution function for a wavelength interval is determined by statistically analyzing the total line opacity spectrum in that interval. The intervals vary from 1 nm in the ultraviolet to 10 nm in the far infrared.

In late 1988 I used the line data described above to compute new solar abundance opacity tables for use in my modelling. The calculations involved 58,000,000 lines, 3,500,000 wavelength points, 56 temperatures from 2000K to 200,000K, 21 log pressures from -2 to $+8$, and 5 microturbulent velocities 0, 1, 2, 4, and 8 km/s, and took a large amount of computer time. The opacity was tabulated twice, as 12-step distribution functions for 328 "big" intervals for speed, and for 1212 "little" intervals for resolution.

Since the beginning of 1990 I have been able to take tremendous advantage of the new Cray YMP at the San Diego Supercomputer Center. I computed opacities ranging from 0.00001 solar to 10 times solar, enough to compute model atmospheres ranging from the oldest Population II stars to high abundance Am and Ap stars. The exact abundances are [+1.0], [+0.5], [+0.3], [+0.2], [+0.1], [+0.0], [-0.1], [-0.2], [-0.3], [-0.5], [-1.0], [-1.5], [-2.0], [-2.5], [-3.0], [-3.5], [-4.0], [-4.5], and [-5.0] where [x.x] stands for log difference from solar abundance of metals. I have just completed opacities with +0.4 dex enhanced α-process elements (O, Ne, Mg, Si, S, Ar, Ca, Ti) for abundances [+0.5], [+0.0], [-0.5], [-1.0], [-1.5], [-2.0], [-2.5], [-3.0], [-3.5], [-4.0], [-4.5], and [-5.0]. These opacities are distributed on CD-ROMs as listed in Table 2 or on VAX backup tapes.

Rosseland mean opacities have been computed for each abundance for use in interiors and envelope calculations.

3 New Grids of Models

I rewrote my model atmosphere program to use the new line opacities, additional continuous opacities, and an approximate treatment of convective overshooting. The revised version is called ATLAS9. The opacity calculation was checked by computing a realistic model for the Sun. See Kurucz (1992b) for details and figures.

Thus far, I have computed a grid of more than 7000 models for [+1.0], [+0.5], [+0.3], [+0.2], [+0.1], [+0.0], [-0.1], [-0.2], [-0.3], [-0.5], [-1.0], [-1.5], [-2.0], [-2.5], [-3.0], [-3.5], [-4.0], [-4.5], and [-5.0] for 2 km/s microturbulent velocity. The effective temperature range is 3500K to 50,000K, i.e. from K stars to O stars. Note that since triatomic molecules are not included in the opacity, the models cannot be used for M stars. Gravities range from log g = 5 down to 0 or the radiation pressure limit in steps of 0.5. Models cooler than 9000K are convective with $l/H = 1.25$. For each model the flux was computed at 1221 wavelengths in the range 0.01 to 160 μm, enough to treat ionization in H II regions and to calibrate the infrared. The temperature range of this grid should allow photometric calibrations consistent for both cool and hot stars.

For each model in the grid I have computed the fluxes, photometry in several systems, Balmer line profiles, and limb-darkening. The models, fluxes, colors, and Balmer line profiles are available on Kurucz CD-ROM No.13 or on VAX backup tapes. Earlier versions of these tapes had errors that have been corrected. The limb-darkening requires a separate tape for each abundance. The CD version is in preparation.

The microturbulent velocity is an important parameter that must be considered. Opacity and model structure vary with v_{turb}. If there is diffusion, v_{turb} must be small. In giants it can be large. In pulsating stars, such as Cepheids and RR Lyraes, it varies with phase. I am presently extending the grid calculations to other microturbulent velocities.

I am also computing grids of models using the new α-enhanced opacities, working inward both from the high abundance bulge stars and from the extremely low abundance halo stars.

4 Opacity Sampling and ATLAS12

In Am and Ap stars, or, perhaps, in all A stars, and in cool evolved giants the atmospheric abundances are not scaled-solar. Scaled-solar grids can produce only an approximate representation of the atmosphere. The cost of computing opacities is so high that it is not practical to pretabulate opacities for individual stars, although it might be possible for subclasses. The alternative is to use opacity sampling to compute individual models for each star.

I have developed a new version of my model atmosphere program, called ATLAS12, that samples from my whole line list to compute opacity. It essentially combines my spectrum synthesis program SYNTHE (Kurucz & Avrett 1981) with my model atmosphere program ATLAS (Kurucz 1970). It recognizes more than 1000 species, each in up to 10 isotopic forms, including all ions of the elements up through Zn and the first 5 ions of heavier elements. The elemental abundances are treated as variable with depth.

At present ATLAS12 has 6 input files containing 58,000,000 lines:

NLTELINES: H, He, C, Mg, Al, Si, Ca, etc., the file used by SYNTHE, species that can be treated in non-LTE by Avrett & Loeser (1992);

LOWLINES: first 5 stages of ionization of all elements not in NLTELINES;

HILINES: ion stages 6 and higher of elements up through Zn;

DIATOMICS: H, C, N, O, Mg, Si, etc combinations;

TIOLINES: TiO now, VO, ZrO, etc, in the future;

TRIATOMICS: empty now, H_2O, etc., in the future.

For each line the wavelength, identification, lower energy level, gf value, and the radiative, Stark, and van der Waals damping constants are packed into 16 bytes. At each wavelength point in a frequency integration, the Voigt profiles of all the significant nearby lines are computed and summed. The lines in file NLTELINES can be treated with more complicated profile functions. These files are available on Kurucz CD-ROMs Nos. 1 and 15. ATLAS12 is still not working reliably at this writing. I hope to distribute it in the fall of 1993. It has been used to make models with 25000 sampled frequencies for Vega (Kurucz 1993; Castelli & Kurucz 1993), which is an easy case.

It will be possible to use ATLAS12 to compute Rosseland opacities for arbitrary abundances for use in envelope and interior calculations.

5 Complete Spectrum Synthesis

It is now possible simply to compute the whole spectrum using all 58 million lines without resorting to sampling or statistical approximations. Such a calculation takes a few days on a very fast workstation. Figure 1, on the six pages following Table 2, shows a sample calculation for Arcturus using an improved version of the model atmosphere from Peterson et al. (1993) that was computed with ATLAS9. The spectrum calculation shown in the top panel of Fig. 1a,1c,1e was repeated 11 times with subsets of the line list. The resolution was 500,000 with 1,864,857 points between 200 and 10000 nm. A full dispersion plot that displays the resolution requires a case of paper. It is now also possible to compute models by integrating over the whole spectrum for each iteration, but even I am daunted by this prospect, for the moment.

References

Anders E., Grevesse N., 1989, *Geochimica et Cosmochimica Acta*, **53**, 197

Avrett E.H., Kurucz R.L., Loeser R., 1984, *Bull. Amer. Astron. Soc.*, **16**, 450

Avrett E.H., Loeser R., 1992, in *7th Cambridge Workshop on Cool Stars, Stellar Systems, and the Sun*, M.S.Giampapa & J.A.Bookbinder (eds.), ASP Conference Series, **26**, 489

Castelli F., Kurucz R.L., 1993, *Astron. Astrophys.*, in press

Cowan R.D., 1968, *J. Opt. Soc. Am.*.**58**, 808

Farrenq R., Guelachvili G., Sauval A.J., Grevesse N., Farmer C.B., 1991, *J. Mol. Spectr.*, **149**, 375

Jørgensen U.G., Jensen P., 1993, *J. Mol. Spectr.*, **161**, 219

Kurucz R.L., 1970, *SAO Spec. Rep.*No. 309, 291 pp.

Kurucz R.L., 1979a, *Astrophys. J. Supp.*, **40**, 1

Kurucz R.L., 1979b, *Dudley Observ. Rep.* No. 14, A.G.Davis Philip (ed.), p. 363

Kurucz R.L., 1992a, *Rev. Mexicana Astron. Astrof.*, **23**, 45

Kurucz R.L., 1992b, *Rev. Mexicana Astron. Astrof.*, **23**, 181

Kurucz R.L., 1993a, in *Peculiar versus Normal Phenomena in A-type and Related Stars*, M.M.Dworetsky, F.Castelli, & R.Faraggiana (eds.), ASP Conference Series, **44**, 87

Kurucz R.L., 1993b, *Physica Scripta*, **T47**, 110

Kurucz R.L., Avrett E.H., 1981, *SAO Spec. Rep.*No. 391, 145 pp.

Kurucz R.L., Peytremann, E., 1975, *SAO Spec. Rep.*No. 362, 1219 pp.

Peterson D.M., 1979, Personal communication

Peterson R.C., dalle Ore C.M., Kurucz R.L., 1993, *Astrophys. J.*, **404**, 303

Vidal C.R., Cooper J., Smith E.W., 1973, *Astrophys. J. Supp.*, **25**, 37

Table 2. Kurucz CD-ROMs

Available from the author (Kurucz@cfa.harvard.edu)

No.01 Atomic data for opacity calculations.
No.02 Opacities for Stellar Atmospheres: [+0.0],[+0.5],[+1.0].
No.03 Opacities for Stellar Atmospheres: [-0.5],[-1.0],[-1.5].
No.04 Opacities for Stellar Atmospheres: [-2.0],[-2.5],[-3.0].
No.05 Opacities for Stellar Atmospheres: [+0.1],[+0.2],[+0.3].
No.06 Opacities for Stellar Atmospheres: [-0.1],[-0.2],[-0.3].
No.07 Opacities for Stellar Atmospheres: [-3.5],[-4.0],[-4.5].
No.08 Opacities for Stellar Atmospheres: [-5.0],[+0.0,noHe],[-0.5,noHe].
No.09 Opacities for Stellar Atmospheres: [+0.5a],[+0.0a],[-0.5a] +.4 alpha.
No.10 Opacities for Stellar Atmospheres: [-1.0a],[-1.5a],[-2.0a] +.4 alpha.
No.11 Opacities for Stellar Atmospheres: [-2.5a],[-3.0a],[-3.5a] +.4 alpha.
No.12 Opacities for Stellar Atmospheres: [-4.0a],[-4.5a],[-5.0a] +.4 alpha.
No.13 ATLAS9 Stellar Atmosphere Programs and 2 km/s Grid.
No.14 Opacities for Stellar Atmospheres: Abundance Sampler.
No.15 Diatomic molecular data for opacity calculations.

In preparation:

Replacement for Kurucz-Peytremann atomic line list for all elements.
Molecular line list with all the old data (Table I).
ATLAS12 opacity sampling model atmosphere program and
SYNTHE spectrum synthesis programs.
Model Atmosphere Grids: [+0.0] for 0,1,2,4,8 km/s Vturb.
Model Atmosphere Grids: [-0.5a],[+0.0a],[+0.5a] 2 km/s for bulge stars.
Model Atmosphere Grids: Limbdarkening for 2 km/s Grid (No.13).

Coming:

All new atomic and molecular calculations as they are made.
Improved atomic data for opacity calculations.
Improved molecular data for opacity calculations.
Model Atmosphere Grids: [-1.0a],[-1.5a],[-2.0a] 2 km/s.
Model Atmosphere Grids: [-2.5a],[-3.0a],[-3.5a] 2 km/s.
Model Atmosphere Grids: [-4.0a],[-4.5a],[-5.0a] 2 km/s.
Model Atmosphere Grids: [-0.5] for 0,1,2,4,8 km/s Vturb.
Model Atmosphere Grids: [-1.0] for 0,1,2,4,8 km/s Vturb.
Model Atmosphere Grids: [-1.5] for 0,1,2,4,8 km/s Vturb.
Model Atmosphere Grids: [-2.0] for 0,1,2,4,8 km/s Vturb.
Model Atmosphere Grids: [+0.5] for 0,1,2,4,8 km/s Vturb.
Model Atmosphere Grids: [+0.0a] for 0,1,2,4,8 km/s Vturb.
Model Atmosphere Grids: [-0.5a] for 0,1,2,4,8 km/s Vturb.
Model Atmosphere Grids: [-1.0a] for 0,1,2,4,8 km/s Vturb.
Model Atmosphere Grids: [-1.5a] for 0,1,2,4,8 km/s Vturb.
Model Atmosphere Grids: [-2.0a] for 0,1,2,4,8 km/s Vturb.
Model Atmosphere Grids: [+0.5a] for 0,1,2,4,8 km/s Vturb.

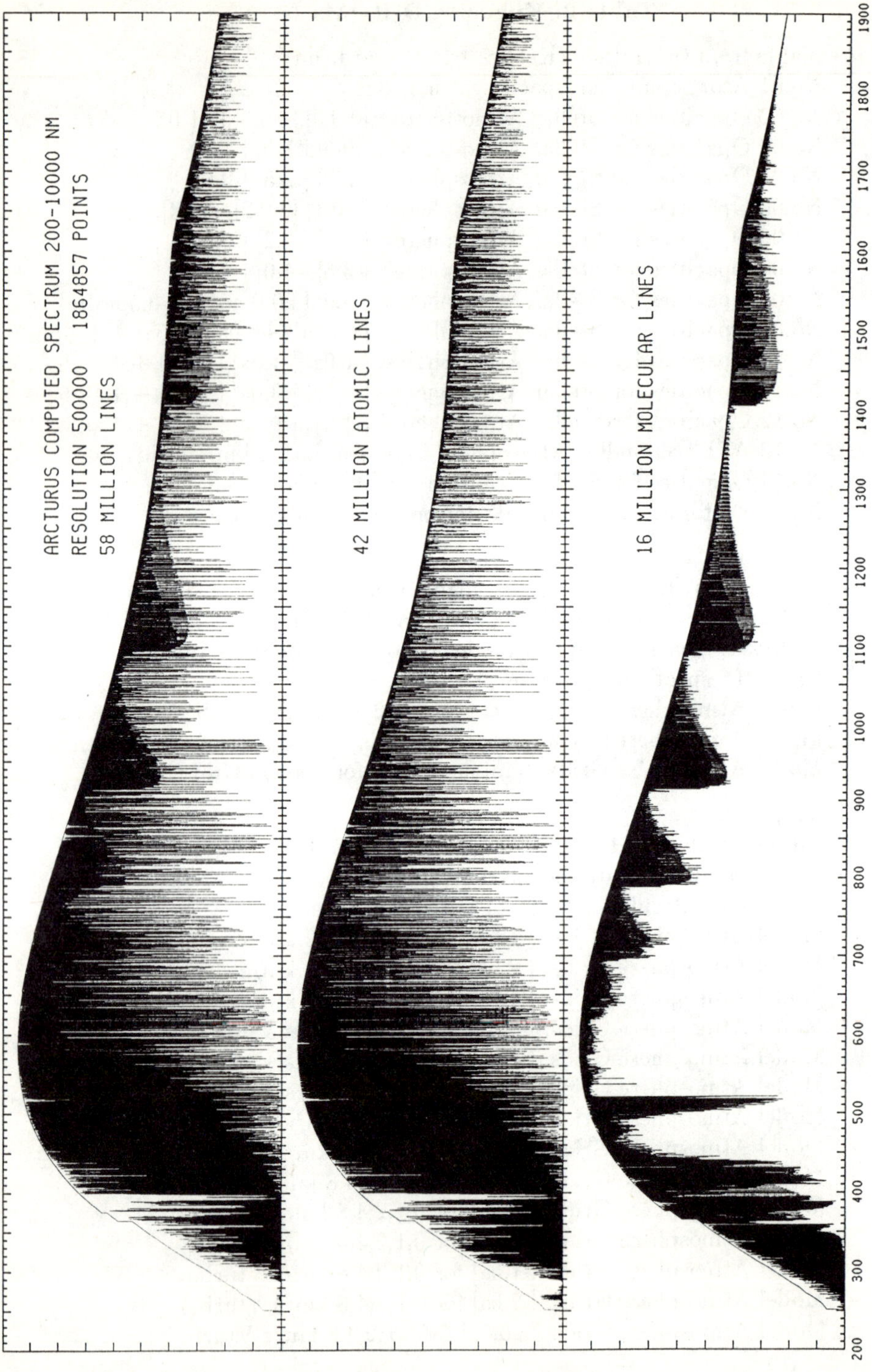
ARCTURUS COMPUTED SPECTRUM 200-10000 NM
RESOLUTION 500000 1864857 POINTS
58 MILLION LINES
42 MILLION ATOMIC LINES
16 MILLION MOLECULAR LINES
200
300
400
500
600
700
800
900
1000
1100
1200
1300
1400
1500
1600
1700
1800
1900

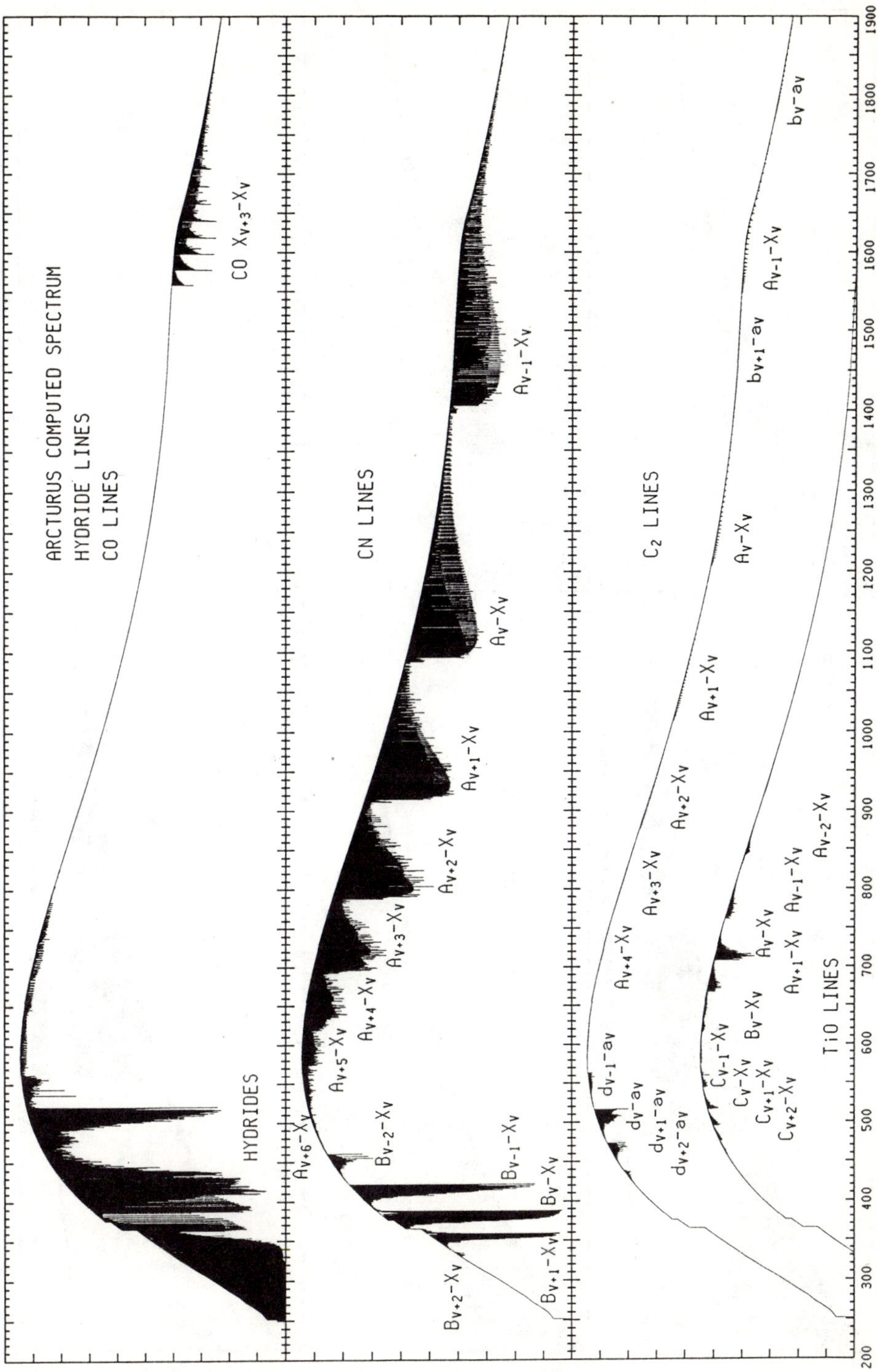
ARCTURUS COMPUTED SPECTRUM
HYDRIDE LINES
CO LINES
HYDRIDES
CO Xv+3-Xv
CN LINES
Bv+2-Xv
Bv+1-Xv
Bv-Xv
Bv-1-Xv
Bv-2-Xv
Av+6-Xv
Av+5-Xv
Av+4-Xv
Av+3-Xv
Av+2-Xv
Av+1-Xv
Av-Xv
Av-1-Xv
C2 LINES
dv+2-av
dv+1-av
dv-av
dv-1-av
Av+4-Xv
Av+3-Xv
Av+2-Xv
Av+1-Xv
Av-Xv
Av-1-Xv
bv+1-av
bv-av
TiO LINES
Cv+2-Xv
Cv+1-Xv
Cv-Xv
Cv-1-Xv
Bv-Xv
Av+1-Xv
Av-Xv
Av-1-Xv
Av-2-Xv
200
300
400
500
600
700
800
900
1000
1100
1200
1300
1400
1500
1600
1700
1800
1900

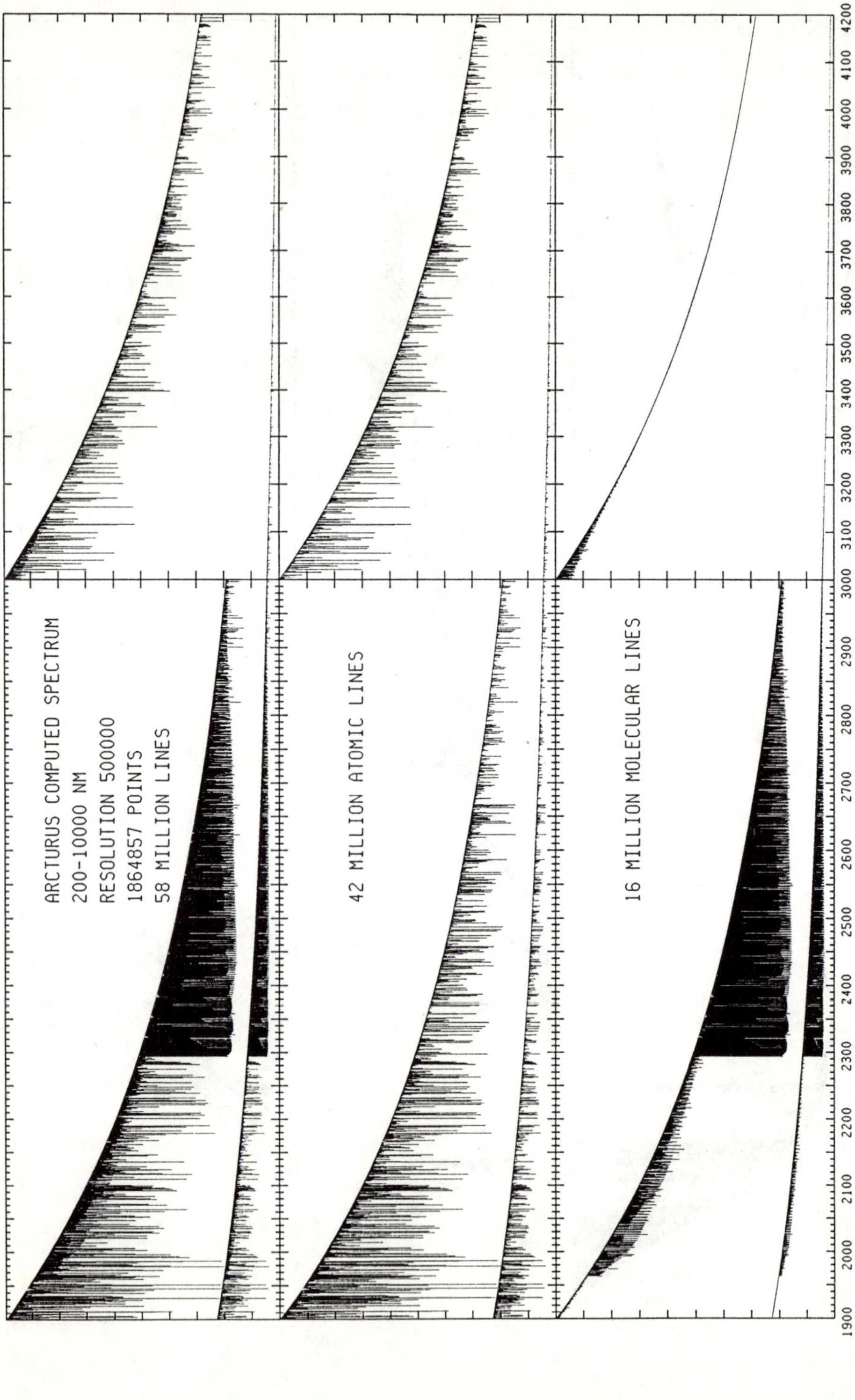
ARCTURUS COMPUTED SPECTRUM
200-10000 NM
RESOLUTION 500000
1864857 POINTS
58 MILLION LINES
42 MILLION ATOMIC LINES
16 MILLION MOLECULAR LINES
1900
2000
2100
2200
2300
2400
2500
2600
2700
2800
2900
3000
3100
3200
3300
3400
3500
3600
3700
3800
3900
4000
4100
4200

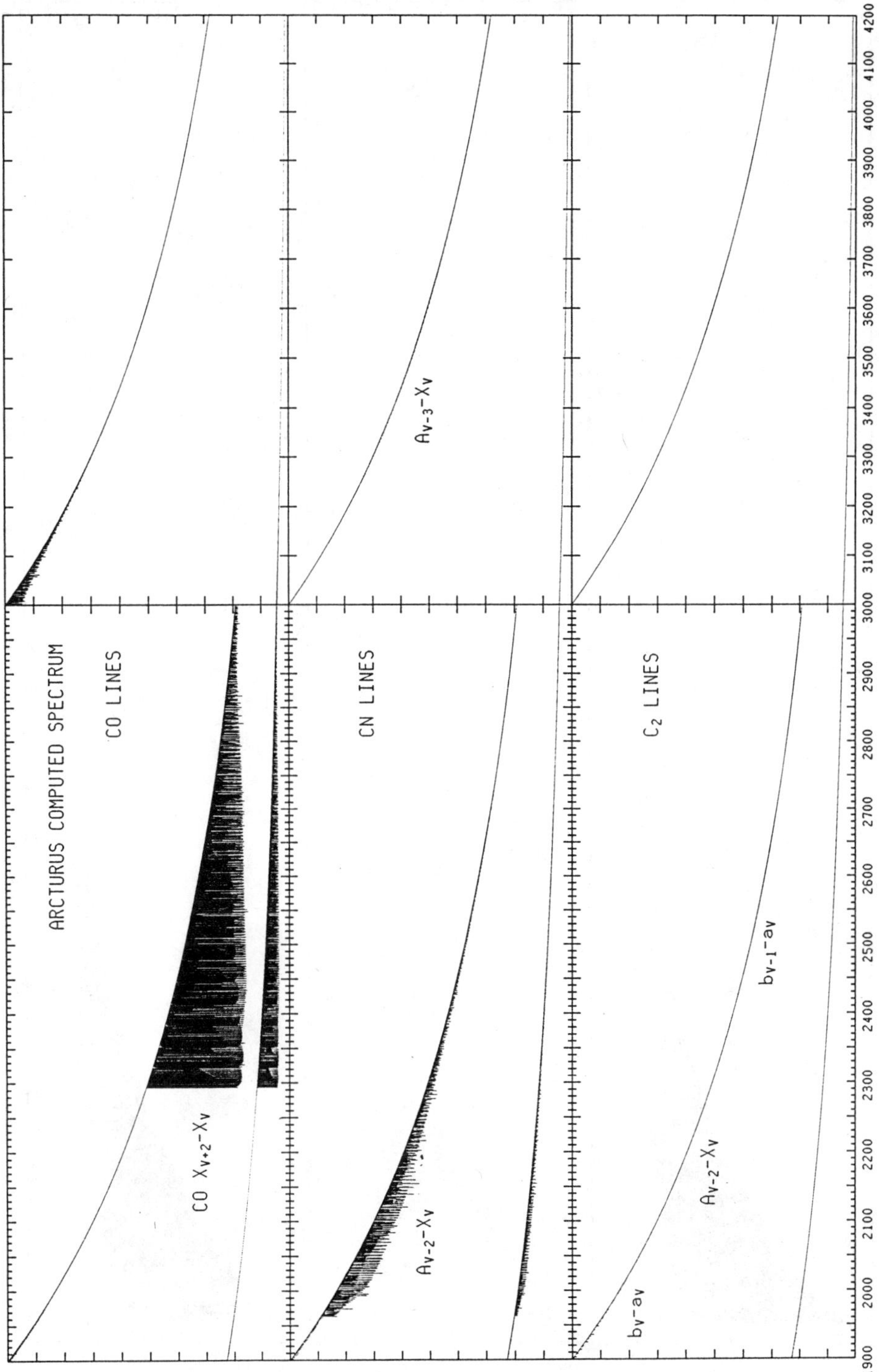
ARCTURUS COMPUTED SPECTRUM
CO LINES
CO $X_{v+2}-X_v$
CN LINES
$A_{v-2}-X_v$
$A_{v-3}-X_v$
C_2 LINES
b_v-a_v
$A_{v-2}-X_v$
$b_{v-1}-a_v$
1900 2000 2100 2200 2300 2400 2500 2600 2700 2800 2900 3000 3100 3200 3300 3400 3500 3600 3700 3800 3900 4000 4100 4200

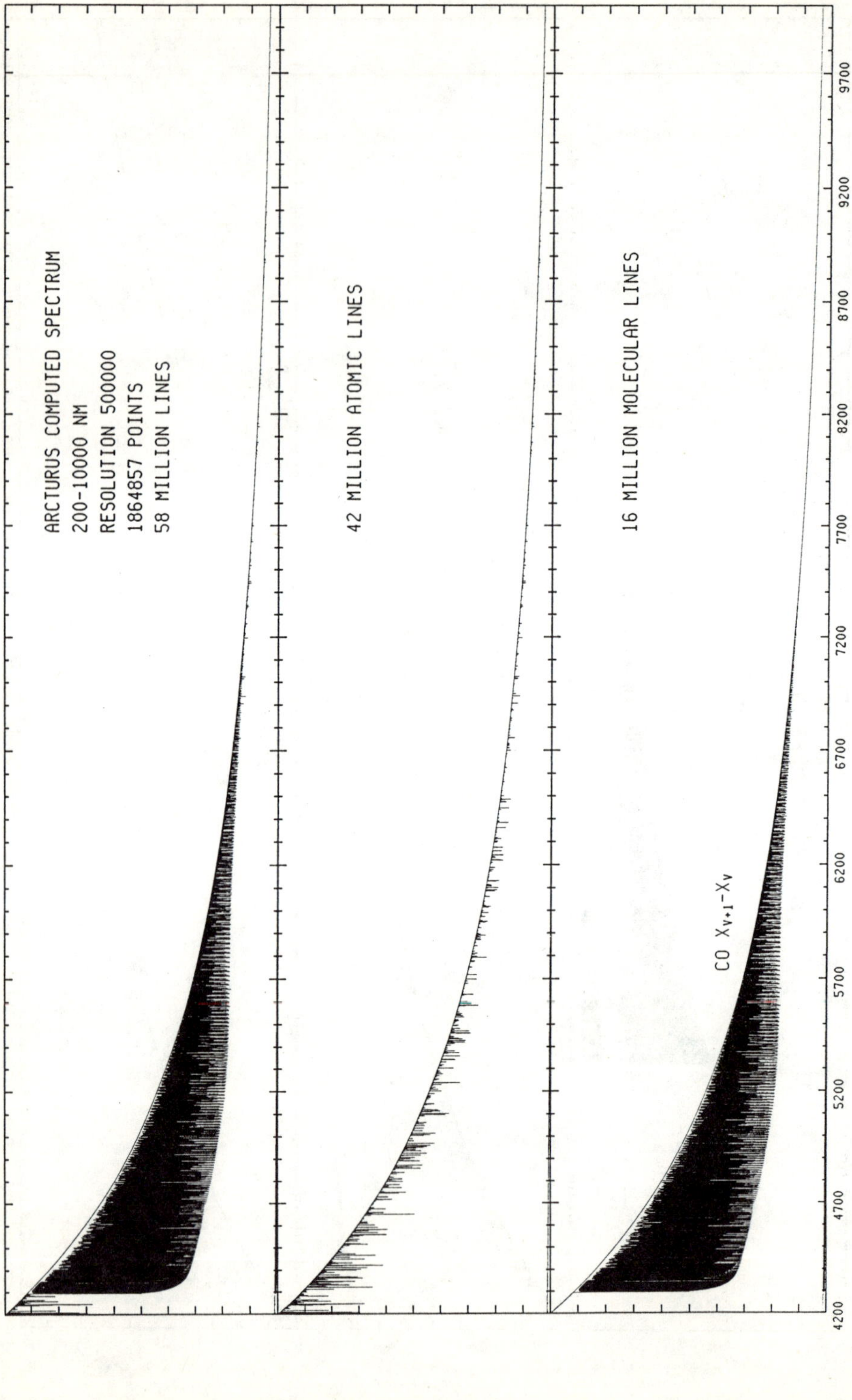
ARCTURUS COMPUTED SPECTRUM
200-10000 NM
RESOLUTION 500000
1864857 POINTS
58 MILLION LINES
42 MILLION ATOMIC LINES
16 MILLION MOLECULAR LINES
CO $X_{v+1}-X_v$
4200
4700
5200
5700
6200
6700
7200
7700
8200
8700
9200
9700

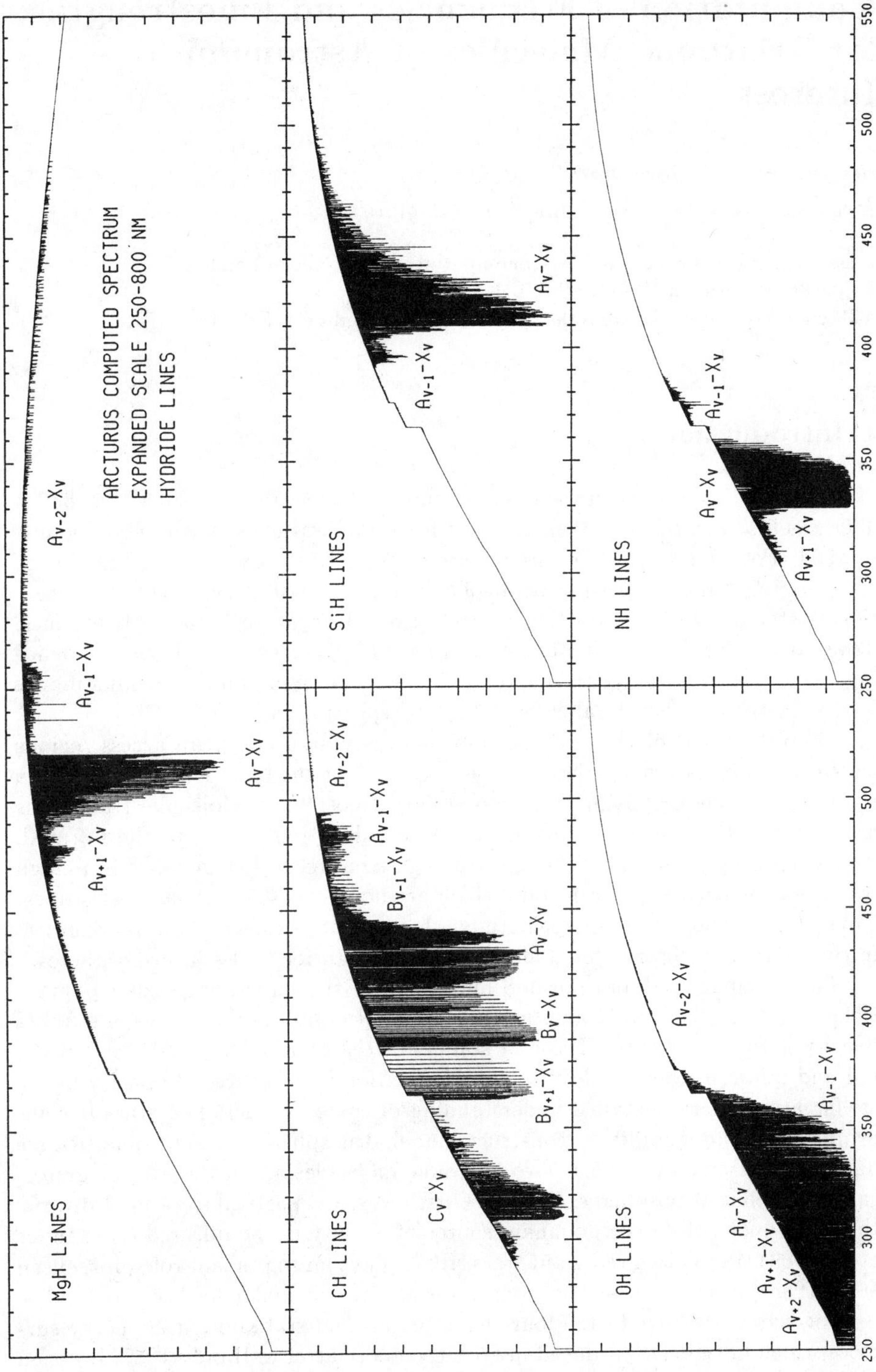
ARCTURUS COMPUTED SPECTRUM
EXPANDED SCALE 250-800 NM
HYDRIDE LINES
MgH LINES
$A_{v+1}-X_v$
A_v-X_v
$A_{v-1}-X_v$
$A_{v-2}-X_v$
CH LINES
C_v-X_v
$B_{v+1}-X_v$
B_v-X_v
A_v-X_v
$B_{v-1}-X_v$
$A_{v-1}-X_v$
$A_{v-2}-X_v$
OH LINES
$A_{v+2}-X_v$
$A_{v+1}-X_v$
A_v-X_v
$A_{v-1}-X_v$
$A_{v-2}-X_v$
SiH LINES
$A_{v-1}-X_v$
A_v-X_v
NH LINES
$A_{v+1}-X_v$
A_v-X_v
$A_{v-1}-X_v$
250
300
350
400
450
500
550

Computation of Frequencies and Linestrengths for Triatomic Molecules of Astronomical Interest

Steven Miller[1], *Jonathan Tennyson*[1],
Hugh R.A. Jones[2] *and Andrew J. Longmore*[2]

[1] Department of Physics and Astronomy, University College London, Gower St., London WC1E 6BT, U.K.
[2] Royal Observatory Edinburgh, Blackford Hill, Edinburgh EH9 3HJ, U.K.

1 Introduction

Molecular bands have been used as a diagnostic of spectral type since the 1860's. TiO was first identified as the dominant feature in optical spectra of cool giants in 1904 (Fowler 1904). Many diatomic molecules have now been included (e.g. Jørgensen & Larsson 1990) in detailed calculations of stellar opacity and attention is starting to focus on the incorporation of accurate data for triatomics. Triatomic molecules have a much greater density of states than diatomics; bending vibrational bands – generally lower in frequency than bond-stretching modes – may contribute considerably to "filling in" spectral gaps.

From the point of view of quantum chemistry, molecules can access regions which would be described classically as "chaotic" at the temperatures (usually a few thousand degrees) available in cool stellar atmospheres. Molecular transitions associated with such behaviour are to be found, in particular, in the infrared. Astronomically the study of this spectral region has recently benefitted from high resolution, sensitive spectrometers which are now available on many telescopes. Thus there is mutual advantage both for chemical physicists and for astronomers in the sharing of observational and theoretical data for cool star atmospheres.

The dominant red and infrared luminosity of the underlying stellar populations of galaxies is from M giants. The dominant population in terms of number density is from M dwarfs. These M stars emit the bulk of their radiation in the red and infrared where molecules provide the dominant source of opacity in the stellar atmosphere. Detailed understanding of the evolution of M stars is compromised by the inability to construct model atmospheres without quantitative data for triatomic molecules. Two triatomic molecules are of the utmost importance in stellar atmospheres. For cool O-rich M stars, particularly the M dwarfs, the water molecule is the dominant source of opacity in the infrared (Alexander et al. 1989). For C-rich red giant stars, HCN plays an important role (Jørgensen et al. 1985). Accurate potential energy surfaces are available for both molecules.

For water, we have tested four electronic ground-state potential energy surfaces which have been produced in recent years (Carter & Handy 1987, Halonen & Carrington 1988, Jensen 1989 and Kauppi & Halonen 1990). In this paper we present only calculations using the Jensen (1989) potential, which was found to be in significantly better agreement with experimental data than the other surfaces, for both vibrational band origins and rotational term values (Fernley et al. 1992. Transition strengths are obtained from an accurate dipole surface computed by Wattson & Rothman (1993). For HCN we have used a recent potential developed for the whole HCN/HNC system (Gadzy & Bowman 1991), which puts the barrier to isomerisation at around 17000cm^{-1}. Band intensities, using a dipole surface valid only for the HCN end of the molecule (Sebald 1991), are presented.

For temperatures applicable to cool star atmospheres (up to 5000K) we consider it will be necessary to compute all levels up to 30000cm^{-1} above the ground state. Calculating all transition frequencies and line strengths up to 14000 to 15000cm^{-1} (wavelengths longer than 0.7μm) will cover the near to far infrared spectral region for which these triatomics are likely to be the main source of opacity. In this paper, however, we report only prelimimary calculations covering a much more restricted spectral range.

2 Ro-vibrational line positions and intensities

Spectroscopists often use techniques based on perturbation expansions to calculate vibrational levels close to the ground state (e.g. Flaud et al. 1983). This involves fitting measured levels using a perturbational Hamiltonian. The parameters obtained from the fit are then used to predict higher, unobserved, levels. Rotational levels belonging to each vibrational manifold may similarly be computed for moderate angular momentum, J, levels. Such techniques, however, become inappropriate when the vibrational modes involve large amplitude motions and when the angular momentum is such that centrifugal distortion plays a major role.

Additionally, perturbational approaches are based on an idealised model in which rotational and vibrational motions are uncoupled to first order. Rovibrational coupling is then added through coriolis contributions, again considered to be perturbations of the ideal rigid-rotor, harmonic oscillator system. The distribution of line intensities is similarly approximated by assuming there is an intrinsic vibrational band intensity, which is then distributed across the rotational structure using Hönl-London factors (Herzberg 1991).

Present techniques of solving the radiative transfer equation in model atmosphere computations make use of a statistical treatment of the computed monochromatic absorption coefficient, as in for example the Opacity Distribution Functions (Jørgensen & Larsson 1990). It might be argued that very accurate calculations of molecular data, requiring large amounts of super-computer processing time, are not worthwhile for computing ODFs and that the approach outlined above is sufficient. For many molecules which have large anharmonicity coefficients and centrifugal distortion constants, however, perturbational expansions of the energy levels become divergent at higher ro-vibrational energies and line frequencies calculated from such Hamiltonians may be out by tens or even hundreds of cm^{-1}. The water molecule, in particular, is often taken as the classic example for which the perturbation expansion of its rotational motion diverges. For the ground state, the rigid rotor approximation computes the J=20, K_a=20, level to be some 2000cm^{-1} too high; addition of the first correction, the centrifugal distortion, makes the level ~3000cm^{-1} too low; the next term in the series causes an error of approximately +4000cm^{-1} (Polyansky 1993). ODFs are known to be sensitive to changes in the monochromatic absorption coefficient at well below this level.

These effects will also cause the distribution of intensity between individual lines using Hönl-London factors to be a poor representation of the actual system. Additionally, effects such as "intensity stealing" can cause whole "dark bands" to become intense; the magnitude of this effect may not be uniform across the entire band, but may depend on individual rotational states. Thus a combination of poor line positions and inappropriately distributed intensity may cause the computed wavelength dependence of the opacity to be very inaccurate. *A fortiori*, any model atmosphere which predicts synthetic spectra for stars needs to have accurate molecular data as input if significant conclusions about physical conditions within stellar atmospheres are to be drawn.

In the last decade, a number of groups (e.g. Tennyson 1986, Spirko & Jensen 1985, Carter & Handy 1986) have developed techniques for calculating triatomic energy levels based on variational techniques. In particular, Sutcliffe & Tennyson (1986) have produced a ro-vibrational Hamiltonian which is exact within the constraints of the Born-Oppenheimer approximation (which decouples **electronic** from **nuclear** motion in a molecule) and the limitations of the electronic potential energy surface available.

The Sutcliffe-Tennyson kinetic energy operator directly relates the cartesian positions of the nuclei of a triatomic molecule to a set of internal co-ordinates comprising two radial co-ordinates and the included angle. Rotation of the molecule-fixed internal co-ordinates in the laboratory frame is carried by the usual Wigner rotational matrices. This operator makes **no** *a priori* assumptions about ro-vibrational separation or equilibrium geometry. In addition, the operator includes parameters which relate the radial co-ordinates to the bonds of the molecule in a flexible way, leading to computational advantages, especially for the accurate calculation of high ro-vibrational energy levels. Other features of the method are described elsewhere (Tennyson et al. 1993).

Ro-vibrational wavefunctions may be calculated by two essentially separate techniques. An approach using model basis functions (Finite Basis Representation or FBR) as building blocks has been developed into a suite of "user-friendly" programs, the latest version of which is described in Tennyson et al. (1993). This suite, known as TRIATOM, contains modules to calculate ro-vibrational transition frequencies and linestrengths from dipole moment surfaces as well as synthetic spectra.

A second, more recent, approach has been pioneered by Light and coworkers (see e.g. Bacic & Light 1989). Known as the Discrete Variable Representation (DVR), this technique calculates wavefunctions on a grid in co-ordinate space (as opposed to the FBR method, which can be considered to calculate wavefunctions on a grid in **function** space). The DVR method is very efficient at building up wavefunctions for high-lying levels. Henderson et al. (1990a,b) have used this method to calculate all the bound vibrational states of H_3^+ (dissociation energy $\sim$ 36000cm^{-1}). A program suite using the DVR technique is also now available (Henderson et al. 1993), and additional developments, required to take full account of molecular symmetry, are currently nearing completion.

3 Water

3.1 Ro-vibrational calculations

At equilibrium, the water molecule has C_{2v} symmetry, with an angle of about 105° between the two O–H bonds. However, the molecule can access linear geometries at energies around 12000cm^{-1} above the minimum, well below the maximum energy level we consider necessary for cool star opacities. Figure 1 shows a contour plot of the wavefunction of the 162nd vibrational level of water (calculated using the Jensen potential (1989); E = 25941cm^{-1}). The x-axis is the inter-bond angle, θ, (in radians) and the y-axis the sum of the two bonds. In this representation, 18 nodes may be counted in the θ-co-ordinate, indicating that there are 18 quanta of bend in this level. The wavefunction has appreciable amplitude for almost all values of θ from 27° to 180°, showing the large displacements of the bond angle from equilibrium.

We have carried out rigorous testing of the four recent potential energy surfaces mentioned above (Fernley et al. 1992) using the Sutcliffe-Tennyson Hamiltonian. All of them perform reasonably well when compared with experimental data for vibrational levels lying below or just above the energy region at which linear geometries become accessible. The surface due to Jensen (1989) gives a 6.4cm^{-1} standard deviation from experiment for the band origins up to $\sim$ 21100cm^{-1} and a 0.14cm^{-1} standard deviation from experiment for the rotational term values up to J=2 for the first ten bands.

We have used this potential in an FBR calculation of a limited number of rovibrational energy levels and wavefunctions: for J=0 to 14, we computed (J+1) $\times$ 25 levels for the states with even parity, and 20 less for those of odd parity; for J=15 to 20, 375 even and 355 odd states were produced. This gave a coverage of just over 11000cm^{-1} for each J-parity block above the lowest energy in the block; at J=20 the lowest level is $\sim$ 4400cm^{-1} above the J=0 zero point energy.

The present calculations were carried out using Radau co-ordinates (Tennyson et al. 1993) for the radial co-ordinates. This method has been shown to produce very accurate and efficient results for rotationally excited states of H_2S and its isotopomers (Miller et al. 1990). Our calculations represent the radial co-ordinate stretching basis functions by Morse-like oscillators, $\Phi(r)$, (Sutcliffe & Tennyson 1986), up to a total of 17 functions for each radial co-ordinate, and the bending co-ordinate basis functions by associated Legendre functions, $P_\lambda^k(\cos\theta)$, with $\lambda \leq 40$. The rotational motion of the molecule is represented by Wigner rotation matrices, $D^J_{M,k}(\alpha\beta\gamma)$. The energy of each $\Phi(r_1) \times$

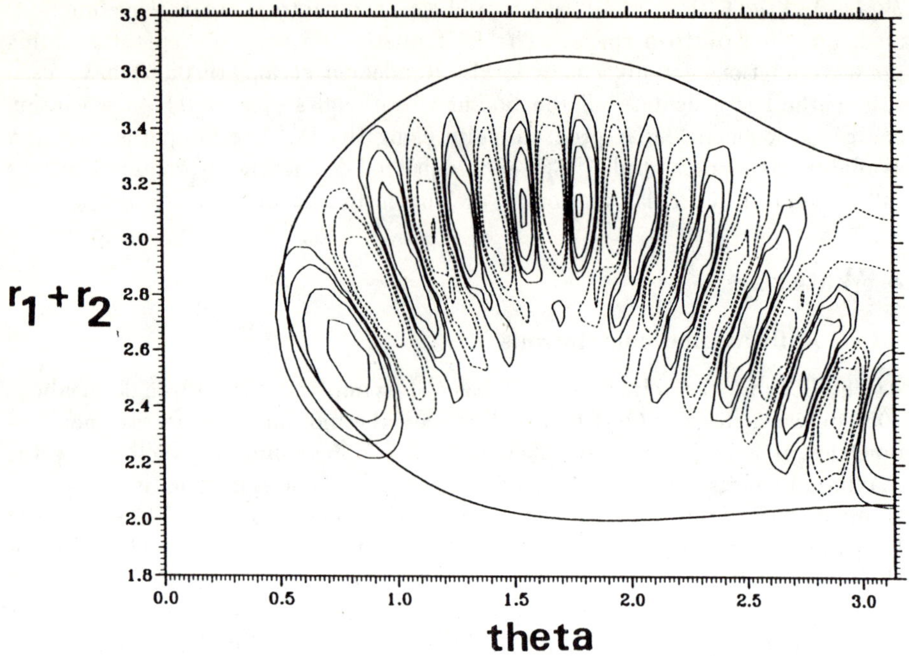

Fig. 1. Contour plot of wavefunction for vibrational level 162

$\Phi(r_2) \times P^0_\lambda(\cos\theta) \times D^0_{M,0}(\alpha\beta\gamma)$ product basis function was calculated, and the lowest 800 taken to define the intermediate basis functions, $\Phi(r_1) \times \Phi(r_2) \times P^k_\lambda(\cos\theta) \times D^J_{M,k}(\alpha\beta\gamma)$. Of these, a total of (J+1) ×650 or 9500 for J≥ 15, ordered on the basis of energy, were used in the final ro-vibrational calculation.

Comparison of the 25 lowest J=0 vibrational energy levels with the highly converged DVR calculation of Fernley et al. (1992) showed that the band origins were all converged to within 0.05cm^{-1}. The rotational levels were also converged to within 0.05cm^{-1} for the ground state and fundamentals rising to a maximum of between 5–15cm^{-1} for some of the higher lying states. As convergence does not worsen uniformly with level number, many higher lying states are better converged than the stated limit.

Ro-vibrational transition moments were calculated using the dipole moment surface of Wattson & Rothman (1993). In table 1, we present a comparison between calculated ro-vibrational transition frequencies and relative intensity for transitions involving J=0 and 1 and those listed in the HITRAN (Rothman 1992) database for 19 low-lying levels. Lines are listed relative to the [J=1, K_a = 0, K_c = 1 ← J=1, K_a = 1, K_c = 0] (101–110 in Table 1) transition of the ν_2 bending fundamental, which has a calculated intensity of 2.60 × 10^{-19} cm/molecule in good agreement with the measured value of 2.47×10^{-19} cm/mol. For these transitions, the agreement with experiment is reasonable. The one serious disagreement is with the $\nu_3 - \nu_2$ difference band, which our calculation

Table 1. H_2O: observed and calculated frequencies and relative intensities

Band	Transition	ω_{obs}	I^a_{obs}	ω_{cal}	I^a_{cal}
000–000	110–101	18.58	.213(0)	18.53	.200(0)
	111–000	37.14	.204(0)	37.09	.191(0)
010–010	110–101	21.95	.122(-3)	21.91	.129(-3)
	111–000	40.22	.980(-4)	40.18	.103(-3)
020–020	110–101	26.48	.583(-7)	26.40	.101(-6)
	111–000	44.46	.467(-7)	44.39	.687(-7)
010–000	101–110	1576.19	.100(+1)	1575.83	.100(+1)
	111–000	1634.97	.235(0)	1634.52	.226(0)
020–010	101–110	1534.93	.822(-3)	1535.81	.977(-3)
	111–000	1601.35	.232(-3)	1602.10	.219(-3)
030–020	101–110	1488.67	.462(-6)	1489.29	.858(-6)
	111–000	1565.84	.140(-6)	1566.23	.189(-6)
020–000	101–110	3133.07	.591(-2)	3133.54	.610(-2)
	111–000	3196.09	.150(-2)	3196.44	.150(-2)
030–010	101–110	3051.07	.660(-5)	3051.50	.759(-5)
	111–000	3122.72	.172(-5)	3123.94	.190(-5)
030–000	101–110	4648.21	.143(-4)	4649.23	.346(-4)
	111–000	4717.47	.453(-5)	4718.28	.977(-5)
040–000	101–110	6155.39	.294(-6)	6115.64	.114(-6)
	111–000	6194.79	.813(-7)	6194.61	.300(-7)
100–000	101–110	3638.08	.303(-1)	3637.63	.323(-1)
	111–000	3693.29	.623(-2)	3892.78	.597(-2)
200–000	101–110	7182.15	.433(-2)	7183.51	.514(-2)
	111–000	7236.75	.106(-2)	7238.09	.110(-2)
001–000	111–110	3749.33	.656(0)	3749.43	.510(0)
	101–000	3779.49	.171(0)	3779.57	.149(0)
002–000	101–110	7425.44	.317(-3)	7426.05	.384(-3)
	111–000	7479.66	.112(-3)	7479.70	.144(-3)
101–000	111–110	7242.44	.555(-1)	7243.71	.258(-1)
	101–000	7273.13	.152(-1)	7274.24	.760(-2)
100–010	101–110	2039.95	.140(-4)	2039.90	.118(-4)
	111–000	2098.55	.400(-5)	2098.43	.319(-5)
001–010	111–110	2151.19	.228(-4)	2151.70	.152(-7)
	101–000	2184.75	.660(-5)	2185.23	.757(-8)
110–010	101–110	3617.89	.115(-4)	3617.73	.112(-4)
	111–000	3679.41	.238(-5)	3679.22	.189(-5)
011–010	111–110	3729.26	.275(-3)	3730.63	.276(-3)
	101–000	3760.13	.733(-4)	3761.41	.820(-4)

[a] Powers of ten in brackets.

predicts to be four orders of magnitude weaker than is observed.

One limitation of the present calculations is that we were not able to use the properly symmetrised wavefunctions necessary to account automatically for spin statistics. For the construction of low temperature spectra, it is clear that this is a serious problem; agreement with experiment can only be achieved by hand-labelling ortho and para levels. At temperatures above about 500K, however, the most important ortho and para transitions (nearly) coincide and spin statistics may be ignored to a good approximation.

3.2 Comparison with M-dwarf spectra

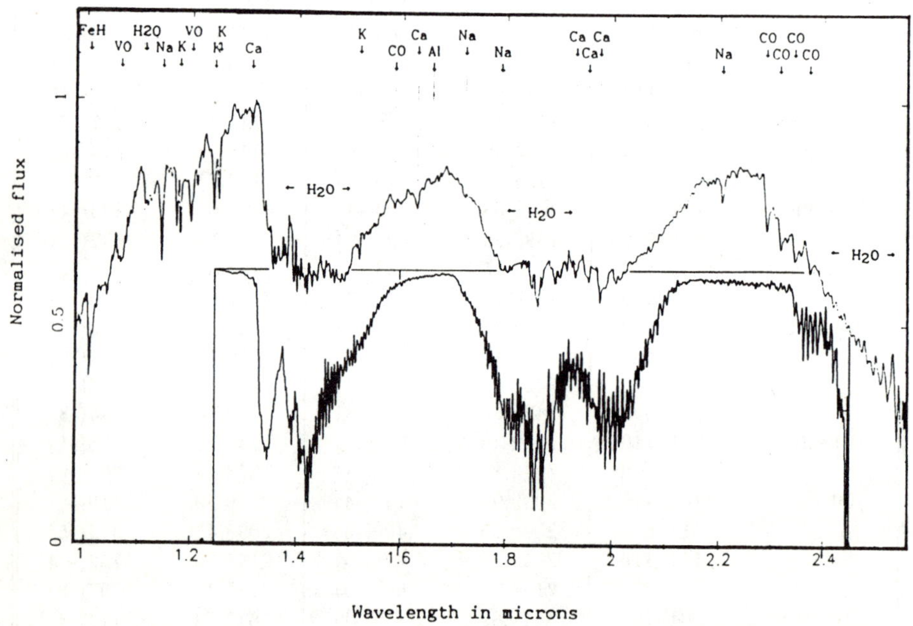

Fig. 2. Comparison of spectrum of VB10 (upper trace) with calculation (lower boxed trace)

The infrared spectra of cool dwarf stars are dominated by strong water absorption features. Jones et al. (1993a) have data on a number of M-dwarfs which show that it should be possible to derive an independent check on the temperatures derived from model atmosphere codes (e.g. Allard et al. 1993) by fitting the spectra of these water absorptions to accurate molecular data. The present calculation is too limited, particularly in terms of the maximum J cutoff, to be able to do this with certainty. But it is possible to test both the ability of the calculated spectrum to reproduce the overall features of the stellar spectrum and to see how sensitive this is to variations in the temperature.

In Fig. 2, we present the spectrum of the M-dwarf VB10 from 1 to 2.5μm. Allard et al. (1993) found that the best fit to the observed spectrum was obtained by use of a model atmosphere with an effective temperature of 2800. The spectrum shown was produced by combining spectra obtained on June 20, 1991 and May 5, 1992 using the cooled grating spectrometer CGS4 on the United Kingdom InfraRed Telescope (Jones et al. 1993a). The spectral range displayed shows two strong absorption features, centred on 1.4 and 1.9μm (as well as the commencement of another at 2.4μm). In the insert box we show the absorption spectrum (assuming an optically thin gas) calculated for a temperature of 3000K, with individual lines computed as a Gaussian profile of full-width half-maximum of about half the CGS4 instrument function to show up the calculated spectral detail. No attempt to model the continuum present in VB10 has been made.

The calculated spectrum shows clearly the main features of that of VB10; in particular, the sharp band head at 1.34μm is well reproduced. Analysis of the calculated spectrum shows that the 1.4μm absorption is due, mainly, to the $\nu_1 + \nu_3$ combination band, with weaker contributions from $2\nu_1$, $2\nu_3$, $2\nu_2 + \nu_1$ and $2\nu_2 + \nu_3$, as well as several hot-band transitions. The longer wavelength absorption feature results mainly from the $\nu_2 + \nu_1$ and $\nu_2 + \nu_3$ combination bands, again with a large number of hot-band lines.

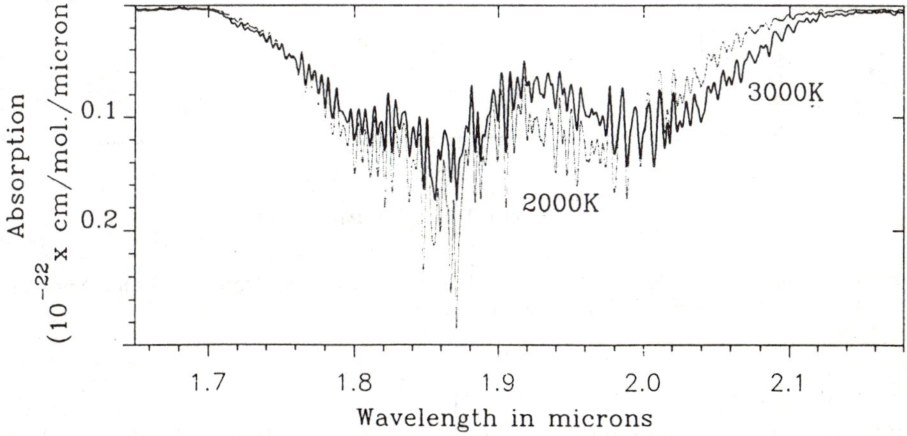

Fig. 3. Temperature sensitivity of calculated $\nu_2 + \nu_1/\nu_3$ spectral region: T=3000K full line; T=2000K broken line

In Fig. 3, the calculated absorption spectra in the $\nu_2+\nu_1/\nu_3$ region for 3000K and 2000K are shown. Both show the R and P branch peaks. The spectra show sensitivity to temperature: the 2000K (dotted line) spectrum shows a sharper R branch peak, and the maximum in the P branch is at shorter wavelengths (corresponding to lower values of J); the 3000K spectrum (full line) shows considerably more intensity in the wings, particularly in the P branch, with the maximum shifted to longer wavelengths (higher J values).

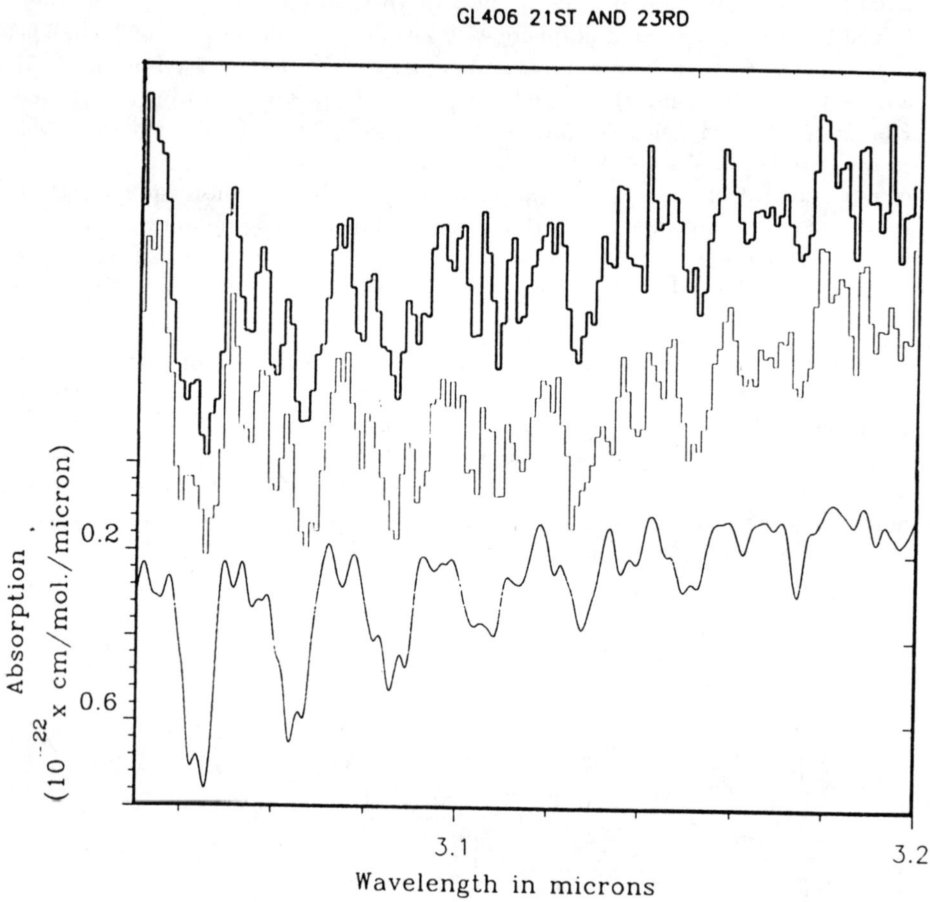

Fig. 4. Comparison of spectrum of GL406 for April 21, 1993 (top trace) and April 23, 1993 (middle trace) with calculation (bottom trace)

More recently, Jones et al. (1993b) have obtained spectra in the 2.85 to 3.4 micron region, corresponding to the P branch of the ν_3 and the ($\sim 20\times$ weaker) ν_1 bands, for a number of M-dwarf stars. One part of these observations taken on 21st April 1993 for the M dwarf GL406 is shown in Fig. 4. Although this spectrum was taken by observing through atmospheric water vapour at 250K the effect of atmospheric transmission is corrected by dividing the observed spectrum by an appropriate standard star very close in airmass to the target object. The observations were repeated on a subsequent night to check the cancellation of atmospheric features and the reproducibility was found to be excellent. Below these two traces we plot the spectrum calculated at 3000K; once more the general agreement with the stellar spectrum is very good.

3.3 Future work

The calculations presented here for water may be used for astrophysical purposes with four main restrictions:

1) The limited range of energies covered means that total opacity will be underestimated. This effect will become increasingly important at shorter wavelengths and for higher temperatures;

2) The convergence of the energy levels, which becomes generally – though not uniformly – worse with higher vibrational and rotational states, means that opacity is not distributed accurately throughout the band. However, the present calculation will be a considerable improvement on any assuming the rigid rotor approximation (for the same v,J states). Moreover, transition frequencies which depend on the difference between energy levels, as opposed to the individual energies themselves, are less affected by the problem of convergence since there is a certain amount of cancellation of errors ;

3) The absence of spin statistics means that both the ro-vibrational partition function and the detailed intensity distribution will be incorrect. Dependent on the resolution required, this problem will probably only be important for temperatures below 500K, as has been shown to be the case for H_3^+ (Sidhu et al. 1992). Systematic tests to determine how low in temperature one can reliably go are in progress;

4) The Jensen potential, while very good, is not perfect in its prediction of vibrational band origins and rotational term values (Fernley et al. 1992). Similarly, the Wattson dipole surface has some short-comings (see Table 1).

None of these problems is insuperable, given time – human and computer – and ingenuity. In collaboration with several co-workers, we have a program to remedy the current defects:

1) A new set of calculations is planned to cover band origins up to 30000cm^{-1} and angular momentum states up to 55, at which value of J the lowest energy level is estimated to be approximately 30000cm^{-1} above the zero point energy;

2) The present calculation has been carried out using an FBR and has been designed to fit into 10 megawords of memory. Convergence problems may be almost totally overcome (to within 1cm^{-1}) by switching to the DVR, probably without the need to use more memory;

3) New program developments not only mean that spin statistics can be explicitly included in the calculation but that computational efficiency can be enhanced by *a priori* excluding spin-forbidden transitions;

4) There is already a program underway to modify the Jensen potential; preliminary results show a significant improvement in both band origins and rotational term values (Paulse & Tennyson 1993). Other work is underway to improve the dipole surface.

4 The HCN/HNC system

4.1 DVR calculations for J = 0

Until recently, ro-vibrational calculations of the HCN/HNC system have made use of an empirically fitted potential energy surface due to Murrell et al. (1982) (MCH) (Bacic & co-workers 1986, 1987, 1990). The MCH surface, however, considerably underestimates the barrier to isomerisation ($E_{isom} = 12168\text{cm}^{-1}$) and Gazdy & Bowman (1991) fitted a new surface, based on the 1982 potential, which incorporated the *ab initio* saddle point ($E_{isom} =\sim 16500\text{cm}^{-1}$). The production of the final surface, which they labelled MCH-M3, also involved scaling the C–H and C–N stretching terms. The agreement between calculated and observed vibrational band origins was considerably improved by the MCH-M3 potential, especially for levels above 10000cm^{-1}. Although a new *ab initio* surface is now available (Bentley et al. 1992), MCH-M3 is still the global potential which gives the best representation of existing experimental vibrational data.

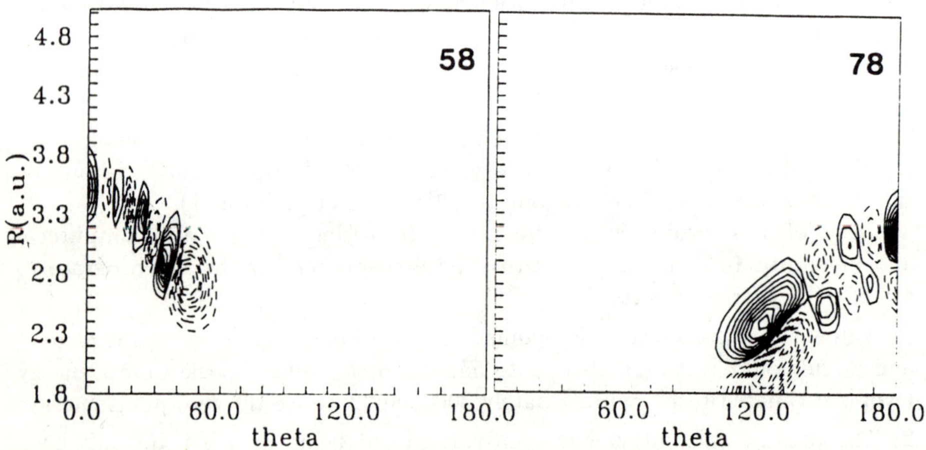

Fig. 5. Contour plots of R_{H-CN} versus $\cos(\theta)$ for vibrational levels 58 (E = 9585cm^{-1}) and 78 (E=10630cm^{-1})

To our knowledge, no global dipole surface for the HCN/HNC system has been published. A recent *ab initio* dipole, expressed in terms of the bending co-ordinate and the C–H and C–N stretches, and valid for H–C–N angles up to 90°, has been produced by Sebald (1991). For band origins below 15000cm^{-1}, this surface gives a fair representation of the measured band intensities when coupled with the corresponding potential energy surface.

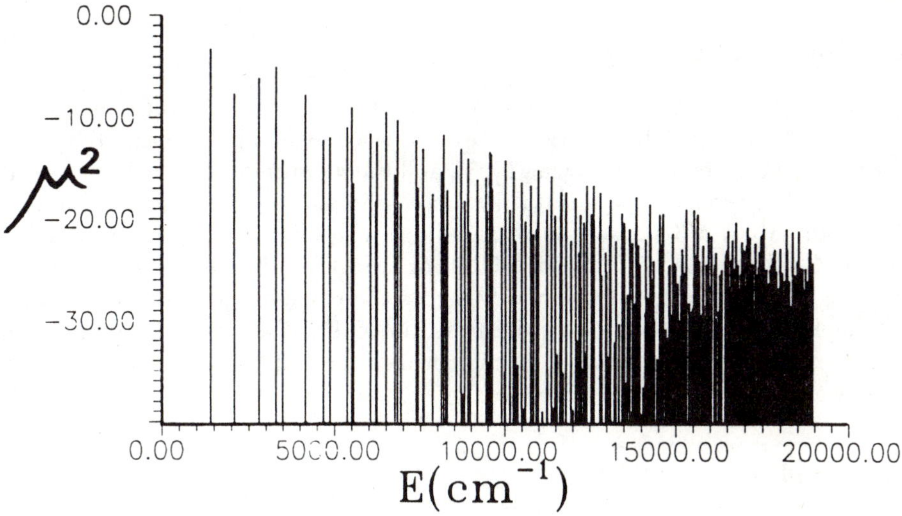

Fig. 6. Log$_{10}$ μ^2 versus transition energy for HCN

We have carried out preliminary calculations of the lowest 500 J=0 vibrational levels using the MCH-M3 potential, which extend up to 20000cm^{-1} above the minimum. The calculation, perfomed in scattering co-ordinates (Tennyson 1993) was done using a DVR in which intermediate two-dimensional r_1, r_2 grid-points, with energies below 25000cm^{-1}, were used to generate a final basis-grid of 2000 r_1, r_2, θ points to be diagonalised (Henderson et al. 1993). By comparing this calculation with one that used only 1500 final grid-points, convergence at the top of the energy range was estimated to be between 5 and 10cm^{-1}. We have also calculated J=0 band dipoles using the Sebald dipole for the HCN transitions.

Details of this work and our results will be published in full elsewhere (Seidel et al. 1993). It is clear that – as in the case of water – very large amplitude motions will be important in calculating levels necessary to model stellar atmospheres. Figure 5 shows contour plots for wavefunctions for two energy levels around 10000cm^{-1} above the zero point energy. Level 58 shows seven quanta of ν_2 for the HCN conformer; level 78 is assigned to the HNC conformer, with four quanta of ν_2 and one of N–H stretch. Although both states are well below the saddle point, large amplitude bending is apparent.

In Fig. 6, we show the band strengths, in terms of μ_b^2, for the HCN conformer up to 18000cm^{-1}. As expected, band strength declines with increasing frequency. At higher energies there is a bath of very low intensity bands, with several more intense transitions, a feature noted by Le Sueur et al. (1993) for H_3^+. It is clear that the combination of increasing density of states at high energies and the existence of several "intense" transitions means that HCN/HNC ro-vibrational opacity will extend right through the 0.5 to 1μm region.

4.2 Future calculations

We intend to carry out a similar project to that outlined for the water molecule. Calculations are already underway to extend the Sebald dipole to the HNC conformer of the system (Botschwina 1993), and the *ab initio* surface will then need to be checked against experimental data (see Lehmann & Romanini this volume). Using the MCH-M3 surface, our calculations will automatically produce levels for HNC mixed in with those for HCN. For energies sufficiently high above the barrier to isomerisation this is clearly a correct procedure; however, HNC transitions have apparently not been identified in cool star atmospheres (Jørgensen 1993) and it may be necessary to implement some form of separation between the two conformers at lower energies.

5 Conclusions

The development of powerful techniques for the computation of large number of ro-vibrational transitions has fortuitously coincided with the advent of high-resolution, sensitive infrared astronomical spectrometers. Our preliminary calculations for H_2O transitions up to J=20 provide a satisfactory fit to the observed features in M dwarfs both at low resolution from 1.3 to 2.5 microns and at higher resolution from 2.85 to 3.05 microns. For molecules of fundamental astronomical and chemical interest, such as water and HCN/HNC, it is both feasible and worthwhile to undertake major computational projects to produce the necessary molecular data.

Acknowledgements. The work presented is a progress report of long-term detailed studies involving a large number of co-workers. Oleg Polyansky, John Fernley and Nic Fulton are contributing to the work on the water molecule, and are gratefully acknowledged for their contribution. Luis Seidel, Rosa Benito, Florentino Borondo and James Henderson are thanked for their work on HCN/HNC. Richard Jameson and Matthew Mountain are valued collaborators in the observation of cool star spectra. The U.K. SERC and British Council and the Spanish Accion Integrada provided financial resources for this work. The staff of UKIRT are thanked for their help in obtaining the stellar spectra included.

References

Alexander D.R., Augason G.C., Johnson H.R., 1989, *Ap. J.*, **345**, 1014
Allard F., Jones H.R.A., Longmore A.J., 1993, *Astrophys. J.*, in preparation
Bacic Z., Light J.C., 1986, *J. Chem. Phys.*, **85**, 4594
Bacic Z., Light J.C., 1987, *J. Chem. Phys.*, **86**, 3065
Bacic Z., Light J.C., 1989, *Ann. Rev. Chem. Phys.*, **40**, 469
Bentley J.A., Bowman J.M., Gazdy B., Lee T.J., Dateo C.E., 1992, *Chem. Phys. Lett.*, **198**, 563
Botschwina P., 1993, private communication.
Carter S., Handy N.C., 1986, *Comp. Phys. Reports*, **5**, 115
Carter S., Handy N.C., 1987, *J. Chem. Phys.*, **87**, 4294
Fernley J.A., Miller S., Tennyson J., 1992, *Astron. Astrophys.*, in press
Flaud J.-M., Camy-Peyret C., Johns J.W.C., 1983, *Can. J. Chem.*, **61**, 1462
Fowler A., 1904, *Proc. R. Soc.*, **73**, 219
Gadzy B., Bowman J.M., 1991, *J. Chem. Phys.*, **95**, 6309
Halonen L., Carrington T. Jnr., 1988, *J. Chem. Phys.*, **88**, 4171
Henderson J.R., Tennyson J., 1990a, *Chem. Phys. Lett.*, **173**, 133
Henderson, J.R., Tennyson, J., Sutcliffe, B.T., 1990b, *J. Chem. Phys.*, **98**, 7191
Henderson J.R., Le Sueur C.R., Tennyson J., 1993, *Comp. Phys. Commun.*, **75**, 379
Herzberg G., 1991, *Molecular Spectra and Molecular Structure*, **III**, Krieger, 226
Jensen P., 1989, *J. Mol. Spectrosc.*, **133**, 438
Jones H.R.A., Longmore A.J, Jameson R.F., Mountain C.M., 1993, *Mon. Not. Roy. Astr. Soc.*, submitted
Jones H.R.A., Longmore A.J., Miller S., Tennyson J., 1993, *Astrophys. J.*, in prep.
Jørgensen U.G., Almlöf J., Gustaffson B., Larsson M., Siegbahn P., 1985, *J. Chem. Phys.*, **83**, 3034
Jørgensen U.G., Larsson M., 1990, *Astron. Astrophys.*, **238**, 424
Jørgensen U.G., 1993, private communication
Kauppi E., Halonen L., 1990, *J. Chem. Phys.*, **94**, 5799
Miller S., Tennyson J., Rosmus P., Senekowitsch J., Mills I.M., 1990, *J. Mol. Spectrosc.*, **143**, 61
Mladenovic M., Bacic Z., 1990, *J. Chem. Phys.*, **93**, 3039
Murrell J.N., Carter S., Halonen L.O., 1982, *J. Mol. Spectrosc.*, **93**, 307
Paulse J.D., Tennyson J., 1993, *48th Ohio State University Symposium on Molecular Spectroscopy*, Abstract WE12
Polyansky O.L., 1993, private communication
Rothman L.S., et al., 1992, *J. Quant. Spectrosc. Radiat. Transfer*, **48**
Sebald P., 1991, *PhD. Thesis*, Universitat Kaiserslautern
Seidel L., Benito R.M., Borondo F., Miller S., Henderson J.R., Tennyson J., 1993, in preparation
Sidhu K.S., Miller S., Tennyson J., 1992, *Astron. Astrophys.*, **255**, 453
Spirko V., Jensen P., Bunker P.R., Cejchan A., 1985, *J. Mol. Spectrosc.*, **112**, 183
Le Sueur C.R., Henderson J.R., Tennyson J., 1993, *Chem. Phys. Lett.*, **206**, 429
Sutcliffe B.T., Tennyson J., 1986, *Mol. Phys.*, **58**, 1053
Tennyson J., 1986, *Comp. Phys. Reports*, **4**, 1
Tennyson J., Miller S., Henderson J.R., 1993, in *Methods in Computational Chemistry*, **4**, (Plenum, New York 1993), 91
Tennyson J., Miller S., Le Sueur C.R., 1993, *Comp. Phys. Commun.*, **75**, 339
Wattson R.B., Rothman L.S., 1993, *J. Quant. Spectrosc. Radiat. Transfer*, in press

Computational Approaches to Determining Accurate Band Strengths

Stephen R. Langhoff and Charles W. Bauschlicher, Jr.

NASA Ames Research Center, Moffett Field, CA 94035, USA

1 Introduction

With the advent of high-speed computing and improved algorithms, computational chemistry techniques have become competitive with the best experimental techniques for determining line strengths for rovibrational and electronic transitions in molecular systems (Bauschlicher & Langhoff 1991). This is particularly the case at high temperatures where the molecules are highly rotationally excited. While it is difficult to measure line strengths at stellar temperatures, the theoretical values can be very reliable if a globally accurate transition moment function (TMF) is used in conjunction with high-quality experimental or theoretical potentials.

In this review we demonstrate the current state-of-the-art in theoretically determining accurate transition probabilities for a variety of astrophysically interesting molecules. The quality of the results is shown to be strongly dependent on the system studied. For example, transition probabilities for first-row diatomic systems rival the best experiments in accuracy, whereas calculations for transition metal systems are much more difficult.

To illustrate a molecular transition where very accurate benchmark studies have been performed (Bauschlicher & Langhoff 1987), we first consider the ultraviolet ($A^2\Sigma^+$–$X^2\Pi$) band system of the hydroxyl radical (OH). Besides being of interest in a variety of astrophysical sources, this band system has been used extensively for combustion diagnostics. To deduce the temperature in a flame or to determine the opacity in the stellar environment, accurate line strengths are required. The lifetimes in OH vary markedly with rotation owing to the rapid decrease in the transition moment with internuclear distance (r), the high degree of centrifugal distortion, and predissociation of higher rotational levels of v=0−2 from a crossing on the outer limb of the $A^2\Sigma^+$ state by the repulsive $^4\Sigma^-$ state. The experimental lifetimes for the lowest ro-vibrational level vary significantly. To determine which of these experimental values is best, very extensive and carefully benchmarked calculations are required. We demonstrate that 5% accuracy can be achieved in the calculated line strengths by carrying out full configuration-interaction (FCI) calculations to benchmark the electron correlation methods in conjunction with a systematic saturation of the one-particle basis.

We next consider the calculation of the line strengths for the ro-vibrational bands in the $X^1\Sigma^+$ ground state of SiO (Langhoff & Bauschlicher 1993). The Δv=1 and 2 sequences have been observed in absorption in the infrared spectra of cool stars (Cohen et al. 1992; Rinsland & Wing 1982; Ridgway et al. 1984). The line strengths that are in current use are based on the electric dipole moment function (EDMF) of Tipping & Chackerian (1981), which is in the form of a Padé approximant, chosen to reproduce the accurate experimental dipole moments for v=0 − 3, to have the proper united and separated atom limits, and to have the correct long-range dependence on internuclear separation. We study the convergence of the EDMF with improvements in the calculations. Line strengths based on our most accurate *ab initio* EDMF are larger than Tipping & Chackerian by a factor of about 1.2 and 2.0 for the fundamentals and first overtone transitions, respectively.

We illustrate calculations on transition metal systems by discussing the spectroscopy of ZrO. Since the absorption bands of ZrO have been used to classify S stars (Johnson 1982; Littleton & Davis 1985), it is important to have accurate electronic transition probabilities for the important band systems of this molecule. Theoretical calculations (Langhoff & Bauschlicher 1990a) have characterized both the singlet and triplet band systems of this molecule up to an energy of about 30 000 cm^{-1}. The theoretical radiative lifetimes are in reasonable accord with the experimental lifetimes that are available for three states. In addition, recent absorption experiments give line strengths that are also in good accord with theory. Thus, overall the band systems of this molecule are relatively well known, although some uncertainties remain. For example, theory predicts a strong $E^1\Phi - A^1\Delta$ transition in the 16 000-17 000 cm $^{-1}$ region, but this transition has not been observed in the laboratory or in stellar spectra. Possibly this transition is obscurred by the γ bands of ZrO or the upper state lies too high to be significantly populated. Analogous calculations for TiO have proven to be very challenging. We conclude this review by discussing the complexity of studying first transition row metal systems such as FeH and TiO.

2 Methods

The ultimate goal of quantum chemistry is to solve the electronic Schrödinger equation, which would give the exact nonrelativistic electronic wave function within the Born-Oppenheimer approximation. However, the exact solution of this integral differential equation is not achievable for more than two electrons. Thus most theoretical approaches (excluding methods such as Monte Carlo (Diedrich & Anderson 1992) and density functional (Parr & Yang 1989)) formulate the solution as a double basis set expansion. The first expansion (or so-called one-particle basis) determines the molecular orbitals as a linear combination of atom centered functions usually taken to be Gaussian functions. The best single Slater determinant (antisymmeterized product of molecular orbitals) gives the so-called self-consistent-field (SCF) approximation. This ignores the instantaneous correlation of the electrons. The second expansion accounts for the electron correlation by formulating the wave function as a linear combination of determinants with one or more electrons promoted from occupied orbitals into virtual orbitals (single, double, triple excitations etc.). This expansion is generally referred to as the n-particle expansion and these various approaches of including electron correlation are often referred to as configuration-interaction (CI) methods (see, for example, Shavitt 1977; Saunders & van Lenthe 1983). An alternative approach is to start from an exponential ansatz, which leads to the singles and doubles coupled cluster (CCSD) approach (see for example Bartlett 1981).

If all excitations (up to n-tuple excitations for n electrons) are included in the wave function, we have the special case of a full configuration-interaction (FCI) calculation. This gives the exact result for the one-particle basis set employed and is an unambiguous standard with which to compare approximate electron correlation approaches that truncate the configuration expansion. Since the size of the expansion increases factorially with the number of electrons correlated, it is computationally intensive to carry out FCI calculations in a high-quality one-particle basis set if more than about six electrons are correlated. However, improvements in algorithms (Siegbahn 1984; Knowles & Handy 1984) and high-speed computers have allowed a rather complete set of benchmark calculations (Bauschlicher et al. 1990a) to be performed. The goal is to find a truncated electron correlation procedure that reproduces the FCI result in a realistic one-particle basis set (one that obtains a significant fraction of the correlation energy) and then carry out a systematic study of the property as a function of improvements in the one-particle basis set at the truncated correlation level. The property should approach the exact non-relativistic result as the one-particle basis approaches completeness if there is not an unexpected strong coupling between the one- and n-particle expansions. We demonstrate this approach for the electronic transition moment of the ultraviolet system of OH in the next section.

To compute accurate spectroscopic constants and transition probabilities it is necessary to describe the ground and excited states for a range of internuclear distances that encompass the Franck-Condon region. Configuration-interaction methods based on a single reference (SCF wave function) are not adequate for this purpose. To describe the multireference character and near degeneracy effects in the wave functions it is necessary to begin with a multiconfigurational SCF or MCSCF description, where the molecular orbitals are optimized simultaneously with the CI coefficients in the MCSCF wave function. One particular choice of an MCSCF wave function that is widely used is the so-called complete-active-space SCF (CASSCF) wave function (Roos 1987), where an FCI calculation is performed within some specified active space (subset of orbitals that describe the important correlation effects). The active space is designated by the number of orbitals in each of the symmetries. For example, an active space of (3221) denotes the number of orbitals in the a_1, b_1, b_2, and a_2 C_{2v} symmetries in which the calculations for a heteronuclear diatomic molecule are carried out. Since it is computationally much easier to use one set of orthonormal molecular orbitals, the energy optimization is generally carried out for the average energy of all states treated in the subsequent CI calculations. The active space must be large enough to provide an accurate description of each of the states over the relevant range of internuclear distances. A CASSCF calculation where the average energy of several states is optimized is generally referred to as a state-averaged or SA-CASSCF treatment. A CI calculation based on a CASSCF zeroth-order wave function (i.e. single and double excitations from all configurations in the CASSCF wave function) is generally referred to as a multireference configuration-interaction (MRCI) treatment to distinguish it from one based on the SCF wave function (generally referred to as a singles and doubles CI or SDCI calculation).

Unfortunately the length of the MRCI expansion grows very rapidly with the size of the active space in the CASSCF calculation. Several procedures have been used to reduce the size of this expansion. One procedure is to not include all excitations in the wave function, but to discard the less important configurations based on their contribution to the energy using second-order perturbation theory (Shavitt 1977; Buenker & Peyerimhoff 1974). A second approach is to generate single and double excitations not from the full CASSCF wave function, but from a subset of important configurations (based on their coefficients in the CASSCF wave function). For example, the notation SA-CASSCF/MRCI(0.05) denotes that a state-averaged CASSCF calculation is performed followed by an MRCI calculation where the zeroth-order reference contains configurations that have a coefficient larger than 0.05 (for any state at any relevant internuclear distance) in the CASSCF wave function. Finally, a relatively new procedure called the internally-contracted MRCI (Werner & Knowles 1988) or IC-MRCI has been introduced as an efficient means of greatly reducing the number of variational parameters. Here the ratio of the coefficients in the final CI wave function are constrained to be those in the zeroth-order wave function. This method closely reproduces the fully variational MRCI calculation when the zeroth-order wave function is adequate. This approach can greatly extend the size of configuration spaces (and thus CASSCF active spaces) that can be employed.

There are electron correlation methods that attempt to incorporate the effects of higher than double excitations into the wave function. A simple empirical approach is to use the so-called Davidson correction (Langhoff & Davidson 1974; Blomberg & Siegbahn 1983) (denoted +Q). However, there are other single-reference based methods, such as the coupled-electron pair approximation (CEPA) (Ahlrichs 1979) and coupled-pair functional (CPF) (Ahlrichs et al. 1985) methods, that incorporate the effect of higher excitations in a more rigorous manner. The accuracy of the CCSD method can also be greatly enhanced by incorporating a perturbational estimate of connected triple excitations (Raghavachari et al. 1989), denoted CCSD(T). A multireference variation of CPF is the averaged coupled-pair functional (ACPF) (Gdanitz & Ahlrichs 1988) method. Our experience (Bauschlicher et al. 1990b) is that this method provides superior properties such as dipole moments compared with the analogous MRCI method. A further discussion of these methods is outside the scope of this review.

We close the discussion of methods by describing the one-particle basis sets currently employed. The FCI benchmark studies (Bauschlicher et al. 1990a) showed that CASSCF/MRCI calculations accounted for most of the electron correlation. Thus the principal limitation of theoretical calculations could often be ascribed to incompleteness in the one-particle basis set. Recently, considerable progress has been made in improving the quality of these bases. For example, the atomic natural orbital (ANO) basis sets developed by Almlöf & Taylor (1987) and the correlation-consistent basis sets developed by Dunning and co-workers (Woon & Dunning 1993 and references therein) have contraction coefficients determined at the correlated level. We discuss basis sets here only to the extent to define the notation and to indicate what is required for high accuracy. For example, an oxygen atom basis set containing four s and two p functions, [$4s\ 2p$], would be considered to be of double-zeta quality, since it contains two functions for each of the oxygen $1s$, $2s$, and $2p$ orbitals. Basis sets of triple-zeta or larger are required for quantitative accuracy. In addition, polarization functions (in this case d, f and higher angular momentum functions) are required to reach basis set convergence. In general, basis sets saturated up through g functions are necessary for very accurate calculations on first-row atoms. For second-row and transition metal atoms the basis set requirements become even more severe. More details of the actual basis sets employed are given in the following sections.

3 The Ultraviolet System of OH

In this section we describe our work (Bauschlicher & Langhoff 1987) on the $A^2\Sigma^+ - X^2\Pi$ electronic transition moment function (TMF) of OH that included FCI benchmark calculations. While it is not feasible to place meaningful rigorous error bounds on calculated properties, an estimate of the accuracy can be obtained by studying the convergence of the property with respect to a systematic expansion of the one- and n-particle spaces. The utility of this procedure is enhanced if FCI calculations can be performed in a meaningful one-particle basis. Thus our approach is to find a truncated correlation method that reproduces the FCI in a relatively small (but realistic) one-particle basis, and then carry out this truncated CI in a nearly complete one-particle basis.

For OH the FCI benchmark calculations were carried out in a [$4s3p2d$] contracted set for oxygen and a [$2s1p$] contracted set for hydrogen described in more detail in the original reference. The subsequent MRCI calculations used to generate the TMF utilized a [$6s\ 5p\ 4d\ 2f\ 1g$] basis for oxygen and a [$4s\ 3p\ 2d$] basis for hydrogen. The oxygen basis set provides an accurate description of both O and O^- character in the wave function. These basis sets were contracted using the ANO approach and should give results close to the basis set limit. Calculations were carried out with several extensions to this basis set to ensure that the TMF was converged with respect to adding additional functions.

In Table I we compare the FCI $A^2\Sigma^+ - X^2\Pi$ transition moment with those obtained from a variety of both single-reference and multireference-based corre-

Table 1. Study of the $A - X$ transition moment of OH with level of correlation treatment at r=1.80 a_0 in the [$4s3p2d/2s1p$] Gaussian basis set

Calculation	moment
FCI 5-electron	0.100959
FCI 7-electron	0.119530
SCF	0.173121
SDCI	0.125307
MCPF	0.121417
(222) CASSCF	0.161812
(222) CASSCF[a]	0.169252
(222) MRCI(0.0)	0.122965
(222) MRCI(0.05)	0.123246
(322) MRCI(0.0)	0.122522
(3221) CASSCF	0.123482
(3221) MRCI(0.0)	0.119841
(3221) MRCI(0.01)	0.120024
(3221) MRCI(0.02)	0.122090

[a] Separate CASSCF calculations are performed for the two states and a non-orthogonal transition moment is evaluated.

lation methods that truncate the n-particle space. All values were obtained at r=1.80 a_o using the [$4s3p2d/2s1p$] Gaussian basis. Since oxygen $2s$ correlation increases the transition moment by 20% at the FCI level, all calculations were carried out correlating all but the oxygen $1s$ electrons.

We first consider the methods based on a single reference treatment given in the upper portion of Table I. The SCF moment is significantly too large, but the magnitude decreases substantially with the addition of electron correlation. The discrepancy with the FCI result is nearly an order of magnitude less at the SDCI level than at the SCF level. Furthermore, the error at the modified CPF (MCPF) (Chong & Langhoff 1986) level is a factor of three less than at the SDCI level, giving a lifetime only 3% shorter than at the FCI level. It is expected, however, that this approach will become less satisfactory as the molecule is displaced further from its equilibrium value.

The $A - X$ transition moment is also significantly too large at the (222) CASSCF level. Note that the moment is only slightly different if instead of the SA-CASSCF procedure, which employs a common set of orthogonal orbitals, we carry out separate (222) CASSCF calculations for each state and compute the moment using a non-orthogonal transition moment method. Hence, the differences with the FCI are *not* principally due to using a common set of averaged orbitals. Significant improvements are also observed as the level of correlation is increased for the multireference-based approaches. The (222) MRCI moment is about 3% larger than the FCI, whether reference configurations are selected using a threshold of 0.05 or all CASSCF configurations are included. The addition of the 5σ orbital to the CASSCF and MRCI active space has virtually no effect on the computed transition moment, as was the case for the $X^2\Pi$ dipole moment of OH (Langhoff et al. 1987). However, the addition of the 1δ orbital to the active space results in a CASSCF transition moment that is superior to the SDCI, and virtually identical to the (222) MRCI results. The addition of more extensive correlation to the (3221) CASSCF wave function yields a transition moment in excellent agreement with the FCI value. This (3221) MRCI calculation is sensitive to the reference selection threshold, with 0.01 reproducing the unselected results, but a threshold of 0.02 giving a TM more similar to the (222) MRCI result, because most of the references involving the 1δ orbital have coefficients less than 0.02. Since the (3221) MRCI(0.01) transition moment is in excellent agreement with the FCI result, a very accurate transition moment is expected at this level of correlation treatment in the limit of a complete one-particle Gaussian basis. The moment should be globally accurate, because the (3221) CASSCF calculation is sufficiently flexible to provide a uniform treatment at all r values.

In our theoretical study of the $A-X$ transition moment we carried out the (3221) CASSCF/MRCI(0.01) calculation using the very complete basis sets for oxygen and hydrogen described above. Basis set saturation studies indicated that this basis was near the basis set limit. Using either the theoretical or RKR potentials in the vibrational analysis produced ratios of Einstein coefficients that were in reasonable accord with the limited experimental data. Experimental lifetime measurements for the $v=0$, $N=1$ level vary from about 625±25 ns for experiments based on the Hanle effect (German et al. 1970) to 760±20 ns measured by Brzozowski et al. (1978) using the high-frequency deflection technique. The lifetimes measured by laser excitation fluorescence range from 721±5 ns measured by McDermid & Laudenslager (1982) to 686±14 ns and 693±10 ns determined by Dimpfl & Kinsey (1979) and German (1976), respectively. Our best estimate for the lifetime of the $v'=0$ $N'=1$ level of the $A^2\Sigma^+$ state of OH is 685±20 ns. This value is in excellent agreement with the LEF experiments of German (1976) and of Dimpfl & Kinsey (1979). This would imply that lifetimes deduced by McDermid & Laudenslager (1982) using LEF techniques are somewhat too long and those based on the Hanle effect too short. In addition, the lifetimes based on the high-frequency deflection technique appear to be too long at least for $N'<5$. Since the r-centroid of the transition increases with N' and the transition moment decreases with r, the vibrationally averaged moment decreases (lifetime increases) monotonically with increasing r (and increasing N'). Thus we have suggested that there is systematic error in the high-frequency deflection lifetimes for small N', such as collisional quenching, which would preferentially increase the population of the lower N' levels giving an apparent increase in the lifetimes.

In our opinion the best transition probabilities for an opacity calculation would be obtained using the *ab initio* transition moment function in Bauschlicher & Langhoff (1987). The transition probabilities could be computed using RKR potentials, which are now known with sufficient accuracy to eliminate any substantial error from this source. Naturally, some estimate of the effect of predissociation is required for those levels affected by the crossing of the $^4\Sigma^-$ and other states on the outer limb of the $A^2\Sigma^+$ state. It is our understanding that these calculations have been carried out to high rotational quantum numbers (Crosley 1993). These data should be accurate to 5% or better for states that have only a radiative component to the lifetime.

4 The electric dipole moment function of SiO

We next consider the calculation of the line strengths for the ro-vibrational bands in the $X^1\Sigma^+$ ground state of SiO. Here it is not possible to carry out FCI calculations, so we estimate the accuracy of the resulting EDMF by studying how it converges with improvements in the theoretical calculation. We employ large contracted Gaussian basis sets: [$8s$ $7p$ $4d$ $2f$ $1g$] and [$7s$ $6p$ $4d$ $2f$ $1g$] for silicon and oxygen, respectively, to study valence correlation. The silicon basis was contracted very flexibly in both the valence and inner-shell regions. Further details of the basis sets are given in the original reference (Langhoff & Bauschlicher 1993).

While it is possible to obtain near convergence of the EDMF with respect to improvements in the one-particle basis, it is much more difficult to demonstrate convergence with respect to the n-particle treatment. Calculations showed that a (433) CASSCF active space is required for an adequate zeroth-order description of the wave function when the oxygen $2s$ electrons are not correlated in the CASSCF. To obtain an equivalent description when the oxygen $2s$ electrons are correlated requires a (633) active space. Larger active spaces did not give significantly different results and the IC-MRCI calculations based on (633) and larger active spaces are exceedingly computationally intensive. The dipole moments are evaluated as expectation values, but the results are not significantly different when formulated as an energy derivative using the finite-field method. The effects of inner-shell (Si $2s$ and $2p$) correlation are not included in the final calculations, but have a relatively small (0.01 Debye) contribution to the permanent moment. Finally, relativistic effects are found to have only a very small effect on the dipole moment (Kellö & Sadlej 1993). Thus IC-MRCI and IC-ACPF calculations based on the (433) and (633) active spaces should provide globally accurate EDMFs for SiO.

In Table 2 we show the variation in the calculated dipole moments for v=0−3 and Einstein coefficients for the fundamental and first two overtones with the level of correlation treatment. At the CASSCF level the effect of including the oxygen $2s$ electrons in the active space is to increase the dipole moment by about 0.21−0.22 Debye. At the IC-MRCI level the difference decreases to about 0.07−0.08 Debye and at the IC-ACPF level the difference is only about 0.03−0.04 Debye. The convergence with level of theory is more rapid for the

Table 2. Theoretical dipole moments (Debye), Einstein coefficients (sec^{-1}), and dipole derivatives (a.u./a_0) as a function of the level of correlation treatment.

	(433) CAS	(633) CAS	(433) MRCI	(633) MRCI	(433) ACPF[a]	Expt[b]
v=0	2.9998	3.2165	3.0541	3.1293	3.0436(3.0983)	3.0982
v=1	3.0210	3.2379	3.0746	3.1500	3.0639(3.1184)	3.1178
v=2	3.0420	3.2590	3.0951	3.1706	3.0836(3.1380)	3.1372
v=3	3.0628	3.2798	3.1153	3.1911	3.1028(3.1572)	3.1574
A_{10}	7.25	7.77	6.92	6.97	6.73(6.68)	
A_{20}	0.281	0.329	0.273	0.270	0.264(0.266)	
A_{30}	0.004	0.004	0.004	0.004	0.005(0.004)	
$(d\mu/dr)_{r_e}$	0.635	0.657	0.621	0.623	0.613(0.610)	

[a] The values in parentheses correspond to a dipole moment that has been shifted in r by $-0.010\ a_0$ and increased in magnitude by 0.0391 Debye.

[b] Raymonda et al. 1970.

(433)-based treatment and the (433) IC-ACPF results are expected to be superior, especially since the calculation can be carried out without reference selection. By shifting the (433) IC-ACPF EDMF globally by $-0.010\ a_0$ in r and 0.0391 Debye in magnitude, we are able to reproduce the experimental dipole moments (Raymonda et al. 1970) for $v=0-3$ to within 0.0008 Debye. The shift in r, which is equal to the error in our r_e at this level of theory, can be partially rationalized in terms of the neglect of Si inner-shell correlation. The shift in magnitude is due primarily to neglect of oxygen 2s correlation. It is clear that an extensive treatment of electron correlation is required to compute the dipole moment to an accuracy of better than a few hundreths of a Debye.

The (433) IC-ACPF and (633) selected reference IC-MRCI EDMFs are compared with the Padé approximant EDMF of Tipping and Chackerian (TC) in Fig. 1, upon which the currently accepted line strengths are based. The true EDMF is expected to lie between the two theoretical ones, but closer to the (433) IC-ACPF EDMF. The present line strengths are about a factor of 1.2 larger than TC for the fundamentals and about a factor of two larger than TC for the overtones. The present line strengths should be accurate to at least 10%.

To aid in spectroscopic and astronomical studies we have generated line strengths in the form of gf values. These are the oscillator strengths (f) multiplied by the Hönl-London factors ($J/(2J+1)$ for the P branch and $(J+1)/(2J+1)$ for the R branch where J is the lower state rotational quantum number). The values are based on our "shifted" (433) IC-ACPF EDMF, which reproduces the experimental dipole moments for $v=0-3$. The gf values were generated for all P and R branch lines for which $J \leq 250$, $v \leq 15$ and $\Delta v \leq 4$. These data are on a file that is available upon request. We recommend using these values in conjunction with an SiO dissociation energy of about D_0=8.26 eV (Hildenbrand

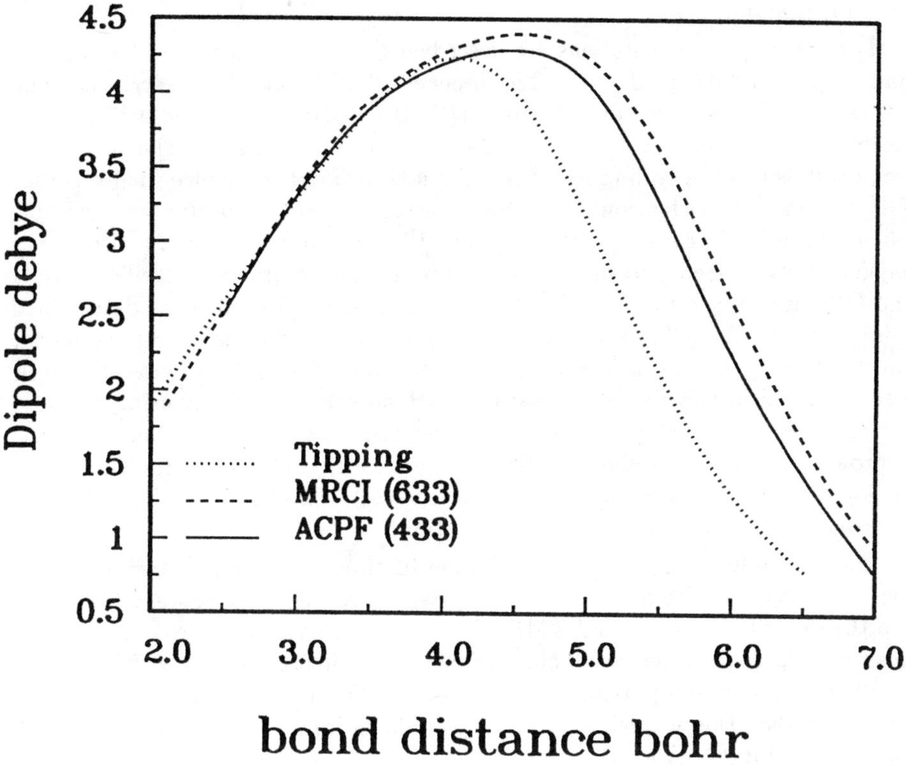

Fig. 1. Three electric dipole moment functions (EDMFs) for the $X^1\Sigma^+$ state of SiO. The solid and dashed lines denote the (433) IC-ACPF and (633) IC-MRCI EDMFs of the present work. The EDMF denoted by the dotted line is from the work of Tipping and Chackerian (1981).

& Murad 1974) to assess the abundance of this molecule in stellar atmospheres.

5 Spectroscopy of ZrO

Since the absorption bands of ZrO have been used to classify S stars (Johnson 1982; Littleton & Davis 1985), it is important to have accurate electronic transition probabilities for the band systems. Although the ground state is $X^1\Sigma^+$, the $a^3\Delta$ state is low-lying (Hammer & Davis 1980), so that transitions in both the singlet and triplet manifolds must be considered. Thus we have carried out calculations (Langhoff & Bauschlicher 1990a) to determine the singlet and triplet potentials in the Franck-Condon region up to an energy of about 30 000 cm^{-1}. Since calculations for transition metal systems are more difficult than for first- and second-row molecules, some compromises were necessary to be able to treat a large number of states over a large range of internuclear distances. Nevertheless, the theoretical results are generally in very good agreement with the available experimental data.

For brevity, the calculations are described only briefly here (see Langhoff & Bauschlicher 1990a for details). The inner shell of Zr atom is described using the relativistic effective core potential (RECP) of Hay & Wadt (1985), which incorporates the mass-velocity and Darwin relativistic effects. The outermost core orbitals (corresponding to $4s^2 4p^6$) are included in the valence shell together with the $4d$ and $5s$ electrons. This RECP is used in conjunction with a flexible valence basis for Zr consisting of [$5s$ $4p$ $4d$ $3f$] contracted Gaussian functions. For oxygen atom we used a [$5s$ $5p$ $2d$ $1f$] ANO basis that accurately describes both O and O^- character in the wave function. The molecular orbitals were determined using the SA-CASSCF approach, averaging separately for the lowest six singlet and triplet states. The six electrons in the Zr $4d$ and $5s$ and oxygen $2p\pi$ orbitals were included in the CASSCF treatment, which was carried out using a (3331) active space. More extensive electron correlation was included using an MRCI approach. The effect of higher excitations was estimated using a multireference analog of the Davidson correction (denoted +Q). This correction significantly improves the $X^1\Sigma^+ - a^3\Delta$ energy separation.

The Einstein coefficients and radiative lifetimes ($\tau_{v'}$) were determined numerically (using finite difference techniques) based on spline representations of the theoretical potentials and TMFs. The experimental T_e values were used to position the states when available (otherwise the theoretical values were used). In all cases the MRCI potentials encompass a sufficiently large range of r values to ensure that this formalism gives a reliable estimate of the lifetime for the lowest few vibrational levels.

The MRCI potentials for the $X^1\Sigma^+$, $A^1\Delta$, $B^1\Pi$, $C^1\Sigma^+$, $D^1\Gamma$, and $E^1\Phi$ states and for the $a^3\Delta$, $b^3\Pi$, $c^3\Sigma^-$, $d^3\Phi$, $e^3\Pi$, and $f^3\Delta$ states are shown in Figs. 2 and 3, respectively. We have adopted the spectroscopic designations based on the ordering of the states shown in these two figures.

The spectroscopic constants (r_e, ω_e, T_e) at the MRCI and MRCI+Q levels are in good agreement with experiment. For example, the MRCI+Q T_e values are within 900 cm^{-1} of experiment for all of the singlet and triplet states that have been characterized experimentally (Huber & Herzberg 1979; Hammer et al. 1981; Simard et al. 1988; Hammer & Davis 1980; Phillips et al. 1979) The MRCI+Q r_e values are uniformly about 0.03 a_0 too long, probably as a result of neglecting Zr inner-shell ($4s$ $4p$) correlation. The calculated moments plotted in Langhoff & Bauschlicher (1990a) for the strong transitions such as $B^1\Pi - X^1\Sigma^+$, $e^3\Pi - a^3\Delta$, and $d^3\Phi - a^3\Delta$ are probably accurate to about 10%. Our calculations indicate that the as yet unobserved $E^1\Phi - A^1\Delta$ transition is quite strong, which could provide a means of characterizing the $E^1\Phi$ state.

In Table 3 we have tabulated the theoretical Franck-Condon factor (q_{00}), the transition moment squared (r_e^2), and the Einstein coefficient (A_{00}) for the $0-0$ band of the strong singlet and triplet band systems. Included for comparison are the transition moment squared values deduced from a laboratory absorption

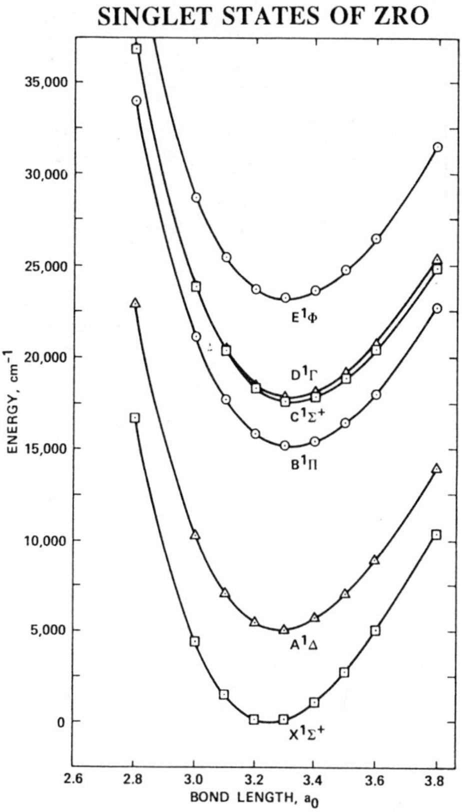

Fig. 2. The MRCI potential curves for the singlet states of ZrO

spectrum by Littleton et al. (1993). These values are determined as ratios to the β system, and given absolute magnitude by requiring them to be consistent with the lifetime measurement (Hammer & Davis 1979) for the v=0, J=77 level of the $e^3\Pi$ state. These "experimental" values agree with the theoretical values to within the combined uncertainty of the calculations and measurements. This lends considerable credence to the accuracy of the theoretical values for the other band systems. The $E^1\Phi - A^1\Delta$ transition has not been characterized experimentally. This band system should probably be included in opacity models, because it should be prominant in emission at transition energies of $\approx$16 000$-$17 000 cm^{-1}, if the $E^1\Phi$ state, which lies $\approx$22 000 cm^{-1} above the ground state, is populated at stellar temperatures.

Radiative lifetimes based on the MRCI transition probabilities are given in Table 4. Although these lifetimes include all radiative channels, there is generally a dominant channel, which is noted in the last column of the table. Experimental values for the $e^3\Pi$, $B^1\Pi$, and $C^1\Sigma^+$ states are included in Table 4 for comparison. The theoretical values are based on rotationless potentials, but

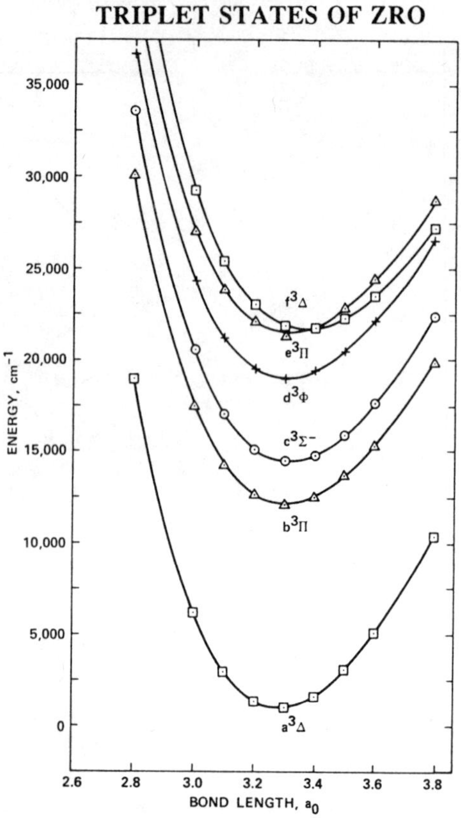

Fig. 3. The MRCI potential curves for the triplet states of ZrO

have a very weak dependence on J. For example, increasing J from 0 to 77 for the $e^3\Pi$ state increases the lifetime from 24.4 to 25.1 ns. The calculated value is about 25% less than the experimental value obtained by Hammer and Davis (1979) using the technique of resonant fluorescence decay. This difference is about the expected accuracy of the theoretical calculations. The discrepancy for the $B^1\Pi$ state lifetime is a little larger. However, our value is very stable to refinements in the theoretical treatment. The agreement is excellent between theory and experiment for the $C^1\Sigma^+$ state lifetime. Thus overall the agreement between the theoretical and experimental lifetimes is excellent. The $d^3\Phi$ and $E^1\Phi$ states are also predicted to have relatively short lifetimes. We recommend using the theoretical values in Table 4 until more reliable experimental values become available.

The isovalent TiO molecule is a very important opacity source in many stars. Thus analogous calculations for this system would be worthwhile. Preliminary calculations (Langhoff unpublished) indicate, however, that the characterization of the states is more difficult due to the large number of low-lying excited states

Table 3. Transition probabilities for selected ZrO band systems

Band system	q_{00}	$r_e^2(e^2 a_0^2)^a$		$A_{00}(s^{-1})$
	MRCI	MRCI	Expt.	MRCI
$b^3\Pi - a^3\Delta$	0.981	0.370	...	9.27(+5)
$d^3\Phi - a^3\Delta(\gamma)$	0.973	2.702	3.5±0.6[b]	2.15(+7)
$e^3\Pi - a^3\Delta(\beta)$	0.949	3.552	3.8±0.2[c]	3.98(+7)
$f^3\Delta - a^3\Delta(\alpha)$	0.672	1.443	...	2.51(+7)
$B^1\Pi - X^1\Sigma^+$	0.796	1.724	2.1±0.4[b]	1.28(+7)
$B^1\Pi - A^1\Delta$	0.946	1.519	...	2.65(+6)
$C^1\Sigma^+ - X^1\Sigma^+$	0.763	0.567	...	5.70(+6)
$E^1\Phi - A^1\Delta$	0.986	2.635	...	3.16(+7)
$E^1\Phi - C^1\Gamma$	0.989	2.250	...	6.76(+5)

[a] The value of the transition moment squared for the $(0-0)$ band.
[b] Values from a laboratory absorption spectrum, Littleton et al. (1993).
[c] Littleton and Davis (1985).

and numerous avoided crossings. This makes it difficult to obtain both an accurate and precise description of the electronic states over a wide range of internuclear distances using the SA-CASSCF approach. Previous calculations (Langhoff & Bauschlicher 1990b) on the FeH molecule also underscored the limitations of the SA-CASSCF method to provide an adequate zeroth-order description, primarily due to the large difference in correlation energy between atomic states with different numbers of *d* electrons. These problems are much more severe in the first transition row, because of the larger correlation energies and smaller radial extents of the orbitals. Nevertheless, it would be very interesting to perform high-level theoretical calculations for TiO to compare with the available experimental data and to ensure that all important band systems have been included in the opacity models.

6 Conclusions

We have discussed three applications of computational chemistry to the determination of accurate transition probabilities in diatomic molecules. For first-row molecules such as OH, the theoretical values rival the best experimental values in accuracy. An advantage of theory is that globally correct transition moment functions can be obtained using MRCI techniques. Thus if theory agrees with observed line strengths at relatively low temperatures, it is capable of providing results of nearly equal accuracy for lines where there is a high degree of vibrational and/or rotational excitation. For transition-metal systems such as ZrO, it

Table 4. Radiative lifetimes for the low-lying states of ZrO

State	Lifetime(v'=0) Theory	Expt.	Dominant transition
$b^3\Pi$	1.07 μs		$b^3\Pi - a^3\Delta$
$d^3\Phi$	45.9 ns		$d^3\Phi - a^3\Delta$
$e^3\Pi$	24.4 ns	32.5±2 ns[a]	$e^3\Pi - a^3\Delta$
$f^3\Delta$	28.3 ns		$f^3\Delta - a^3\Delta$
$B^1\Pi$	56.4 ns	83 ns[b]	$B^1\Pi - X^1\Sigma^+$
$C^1\Sigma^+$	136 ns	126±9[b] ns	$C^1\Sigma^+ - X^1\Sigma^+$
$E^1\Phi$	30.8 ns		$E^1\Phi - A^1\Delta$

[a] The experimental value for v'=0, J'=77 from Hammer & Davis (1979). The theoretical value for J'=77 is 25.1 ns.

[b] The experimental values for v'=0, from Simard *et al.* (1988).

is more difficult to obtain reliable line strengths, but in many cases the corresponding experiments are difficult or impossible. Thus theory may still provide the best method of obtaining opacity data. Clearly these *ab initio* methods can be applied to triatomic and larger systems (see, for example, Jørgensen & Jensen (1993) or Jørgensen this volume on the opacity of water vapor). Although it is much more difficult to solve the nuclear problem for triatomic molecules and the number of accessible levels increases rapidly with energy, it is clear that theoretical methods will play an important role in determining the opacity of these molecules as well.

References

Ahlrichs R., 1979, *Comput. Phys. Commun.*, **17**, 31

Ahlrichs R., Scharf P., Ehrhardt C., 1985, *J. Chem. Phys.*, **82**, 890

Almlöf J., Taylor P. R., 1987, *J. Chem. Phys.*, **86**, 4070

Bartlett R. J., 1981, *Annu. Rev. Phys. Chem.*, **32**, 359

Bauschlicher C. W., Langhoff S. R., 1987, *J. Chem. Phys.*, **87**, 4665

Bauschlicher C. W., Langhoff S. R., 1991, *Chem. Rev.*, **91**, 701

Bauschlicher C. W., Langhoff S. R., Taylor, P. R., 1990a, *Adv. Chem. Phys.*, **77**, 103

Bauschlicher C. W., Langhoff S. R., Komornicki A., 1990b, *Theor. Chim. Acta.*, **77**, 263

Blomberg M. R. A., Siegbahn P. E. M., 1983, *J. Chem. Phys.*, **78**, 5682

Brzozowski J., Erman P., Lyyra M., 1978, *Phys. Scr.*, **17**, 507

Buenker R. J., Peyerimhoff S. D., 1974, *Theor. Chim. Acta.*, **35**, 33

Chong D. P., Langhoff S. R., 1986, *J. Chem. Phys.*, **84**, 5606

Cohen M., Witteborn F. C., Carbon D. F., Augason G., Wooden D., Bregman J., Goorvitch D., 1992, *Astron. J.*, **104**, 2045

Crosley D. R., 1993, personal communication

Diedrich D. L., Anderson J. B., 1992, *Science*, **258**, 786
Dimpfl W. L., Kinsey J. L., 1979, *J. Quant. Spectrosc. Radiat. Transfer*, **21**, 233
Gdanitz R. J., Ahlrichs R., 1988, *Chem. Phys. Lett.*, **143**, 413
German K. R., 1976, *J. Chem. Phys.*, **62**, 2584, 4065 and references therein
German K. R., Bergeman T. H., Weinstock E. M., Zare R. N., 1970, *J. Chem. Phys.*, **58**, 4304
Hammer P. D., Davis S. P., 1979, *J. Mol. Spectrosc.*, **78**, 337
Hammer P. D., Davis S. P., 1980, *Ap. J.*, **237**, L51
Hammer P. D., Davis S. P., Zook A. C., 1981, *J. Chem. Phys.*, **74**, 5320
Hay P. J., Wadt W. R., 1985, *J. Chem. Phys.*, **82**, 299
Hildenbrand D. L., Murad E., 1974, *J. Chem. Phys.*, **61**, 1232
Huber K. P., Herzberg G., 1979,*Constants of Diatomic Molecules*, (Van Nostrand Reinhold, New York)
Johnson H. R., 1982, *Astrophys. J.*, **260**, 254
Jørgensen U.G., Jensen P., 1993, *J. Mol. Spectrosc.*, **161**, 219
Kellö V., Sadlej A. J., 1993, *J. Chem. Phys.*, **98**, 1345
Knowles P. J., Handy N. C., 1984, *Chem. Phys. Lett.*, **111**, 315
Langhoff S. R., Bauschlicher C. W., 1990a, *Astrophys. J.*, **349**, 369
Langhoff S. R., Bauschlicher C. W., 1990b, *J. Mol. Spectrosc.*, **141**, 243
Langhoff S. R., Bauschlicher C. W., 1993, *Chem. Phys. Lett.*, in press
Langhoff S. R., Bauschlicher C. W., Taylor P. R., 1987, *J. Chem. Phys.*, **86**, 6992
Langhoff S. R., Davidson E. R., 1974, *Int. J. Quantum Chem.*, **8**, 61
Littleton J. E., Davis S. P., 1985, *Astrophys. J.*, **296**, 152
Littleton J. E., Davis S. P., Song M., 1993, *Astrophys. J.*, **404**, 412
McDermid I. S., Laudenslager J. B., 1982, *J. Chem. Phys.*, **76**, 1824
Parr F. G., Yang W., *Density-Functional Theory of Atoms and Molecules*, (Oxford University Press, New York, 1989)
Phillips J. G., Davis S. P., Galehouse D. C., 1979, *Astrono. J.*, **234**, 401
Raghavachari K., Trucks G. W., Pople J. A., Head-Gordon M., 1989, *Chem. Phys. Lett.*, **157**, 479
Raymonda J. W., Muenter J. S., Klemperer W. A., 1970, *J. Chem. Phys.*, **52**, 3458
Ridgway S. T., Carbon D. F., Hall D. N. B., Jewell J., 1984, *Astrophys. J. Suppl.*, **54**, 177
Rinsland C. R., Wing R. F., 1982, *Astrophys. J.*, **262**, 201
Roos B. O., 1987, *Adv. Chem. Phys.*, **69**, 399
Saunders V. R., van Lenthe J. H., 1983, *Mol. Phys.*, **48**, 923
Shavitt I., *in Methods of Electronic Structure Theory (Ed. H. F. Schaefer)*, **p. 189**, (Plenum, New York, 1977)
Siegbahn P. E. M., 1984, *Chem. Phys. Lett.*, **109**, 417
Simard B., Mitchell S. A., Humphries M. R., Hackett P. A., 1988, *J. Mol. Spectrosc.*, **129**, 186
Tipping R. H., Chackerian Jr. C., 1981, *J. Mol. Spectrosc.*, **88**, 352
Werner H.-J., Knowles P. J., 1988, *J. Chem. Phys.*, **89**, 5803
Woon D. E., Dunning T. H., 1993, *J. Chem. Phys.*, **98**, 1358

Ab initio Calculations of Absorption Coefficients of Small Carbon Hydride Molecules

S.D. Peyerimhoff

Institut für Physikalische und Theoretische Chemie, Universität Bonn, Wegelerstraße 12, 53115 Bonn, Germany

1 Introduction

Quantum chemical methods have become a powerful tool for investigating the electronic structure of small molecules in ground and excited states. This is particularly true for species which are not easily accessible by experiment, either because they are short-lived, for example, or unstable under laboratory conditions or difficult to detect since their signals are hidden by other, prevailing molecules. In this sense theoretical methods are of special importance for molecules in atmospheric, stellar and interstellar environments.

The present contribution will focus on the use of ab initio computations to determine the absorption of light by molecules and will exemplify some aspects on the spectra of CH_2^+, C_2H and their deuterated species.

2 General considerations on the ab initio calculation of radiative transitions in molecules

Calculations for molecules are generally based on the Born-Oppenheimer approximation, i.e. the separability of nuclear and electronic motion is assumed and the coupling terms between the two types of motion are neglected. The total wavefunction is then a product of an electronic part $\phi_e(\mathbf{r}, Q)$ and a function $\chi(\mathbf{Q})$ describing the nuclear motion; the latter depends on the nuclear coordinates $\mathbf{Q}$, the former on the electronic coordinates $\mathbf{r}$ and is evaluated at a fixed nuclear arrangement, i.e. it depends only parametrically on the nuclear position Q. Neglect of the coupling terms $\partial\phi_e/\partial\mathbf{Q}$ and $\partial^2\phi_e/\partial\mathbf{Q}^2$ (which arise from the nuclear kinetic energy operator) is justified if the electronic wavefunction does not change significantly with the nuclear coordinates; if ϕ_e changes considerably along a nuclear coordinate the validity of the Born Oppenheimer approximation is questionable.

In order to obtain the potential energy surface $U(\mathbf{Q})$ one has to solve the electronic Schrödinger equation

$$\hat{H}_e\phi_e = E_e\phi_e \tag{1}$$

for a number of nuclear positions Q and then obtain $U(Q) = E_e(Q) + V_{NN}$ point by point, where V_{NN} is the nuclear repulsion energy. In order to obtain vibrational levels, it is necessary to solve the nuclear Schrödinger equation

$$(\hat{T}_N + U)\chi_\nu = E\chi_\nu \tag{2}$$

where $\hat{T}_N$ is the vibrational kinetic energy operator and are χ_ν, the vibrational wavefunctions. More details of potential energy calculations on ground and excited states can be found elsewhere (Bruna & Peyerimhoff 1987).

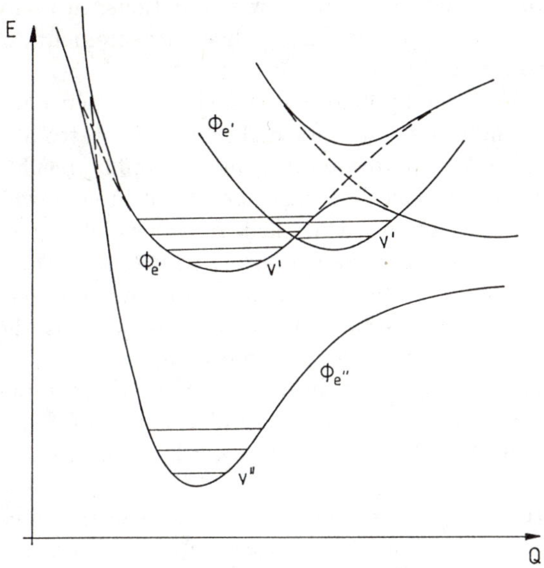

Fig. 1. Typical examples for potential curve interactions: dotted lines refer to diabatic states of the same symmetry, their adiabatic counterparts show avoided crossings.

A typical example of potential energy surfaces is presented in Fig. 1. If transitions occur between two separate curves, the oscillator strength as a measure of the absorption strength is simply evaluated as

$$\begin{aligned} f_{e''e'\nu''\nu'} &= \frac{2}{3} | < \phi_{e''}\chi_{\nu''} | \boldsymbol{\mu} | \phi_{e'}\chi_{\nu'} > |^2 \Delta E_{e''e'\nu''\nu'} \\ &\equiv \frac{2}{3} | < \chi_{\nu''} | R_{e'e''}(Q) | \chi_{\nu'} > |^2 \Delta E_{e''e'\nu''\nu'} \end{aligned} \tag{3}$$

In this formula quantities with double and single prime refer to the ground and the electronically excited state, respectively. The quantities ϕ_e and χ_ν are obtained from Eqs. 1 and 2; $\boldsymbol{\mu}$ is the dipole operator and integration is over $\mathbf{r}$ and $\mathbf{Q}$. The quantity $R_{e'e''}$ is the electronic transition moment. The relations between oscillator strength, lifetime and other quantities measuring the absorption strength are well documented (Herzberg 1950, Peyerimhoff 1992, Larsson this volume).

Frequently Eq. 3 is simplified by taking the average of $R_{e'e''}$, assuming it is independent of the nuclear coordinate (Franck-Condon principle)

$$f_{e''e'v''v'} = \frac{2}{3}|R_{e'e''}|^2| < \chi_{v''}|\chi_{v'} > |^2 \tag{4}$$

Hence, very often only the electronic part $R_{e'e''}$ and the vibrational overlap $< \chi_{v''}|\chi_{v'} >$ are considered when discussing absorption strengths.

Numerous calculations of absorption strengths according to Eq. 4 are in the quantum chemistry literature. Generally, errors are of the order of 10 % or less if suitable treatment is chosen to obtain the electronic wavefunctions. In many instances computed values are the only ones available when measurements of absolute absorption strength have yet to be carried out.

A more extended treatment is required if potential surface interaction occurs, a situation which becomes increasingly important for higher excited states and polyatomic molecules with several degrees of freedom. Frequent coupling mechanisms are the coupling due to the nuclear kinetic energy operator ("breakdown" of the Born-Oppenheimer approximations, as mentioned above) which generally couples states of the same symmetry and the spin-orbit and Renner-Teller interactions coupling states of different symmetry. Furthermore, the zero-field splitting of the spin multiplets of a single state may become as large as the energy separation between different states in molecules containing heavy atoms (large spin-orbit effects) and must also be taken into account for the computation of the absorption strength. In such cases the simple quantity of Eq. 4 serves as a guideline for the magnitude of the absorbing power but may deviate considerably from reality.

A typical example for the interaction of many potential surfaces of the same symmetry is given in Fig. 2 which shows the bending curves for excited states in HCO. Two valence states (denoted by V), one with its minimum at linear geometry and the other preferring a bent HCO molecule with an angle around 90°, couple with the series of Rydberg states of ns, npπ and npσ symmetry, all of which prefer the linear HCO arrangement; their bending curves run fairly parallel to the Rydberg state limit, i.e. the HCO^+ ion. The perturbed states (solid curves in Fig. 2) show a complicated pattern for the bending potential curves with various shoulders, barriers and wells which can be rationalized based on the dotted unperturbed (diabatic) curves. Note that only the Rydberg states up to $n = 4$ are considered in the figure and that the nd series has not explicitly been taken into account. The first excited state, for example, is characterized as a valence state in the region between 180° and 160°, has Rydberg character for bending angles between 160° and 100° and regains more compact valence character for smaller bond angles. The intensity of a transition from the ground state to this first excited state thus changes drastically with the angular coordinate (the intensity is generally weak for Rydberg-like transitions and stronger for suitably allowed valence-like transitions); a very irregular spectrum in the entire wavelength region is the consequence of such interactions. Such couplings occur not only in the bending but also in the stretching coordinate as discussed by Lorenzen-Schmidt et al. (1993) and Perić & Peyerimhoff (1993a,b).

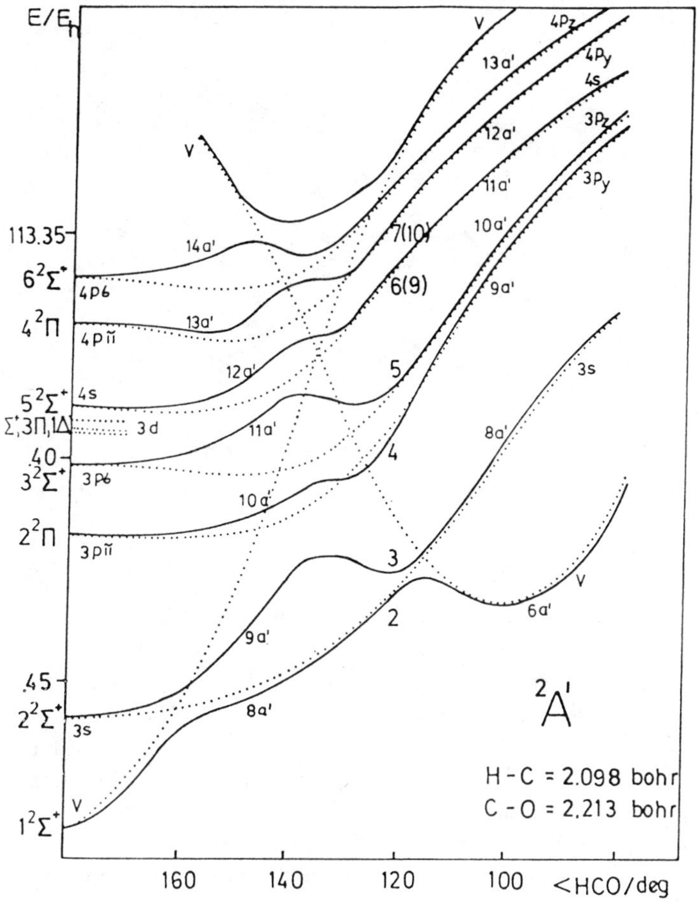

Fig. 2. Calculated potential curves for the excited states of HCO as a function of the bending angle (180° ≡ linear). The dominant molecular orbital in a given state is indicated. Dotted lines refer to the diabatic representation (same character) of states.

Such "non-adiabatic" interactions are often also the cause for non-radiative deactivation. In rare gas hydrides such as ArH and HeH, for example, interactions between the repulsive inner part of the potential curves as indicated in Fig. 1 lead to much faster deactivation of the excited states to the ground state than do the radiative transitions (Hemert et al. 1986, Hemert & Peyerimhoff 1991); this mechanism explains why the search for the corresponding emission spectrum was unsuccessful.

A representative example for the coupling due to spin-orbit interaction is the interaction of the $X^1\Sigma^+$ and $a^3\Pi$ states in MgO which is isovalent to the C_2 molecule. The situation is shown schematically in Fig. 3. The strength of the vibronic coupling depends on: (a) the magnitude of the spin-orbit coupling between $^1\Sigma^+$ and $^3\Pi$, (b) on the form of the unperturbed vibrational wavefunctions in

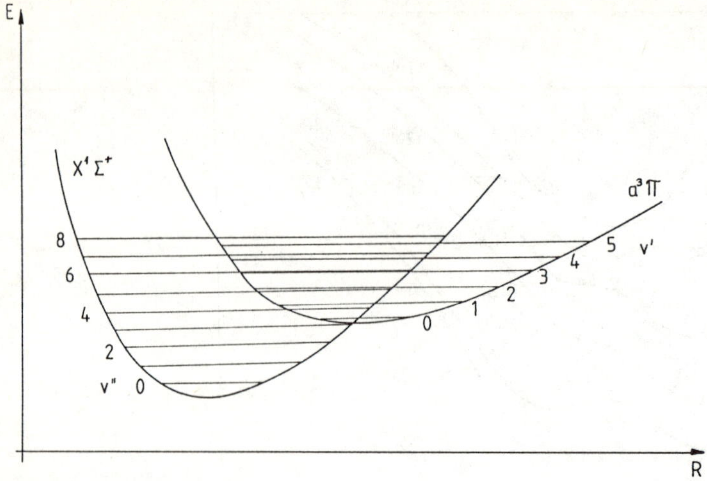

Fig. 3. Schematic representation of the interaction of the $X^1\Sigma^+$ and $a^3\Pi$ states in MgO.

the $^1\Sigma^+$ and $^3\Pi$ potential wells, in particular their possible spatial overlap and (c) on the relative energy of the vibrational levels in the two potential wells.

Fig. 4 shows the computed vibronic functions. The $^3\Pi(v = 0, 1)$ and $X^1\Sigma^+(v = 3, 4)$ are essentially unperturbed (Thümmel et al. 1989). The $^3\Pi(v = 2)$ function already shows admixture of the $^1\Sigma^+(v = 5)$ function in the area of the inner turning point; $^3\Pi(v = 3)$ and $^1\Sigma^+(v = 6)$ are in near energetic resonance and hence the two vibronically mixed functions have very different character as compared to their unperturbed counterpart. For higher vibrational levels the degree of mixing becomes small again. Absorbtion characteristics for "infrared" transitions (within the $^1\Sigma^+$ and the $^3\Pi$ vibrational states) or "electronic" transitions (between $^1\Sigma^+$ and $^3\Pi$) are thus markedly changed as soon as the strongly perturbed levels are involved. Ab initio treatments can fairly easily compute the necessary spinorbit matrix element, the unperturbed vibrational functions and the actual coupling. However, their accuracy is generally not high enough to determine the relative location of the interacting levels to within a few wavenumbers so that information about the relative location of potential surfaces to this accuracy has to come from other sources (spectroscopy).

Vibronic coupling is not very effective if the overlap between the unperturbed vibrational wavefunctions of the two coupling states is small, even if near energy resonance of levels and strong coupling is present. This has been shown by Gemein & Peyerimhoff (1991) for the $X^1\Sigma^+$ and $a^3\Pi$ and $a'^3\Sigma^+$ states in CO; the ground state vibrational levels which could possibly mix possess quantum numbers $v > 30$ and therefore have so many "wiggles" that the overlap integral with the $^3\Pi$ or $^3\Sigma^+$ wavefunctions having small vibrational quantum numbers is very small. This observation that vibronic coupling is not very efficient if the difference in the vibrational quantum numbers of the unperturbed states is very large can be considered as a general rule.

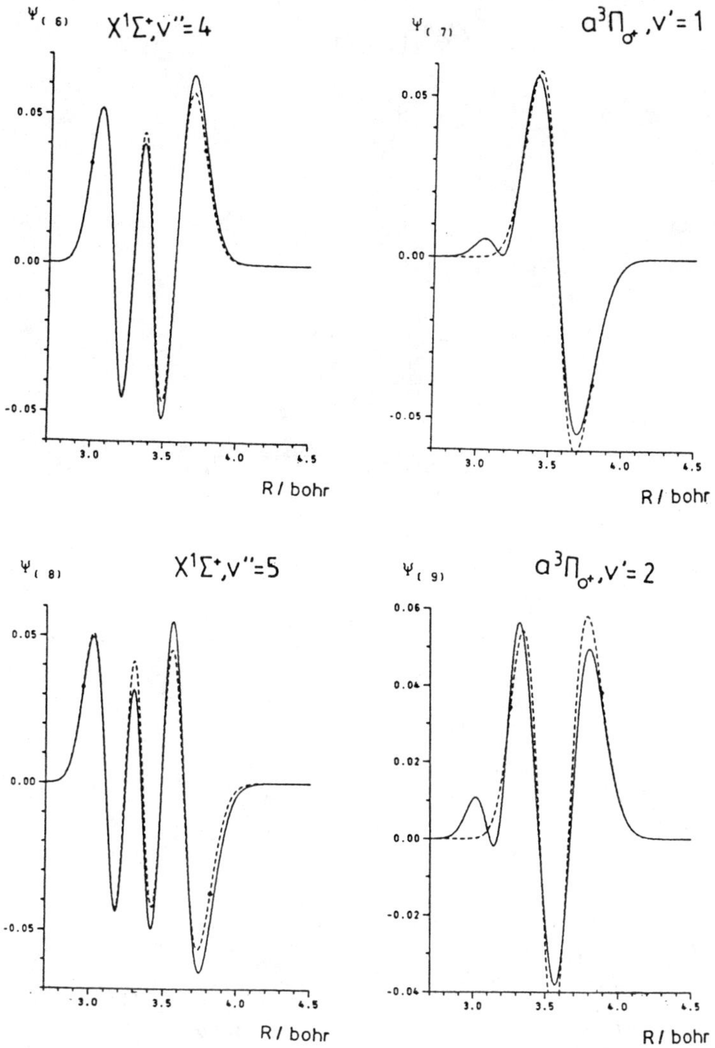

Fig. 4. Calculated vibronic wavefunctions in the $X^1\Sigma^+ - a^3\Pi$ interaction zone of MgO. For labelling of states see Fig. 3. The dotted lines refer to the unperturbed vibrational functions.

Finally, an example for the importance of zero-field splitting can be found in the work of Marian (1990) on NiH, which is also briefly discussed by Peyerimhoff (1992). The NiH ground state $^2\Delta$ is followed by $^2\Sigma^+$ and $^2\Pi$ states. Spin-orbit interaction (L+S) mixes components of $^2\Delta$ and $^2\Pi$, as well as of $^2\Pi$ and $^2\Sigma^+$, so that the correct potential curves are $^2\Delta_{5/2}$, $^2\Delta_{3/2}$, $^2\Pi_{3/2}$, $^2\Pi_{1/2}$ and $^2\Sigma^+_{1/2}$. The vibronic functions then also have character of the different electronic states.

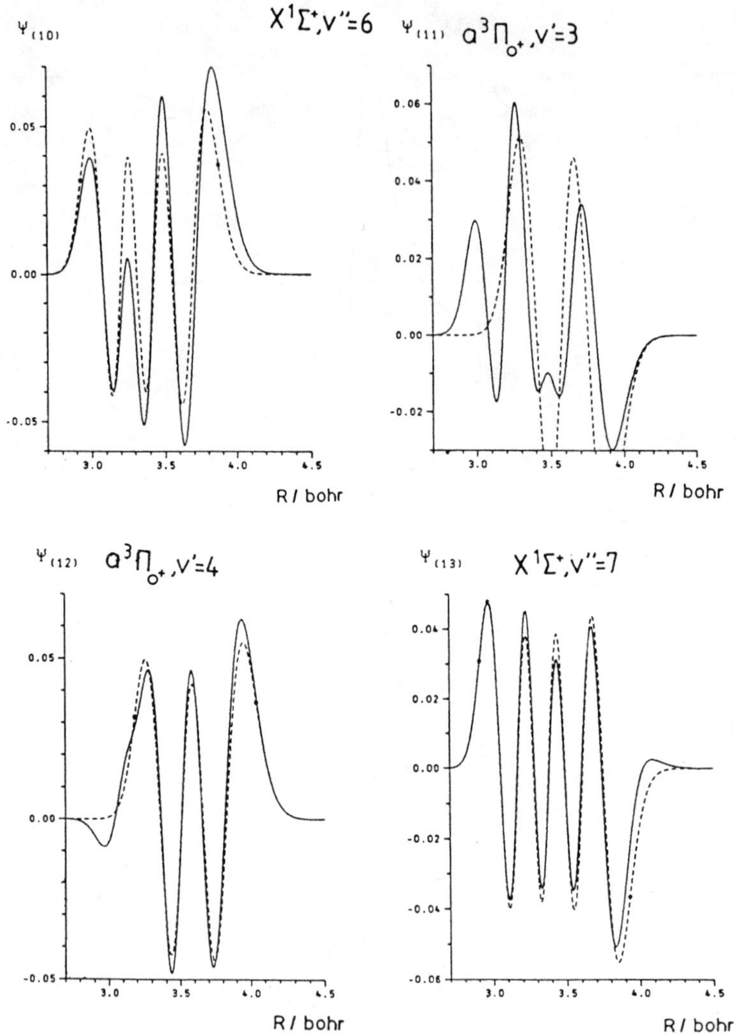

Fig. 4. –continued. Calculated vibronic wavefunctions in the $X^1\Sigma^+ - a^3\Pi$ interaction zone of MgO. For labelling of states see Fig. 3. The dotted lines refer to the unperturbed vibrational functions.

3 The spectrum of CH_2^+

The occurrence of the CH_2^+ ion in mass spectroscopy is well known since its detection (Hogness & Kvalness 1928) 65 years ago; this ion is furthermore assumed to be an important partner in chemical reactions. The first spectroscopic indication for its existence came in 1961 from Herzberg (1961) and since then various efforts to characterize this ion spectroscopically have failed to be successful. Only very recently (Rösslein et al. 1992) were the infrared transitions of the ν_3 band seen. The best characterization of this molecule comes from various ab initio calculations (Bartholomae et al. 1981, Reuter & Peyerimhoff 1992). The bending potential curves for the two lowest states are seen in Fig. 5 as taken from our own work (Reuter & Peyerimhoff 1992).

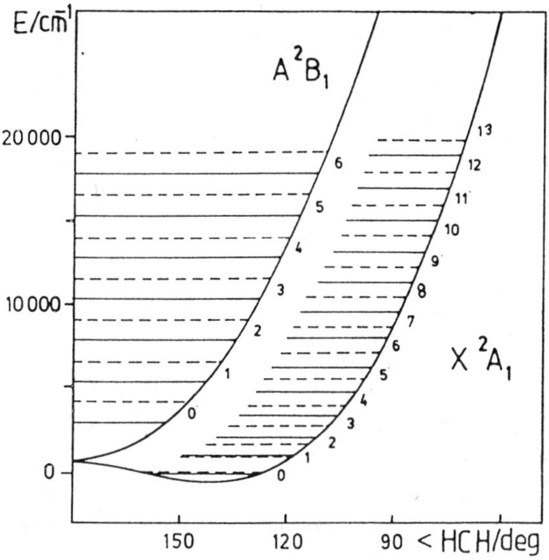

Fig. 5. Calculated potential curves for CH_2^+ and vibrational levels: full lines $K = 0$, dashed lines $K = 1$ quantum numbers.

The lowest CH_2^+ state is a $^2\Pi$ and is split by the Renner-Teller effect into the X^2A_1 state in which the molecule is slightly bent (around 140°) and the A^2B_1 state which prefers the linear geometry. The barrier to linearity is predicted to be of the order of 1000–1100 cm^{-1} (Bartholomae et al. 1981, Reuter & Peyerimhoff 1992).

In principle the three steps for a purely ab initio treatment of the spectrum are (a) to compute the potential energy surface of the Renner-Teller coupled pair according to Eq. 1, (b) to generate the vibrational wavefunctions (considering large-amplitude motion) for the three modes on the two "unperturbed" surfaces and to add rotational functions for motion about the z-axis (linear axis) when solving Eq. 2, and (c) to couple the functions describing nuclear motion on the two surfaces to produce the vibronic functions in a similar manner as shown pictorially for the diatomic molecule MgO in Figs.3 and 4. Since we found very little coupling between the symmetric stretching and bending modes in CH_2^+ we have explicitly treated only the large-amplitude bending motion in the effective one-dimensional potential which includes the optimized CH distance at each point (for further discussion see Reuter & Peyerimhoff 1992). Details of the vibronic mixing will again depend substantially on the relative energetic location of the unperturbed levels. The calculated intensity distribution (i.e. the square of the vibronic transition moment which is the quantity in Eq. 3 without the factor 2/3 and the energy) for transitions from the lowest ground state level v_2"$= 0$, $K = 0$ to $v_2' = n$, $K = 1$ is plotted in Fig. 6.

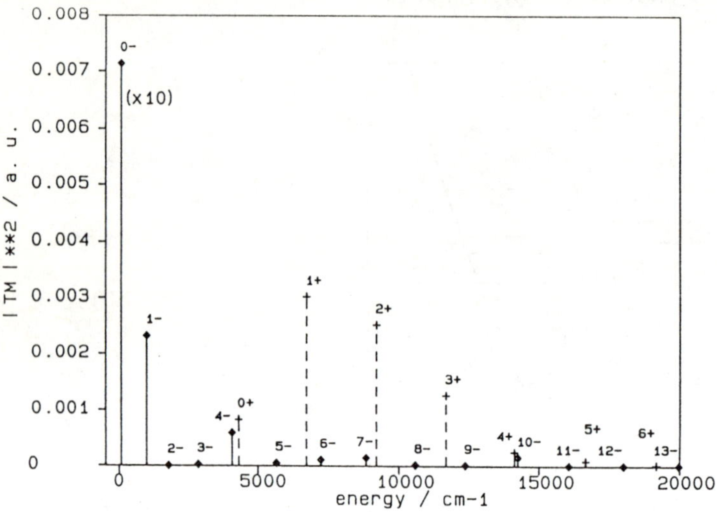

Fig. 6. Calculated intensity distribution (square of vibronic transition moment) from $v'' = 0$, $K =$ to $v' = n$, $K = 1$ levels in CH_2^+. Levels which have dominant lower-state character are denoted by a minus, those with dominant upper state $^2B_1$ character by a plus sign.

For low quantumnumbers the typical "infrared" intensity pattern is seen, i.e. high intensity for the $\Delta v = 1$ transition, rapidly decreasing for the $\Delta v = 2, 3$ transitions. The first intensity borrowing is seen for the combination $v' = 0$ of the upper and $v''=4$ of the lower state (levels of essentially upper state character are denoted by a plus, those with dominant lower state character by a minus) so that formally the v"$= 0$ to v"$= 4$ becomes a much stronger absorber while $v'' = 0$ to $v' = 0$ is weaker than expected. Similar calculations have been carried out for the other isotopomers CHD^+, CD_2^+, as well as for transitions from v"$= 0$, $K = 1$ to $v' = n$, $K = 0$ (Reuter & Peyerimhoff 1992).

Furthermore, this vibronic coupling is apparent in other properties; predictions have been made for the spin-orbit splitting of the vibronic levels (Reuter & Peyerimhoff 1992) and the vibronically averaged hyperfine coupling constants (Engels & Perić 1992). All these results can be used as guidelines to interpret and understand the overall structure of the CH_2^+ spectrum; numerically accurate details require higher accuracy in the ΔE values of the two states than are normally achieved by the computations. Therefore an experiment from which this difference could be determined to wavenumber accuracy would be very helpful for calibration of the potential surfaces which would then allow the accurate prediction of many properties of the CH_2^+ spectrum.

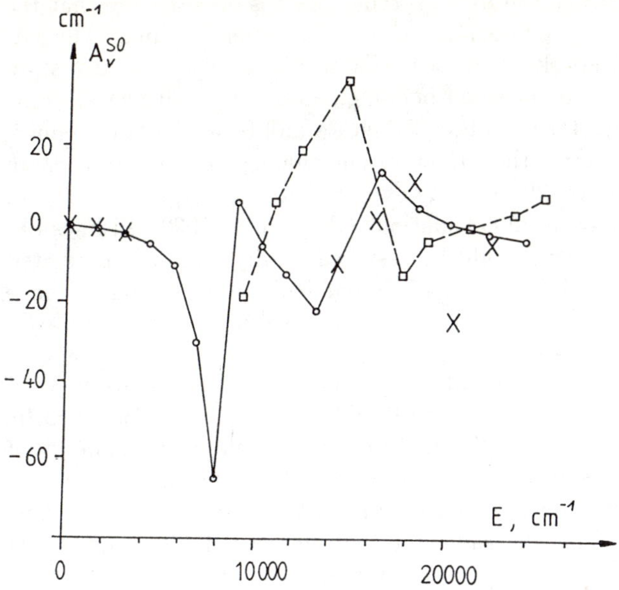

Fig. 7. Calculated spin-orbit splitting of the vibronic levels of H_2O^+. The crosses are experimental data. Solid lines connect the vibronic levels of dominant X^2A_1 character, dashed lines those of dominant A^2B_2 character.

A comparison with the H_2O^+ spectrum, treated in an equivalent manner (Reuter et al. 1991), shows convincingly good correspondence between the measured and the theoretically predicted results. Fig. 7 compares the measured and calculated spin-orbit splitting of the various vibronic levels. As a result of the computational analysis the suggestion made earlier (Weis et al. 1989), namely that the first observed upper state vibronic level should be assigned to $v_2' = 1$ rather than to $v_2' = 0$, is strongly supported. Thus we are confident that the numbering of the $^2A_1$ vibronic levels in earlier analyses (Jungen et al. 1980) should be corrected by one unit. The calculated barrier to linearity is then 8235 cm^{-1} (Reuter et al. 1991) or 7948 cm^{-1} (Weis et al. 1989) and not 1000 wavenumbers higher (9187 cm^{-1}) as assumed in empirical fits (Jungen et al. 1980).

4 The spectrum of C_2H

The spectrum of C_2H is even more complicated than that of CH_2^+ because (a) it results from a three-state problem and (b) the CC-stretching motion and the bending motion are strongly coupled. An extended ab initio study was performed by Perić et al. (1992).

C_2H has two low-energy states: $^2\Sigma^+$ and $^2\Pi$. At small internuclear CC separations (1.2Å) the $^2\Sigma^+$ state lies below the $^2\Pi$, around 1.3Å the two states are almost isoenergetic and at larger separations (> 1.4Å) the $^2\Pi$ state is the lowest energy state. In addition, the Renner-Teller effect is operative so that the $^2\Pi$ state splits into the two components ^{2}A' and ^{2}A" upon bending. The ^{2}A' component interacts (analogously to the situation in Fig. 2) with the $^2\Sigma^+$ state which also takes on ^{2}A' symmetry upon bending. As a net result the vibronic wavefunctions possess character from the $^2\Sigma^+$ state and from the two Renner-Teller components derived from the $^2\Pi$ state (in analogy to the situation in CH_2^+) with variable mixing coefficients. As expected, the absorption spectrum is very complex. Based on ab initio calculations (Perić et al. 1992) all lines observed to date in ^{12,13}C and ^{1,2}H could be assigned. Many more are predicted so that the experimental search is facilitated. In particular a few "puzzles" have been solved. Examples are:

(a) In several experiments the energy difference between the (011) vibronic level (T = 2091 cm^{-1}) and the (001) level (T = 1841 cm^{-1}) was found to be 250 cm^{-1}. This is significantly different from the ν_2 fundamental energy of 372 cm^{-1}. These findings are almost exactly reproduced by the calculations which show an energy difference of 262 cm^{-1} (T = 2142 cm^{-1} and T = 1880 cm^{-1}) and show that the cause for the unexpected (001) and (011) difference is not an unusually large anharmonicity but a level shift due to vibronic mixing of the $X^2\Sigma^+$ and $A^2\Pi$ states.
(b) The lowest $K = 1$ vibronic state shows a measured spin-orbit splitting of -0.34684 cm^{-1}, unexpected for a $^2\Sigma$ state. The computed value for the vibronically coupled $^2\Sigma - ^2\Pi$ level of -0.35 cm^{-1} agrees excellently with the measurements.
(c) The measured ratio of the splittings of the lines in the ESR spectra of C_2H and C_2D attributed to the hyperfine structure of the hydrogen isotope is 6.33 which differs significantly from the expected value of 6.51. The calculations predict a value of 6.32 and offer a simple explanation: since the ground state vibronic levels in C_2D lie somewhat lower than in C_2H they have a somewhat smaller percentage of the electronic $^2\Pi$ character ($^2\Pi$ is the higher state in this case). Since the computed values for the isotropic hyperfine coupling constant at linear geometry are about 45 MHz for the Σ state and about -43 MHz for the Π state, the actual percentage of $\Sigma - \Pi$ mixing in the vibronic level is important and the calculations show that vibronically averaged value for the isotropic hfcc should be slightly larger in C_2D than in C_2H.

5 Summary

The present contribution has shown in a few representative examples that modern ab initio calculations are far enough advanced to not only yield potential energy surfaces but also to predict many other quantities of spectroscopic interest with quite high accuracy. Results of this quality require that the electronic wavefunctions be of high accuracy by using at least doublezeta atomic orbital basis sets (see also Langhoff & Bauschlicher this volume) with additional polarization-type functions; furthermore, a considerable amount of electron correlation must be accounted for in the theoretical treatment employed. For all examples the multireference configuration-interaction method (MRD–CI) was chosen (Buenker et al. 1978, Buenker & Peyerimhoff 1983). Among the properties discussed (other than energies) the hyperfine coupling seems to be the most sensitive to the CI expansion length, while spin-orbit coupling or electronic transition probabilities seem to show less dependence on the rather large tail (several 100000 compared to first few 10000 configurations) of the CI expansion.

References

Bruna P.J., Peyerimhoff S.D., 1987, in: *Ab initio Methods in Quantum Chemistry I* (ed. K. P. Lawley), **1**

Herzberg G., 1950, in: *Molecular Spectra and Molecular Structure*, Vol.1, van Nostrand, **20**, 382

Peyerimhoff S.D., 1992, In: E.Kochanski (ed.), *Photoprocesses in Transition Metal Complexes, Biosystems and Other Molecules*, Kluwer Acad. Publ., **89**

Lorenzen-Schmidt H., Perić M., Peyerimhoff S.D., 1993, *J. Chem. Phys.*, **98**, 525

Perić M., Peyerimhoff S.D., 1993, *J. Chem. Phys.*, **98**, 3587

Perić M., Peyerimhoff S.D., 1993, *J. Mol. Structure*, in press

van Hemert M., Dohmann H., Peyerimhoff S.D., 1986, *Chem. Phys.*, **110**, 55

van Hemert M., Peyerimhoff S.D., 1991, *J. Chem. Phys.*, **94**, 4369

Thümmel H., Klotz R., Peyerimhoff S.D., 1989, *Chem. Phys.*, **135**, 229

Gemein B., Peyerimhoff S.D., 1991, *Chem. Phys. Letters*, **184**, 45

Marian C.M., 1990, *J. Chem. Phys.*, **93**, 1176

Hogness T.R., Kvalness H.M., 1928, *Phys. Rev.*, **32**, 942

Herzberg G., 1961, *Can. J. Phys.*, **39**, 1511

Rösslein M., Gabrys C.M., Jagod M.-F., Oka T., 1992, *J. Mol. Spectry*, **153**, 738

Bartholomae R., Martin D., Sutcliffe B.T., 1981, *J. Mol. Spectry*, **87**, 367

Carter S., Handy N.C., 1984, *Mol. Phys.*, **52**, 1367

Sutcliffe B.T., Tennyson J., 1987, *J. Chem. Soc. Faraday Trans.II*, **83**, 1663

Lee J.S., Secrest D., 1988, *J. Phys. Chem.*, **92**, 1821

Reuter W., Peyerimhoff S.D., 1992, *Chem. Phys.*, **160**, 11

Engels B., Perić M., 1992, *J. Chem. Phys.*, **97**, 7629

Reuter W., Perić M., Peyerimhoff S.D., 1991, *Mol. Phys.*, **74**, 569

Weis B., Carter S., Rosmus P., Werner H.-J., Knowles P.J., 1989, *J. Chem. Phys.*, **91**, 2818

Jungen Ch., Hallin K.-E., Merer A.J., 1980a, *Mol. Phys.*, **40**, 25

Jungen Ch., Hallin K.-E., Merer A.J., 1980b, *Mol. Phys.*, **40**, 65

Perić M., Peyerimhoff S.D., Buenker R.J., 1992, *Z. Phys. D.*, **24**, 177

Buenker R.J., Peyerimhoff S.D., Butscher W., 1978, *Mol. Phys.*, **35**, 771

Buenker R.J., Peyerimhoff S.D., 1983, *in: New Horizons of Quantum Chemistry*, (eds. P. O. Löwdin, B. Pullmann), **183**

The RASSCF, RASSI, and CASPT2 Methods Used on Small Molecules of Astrophysical Interest

Per-Åke Malmqvist

Theoretical Chemistry, Chemical Center,
P.O. Box 124, 221 00 Lund, Sweden

1 Introduction

To a quantum chemist with no particular background in astrophysics or astronomy, a brief glance at journals and textbooks in these fields shows at least three areas where computational quantum chemistry has had a valuable impact: Interstellar cloud chemistry; stellar atmosphere modelling; and chemistry in extreme conditions, such as at the surface of a neutron star. The first two uses are particularly suitable, since standard methods are directly applicable.

For such problems, good calculations of potential energy as well as expectation values and matrix elements of dipole and other operators appears to be in demand. Many electronic states may be involved, at least a broad range of problems involve fairly small molecules, often radicals, and conformation regions far from equilibrium. Such problems are addressed by three methods originated in our laboratory, and known by the acronyms RASSCF (Restricted Active Space Self-Consistent Field, Malmqvist et al. 1990), RASSI (RAS State Interaction) and CASPT2 (Complete Active Space Perturbation Theory to Second Order-Complete Active Space Perturbation Theory to Second Order, Andersson et al. 1990; Andersson et al. 1992).

The first method is an extension of the well-known CASSCF (Complete Active Space SCF) method (Roos et al. 1980; Roos 1980). The RASSI is a versatile analysis method, (an extension of CASSI; Malmqvist & Roos 1989), whereby separately optimized RASSCF wave functions can be used as basis functions, for computing transition moments, defining diabatic states, etc. Finally, the CASPT2 method uses a CASSCF wave function as a multiconfigurational zeroth order wave function, and corrects for dynamic correlation to second order in energy. In contrast to most conventional perturbation techniques, it works well for arbitrary spin and symmetry, for excited states, etc., without any particular demands on the electronic structure of the root function.

A rough sketch of computational quantum chemistry methods will be given, and against this background the methods will be briefly described with demonstration applications.

1.1 The Basic Problem: The Adiabatic Electronic Structure

Quantum chemical problems are concerned with some aspect of the solution of the Schrödinger equation applied to a system of nuclei and electrons. The electrodynamic field is imposed only as an external field. Approximate relativistic effects, and nuclear size effects, are incorporated by modifying the interactions. Any more sophisticated effects are added as corrections computed by perturbation theory. The time-dependent Schrödinger equation, with nuclear and electronic coordinates **Q** and **q**, resp., reads

$$\left.\begin{aligned} i\tfrac{\partial}{\partial t}\Psi(\mathbf{q},\mathbf{Q},t) &= \hat{H}\Psi(\mathbf{q},\mathbf{Q},t) \\ \hat{H} &= \hat{T}_n + \hat{T}_e + \hat{V}_{nn} + \hat{V}_{ee} + \hat{V}_{en} \end{aligned}\right\} \tag{1}$$

where the second line simply sums the contributions from kinetic energy of nuclei and electrons, the interaction among nuclei, among the electrons, and between nuclei and electrons, respectively. Those terms may include relativistic corrections, and are believed to account for the physics of this finite system to a precision well in excess of that needed for all chemical and almost all spectroscopic applications.

Practical solution of this equation preceeds in two steps. The hamiltonian is split into $\hat{H} = \hat{T}_n + \hat{H}_e$. All terms (including nuclear-nuclear repulsion) that commute with the nuclear position operators have been included in the electronic hamiltonian $\hat{H}_e(\mathbf{Q})$, which is thus an operator on electronic wave functions as function of the nuclear positions. Expanding $\hat{H}_e(\mathbf{Q})$ in a complete set of electronic basis functions, which depend on the nuclear coordinates, the full Schrödinger equation takes a form that involves only nuclei.

A complete basis is never available. However, a practical and almost always sufficient basis is provided by the few lowest eigenfunctions of $\hat{H}_e(\mathbf{Q})$; this is called the adiabatic basis, and the Schrödinger equation becomes

$$\begin{aligned} i\frac{\partial}{\partial t}\chi_K = -\frac{1}{2M}\frac{\partial^2}{\partial Q^2}\chi_K + V_K(Q)\chi_K - \frac{1}{M}\sum_L \langle \Phi_K | \frac{\partial}{\partial Q}\Phi_L\rangle \frac{\partial}{\partial Q}\chi_L \\ -\frac{1}{2M}\sum_L \langle \Phi_K | \frac{\partial^2}{\partial Q^2}\Phi_L\rangle \chi_L \end{aligned} \tag{2}$$

where $\Phi_K(\mathbf{q},\mathbf{Q})$ and $\chi_K(\mathbf{Q},t)$ are the electronic and nuclear wave functions. The potential energy functions for the nuclear motion are just the energy eigenvalues of $\hat{H}_e(\mathbf{Q})$.

So far, the treatment is formally exact. If the two matrix elements involving derivatives with respect to nuclear coordinates are neglected, we obtain the Born-Oppenheimer (BO) approximation. When this is not valid, the equation can in practice still be solved using a few selected adiabatic wave functions as basis.

Disregarding exceptional cases where non-BO effects must be included already from the beginning, we see that the problem has a very nice modularity: We first solve the electronic problem for relevant eigenstates. The result provides input for a number of other methods, which perform e.g. reactive scattering or ro-vibrational analysis. Such calculations have their own broad range of methodology and is also a part of the quantum chemistry repertoir. However, we will keep within the scope of electronic eigenstates.

The basic requirement of quantum chemistry methods is then the ability to obtain, for *any* conformation of nuclei, the adiabatic electronic energy, and electronic matrix elements of various fairly simple operators, of *any* of the lower electronic eigenstates. The required differential accuracy depends on the problem, but is typically a few kJ/kmol in potential energy functions. Much higher accuracy can be demanded in spectroscopic applications, but may still be reached when errors cancel, or when matrix elements for small perturbations are wanted. Spectroscopy usually requires matrix elements of a broader range of operators. From a quantum chemist's viewpoint, the basic problems are similar.

The general applicability, with a well-balanced description of different electronic structures, is a tall order. One cannot expect any single method to solve all problems. As a consequence, a large number of methods are used to obtain electronic energies. In the following, a brief overview is offered, on the understanding that it is not complete, and in some cases not well-informed as regards latest research status.

2 Quantum Chemistry Methods in General

Those methods which compute electronic eigenstate energies and properties can be subdivided into density functional and wave function methods, and the latter into configuration based methods (single- vs. multi-configurational) and explicitly correlated ("r_{12}"-) methods. This subdivision is ordered by an increasingly detailed description of the system, and thus an increasing accuracy but also a rapidly increasing computation cost.

Density functional theory expresses the minimum electronic energy for any specified electronic charge density function in the form of a so-called energy functional. The exact functional is not known in any practically useful formulation, and its existence is just used as a justification for a variety of pragmatic approximations. Recent improvements have made such methods useful for molecules, and since they are cheap and scale very favourably with molecular size, they may well be the only practical alternative for many problems. A number of useful references can be found, e.g., in Clementi et al. 1991. A drawback is the lack of any systematic way to improve the result by larger calculations, and of a good indicator of the reliability of any specific result.

Response and transition properties can also be computed by a range of other methods, which will only be mentioned in passing: In general, they use wave functions calculated by a specific wave function method, and use some of the theoretical and computational machinery of that method to obtain transition strengths and energies, ionization potentials etc. The accuracy in transition properties can surpass that obtained from simple matrix elements over the separately computed state functions.

The end result of methods described below is in general a set of potential energies and various electronic matrix elements, which are then expressed as parametrized functions of internal coordinates.

2.1 Wave Function Methods

Excepting density functional methods, we compute an approximation to the wave function. This is defined in terms of a large number of parameters, whose meaning are defined by some wave function model. The parameter values are optimized by requiring key properties of the exact wave function to be exactly represented in the model: usually the stationarity criterion for an energy eigenstate. The model should be able to reproduce the most important features of the exact wave function. A list of such features, in approximate order of diminishing importance, could be:

1. A major part (overlapping by about 0.985^n, if n is the number of electrons, with the exact wave function) requires a few (typically 1–100) so-called determinant functions, if those are properly chosen. At specific conformations, non-radicals may need only 1 (SCF). Dissociation or reaction paths require more. The entire chemistry, i.e. the set of all reaction and dissociation paths, for even a small number of atoms is rarely done, but may require thousands of determinants. A couple of hundred are needed to dissociate the Cr_2 molecule. When more than one configuration is important, one speaks of static correlation, or near-degeneracy effects. Other correlation effects are then called dynamic correlation.
2. When an electron is close to a nucleus, the wave function has an electron-nuclear cusp.
3. When it is far away from the rest of the molecule, the wave function goes asymptotically to describe the ground state of the kation, properly symmetry coupled to an orbital with exponential tail describing the remote electron.
4. The major effect of varying the position of one electron is to polarize the remaining electron system. This effect may be called dynamic polarization, or intramolecular dispersion. It affects most strongly the immediate vicinity of the electron, but is fairly long-range. It affects quite strongly also the core electrons, and the sigma skeleton of conjugated molecules, which are traditionally though of as rather inactive.
5. Apart from this polarization, there is also an inter-electron cusp. This is of small importance to the energy, but affects some two-electron properties.

In addition, it is usually quite important that dissociation gives asymptotically a properly spin- and symmetry-coupled sum of products, with factors that are energy eigenstates of the respective fragments (Size consistency).

There are of course methods designed to solve one-, two-, or three-electron problems, which use specialized forms for the wave function. Excepting these, the universal ansatz is an expansion in Slater Determinant Functions (SDs) or Configuration State Functions (CSFs). The SDs are antisymmetrized products of one-electron functions; the CSFs are linear combinations of SDs which have desired total spin and spatial symmetry. The expansion coefficients can be explicitly used as variation parameters, or implicitly defined as in Coupled Cluster or perturbation theory.

On top of this expansion, correlation factors which depend explicitly on the interelectron distances can be used to introduce the interelectronic cusp, which improves the convergence rate of CSF or perturbation expansions, and which is crucial to sampling methods such as Quantum Monte Carlo. The latter type of methods are able to generate millions of electron positions statistically drawn from an ensemble that accurately represents the ground state wave function, and form a Monte-Carlo estimate of the energy. However, mainstream methods rely only on CSF expansions, where local features are very poorly represented, but where energies and most properties are obtained by integrals which are not very sensitive to these defects.

2.2 Configuration Expansion Methods: Orbital Generation

All methods considered here rely on a one-electron basis set, which must be able to reproduce the cusp and tail of items 2 and 3 above. This would seem to favor STO basis sets, with exponential functions similar to hydrogenic wave functions. However, contracted GTO bases, which are sums of gaussian functions instead, can easily fulfill these requirements in the $\mathcal{L}^2$ sense, and are much easier to handle. For accurate pointwise evaluation of wave function (QMC) or density there is some justification for other bases in molecular problems, and then a better approach is to use some modifying factor times a standard gaussian basis set. The selected basis set, the "AO basis", is used to express molecular orbitals, or MO's. Modern basis sets are usually general contractions of the ANO type (Atomic Natural Orbitals, Almlöf & Taylor 1987).

Determination of suitable MO coefficients is done by SCF or MCSCF. One or more Slater Determinant functions (SD), or Configuration State Functions (CSF) are formed from the MOs. For any given such function, those MOs which are used to define it are called occupied, the other are unoccupied. The MO coefficients, describing the orbitals, are used as variational parameters. If more than one SD or CSF is used, there are in addition CI (Configuration Interaction) coefficients to express the wave function. The MO and CI coefficients are varied to find a stationary energy expectation value. The usual orbital generation methods are:

1. **UHF ("Unrestricted" HF)**. The spin-orbitals are restricted to have either spin up or down (α or β).
2. **RHF (Restricted HF)**: All orbitals come in pairs, with α and β spin but the same spatial dependence. They do not have to be occupied in pairs:
3. **High-Spin RHF**: All singly occupied orbitals have α spin.
4. **OSHF (Open-Shell HF)**: A spin- and symmetry-coupled CSF is being optimized. This is rarely done nowadays, except in the High-Spin special case, where the CSF is a single SD.

5. **MCSCF (Multi-Configurational Self-Consistent Field)**: A wave function composed of several or many CSFs is optimized.

In a basis set approach, the orbital generator produces not only the set of occupied orbitals, but a complete set within the basis. Those which are doubly occupied in all configurations are *inactive* orbitals. Those that are never used are *virtual*, or *secondary*, orbitals. Those with varying occupancy in MCSCF, or open shells, are *active* orbitals.

2.3 Correlation by CSF Expansion

The SCF and MCSCF techniques give energies and properties which are often useful in their own right. In particular, for few-electron systems the MCSCF can include enough configurations to account not only for static correlation, but for intramolecular dispersion and some close-range dynamic correlation as well. The advantage is then that the orbitals are optimized together with the correlation treatment. To afford large-scale MCSCF for more than a few electrons, one must select the CSFs used as economically as possible. However, this is not easy when the selection is to be valid for several or many conformations and states. Moreover, such general MCSCF is plagued with existence of multiple solutions to the stationarity conditions. The most popular form of MCSCF is therefore the CASSCF (Complete Active Space) method, where all the CSFs possible with a given active space are included. Their number increases as a polynomial with the number of active orbitals, with leading degree equal to the number of active electrons. This dependence sets a sharp limit to the possibility of including dynamic correlation by merely increasing the active space. The CASSCF is thus followed by a calculation of remaining correlation energy by CI-like or non-CI-like methods. With the former, I mean the well-known SDCI (Singles and Doubles CI) with single- or multi-configuration reference (MR-SDCI, or MRCI for short), and to methods which technically are simple modifications to SDCI: CPF and MCPF ([Modified] Coupled Pair Functional) in the single-configuration case, ACPF (Average CPF) in the multi-configuration case (Ahlrichs et al. 1985; Chong & Langhoff 1986; Gdanitz & Ahlrichs 1988).

With these methods, wave functions are obtained in the CSF space generated by substituting at most two orbitals in the reference configuration(s). The SDCI is not size-consistent, so the modifications are preferred if many electrons are correlated. With the above methods, properties are easily obtained. Transition matrix elements are easily obtained only by the SDCI method, and assuming a common orbital basis for the states.

With non-CI-like methods, there is a large difference between the single- and multi-configuration treatments. In the single-configuration case, there is a wide range of different methods, with programs available and useful also for non-specialists. This is especially true for RHF-based methods for non-radicals (which could, technically speking, also be considered as CI-like). High-spin open shell cases can be treated as UHF, with some annoying loss of efficiency, while OSHF references are on their way but not yet firmly established. The most important methods are MP2 (Second-order Møller-Plesset), MBPT4 (Fourth-order Many-Body Perturbation Theory), and CC (Coupled Cluster). The latter usually includes single and double excitations in its parameter space (CCSD), but unless the number of orbitals is large, triples can be included approximately or exactly (CCSD(T), CCSDT) to yield very accurate results, whenever a single closed-shell reference is adequate. A good application example with important references is the article by Watts et al. 1991. By contrast, the MP2 is a very popular method for larger molecules, since it is fairly cheap and can be used without storing the electron repulsion integrals.

With a multi-configuration reference space, the only successful general scheme so far is to use a single reference function composed of many configurations, of the CASSCF type, as the root function of a perturbation expansion.

3 The RASSCF, RASSI, and CASPT2 Methods

3.1 RASSCF

The major drawback of CASSCF is the steep increase in the number of CSFs with the number of active orbitals, except when very few electrons are correlated. With twelve active electrons, there is a limit at around twelve or thirteen active orbitals, which give 226 512 and 736 164 singlet CSFs, respectively. There are many examples where such an active space is not sufficient for obtaining good CASSCF energies and properties, and quite a few examples where not even the CASSCF orbitals are good enough to be used in subsequent MRCI or similar methods. This happens when the dynamic correlation, lacking in the CASSCF, is important for determining the shape of the orbitals. A typical example is negative ions, where lack of dynamic correlation gives too diffuse orbitals, or when dynamic correlation alters the balance between competing electronic structures with large and different dipole moments. General MCSCF is not an attractive scheme, since it is difficult to preselect the important configurations, and to avoid multiple minima.

The name Restricted Active Space was given by Olsen (Olsen et al. 1988) to a scheme where the active orbitals are subdivided into RAS-1, RAS-2, and RAS-3 orbitals. All CSFs are included, which have at most a specified number of electrons in the RAS-3 space, and at most a specified number of electrons taken out from the RAS-1 space. In RAS-2 orbitals, any occupation is allowed. The restrictions on RAS-1 and RAS-3 can cut down the number of CSFs quite drastically, thus allowing a larger active space, and are based entirely on an orbital selection which is easier and more natural than a general CSF selection.

The RAS expansion space is closed under deexcitation, which is a technically important property. It ensures that any matrix elements expressed by spin-free second quantization formulae can, after minor rearrangement, be expressed by means of vector transformations within the expansion space. This makes it efficient to regard the application of a linear operator as a programming primitive. Our RASSCF code first used the Split-Graph GUGA method for operations on the CSF expansion (Malmqvist et al. 1990). In a later version such operations are performed by temporarily reexpanding in SDs and, after the operations, transform back to CSFs.

Leaving out any technicalities, the program is simple: Given a set of starting orbitals, the CI equations are solved for the state of interest. For this given CI expansion, the orbitals are refined so that the energy is stationary. If this scheme is repeated as it stands, convergence may suffer from neglect of the coupling between the CI- and the orbital-optimization steps. However, this coupling is introduced in all subsequent iterations by a correction, which is refined at each iteration. A number of variants are possible, such as a weighted optimization of several states simultaneously. The end result is orbitals, wave functions, density matrices, properties and matrix elements for the RASSCF states.

3.2 RASSI

Since its CSF space is closed under deexcitation, a RASSCF wave function can also be reexpressed in any new orbital set, provided that the transformation matrix is triangular (or even block-triangular with the subdivision into inactive,RAS-1, etc. orbitals). The CI expansion coefficients are easily and efficiently recomputed to express identically the same wave function in the new orbital set. The new orbitals will no longer be orthonormal. The RASSI program puts this to use by accepting as input two different RASSCF wave functions, each with different orbitals, and proceeds by performing such non-orthogonal orbital transformations that the resulting two new orbital sets $\{\psi_p^A\}$ and $\{\psi_q^B\}$ are biorthonormal:

$$\langle\psi_p^A|\psi_q^B\rangle = \delta_{pq}, \quad \forall p,q\,. \tag{3}$$

The point is that with such orbitals, any matrix element can be computed by standard methods: If an operator is expressed by a second quantization formula with integrals over the mixed new basis, the expression for its matrix element between states A and B is identical to the form it would take in an common orthonormal basis.

The RASSI program is extensively used just to compute standard one-electron matrix elements of e.g. the dipole operator, for oscillator strengths (See the description of its precursor, CASSI, in Malmqvist & Roos 1989). With state-specific orbitals, such matrix elements cannot be obtained from dynamic correlation post-RASSCF programs, and the RASSCF values computed by RASSI is then our only source of oscillator strengths. When the states have the same symmetry, the RASSCF states may well be non-orthogonal and interacting. However, the RASSI program can simply solve the Schrödinger eigenvalue problem in the basis of any provided RASSCF wave functions, thus yielding oscillator strenghts for orthogonal and non-interacting RASSI states instead. The difference may be quite important, when two RASSCF states have different permanent dipole moment: The transition dipole, if evaluated for the RASSCF states, is contaminated by the product of the overlap with the difference in dipole moment.

The RASSI program is also very helpful in a variety of situations: Even if the RASSCF is simpler than general MCSCF, it is sometimes difficult to select orbital subspaces properly. Two putative different solutions may actually be misrepresentations of the same physical state, if the optimization has resulted in correlation of either of two bonds for instance, when the subdivision of the active space prevent them from being treated democratically. Occasionally, there are even problems with spontaneous symmetry breaking. This type of problems may be resolved by simply giving all the available RASSCF states as input to RASSI and solve the Schrödinger equation in the space they span. The result will in general be a number of reliable RASSI states, and their natural orbitals will be excellent start orbitals for new state-specific calculations, while their natural occupancies will indicate how to pick the new active orbitals. The original RASSCF wave functions will be reexpressed in the orthonormal, non-interacting RASSI states, which indicates in which way the states have perturbed each other.

3.3 CASPT2

When RASSCF does not give sufficient dynamic correlation, the preferred additional procedure is MRCI, if it can be afforded. If more than a few electrons are correlated, it must be corrected for size consistency or replaced by its cousin ACPF. However, it is necessary that the reference selection of CSFs is large enough. It must contain a large fraction of the wave function, and a similar fraction for all states and conformations. It must be realized that when going from CASSCF or RASSCF to MRCI, we are not just adding a large number of important configurations, but also deleting by selection, and perhaps more severe: One part, with a weight of e.g. around 85%, of the CASSCF wave function, is now improved by dynamic correlation; 15% is not. These two parts may differ considerably in ionizity, dipole moment etc, and are not treated in a balanced way, which will affect the result. In fact, whenever possible, the best procedure is to use several references, and extrapolate the results, as function of a selection threshold. For more than a few atoms, or when large references are necessary, MRCI may be too expensive. The CASPT2 method was developed to account for extra-CAS dynamic correlation, when a satisfactory MRCI treatment is too expensive (Andersson et al. 1990; Andersson et al. 1992). However, it is not an approximation to MRCI, but a (size-consistent) perturbative approximation to Full CI, and uses the complete CASSCF wave function as reference.

The CASPT2 program solves the traditional first-order Rayleigh-Schrödinger perturbation equations with a Møller-Plesset splitting of the Hamiltonian:

$$\left.\begin{aligned} \hat{H} &= \hat{H}^{(0)} + \hat{H}^{(1)} \\ (\hat{H}^{(0)} - E^{(0)})\Psi^{(1)} &= -\hat{H}^{(1)}\Psi^{(0)} \\ \hat{H}^{(0)} - E^{(0)} &= \hat{Q}(\hat{F} - \text{const.})\hat{Q} \\ \hat{F} &= \textstyle\sum_{pq} F_{pq}\hat{E}_{pq} \end{aligned}\right\} \qquad (4)$$

where $\hat{Q}$ is a projector onto the interacting space of $\Psi^{(0)}$, which is a CASSCF wave function. The Fock-like matrix F_{pq} is defined such that it gives the usual Fock matrix for a closed-shell reference function, and fulfills Koopmans' theorem for ionization energy and electron affinity in an average sense for the active orbitals. The equation is solved by expanding in the interacting space:

$$\Psi^{(1)} = \sum_{pqrs} C_{pqrs} \hat{E}_{pqrs} \Psi^{(0)} \tag{5}$$

and the correlation energy is obtained to second order as

$$E^{(2)} = -\langle \Psi^{(1)} | \hat{H}^{(0)} \Psi^{(1)} \rangle \tag{6}$$

In these equations, $\hat{E}_{pqrs}$ is an excitation operator, which effects a spin-coupled excitation of two electrons from orbitals q and s to orbitals p and r. For a CASSCF reference, the meaningful excitations are those where q and s are inactive or active, p and r are active or secondary, but all four indices are not active. Of the eight different cases, the most expensive terms arise from the semi-internal excitations, where three of the orbital indices *pqrs* refer to active orbitals. To solve the resulting equations requires then essentially the three-body density matrix elements for active orbitals, which are reasonably easy to compute but gives equation matrices of the size $n^3 \times n^3$. This causes a calculation bottleneck for more than about $n = 12$ active orbitals. The calculation then scales as fast as n^9 in the present algorithm, which is the main reason why the method has not yet been worthwhile to append to RASSCF rather than CASSCF.

It should be noted that the present implementation is completely independent of the number of CSFs, apart from the initial, rapid evaluation of the density matrix elements. It has been used with equal ease for single configuration root functions as with CASSCF wave functions with a hundred thousand configurations. The interacting space can easily comprise billions of CSFs; the number of variational parameters still scale quite modestly, and is slightly larger than with a similar closed-shell MP-2 calculation.

4 Demonstration Examples

4.1 RASSCF

Ethene V state. The first singlet $\pi\pi^*$ state of ethene, the so-called V state, is an example where a meaningful orbital optimization must include dynamic correlation. The state is dominated by two valence-bond structures, each with one positive and one negative methylene ion. It has been much studied as a prototype example of high ionicity states. Unless enough dynamic correlation is included, the orbitals will be too diffuse. Table 1 illustrates the very rapid growth of the number of CSFs in an ill-advised attempt to obtain good orbitals by CASSCF (In the original study, another approach was taken, of course: MRCI with iterated natural orbitals).

By contrast, in the last line is a RASSCF calculation with 29 active orbitals, requiring a very modest number of CSFs. The total energy is much lower, of course, but this is a dangerous quality measure in this case. The fourth column

Table 1. RASSCF demonstration example: Ethene

Calculation	# CSFs	Energy (a.u.)	$\langle z^2 \rangle$ (a.u.)	CPU (s)
CAS $(10\sigma, 2\pi)$	28 848	-77.8993	22.8	15
CAS $(10\sigma, 3\pi)$	92 406	-77.9239	23.4	60
CAS $(10\sigma, 4\pi)$	268 800	-77.9341	23.0	227
CAS $(10\sigma, 5\pi)$	716 523	-77.9433	21.9	878
RAS (5,2,22)	4 704	-78.0484	20.8	68

is the expectation value of $\sum z_i^2$. Its true value is believed to be around 15 a.u., and the excess is primarily due to the π^* orbital which is forced to be too diffuse. All calculations correlate the 12 valence electrons. The basis set is of ANO type, with 6s, 5p, 3d and 2f functions on carbons, 3s and 2p functions on hydrogens. The CPU time is for an IBM 3090-17S computer.

The radiative lifetime of SH^+ $A^3\Pi$. Here, the potential and transition dipole functions of the ground $X^1\Sigma^+$ and the $A^3\Pi$ states have been computed by RASSCF (González-Luque et al. 1992). The spectroscopic constants derived for the two states are presented in Tabs. 2 and 3, compared to experiment (Rostas et al. 1984).

Table 2. The SH^+ $X\,^3\Sigma^-$ state.

	Normal		Deuterated	
Quantity	RAS	Exp	RAS	Exp
R_e	1.367 Å	1.364	1.367	–
D_e	3.46 eV	3.70	3.46	–
D_0	3.30 eV	3.54	3.35	3.59
ω_e	2540 cm^{-1}	2548	1824	1830
$\omega_e x_e$	51.7 cm^{-1}	49.3	26.8	25.4
B_e	9.24 cm^{-1}	9.28	4.76	4.79
α_e	0.30 cm^{-1}	0.28	0.11	0.11

The end results for the life time is $1.07 \pm 0.13\,\mu$s and $1.14 \pm 0.13\,\mu$s, in excellent agreement with one experiment but in disagreement (but within error bars) with another. For details, we refer to the original article which also quotes other similar examples.

Table 3. The SH^+ $A\,^3\Pi$ state.

	Normal		Deuterated	
Quantity	RAS	Exp	RAS	Exp
R_e	1.502 Å	1.501	1.502	–
D_e	1.64 eV	1.78	1.64	–
D_0	1.53 eV	1.68	1.56	1.71
ω_e	1660 cm^{-1}	1672	1194	1201
$\omega_e x_e$	49.1 cm^{-1}	47.6	26.6	24.5
B_e	7.65 cm^{-1}	7.66	3.95	3.95
α_e	0.40 cm^{-1}	0.36	0.15	0.13
T_e	30 595 cm^{-1}	30 345	–	30 356

4.2 RASSI

The routine use of the RASSI program is for calculation of transition moments and similar quantities, such as the life times reported above. However, the program is very useful in a number of less standard contexts. In a recent study of O_4^+, a strong symmetry-breaking tendency (giving $O_2 + O_2^+$ even in symmetric conformations) was handled by simply using the two symmetry-broken solutions as quasi-diabatic states and computing the hamiltonian coupling elements by RASSI. Rather than reproducing the relevant figures here, we refer the interested reader to the original article (Lindh & Barnes 1993).

4.3 CASPT2

Apart from the two benzene examples presented below, there is a rapidly accumulating body of experience showing that CASPT2 may be one of the most powerful methods to emerge for many years. The interested reader should see also e.g. Roos & Andersson 1993 or Roos et al. 1993.

Benzene pilot study. The valence electronic spectrum of benzene has been used for a very long time as a test of ab initio methods. Only recently has reliable results been obtained with an overall accuracy of better than 1 eV. The key problem is the large variation in ionicity among the valence states, with a resulting variation in the dynamic correlation, primarily in the form of intramolecular dispersion interaction between the π electrons and the σ skeleton. The following pilot study demonstrates the problem (Andersson & Roos 1993). A common misrepresentation is to regard these excited states as primarily single excitations, with an open-shell Hartree-Fock structure. However, in reality, the structure varies strongly, with the $^1E_{2g}$ state as a rather extreme case. In Tab. 4, note that the multi-excited character of the wave functions would be much more pronounced if e.g. ground state Hartree-Fock orbitals were used. The figures here apply to state-specific optimized CASSCF orbitals. In the second part of

Table 4. Pilot study: Singlet excited valence states of benzene.

Weight of excitations from leading conf. Exc. type	$^1A_{1g}$	$^1B_{2u}$	$^1B_{1u}$	1^1E_{1u}	2^1E_{1u}	$^1E_{2g}$
Singly		80%	92%	87%	-	48%
Doubly		15%	3%	5%	-	42%
Triply		2%	-	2%	-	3%

Excitation energies, eV. Method	$^1A_{1g}$	$^1B_{2u}$	$^1B_{1u}$	1^1E_{1u}	2^1E_{1u}	$^1E_{2g}$
CAS(6)		5.0	7.9	9.3	-	8.1
+PT2		4.6	5.9	6.5	-	7.7
CAS(12)		4.9	7.4	8.8	7.7	8.1
+PT2		4.7	6.1	7.1	7.7	7.8
Exp		4.9	6.2	6.9	7.6	7.8

Dynamic correlation contributions, eV. Corr. type	$^1A_{1g}$	$^1B_{2u}$	$^1B_{1u}$	1^1E_{1u}	2^1E_{1u}	$^1E_{2g}$
$\sigma\sigma$	12.6	12.7	12.8	12.8	-	12.7
$\sigma\pi$	7.7	8.0	8.8	9.6	-	8.0
$\pi\pi$	0.9	1.0	1.8	1.8	-	1.0

the table, it is seen how CASPT2 is able to correct for most of the occasionally large errors in the CASSCF energies. Only the six π-electrons are active. Two active spaces are used: The first comprises the valence π orbitals, while the second is twice as large. Note the small improvement to the $^1B_{1u}$ and 1^1E_{1u} states with the larger active space, and how CASPT2 acts to correct them. Note also how, with the larger space, we begin to see Rydberg intruders (The 2^1E_{1u} state). In the last part of the table, the dynamic correlation energies have been broken up into different contributions. The second line can be interpreted as the contributions from the instantaneous response of the sigma electrons to the charge of the π electrons; naturally, the effect is larger (by up to 2 eV!) for the states with predominantly ionic structures.

Benzene application example. We finally offer a recent application example: Table 5 presents some of the results of a recently conducted study of singlet and triplet excited states of benzene up to about 7.9 eV (64 000 cm^{-1}) (Lorentzon et al. 1993). The basis set was basically our standard basis set of ANO type (Widmark et al. 1990), but augmented with a specially prepared one-center basis to represent the Rydberg orbitals correctly. As can be seen in the tables, the accuracy in vertical excitation energies is generally about 0.1 eV.

Acknowledgements. I wish to thank the organizers of this conference for the kind invitation. To me, it seems that this kind of inter-disciplinary symposia is

Table 5. CASPT2 singlet excitation energies for Benzene.

State	E_{CAS} (eV)	$\langle z^2 \rangle$ (a.u.)	E_{PT2} (eV)	Exp. (eV)
Valence $^1\pi\pi^*$				
1^1B_{2u}	4.83	30.1	4.82	4.90 vert
1^1B_{1u}	7.33	31.9	6.28	6.20 vert
1^1E_{1u}	8.45	35.1	7.09	6.94 vert
2^1E_{2g}	7.96	29.8	7.86	7.86 vert
Rydberg $^1\pi\pi^*$				
2^1E_{1u}	6.48	79.1	7.17	7.41
2^1A_{1g}	7.19	88.3	7.76	7.81?
1^1E_{2g}	7.11	88.3	7.77	7.81
1^1A_{2g}	7.10	88.2	7.82	7.81
Rydberg $^1\pi\sigma^*$				
1^1E_{1g}	6.56	46.2	6.34	6.334
1^1A_{2u}	6.68	44.7	6.85	6.932
1^1E_{2u}	6.81	45.1	6.95	6.953
1^1A_{1u}	6.77	45.4	6.99	–
1^1B_{2g}	7.30	64.9	7.54	7.460?
1^1B_{1g}	7.30	65.2	7.54	7.460?
2^1E_{1g}	7.34	63.5	7.54	7.535
3^1E_{1g}	7.39	52.4	7.54	–

a fruitful idea. The RASSCF, RASSI and CASPT2 methods reported in this contribution results from work within a joint study contract with IBM Sweden, and supported by the National Research Council of Sweden. They are part of the MOLCAS program package described by Andersson et al. 1991.

References

Ahlrichs R., Scharf P., Ehrhardt C., 1985, *J. Chem. Phys.*, **82**, 890

Almlöf J., Taylor P. R., 1987, *J. Chem. Phys.*, **86**, 4070

Andersson K., Fülscher M. P., Lindh R., Malmqvist P.-Å., Olsen J., Roos B. O., Sadlej A. S., Widmark P. O., 1991, *MOLCAS version 2.* University of Lund, Sweden

Andersson K., Malmqvist P.-Å., Roos B. O., Sadlej A. S., Wolinski K., 1990, *J. Phys. Chem.*, **94**, 5483

Andersson K., Malmqvist P.-Å., Roos B. O., 1992, *J. Chem. Phys.*, **96**, 1218

Chong D. P., Langhoff S. R., 1986, *J. Chem. Phys.*, **84**, 5606

Clementi E., Corongiu G., Stradella O. G., 1991, *Density Functionals for Molecules*, p321 ff in: Modern techniques in computational chemistry: MOTECC-91 (ed. E. Clementi), Leiden 1991.

Gdanitz, R. J., Ahlrichs, R., 1988, *Chem. Phys. Letters*, **143**, 413

González-Luque R., Merchán M., Roos B. O., 1992, *Molec. Phys.*, **76**, 201

Lindh R., Barnes L. A., 1993, *J. Chem. Phys.*, (in press).
Lorentzon J., Malmqvist P.-Å, Roos B. O., 1993, *J. Chem. Phys.*, (submitted).
Malmqvist P.-Å., Rendell A. P., Roos B. O., 1990, *J. Phys. Chem.*, **94**, 5477
Malmqvist P.-Å., Roos B. O., 1989, *Chem. Phys. Letters*, **155**, 189
Olsen J., Roos B. O., Jørgensen P., Jensen H. J. Aa., 1988, *J. Chem. Phys.*, **89**, 2185
Roos B. O., 1980, *Int. J. Quantum Chem. Symp.*, **14**, 175
Roos B. O., Andersson K., 1993, *Int. J. Quantum Chem.*, **45**, 591
Roos B. O., Serrano-Andrés L., Merchán M., 1993, *Pure Appl. Chem.*, **65**, 1693
Roos B. O., Taylor P. R., Siegbahn P. E. M., 1980, *Chem. Phys.*, **48**, 157
Rostas J., Horani M., Brion J., Daumont D., Malicet J., 1984, *Molec. Phys.*, **52**, 1431
Watts J. D., Stanton J. F., Bartlett R. J., 1991, *Chem. Phys. Letters*, **178**, 471
Widmark P. O., Malmqvist P.-Å., Roos B. O., 1990, *Theor. Chim. Acta*, **77**, 271

The MORBID Method

Per Jensen⋆

FB 9–Theoretische Chemie, Bergische Universität–Gesamthochschule Wuppertal, D-42097 Wuppertal, Germany

1 Introduction

The present communication is concerned with the problem of calculating the rotation-vibration energies and the absorption intensities of a triatomic molecule from the effective nuclear potential energy function and the dipole moment surfaces within the Born-Oppenheimer approximation (Born & Oppenheimer 1927). There are at least two astrophysical applications of such calculations:

- Detailed spectra of triatomic molecules can be computed under the conditions prevailing in stellar atmospheres and compared with observed spectra.
- Such spectra can also be used for constructing a wavenumber-dependent absorption coefficient which can be employed for opacity calculations as outlined, for example, by Jørgensen & Jensen (1993).

In the Born-Oppenheimer approximation (Born & Oppenheimer 1927), the rovibronic energies of a molecule are calculated in two steps:

- In the first step, the nuclei are held fixed in space, and the Schrödinger problem for the electrons (which interact with the "clamped" nuclei and with each other through Coulomb forces) is solved for many different nuclear geometries. The energy eigenvalue obtained in this calculation, taken as a function of the nuclear coordinates (which we denote as $\mathbf{R}_n$ here), is the Born-Oppenheimer potential energy function $V(\mathbf{R}_n)$.
- The second step is the solution of the Schrödinger equation for the nuclear motion,

$$\left[\widehat{T}_n + V(\mathbf{R}_n)\right] \psi_n(\mathbf{R}_n) = E_{ne}\psi_n(\mathbf{R}_n) \tag{1}$$

  where $\widehat{T}_n$ is the operator representing the kinetic energy of the nuclei, $\psi_n(\mathbf{R}_n)$ is a nuclear wavefunction, and E_{ne} is the corresponding energy eigenvalue.

⋆ Guest professor 1992-93. On leave from Physikalisch-Chemisches Institut, Justus-Liebig-Universität Giessen, Heinrich-Buff-Ring 58, D-35392 Giessen, Germany.

One may think of the motion of the nuclei as a superposition of translation (uniform motion of all nuclei along straight lines through space), vibration (motion of the nuclei relative to each other), and rotation. It can be shown that the nuclear kinetic energy operator $\widehat{T}_n$ can be separated into the sum of two commuting operators

$$\widehat{T}_n = \widehat{T}_{\text{Trans}} + \widehat{T}_{\text{Rot-Vib}}, \tag{2}$$

where $\widehat{T}_{\text{Trans}}$ describes the kinetic energy resulting from translation and $\widehat{T}_{\text{Rot-Vib}}$ that resulting from rotation and vibration. In theoretical studies aimed at describing "spectroscopic" phenomena (i.e., the interaction between molecules and radiation) for isolated molecules in a field-free space we can ignore the effects of translation since radiation will not induce transitions between translational states. We are then left with the Schrödinger equation for rotation and vibration

$$\left[\widehat{T}_{\text{Rot-Vib}} + V(\mathbf{R}_{\text{Vib}})\right] \psi_{\text{Rot-Vib}}(\mathbf{R}_{\text{Rot}}, \mathbf{R}_{\text{Vib}}) = E_{\text{Rot-Vib}}\psi_{\text{Rot-Vib}}(\mathbf{R}_{\text{Rot}}, \mathbf{R}_{\text{Vib}}). \tag{3}$$

In Eq. (3), we have indicated that the rotation-vibration wavefunction $\psi_{\text{Rot-Vib}}$ depends on rotational coordinates $\mathbf{R}_{\text{Rot}}$ and vibrational coordinates $\mathbf{R}_{\text{Vib}}$, whereas the potential energy function V depends on the vibrational coordinates only.

We discuss here the solution of Eq. (3). In theoretical spectroscopy, the traditional approach to obtaining the rotation-vibration energies $E_{\text{Rot-Vib}}$ and the corresponding eigenfunctions $\psi_{\text{Rot-Vib}}$ involves perturbation theory: The molecular vibration is modelled by a set of independent harmonic oscillators, and the molecular rotation is modelled as that of a rigid molecule. The deviation between the real molecule and this idealized model is treated as a perturbation.

In recent years it has become apparent that in many cases, the perturbation treatment of molecular rotation and vibration leads to considerable inaccuracies in the calculated energies, and a number of more accurate methods for solving Eq. (3) have been developed for small (mostly triatomic) molecules. In the present communication we shall describe one such method for triatomic molecules, the MORBID (Morse Oscillator Rigid Bender Internal Dynamics) approach (Jensen 1988a, 1988b, 1992). The computer programs developed on the basis of the MORBID model can be used for three different types of calculations:

- If the nuclear potential energy function $V(\mathbf{R}_n)$ for one isolated electronic state is known, the rotation-vibration energies belonging to this electronic state can be calculated.
- The parameters in an analytical representation for $V(\mathbf{R}_n)$ (see below) can be refined in a least squares fitting to experimentally observed rotation-vibration energy spacings.
- If $V(\mathbf{R}_n)$ and the dipole moment components are known as functions of the nuclear coordinates, absorption spectra (i.e., transition wavenumbers and intensities) can be calculated.

We give examples of such calculations below.

2 The MORBID Approach

We give here a very sketchy outline of the ideas behind the MORBID approach. The reader is referred to Jensen (1988a, 1988b, 1992) for details.

2.1 Coordinates

In order to describe a quantum mechanical system such as a triatomic molecule, we must initially choose a set of coordinates defining its instantaneous configuration. In the MORBID approach, we follow the work of Hougen et al. (1970) in that we base the coordinate definitions on a socalled reference configuration (Fig. 1) which follows the rotation and the bending motion. The reference configuration is chosen as having fixed internuclear distances r_{j2}^e [the equilibrium distance between the "outer" nucleus j =1 or 3 (Fig. 1) and the center nucleus 2], but with a variable bond angle α. The coordinate $\rho = \pi - \alpha$ is chosen to describe the large amplitude motion, and the stretching motions are described as displacements from the reference geometry through the coordinates

$$\Delta r_{j2} = r_{j2} - r_{j2}^e, \tag{4}$$

$j = 1$ or 3, defined as the displacement of the instantaneous internuclear distance r_{j2} from its equilibrium value r_{j2}^e. We attach a molecule fixed axis system xyz to the reference configuration (Fig. 1) and describe the rotation of the molecule through usual Euler angles θ, ϕ, and χ (see, for example, Papoušek & Aliev (1982)) which give the instantaneous orientation of the xyz axis system relative to a laboratory fixed axis system XYZ.

2.2 The potential energy operator

We choose the potential energy function for a triatomic molecule to be given by the analytical function

$$V(\Delta r_{12}, \Delta r_{32}, \bar{\rho}) = V_0(\bar{\rho}) + \sum_j F_j(\bar{\rho}) y_j + \sum_{j \le k} F_{jk}(\bar{\rho}) y_j y_k + \sum_{j \le k \le m} F_{jkm}(\bar{\rho}) y_j y_k y_m + \sum_{j \le k \le m \le n} F_{jkmn}(\bar{\rho}) y_j y_k y_m y_n, \tag{5}$$

where all of the indices j, k, m, and n assume the values 1 or 3. In Eq. (5), $\bar{\rho}$ is the supplement of the instantaneous value of the bond angle (which is almost, but not quite, equal to the vibrational coordinate ρ discussed above), and

$$y_j = 1 - \exp(-a_j \Delta r_{j2}), \tag{6}$$

where the Δr_{j2} are defined in Eq. (4) and the a_j are molecular parameters. The $F_{jkm...}$ expansion coefficients of Eq. (5) are functions of $\bar{\rho}$. The function $V_0(\bar{\rho})$ is the pure bending potential energy function, i.e. the potential energy for the molecule bending with its bond lengths fixed at their equilibrium values.

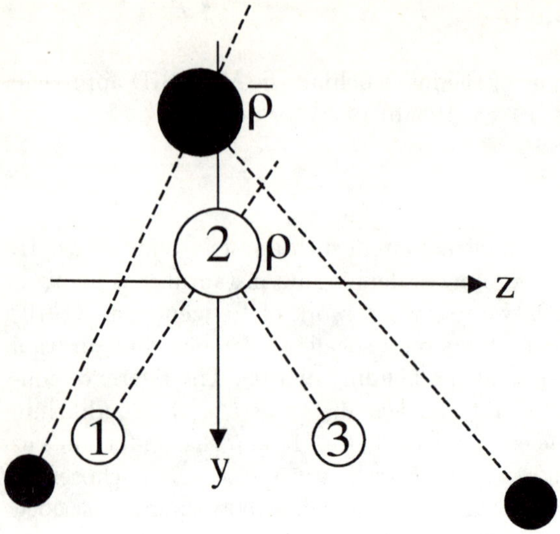

Fig. 1. The numbering of the nuclei and the molecule fixed coordinate system used in the MORBID approach. The molecule fixed x axis is perpendicular to the plane of the molecule. The open circles are the positions of the nuclei in the reference configuration (which has the bond angle supplement ρ) and the filled circles are the instantaneous positions of the nuclei (with bond angle supplement $\bar{\rho}$).

The $\bar{\rho}$-dependent functions entering into Eq. (5) are defined as general cosine expansions:

$$F_j(\bar{\rho}) = \sum_{i=1}^{4} f_j^{(i)}(\cos\rho_e - \cos\bar{\rho})^i,$$

$$F_{jk\ldots}(\bar{\rho}) = f_{jk\ldots}^{(0)} + \sum_{i=1}^{N} f_{jk\ldots}^{(i)}(\cos\rho_e - \cos\bar{\rho})^i \qquad (7)$$

where ρ_e is the equilibrium value of $\bar{\rho}$ and the $f_{jk\ldots}^{(i)}$ are expansion coefficients. The function $F_{jk}(\bar{\rho})$ has $N = 3$, $F_{jkl}(\bar{\rho})$ has $N = 2$, and $F_{jklm}(\bar{\rho})$ has $N = 1$. Symmetry relations exist between the potential energy parameters for the symmetrical ABA molecule to ensure that V is totally symmetric under the interchange of Δr_{12} and Δr_{32}. For $V_0(\bar{\rho})$, we use two different parameterizations,

$$V_0(\bar{\rho}) = \sum_{i=2}^{8} f_0^{(i)}(\cos\rho_e - \cos\bar{\rho})^i, \qquad (8)$$

or

$$V_0(\bar{\rho}) = \sum_{i=1}^{8} f_0^{(i)}(1 - \cos\bar{\rho})^i. \qquad (9)$$

The parameterization given by Eq. (8) has ρ_e as an independent parameter and is appropriate for *bent* molecules, i.e. molecules with $\rho_e \neq 0$. Equation (9) has $f_0^{(1)}$ as an independent parameter, whereas ρ_e is determined as a function of the $f_0^{(i)}$ quantities. This parameterization is appropriate for both linear and bent molecules.

For a potential energy function with a single minimum, the analytical expression given by Eqs. (5-9) has a physically reasonable asymptotic behaviour at all coordinate boundaries: at $\bar{\rho} = 0$ and $\bar{\rho} = \pi$ it has zero slope for all values of r_{12} and r_{32}, at large bond length values it approaches a constant for any value of $\bar{\rho}$, and at short bond lengths it approaches a very large (although not infinite) value.

2.3 The kinetic energy operator

Using the rules of quantum mechanics, we can derive a kinetic energy operator $\widehat{T}_{\mathrm{MORBID}}$ expressed in terms of the coordinates given in Sect. 2.1. This operator is obtained as

$$\widehat{T}_{\mathrm{MORBID}} = \frac{1}{2} \sum_{\delta,\gamma=x,y,z,\rho} (\widehat{J}_\delta - \widehat{p}_\delta)\mu_{\delta\gamma}(\Delta r_{12}, \Delta r_{32}, \rho)(\widehat{J}_\gamma - \widehat{p}_\gamma)$$
$$+ \frac{1}{2} \sum_{j,k=1,3} \widehat{P}_j G_{jk}^{(r)}(\Delta r_{12}, \Delta r_{32}, \rho)\widehat{P}_k + U_1(\Delta r_{12}, \Delta r_{32}, \rho) + U_0(\rho), \qquad (10)$$

where $\widehat{J}_\delta$, $\delta = x, y, z$, are the components of the total angular momentum along the molecule fixed axes, $\widehat{J}_\rho = -i\hbar\partial/\partial\rho$, $\mu_{\delta\gamma}$ ($\delta, \gamma = x, y, z, \rho$) are the elements of the 4 × 4 "inverse inertial" matrix $\boldsymbol{\mu}$ defined by Eqs. (45-50) of Jensen (1988a), $\widehat{P}_j = -i\hbar\partial/\partial\Delta r_{j2}$ is the momentum conjugate to Δr_{j2}, the vibrational angular momentum $\widehat{p}_\delta$ is given by

$$\widehat{p}_\delta = \frac{1}{2} \sum_{j,k=1,3} \left\{ \widehat{P}_j G_{jk}^{(r)} \chi_k^{(\delta)} + \chi_k^{(\delta)} G_{jk}^{(r)} \widehat{P}_j \right\}, \qquad (11)$$

the $\chi_k^{(\delta)}$ being Coriolis coupling coefficients, and the $G_{jk}^{(r)}$ matrix elements are defined by Eq. (40) of Jensen (1988a). The U_1 and U_0 functions are pseudopotential terms defined in Eqs. (42-44) of Jensen (1988a). The dependence of U_0 on the stretching coordinates Δr_{12} and Δr_{32} has been neglected; this possibly explains why sometimes splittings due to Fermi resonances are calculated to be too small by the MORBID program [see Bunker et al. (1990) and Barclay et al. (1993)].

The functions $\mu_{\delta\gamma}$, $G_{jk}^{(r)}$, and $\chi_k^{(\delta)}$ entering into Eq. (10) all depend on the coordinates Δr_{12}, Δr_{32}, and ρ, and we expand them as polynomials in y_1 and y_3 with ρ-dependent coefficients, e.g.

$$\boldsymbol{\mu} = \boldsymbol{\mu}_0(\rho) + \sum_j \boldsymbol{\mu}_j(\rho) y_j + \sum_{j \le k} \boldsymbol{\mu}_{jk}(\rho) y_j y_k$$
$$+ \sum_{j \le k \le m} \boldsymbol{\mu}_{jkm}(\rho) y_j y_k y_m + \sum_{j \le k \le m \le n} \boldsymbol{\mu}_{jkmn}(\rho) y_j y_k y_m y_n \qquad (12)$$

where $\boldsymbol{\mu}_{jk\ldots}(\rho)$ is a 4×4 matrix with elements $\mu_{\delta\gamma}^{(jk\ldots)}(\rho)$, δ, $\gamma = x, y, z, \rho$. The expansion of the $\boldsymbol{\mu}$-tensor elements are truncated after the quartic terms, the expansions of the $\chi_k^{(\delta)}$ functions are truncated after the cubic terms, and the expansions of the $G_{jk}^{(r)}$ matrix elements are truncated after the quadratic terms. In this way, we obtain $\widehat{T}_{\mathrm{MORBID}}$ as an expansion to fourth order in the vibrational operators y_j and $\widehat{P}_k$.

2.4 The calculation of the rotation-vibration energies

The MORBID Hamiltonian is now given as

$$\widehat{H}_{\mathrm{MORBID}} = \widehat{T}_{\mathrm{MORBID}} + V(\Delta r_{12}, \Delta r_{32}, \rho) \tag{13}$$

where $\widehat{T}_{\mathrm{MORBID}}$ is given by Eq. (10) (with the coordinate-dependent functions expressed as the series expansions described above) and the potential function V from Eq. (5) has been transformed so that instead of depending on Δr_{12}, Δr_{32}, and the instantaneous value of the bond angle supplement $\bar{\rho}$ [see Sect. 2.2], it now depends on Δr_{12}, Δr_{32}, and the coordinate ρ. The transformation is carried out using basic geometrical identities as shown in Eqs. (34-36) of Jensen (1988a).

We obtain the solutions of the eigenvalue problem

$$\begin{aligned}&\widehat{H}_{\mathrm{MORBID}}\ \psi_{\mathrm{MORBID}}(\Delta r_{12}, \Delta r_{32}, \rho, \theta, \phi, \chi)\\ &\quad = E_{\mathrm{MORBID}}\ \psi_{\mathrm{MORBID}}(\Delta r_{12}, \Delta r_{32}, \rho, \theta, \phi, \chi)\end{aligned} \tag{14}$$

by "taking the molecule apart". That is, we consider three simplified molecules, each of which only carries out part of the molecular motion.

Stretching. We initially consider a molecule which carries out stretching of its bond lengths r_{j2} with the bond angle fixed at the equilibrium value α_e and the rotation frozen. This molecule has the Hamiltonian

$$\begin{aligned}\widehat{H}_{\mathrm{Stretch}} &= \frac{1}{2}\left[\frac{1}{m_1}+\frac{1}{m_2}\right]\widehat{P}_1^2 + \frac{1}{2}\left[\frac{1}{m_3}+\frac{1}{m_2}\right]\widehat{P}_3^2 - \frac{1}{m_2}\cos\rho_e\,\widehat{P}_1\widehat{P}_3\\ &+ \sum_{j\le k} f_{jk}^{(0)} y_j y_k + \sum_{j\le k\le m} f_{jkm}^{(0)} y_j y_k y_m + \sum_{j\le k\le m\le n} f_{jkmn}^{(0)} y_j y_k y_m y_n\end{aligned} \tag{15}$$

[see Eq. (58) of Jensen (1988a)], where m_j is the mass of nucleus j. We obtain the eigenfunctions $|N_{\mathrm{Vib}}\Gamma_{\mathrm{Sym}}\rangle$ of this Hamiltonian [and the corresponding eigenvalues, which are our zero order stretching energies] by setting up the matrix of $\widehat{H}_{\mathrm{Stretch}}$ in a basis of symmetrized Morse oscillator functions as described in Sect. V of Jensen (1988a). The eigenfunctions are labeled by the index N_{vib} which characterizes the zeroth order stretching state, and by Γ_{Sym} which is the irreducible representation spanned by the function $|N_{\mathrm{vib}}\Gamma_{\mathrm{Sym}}\rangle$ in the appropriate molecular symmetry group.

Bending and K-Type Rotation. We then consider a molecule which carries out bending motion and rotation around the molecule-fixed z axis (Fig. 1) with its bond lengths fixed at the equilibrium values r_{12}^e and r_{32}^e. The molecule fixed z-axis becomes the molecular axis when the molecule is in a linear configuration. It is necessary to consider these two types of motion together since, for a triatomic molecule with a linear equilibrium configuration, the rotation around the z-axis is only made possible through the bending motion. This simplified molecule has the Hamiltonian

$$\begin{aligned}\hat{H}_{\text{Bend}} &= \frac{1}{2}\hat{J}_\rho \mu_{\rho\rho}^{(0)}(\rho)\hat{J}_\rho + \frac{1}{2}\mu_{zz}^{(0)}(\rho)\hat{J}_z^2 \\ &+ \frac{1}{2}\hbar^2\mu_{\rho\rho}^{(0)}(\rho)\; g(\rho) + \frac{1}{2}(\mu^{(0)})^{-1/4}[\hat{J}_\rho, \mu_{\rho\rho}^{(0)}][\hat{J}_\rho, (\mu^{(0)})^{1/4}] + V_0(\rho), \qquad (16)\end{aligned}$$

[see Eq. (63) of Jensen (1988a)] where $\mu^{(0)}$ is the determinant of the matrix $\boldsymbol{\mu}_0$ from Eq. (12), and the function $g(\rho)$ is given by Eq. (43) of Jensen (1988a). We obtain the eigenfunctions $|v_2, K\rangle$ for this Hamiltonian through Numerov-Cooley numerical integration (Cooley 1961) as described in Sect. V of Jensen (1988a). The index v_2 is the bending quantum number for a bent triatomic molecule, and K is the rotational quantum number describing the rotation around the z-axis.

End-over-end Rotation. Finally, we consider a drastically simplified molecule which carries out rotation around the molecule fixed x- and y-axes. We assume this molecule to be a rigid symmetric top, so that its rotational eigenfunctions are the symmetrized symmetric top eigenfunctions $|J, K, M, \tau\rangle$ defined in Eq. (7.1) of Jensen (1983).

The Complete Molecule. We use the products $|N_{\text{vib}}\Gamma_{\text{Sym}}\rangle\;|v_2, K\rangle\;|J, K, M, \tau\rangle$ as basis functions for constructing a matrix representation of the rovibrational Hamiltonian $\hat{H}_{\text{MORBID}}$ for the "complete" molecule. We may label the eigenfunctions ψ_{MORBID} of this Hamiltonian through an index s together with the good quantum numbers J, M, and Γ_{rv}, where $M\hbar$ is the projection of the total angular momentum on the space-fixed Z-axis and Γ_{rv} is the symmetry of the wavefunction, and we approximate such a function as

$$\begin{aligned}\psi_{\text{MORBID}} &= |s; J, M, \Gamma_{rv}\rangle \\ &= \sum_{N_{\text{vib}},\Gamma_{\text{Sym}},v_2,K} c_{N_{\text{vib}},\Gamma_{\text{Sym}},v_2,K}^{(s;J,M,\Gamma_{rv})}|N_{\text{vib}}\Gamma_{\text{Sym}}\rangle|v_2, K\rangle|J, K, M, \tau\rangle. \qquad (17)\end{aligned}$$

By diagonalizing the matrix representation of the Hamiltonian, we can determine its eigenvalues (which, in the approximation adopted here, are the rotation-vibration energies) and the expansion coefficients $c_{N_{\text{vib}},\Gamma_{\text{Sym}},v_2,K}^{(s;J,M,\Gamma_{rv})}$ defining the rotation-vibration wavefunction in Eq. (17).

2.5 Intensity calculations

With the rotation-vibration wavefunctions given by Eq. (17) we can calculate the intensities of absorption transitions between rotation-vibration states. If we consider molecules in thermal equilibrium at an absolute temperature T, the integrated absorption coefficient for a transition from an initial state $|s_i; J_i, M_i, \Gamma_{rv(i)}\rangle$ to a final state $|s_f; J_f, M_f, \Gamma_{rv(f)}\rangle$ is given by (Smith et al. 1985)

$$I(f \leftarrow i) = \frac{8\pi^3 N_A g_{ns} \omega_{if} \exp(-E_i/kT)[1 - \exp(-hc\omega_{if}/kT)]}{3hcQ} S(f \leftarrow i) \quad (18)$$

where N_A is Avogadro's number, g_{ns} is the nuclear spin statistical weight, ω_{if} is the wavenumber of the transition (in cm^{-1}), E_i is the energy of the initial state, k is the Boltzmann constant, h is Planck's constant, c is the vacuum velocity of light and the partition function Q is given by

$$Q = \sum_w g_w \exp(-E_w/kT). \quad (19)$$

In Eq. (19), E_w is the energy and g_w the total degeneracy of the state w and the summation extends over all such states of the molecule. The line strength $S(f \leftarrow i)$ of an electric dipole transition is defined as

$$S(f \leftarrow i) = \sum_{A=X,Y,Z} \sum_{M_i, M_f} \left|\langle s_f; J_f, M_f, \Gamma_{rv(f)} | \mu_A | s_i; J_i, M_i, \Gamma_{rv(i)}\rangle\right|^2 \quad (20)$$

where μ_A is the component of the electronically averaged molecular dipole moment operator along the A axis ($A = X$, Y, or Z) of the space fixed coordinate system and M_i and M_f are the rotational M quantum numbers (quantizing the Z component of the total angular momentum) of the initial and final states, respectively.

As described by Jensen (1988c), the MORBID program can calculate the linestrengths given by Eq. (20) on the basis of the wavefunctions from Eq. (17) and the electronically averaged dipole moment surfaces of the molecule. The dipole moment functions are chosen as follows: A pq axis system is defined for the triatomic molecule. This axis system has its origin at the nuclear center of mass and the p and q axes are in the plane defined by the three nuclei. For an unsymmetrical ABC molecule, the p axis is parallel to the bond between nucleus 2 and nucleus 3 and points so that the p coordinate of nucleus 1 is negative. The q axis is perpendicular to the p axis and points so that the q coordinate of nucleus 1 is positive. For a symmetical ABA molecule, the q axis is the bisector of the angle $\angle$(ABA) and points so that the two A nuclei have positive q-coordinates. The p axis is perpendicular to the q axis and points so that the p coordinate of nucleus 3 is positive. The dipole moment component along the p axis, $\bar{\mu}^{(p)}$, is expanded as:

$$\bar{\mu}^{(p)}(\Delta r_{12}, \Delta r_{32}, \bar{\rho}) = \mu_0^{(p)}(\bar{\rho}) + \sum_j \mu_j^{(p)}(\bar{\rho}) \Delta r_{j2}$$

$$+\sum_{j\le k}\mu_{jk}^{(p)}(\bar\rho)\Delta r_{j2}\Delta r_{k2}+\sum_{j\le k\le m}\mu_{jkm}^{(p)}(\bar\rho)\Delta r_{j2}\Delta r_{k2}\Delta r_{m2}$$
$$+\sum_{j\le k\le m\le n}\mu_{jkmn}^{(p)}(\bar\rho)\Delta r_{j2}\Delta r_{k2}\Delta r_{m2}\Delta r_{n2}, \tag{21}$$

where all of the indices j, k, m, and n assume the values 1 or 3 and

$$\mu_{jk\ldots}^{(p)}(\bar\rho)=\sum_{i=0}^{N}p_{jk\ldots}^{(i)}(\cos\rho_e-\cos\bar\rho)^i. \tag{22}$$

The function $\mu_0^{(p)}(\bar\rho)$ has $N=8$, $\mu_j^{(p)}(\bar\rho)$ has $N=4$, $\mu_{jk}^{(p)}(\bar\rho)$ has $N=3$, $\mu_{jkm}^{(p)}(\bar\rho)$ has $N=2$, and $\mu_{jkmn}^{(p)}(\bar\rho)$ has $N=1$.

The dipole moment component along the q axis, $\bar\mu^{(q)}$, is represented by

$$\bar\mu^{(q)}(\Delta r_{12},\Delta r_{32},\bar\rho)=\sin\bar\rho\,\Big[\mu_0^{(q)}(\bar\rho)+\sum_{j}\mu_j^{(q)}(\bar\rho)\Delta r_{j2}$$
$$+\sum_{j\le k}\mu_{jk}^{(q)}(\bar\rho)\Delta r_{j2}\Delta r_{k2}+\sum_{j\le k\le m}\mu_{jkm}^{(q)}(\bar\rho)\Delta r_{j2}\Delta r_{k2}\Delta r_{m2}$$
$$+\sum_{j\le k\le m\le n}\mu_{jkmn}^{(q)}(\bar\rho)\Delta r_{j2}\Delta r_{k2}\Delta r_{m2}\Delta r_{n2}\Big]. \tag{23}$$

Again, all of the indices j, k, m, and n assume the values 1 or 3. The equations for the $\mu_{jkm\ldots}^{(q)}(\bar\rho)$ functions are obtained when p is replaced by q in Eq. (22).

The values of the parameters $p_{jk\ldots}^{(i)}$ and $q_{jk\ldots}^{(i)}$ in Eqs. (21-23) are obtained by fitting the analytical expressions through ab initio dipole moment values. As explained by Jensen (1988c)

- The dipole moment components along the molecule fixed y and z axes, μ_y and μ_z, can be obtained by rotating the components $\bar\mu^{(p)}$ and $\bar\mu^{(q)}$.
- The functions μ_y and μ_z can be transformed into polynomials in the quantities $y_k=1-\exp(-a_k\Delta r_{k2})$ [with expansion coefficients depending on the bending coordinate ρ].
- The matrix elements of these expansions between vibrational basis functions can be straightforwardly calculated using Morse oscillator matrix elements from Špirko et al. (1985).
- On the basis of the vibrational matrix elements of μ_y and μ_z and the expansion coefficients $c_{N_{\mathrm{vib}},\Gamma_{\mathrm{Sym}},v_2,K}^{(s;J,M,\Gamma_{rv})}$ from Eq. (17), the line strength $S(f\leftarrow i)$ for an individual rotation-vibration transition can be obtained from Eq. (55) of Jensen (1988c).

When the linestrengths $S(f\leftarrow i)$ have been obtained, the integrated absorption coeffients $I(f\leftarrow i)$ can be calculated from Eq. (18) if the absorbing molecules are in thermal equilibrium. For non-equilibrium situations the calculation of $I(f\leftarrow i)$ requires detailed knowledge about the population distribution of the molecules under study. In order to allow direct comparison with experimental results it might be necessary to compute peak absorption coefficients for the rotation-vibration transitions. Such calculations require further knowledge about the line shape function (i.e., about the broadening effects present in the experiment). An example of a calculation of this type is given by Jensen et al. (1992).

3 Applications

We give here a brief survey of results from MORBID calculations.

3.1 H_2O

Calculations from ab initio data. Two calculations of the rotation-vibration energies of the water molecule have been carried out (Jensen 1988b,1988d) on the basis of ab initio potential energy surfaces from Kraemer et al. (1982) and Bartlett et al. (1987), respectively.

Fitted potential energy surfaces. The parameters of Eqs. (5-9) have been optimized (Jensen 1989a) through a least squares fitting to an extensive set of rotation-vibration energy spacings for $H_2{}^{16}O$ and its isotopic species. These input data consisted of 550 experimental energy separations involving rotation-vibration energy levels with $J \leq 2$ in 103 vibrational states [with energies up to 19 000 cm^{-1} above the (0,0,0) state] belonging to the six isotopic molecules $H_2{}^{16}O$, $D_2{}^{16}O$, $HD^{16}O$, $H_2{}^{17}O$, $H_2{}^{18}O$, and $T_2{}^{16}O$. In the fitting 19 parameters were varied, and the standard deviation was 0.63 cm^{-1}. In a careful comparison between four potential energy surfaces for water available in the literature (Jensen 1989a; Carter & Handy 1987; Halonen & Carrington 1988; Kauppi & Halonen 1990) Fernley et al. (1991) found the MORBID potential (Jensen 1989a) to be the most accurate surface currently available for the water molecule. This potential has also been used for calculations of water energies by Wattson & Rothman (1992) and by Choi & Light (1992). On the basis of the MORBID potential energy surface and dipole moment surfaces calculated ab initio, Jørgensen and Jensen (1993) have recently carried out an extensive calculation of the vibrational transition moments for water, which they intend to use as input for an accurate calculation of the absorption coefficient as a function of wavenumber. A new refined potential for H_2O based on experimental data involving $J \leq 10$ is presently being fitted with the MORBID program (Jensen & Tyuterev 1993).

3.2 CH_2

Calculations from ab initio data. Ab initio calculations of the rotation-vibration energies in the vibrational ground state $\widetilde{X}^3B_1$ of the methylene radical CH_2 have been carried out by Jensen (1988a) (who used a potential energy surface calculated by McLean et al. (1987)) and by Comeau et al. (1989), who also did calculations for the first excited electronic state ($\widetilde{a}^1A_1$).

Fitted potential energy surfaces. Jensen & Bunker (1988) have fitted the potential energy surfaces for the $\tilde{X}^3B_1$ and $\tilde{a}^1A_1$ electronic states of CH_2. For the $\tilde{X}^3B_1$ state, they used all extant rotation-vibration data as input for the fitting, whereas only the $J = 0$ data for the $\tilde{a}^1A_1$ state were fitted. On the basis of the fitted potential energy surfaces, the stretching fundamental energies of $\tilde{X}^3B_1$ $^{12}CH_2$ were predicted at $E(\nu_1) = 2992$ cm^{-1} and $E(\nu_3) = 3213$ cm^{-1}. By combining the results of the MORBID fittings with experimental data for the $\tilde{X}$ and $\tilde{a}$ states Jensen & Bunker (1988) obtained the energy splitting between the vibrational ground states of these two electronic states as $T_0\left(\tilde{a}^1A_1\right) = 3147$ cm^{-1} (8.998 kcal/mol) and the splitting between the equilibrium energies of the two states as $T_e\left(\tilde{a}^1A_1\right) = 3223$ cm^{-1} (9.215 kcal/mol) for $^{12}CH_2$.

3.3 C_3

Calculations from ab initio data. The rotation-vibration energy spectrum of C_3 in the electronic ground state have been calculated by Jensen & Kraemer (1988a) on the basis of an ab initio potential energy surface by Kraemer et al. (1984). Recently, more extensive ab initio calculations have been carried out by Jensen et al. (1992). These authors also considered the intensities of the allowed ν_2 and ν_3 fundamental bands of $^{12}C_3$ and showed that their theoretical transition moments were in keeping with recent experimental observations (Schmuttenmaer et al. 1990).

Fitted potential energy surfaces. Jensen (1989b) fitted a very limited set of rovibrational data for $^{12}C_3$ and obtained a potential energy function with a small barrier to linearity (16.5 cm^{-1}). This agreed well with the ab initio results by Kraemer et al. (1984) and apparently supported suggestions by Matsumura et al. (1988) who argued on experimental grounds that C_3 might be slightly bent. A recent fit to a more extended data set by Jensen et al. (1992), however, yields a potential energy surface with a linear equilibrium geometry, in agreement with the ab initio calculation by these authors.

3.4 H_2Se

Jensen & Kozin (1993) have fitted a potential energy surface for the electronic ground state of hydrogen selenide H_2Se. On the basis of this potential energy function, the rovibrational energy structure in the vibrational ground state (Jensen & Kozin 1993) and in the fundamental vibrational states (Kozin & Jensen 1993) has been calculated. In all the vibrational states studied, it was found that four-fold energy clusters (i.e., four-member groups of nearly degenerate energies) form at high rotational excitation. This effect has been experimentally verified for the vibrational ground state of $H_2{}^{80}Se$ (Kozin et al. 1992a, 1992b, 1993).

3.5 Other molecules

Other molecules for which MORBID calculations have been carried out include NH_2^+ (Jensen et al. 1987; Barclay et al. 1993), HOC^+ (Jensen & Kraemer 1988a), CCN^+ and CNC^+ (Jensen & Kraemer 1988b), LiOH (Bunker et al. 1989), NH_2 (Jensen et al. 1990), H_2F^+ (Bunker et al. 1990), HNC^+ (Kraemer et al. 1992), Si_2C (Barone et al. 1992), HO_2 (Bunker et al. 1992), and HNSi (Chong et al. 1993).

Acknowledgments. Since 1989, this work has been supported by the Deutsche Forschungsgemeinschaft. Additional funding has been provided by the the Fonds der Chemischen Industrie, the Dr. Otto Röhm Gedächtnisstiftung, and the Fritz Thyssen-Stiftung. Computer time has been made available by the Computer Center of the Justus Liebig-Universität Giessen, the Computing Center of the Technical University Darmstadt, and the Deutsche Versuchsanstalt für Luft- und Raumfahrt, Oberpfaffenhofen near Munich. B.P. Winnewisser is thanked for critically reading the manuscript and suggesting improvements.

References

Barclay V.J., Hamilton I.P., Jensen P., 1993, *J. Chem. Phys.*, submitted for publication
Barone V., Jensen P., Minichino C., 1992, *J. Mol. Spectrosc.*, **154**, 252
Bartlett R.J., Cole S.J., Purvis G.D., Ermler W.C., Hsieh H.C., Shavitt I., 1987, *J. Chem. Phys.*, **87**, 6579
Born M., Oppenheimer J.R., 1927, *Ann. der Physik*, **84**, 457
Bunker P.R., Hamilton I.P., Jensen P., 1992, *J. Mol. Spectrosc.*, **155**, 44
Bunker P.R., Jensen P., Karpfen A., Lischka H., 1989, *J. Mol. Spectrosc.*, **135**, 89
Bunker P.R., Jensen P., Wright J.S., Hamilton I.P., 1990, *J. Mol. Spectrosc.*, **144**, 310
Carter S., Handy N.C., 1987, *J. Chem. Phys.*, **87**, 4294
Choi S.E., Light J.C., 1992, *J. Chem. Phys.*, **97**, 7031
Chong D.P., Papoušek D., Chen Y.-T., Jensen P., 1993, *J. Chem. Phys.*, **98**, 1352
Comeau D.C., Shavitt I., Jensen P., Bunker P.R., 1989, *J. Chem. Phys.*, **90**, 6491
Cooley J.W., 1961, *Math. Comp.*, **15**, 363
Fernley J.A., Miller S., Tennyson J., 1991, *J. Mol. Spectrosc.*, **150**, 597
Halonen L., Carrington Jr., T., 1988, *J. Chem. Phys.*, **88**, 4171
Hougen J.T., Bunker P.R., Johns J.W.C., 1970, *J. Mol. Spectrosc.*, **34**, 136
Jensen P., 1983, *Comp. Phys. Reports*, **1**, 1
Jensen P., 1988a, *J. Mol. Spectrosc.*, **128**, 478
Jensen P., 1988b, *J. Chem. Soc. Faraday Trans. 2*, **84**, 1315
Jensen P., 1988c, *J. Mol. Spectrosc.*, **132**, 429
Jensen P., 1988d, *J. Mol. Structure*, **190**, 149
Jensen P., 1989a, *J. Mol. Spectrosc.*, **133**, 438
Jensen P., 1989b, *Collection of Czechoslovak Chemical Communications*, **54**, 1209
Jensen P., 1992, in *Methods in Computational Molecular Physics*, S. Wilson and G. H. F. Diercksen, eds., Plenum Press, New York.
Jensen P., Buenker R.J., Hirsch G., Rai S.N., 1990, *Mol. Phys.*, **70**, 443
Jensen P., Bunker P.R., 1988, *J. Chem. Phys.*, **89**, 1327
Jensen P., Bunker P.R., McLean A.D., 1987, *Chem. Phys. Lett.*, **141**, 53

Jensen P., Kozin I.N., 1993, *J. Mol. Spectrosc.*, **160**, 39
Jensen P., Kraemer W.P., 1988a, *J. Mol. Spectrosc.*, **129**, 172
Jensen P., Kraemer W.P., 1988b, *J. Mol. Spectrosc.*, **129**, 216
Jensen P., Rohlfing C.M., Almlöf J., 1992, *J. Chem. Phys.*, **97**, 3399
Jensen P., Tyuterev V.G., 1993, in preparation.
Jørgensen U.G., Jensen P., 1993, *J. Mol. Spectrosc.*, **161**, 219
Kauppi E., Halonen L., 1990, *J. Phys. Chem.*, **94**, 5779
Kozin I.N., Belov S.P., Polyansky O.L., Tretyakov M.Yu., 1992a, *J. Mol. Spectrosc.*, **152**, 13
Kozin I.N., Jensen P., 1993, *J. Mol. Spectrosc.*, **161**, 186
Kozin I.N., Klee S., Jensen P., Polyansky O.L, Pavlichenkov I.M., 1993, *J. Mol. Spectrosc.*, **158**, 409
Kozin I.N., Polyansky O.L., Pripolzin S.I, Vaks V.L., 1992b, *J. Mol. Spectrosc.*, **156**, 504
Kraemer W.P., Bunker P.R., Yoshimine M., 1984, *J. Mol. Spectrosc.*, **107**, 191
Kraemer W.P., Jensen P., Roos B.O., Bunker P.R., 1992, *J. Mol. Spectrosc.*, **153**, 240
Kraemer W.P., Roos B.O., Siegbahn P.E.M., 1982, *Chem. Phys.*, **69**, 305
Matsumura K., Kanamori H., Kawaguchi K., Hirota E., 1988, *J. Chem. Phys.*, **89**, 3491
McLean A.D., Bunker P.R., Escribano R., Jensen P., 1987, *J. Chem. Phys.*, **87**, 2166
Papoušek D., Aliev M.R., 1982, *Molecular Vibrational-Rotatonal Spectra*, Elsevier, Amsterdam.
Schmuttenmaer C.A., Cohen R.C., Pugliano N., Heath J.R., Cooksy A.L., Busarow K.L., Saykally R.J., 1990, *Science*, **249**, 897
Smith M.A.H., Rinsland C.P., Fridovich B., Narahari Rao K., 1985, in *Molecular Spectroscopy: Modern Research*, Vol. III, K. Narahari Rao, ed., Academic Press, Orlando.
Špirko V., Jensen P., Bunker P.R., Čejchan A., 1985, *J. Mol. Spectrosc.*, **112**, 183
Wattson R.B., Rothman L.S., 1992, *J. Quant. Spectrosc. Radiat. Transfer*, **48**, 763

Experimental Measurement of Weak Band Intensities

Kevin K. Lehmann and Daniele Romanini

Department of Chemistry, Princeton University, Princeton NJ 08544, USA

1 Introduction

In modeling the opacity of astronomical objects, such as stellar atmospheres, the most important transitions are those with absorbance[1] $= \alpha L \sim 1$, where α is the absorption coefficient of the atmosphere and L is the pathlength that light travels in leaving the atmosphere. Transitions with much smaller absorbance are not important, while those with much larger absorbance are effectively opaque. The relevant absorption lengths L are typically orders of magnitude larger than can be realized in the laboratory, and thus obtaining the needed laboratory data is a significant experimental challenge. This volume contains several examples of the power of modern *ab initio* methods to predict spectra of simple triatomic molecules, such as H_2O, but it is still necessary to at least calibrate the predictions against laboratory bands of similar oscillator strength. The theoretical treatment of molecules with four or more atoms lags far behind the theoretical treatment of triatomics and present experimental capabilities.

The absorption coefficient α can be decomposed into a product

$$\alpha = N_0 \sigma^2(\lambda) \tag{1}$$

where N_0 is the number density (in atoms or molecules per cm^3) and $\sigma^2(\lambda)$ is the absorption cross section. At pressures low enough to ignore pressure broadening, $\sigma^2(\lambda)$ is a function of only the species responsible for the absorption and the temperature, and is proportional to the molecular transition dipole moment squared. The absorbance of a transition can be small either because of a small number density or a small cross section. Low number density is typically caused by either an unstable or refractory material, or because the molecule in question contains elements of low abundance. Modern chemical physics offers many methods for dealing with these problems, such as electric discharge (see Lovas et al. 1992, for example) or laser oblation sources (Powers et al. 1982) to produce radicals or small clusters. Sensitive indirect spectroscopic detection methods, such as

[1] The absorbance is defined as $\ln T$ where T is the transmission of the atmosphere at a particular wavelength λ. The absorption coefficient is the absorbance per unit length. The extinction is defined as $\log_{10} T$ and thus is proportional to the absorbance.

laser induced fluorescence[2] or resonant enhanced multiphoton ionization[3] allow spectroscopic detection of as little as 10^2 molecules/cm^3/quantum state in favorable cases (Fairbanks et al. 1975). While it is hard to combine these methods with high temperatures, because of thermal radiation and ionization, they are important spectroscopic methods that provide a wide range of data important in astrophysics.

Much harder to overcome is limitation imposed by intrinsically weak transitions. The most important of these are high overtone vibrational transitions. The standard harmonic oscillator approximation predicts that light can only change vibrational modes by ± 1 quanta, and this leads to intense absorption in the IR for all but homonuclear diatomic molecules. Molecules are not, in fact, harmonic oscillators, and real molecules have overtone transitions that change the excitation of vibrational modes by 2, 3, or more quanta, as well as combination bands where two or more modes are simultaneously excited. For hydrogen containing molecules, transitions to levels with between 5 and 8 quanta of the X-H stretching excitation fall in the visible and have been known for many years. See the text by Herzberg (1945) for a compilation of the early work. These transitions, however, have cross sections only 10^{-5}–10^{-8} as large as the allowed IR fundamentals (Burberry et al. 1979). Such weak transitions are important in the calculation of the opacity of both oxygen-rich and carbon-rich stars, since the long column density of species such as H_2O, HCN, and C_2H_2 give even these weak overtone transitions appreciable absorbance. The many weak lines fill in the spectral windows between the strong lines, and in this way effectively block the stellar light flux.

In this chapter, we will attempt to briefly review the methods that have been used to detect weak overtone transitions. These include absorption measurements in traditional long pathlength cells, photoacoustic spectroscopy, and intracavity loss spectroscopy. We will then describe a new method of high sensitivity and flexibility, ring-down cavity spectroscopy. This method was first introduced by O'Keefe & Deacon (1988) and has been further developed in our laboratory and used for the study of extremely weak overtone bands of HCN (Romanini & Lehmann 1993).

2 Traditional Long-path Spectroscopy

The traditional approach to increasing the absorbance pathlength is to use a multipass cell to fold the light path many times through the cell. The most important design is that first reported by White (1942), based upon three mirrors, though many modification on this design have been published (Welsh et

[2] In laser induced fluorescence, molecular absorption of laser radiation is detected by monitoring emission at right angles to the laser beam.

[3] In resonant enhanced multiphoton ionization, one monitors absorption via detection of ions produced by selective photoionization from an excited electronic state.

al. 1951). A key feature of all these designs is the use of low F number[4] optics and a placement of the mirrors at a separation that is on the edge between a stable and unstable optical cavity.[5] This results in a refocusing to a constant spot size on each round trip through the cell, and a running line of spots on one mirror. The pathlength is adjusted by changing the number of spots that fall on this mirror. Large mirrors are needed to keep the light throughput high when such a cell is used with incoherent sources, but also to give ample space to resolve the spots used for coupling the light into and out of the cell. The maximum number of traversals of the cell is limited by both the reflectivity of the mirrors and by the need to physically separate the spots. Practical limits have been on the order of a few hundred passes of the cell, which translates to a maximum pathlength on the order of 0.5 km for cells of a few meters physical length (Smith et al. 1986). Special, very large cells have been constructed that allow pathlengths of several km to be reached (Giver 1978), but these are quite expensive and difficult to control in temperature. Because they are not stable optical cavities, density gradients and turbulence caused by convective flow can easily destroy the optical beam quality and lead to substantial noise on the optical transmission. This is particularly a problem when the cells are used with Fourier Transform instruments, but also limits the sensitivity when used with coherent sources.

In recent years, novel cell designs based upon off-axis coupling into a marginally stable, near concentric optical cavity have been used, and are known as Herriott Cells (Herriott et al. 1964). Light enters and leaves the cell through a single hole in one of the two mirrors. Unless the input light is carefully mode matched, the spot size changes on each round trip, but it returns to its original size when it once again passes through the coupling hole. The stable optical cavity makes the cell less sensitive to both mechanical vibration and convection than a White type cell. One minor drawback is that the pathlength is varied by scanning the physical separation of the two mirrors and this can be somewhat difficult in vacuum. The effective F number of a Herriott cell is much greater than a White cell, and thus they are not as efficient for incoherent sources, but for laser sources they are much superior. Like White cells, the maximum pathlength is limited by both the finite reflectivity of the large mirrors required

---

[4] The F number is the ratio of the focal length to the limiting aperture of an optical system. The light throughput of such a system is proportional to $1/F^2$.

[5] There is not enough room in this chapter to adequately review the theory of optical cavities. See the book by Yariv (1989) for an excellent introduction to this subject. For the uninitiated, it is enough to know that both multipass cells and lasers can be treated as optical cavities which confine light with a certain loss per pass. Cavities are stable or unstable depending on the behaviour of light rays slightly divergent from the optical axis of the cavity. These will remain close to the axis if the cavity in which they propagate is stable, while in an unstable cavity their distance from the axis will diverge exponentially. For a cavity on the edge between stable and unstable, the divergence is linear with the number of passes (think to the case of two plane parallel mirrors). Light is mode matched into a cavity if the light beam has the right size and wavefront to exactly overlap itself on each round trip of the cavity.

(the beam attenuation through the cell is proportional to the mirrors reflectivity raised to the power of the number of cell traversals) and by the requirement that at each round trip the beam forms a distinct, spatially resolved spot on one of the mirrors. For cells of practical size, these limit the total pathlengths to one the order of 1 km or less.

3 Photoacoustic Spectroscopy

Photoacoustic or optoacoustic spectroscopy monitors absorption by its effect on the sample and thus is an indirect detection method such as laser induced fluorescence. Following absorption of laser light, molecules collisionally relax, releasing heat which increases the local temperature. This local temperature rise leads to expansion, followed by contraction as the heat energy is dissipated. Thus chopping of a laser beam at an acoustical frequency (~ 1 kHz), will produce a synchronous sound wave which can be detected by a microphone. Photoacoustic cells are built to minimize sound waves produced by window absorption and may include a resonant inner section which allow the formation of acoustic standing waves at the laser chopping frequency. Microphones are extremely sensitive and, with this careful cell design, allow one can detect on the order of 10^{-8} W/cm of deposited energy (Kritchman et al. 1978). Since one measures the deposited energy, the signal (and thus sensitivity) increases linearly with average laser power. Thus photoacoustic spectroscopy is typically done with the sample cell inside the optical cavity of a 100% amplitude modulated, continuous wave laser to maximize the signal strength. A minimum noise equivalent absorption coefficient of $4 \cdot 10^{-10}\text{cm}^{-1}$ has been obtained in this way (Davidsson et al. 1990). Efficient conversion of the heat to acoustic energy requires pressures of at least 10–100 torr, but this can be made up principally of buffer gas when the sample vapor pressure is too low.

The principal advantage of photoacoustic spectroscopy is its extreme sensitivity, the highest so far obtained for species that do not fluoresce. A number of drawbacks of photoacoustic spectroscopy are: (1) One must have a quiet acoustic environment; thus use of an electric discharge or rapid flow of the sample leads to a substantial increase in noise. (2) The sample is exposed to high average light flux, which can lead to photochemistry in some situations. (3) One needs a high average power continuous wave laser source, which is higher in cost and has less wavelength flexibility than pulsed laser sources (which are easily frequency doubled in nonlinear crystals and shifted by stimulated Raman scattering into the UV and near IR respectively (Yariv 1989)). (4) The indirect nature of the detection results makes determination of absolute absorption strengths difficult. The only practical way to calibrate the strength of the acoustic signal is to use a mixture of a gas which has some transition whose cross section is already known along with the gas of interest. Even with such calibration, uncertainties on the order of 20% have remained (Smith et al. 1987).

4 Intracavity loss absorption spectroscopy (ICLAS)

When a laser with a large gain bandwidth[6] (such as a dye laser[7]) is switched on, light intensity builds up from spontaneous emission. At first, the spectrum matches the gain bandwidth of the laser, but then begins to shift and narrow due to gain narrowing and then mode competition.[8] If the laser contains a weak intracavity (i.e. inside the optical cavity) absorber, whose absorption features are narrow compared to the laser bandwidth, then as the time during which the laser has been pumped (t_g) increases, narrow "holes" are eaten out of the broad band laser emission spectrum. It has been established that for the first 500 μs the strength of these holes have an equivalent absorbance, A, of

$$A = \alpha \cdot f \cdot c \cdot t_g \tag{2}$$

where α is the absorbance coefficient of the gas, f the fraction of the laser cavity filled by the absorber (typically ~ 0.5) and c the speed of light. Thus one has an effective pathlength, L_{eff} given by:

$$L_{eff} = f \cdot c \cdot t_g \tag{3}$$

which can be up to a maximum of $\sim$ 100 km. Beyond this generation time t_g, mechanical vibrations and other sources of dephasing cause the depth of the absorption features to no longer follow a simple Lambert-Beer's law (Stoeckel & Atkinson 1985). Timing is achieved by the use of acousto-optic modulators on the pump and output of the laser, and the laser output is dispersed on a large spectrograph with an array detector. Using this technique, it is possible to measure spectroscopic features with an absorption coefficient as low as a few times 10^{-8}cm^{-1} (Stoeckel et al. 1982).

The sensitivity of this method is not quite as high as photoacoustic spectroscopy, but is much higher than that obtained with traditional long path cells. As long as one is careful to remain in the early time region, the absorption coefficient, and thus the cross section (assuming one knows the gas density and cavity fill factor), can be extracted from the optical depth of the observed absorption features. Because of the pulsed nature of the experiment, and the near threshold operation of the laser, chances for inadvertent photochemistry are reduced compared to intracavity photoacoustic spectroscopy. The chief disadvantages are that this method requires complex instrumentation. One must use a custom designed continuous wave dye laser that has been carefully designed to remove all sources of stray interference which can lead to narrow bandwidth spikes on the laser output spectrum. In order to achieve Doppler limited resolution, one must use a spectrograph of extremely high ($\sim 10^6$) dispersion, which is an expensive piece of custom instrumentation.

---

[6] The gain bandwidth is the spectral region over which a laser has net gain on each round trip.

[7] A dye laser uses a solution of an organic dye that is optically pumped by another laser to produce gain. The principal advantage of these lasers is that they can be tuned over a broad spectral region.

[8] These phenomena are discussed in the text by Yariv (1989).

5 Ring-Down Cavity spectroscopy

Ring down cavity spectroscopy is based upon a simple idea which has become practical due to recent advances in reducing loss in dielectric mirrors. A ring down cavity is made from two highly reflective, concave mirrors aligned as a stable, low loss optical cavity (i.e. the separation less than twice the radius of curvature). In our own work, radii of 1–2 m and separation L = 1.5–3 m are typical. It is now possible to purchase at low cost ($\sim$ \$100 each) small mirrors with reflectivity R > 99.99% over a range of some 60 nm anywhere in the visible, and mirrors with R=99.9998% have been reported near 840 nm (Rempe et al. 1992). Light from a conventional pulsed dye laser is coupled into the ring down cavity through one end. If the length of the laser pulse is less than the round trip time of the cavity ($2L/c = 10 - 20$ nsec), then there can be no interference and a small but stable fraction of the incident light ($\sim 10^{-5}$) enters the cavity. For a typical input pulse energy of a modest 1 mJ, this corresponds to $\sim 3 \cdot 10^{10}$ photons. These photons are trapped between the high reflectivity mirrors and slowly decay due to the combined loss of the mirrors and any molecular absorber located between the mirrors. The empty cell has a decay time $\tau = L/c(1-R)$. On each round trip, something like 10^{-5} of the intracavity light intensity is transmitted through the back mirror and is detected by a photomultiplier tube (PMT). With the time constant of the PMT set long compared to the round trip time, its output current follows a smooth exponential decay. O'Keefe & Deacon (1988) digitized this complete curve and did a least squares fit to extract the decay rate. In our own work (Romanini & Lehmann 1993), we have found that a much simpler system of a double gated integrator, which averages the light intensity in two time windows separated by $\Delta t \sim 2\tau$, works essentially as well. The gated integrator output, S, is proportional to the ln of the ratio of the signals in each of the two windows, which is given by:

$$S = \ln\left[\frac{A}{B}\right] = -\frac{\Delta t}{\tau} - L_{\text{eff}}\alpha \tag{4}$$

where τ is the empty cavity decay time given above, $L_{\text{eff}} = c\Delta t$ is the effective cavity length (how far the light traveled between the two times that the gated integrator samples the light intensity leaving the ring down cavity) and α is the absorption coefficient of the gas in the cell. In our work with $\lambda = 570 - 435$ nm, we have achieved effective pathlengths as large as 60 km, near the maximum attained by the ICLAS method. Figure 1 shows a schematic of the apparatus used. Since the signal is the ln of the ratio of the light intensity at early and late times, shot-to-shot fluctuations in the laser intensity are normalized out. In practice we have achieved noise levels of about 0.005 ($2\sigma_S$), which corresponds to a noise equivalent absorption coefficient of $\sim 8 \cdot 10^{-10}\text{cm}^{-1}$. This is already quite competitive with the best that has ever been achieved by photoacoustic spectroscopy, and significantly better than the claimed sensitivity of ICLAS. We have estimated that with mirrors of 1 ppm loss, such as has been reported by Rempe et al. (1992), a pathlength of 4000 km and noise equivalent absorption coefficient on the order of 10^{-12}cm^{-1} should be attainable, which is orders of

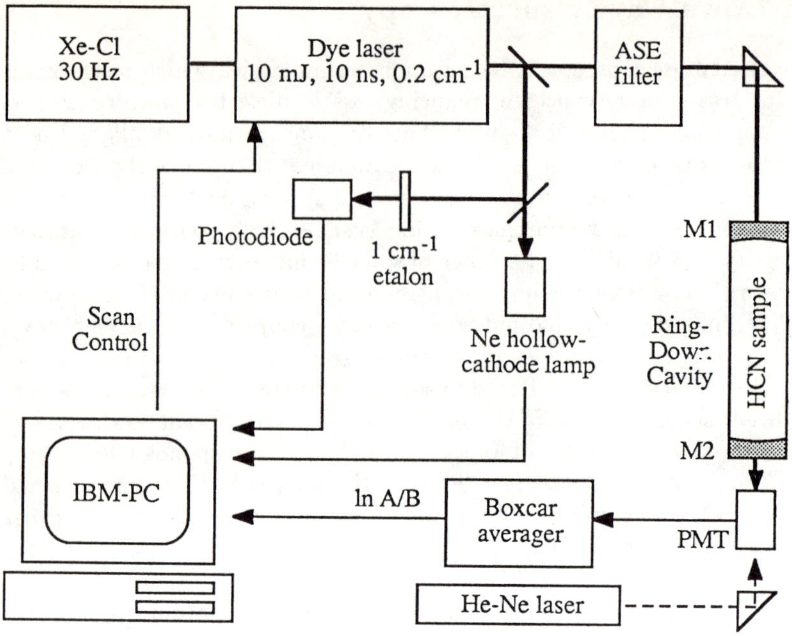

Fig. 1. Schematic of the experimental apparatus. The ASE filter is a stimulated Brillouin scattering cell for filtering the dye laser output of residual amplified spontaneous emission. The HeNe laser is used for rapid alignment of the Ring-Down cavity. Figure taken from Romanini & Lehmann (1993).

magnitude better than has ever been realized in any absorption based detection method. At one atmosphere pressure, this corresponds to an absorption cross section of $3 \cdot 10^{-32}$ cm^2! Since the method is based upon light traveling a known distance through a passive optical cavity, Lambert-Beer's law should hold quantitatively for all pathlengths, and thus the method provides a direct determination of the optical extinction coefficient at a known pathlength.

Compared with the other methods which we have discussed above, the ring down cavity spectroscopy is the simplest and least expensive to implement. It is also the most flexible. It is important to point out that the optics used are of such high reflectivity and modest cost because they are of small size; the coated surfaces are less than 1 cm in diameter. It is also possible to combine this method with pulsed molecular beam sources (O'Keefe et al. 1990), or with transient species produced in a discharge. The method samples only a narrow pencil of the sample, with a cross section less than 1 mm^2. By putting a furnace around the ring down cavity, it should be possible to study high temperature species, something that likely will prove useful for stellar astronomy. Because of the use of a stable optical cavity, small deviations of the light beam due to density gradients will average out over several round trip, and should not contribute to noise as in traditional long pathlength cells. Furthermore, since the cross sectional area of a ring down cell is much less than for a White cell, gas turbulence due to convection is much less likely to occur.

6 HCN Overtone Spectrum

Using the method of photoacoustic spectroscopy, we had previously investigated the spectrum of HCN for λ from 700 – 540 nm (Smith et al. 1987), which included bands with up to six quanta of CH stretch, as well as combination bands of the CH and CN stretching modes. The wavelength region was limited by the tuning range of a dye laser pumped by a visible Ar^+ ion laser. Attempts to determine absolute absorption strengths using the photoacoustic method proved disappointing. We also studied the near-IR spectrum from 1850 – 660 nm by using a 6 meter base length White cell (allowing absorption pathlengths up to 432 m) with the 1-m Fourier transform spectrometer at the McMath National Solar Observatory at Kitt Peak (Smith et al. 1989).

In order to push the HCN measurements to higher energy, and weaker bands, we have measured the overtone spectrum of HCN for $\lambda = 570 - 430$ nm using a ring down cavity (Romanini & Lehmann 1993). Figure 2 shows the rotational

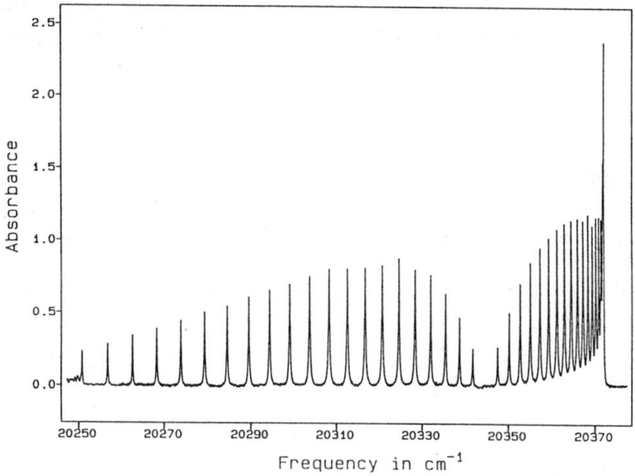

Fig. 2. The spectrum of the $(1,0,6) \leftarrow (0,0,0)$ overtone band of HCN (100 torr, $\Delta t = 80\mu s$, $L = 24$ km). Near the last displayed line of the P-branch is the beginning of the $(1,0,6) \leftarrow (0,1,0)$ hot band. Figure taken from Romanini & Lehmann (1993).

structure of one of the observed bands, the $(1,0,6) \leftarrow (0,0,0)$ band at 491.5 nm, obtained with an effective pathlength of 24 km. In this wavelength region, we have observed and analyzed nine bands of the principal isotopic species (for an analysis on this spectrum see Romanini & Lehmann 1993). Table 1 contains a list of the observed band origin and integrated band intensity [9] for each transition. Also, for comparison, are listed two predictions for the transition intensities. Both are calculated from two dimensional variational wavefunctions using the *ab initio* dipole function of Botschwina (1988). The column labeled Botschwina contain the intensities calculated by Botschwina (1992) using his corrected *ab initio* potential (Botschwina 1988). These calculations used the CEPA-1 method (coupled electron pair approximation). The second column, labeled Carter-Handy-Mills, contains intensities calculated by us, using the recently determined empirical HCN potential reported by these authors (Carter et al. 1993). Overall, the agreement between theory and experiment is excellent, especially considering the weakness of these bands. For comparison, the CH stretching fundamental of HCN has a strength of 6 x 10^6 cm/mol (Smith et al. 1981), nearly 10^8 times larger than the weakest of the bands we have presently observed (Romanini & Lehmann 1993). These measurements provide a critical test of the accuracy of theoretical calculations, which can easily be used to simulate the spectrum of HCN even at temperatures of stellar atmospheres. As this book was going to press, Botschwina (1993) reported new results with a dipole function calculated at a higher level of theory, the CCSD(T) method (coupled clusters with single, double and an approximate treatment of triple exciations). The vibrational wavefunctions are unchanged from those he had calculated with his earlier (CEPA-1) potential. The intensities he calculated with this new dipole function are reported in the last column of the table, under the heading CCSD(T). It is seen that the agreement with experiment is substantially improved, with the remaining dif-

Table 1. Comparison of observed and predicted band intensities of the overtone bands of HCN. Observed band origins are accurate to ±0.03 cm^{-1} intensities to ±5%. The three theoretical predictions are described in the text.

Transition	Frequency cm^{-1}	Exp. intensity cm/mole	Botschwina CEPA	Carter-Handy-Mills	CCSD(T)
105←000	17550.42	3.51	4.53	5.4	3.62
006←000	18377.03	2.61	2.88	3.4	2.61
205←000	19528.57	0.58	0.39	0.56	0.52
116←010	20222.27	0.78			
106←000	20344.51	0.89	0.69	0.94	0.85
007←000	21116.31	0.55	0.28	0.40	0.37
305←000	21486.77	0.092	0.026	0.052	0.073
206←000	22292.02	0.22	0.092	0.16	0.20
107←000	23047.11	0.24	0.098	0.17	0.20

[9] The integrated band intensity S, given in units of cm/mole, is related to the cross section by $S = N_A \int \sigma^2(\nu) d\nu$ where N_A is the Avogadro constant, ν is the frequency in wavenumber units (cm^{-1}), and the integral extends over all the rotational lines in a given vibrational band. The more common unit of integrated band intensity is km/mole = 10^5 cm/mole and is used because typical allowed IR fundamentals have strength of a few tens to hundreds of km/mole.

ferences being in many cases within experimental uncertainties. One can only assume that with vibrational wavefunctions calculated using the higher level of theory that the accuracy will improve further.

7 Conclusions

We have here reviewed the methods for obtaining spectra of molecules in regions of very small cross sections, and compared the established methods with the recently developed method of ring-down cavity spectroscopy. The latter method has many advantages over the others, and will likely prove to be widely useful in obtaining spectra of importance for astrophysics.

References

Botschwina P., 1988, *J. Chem. Soc. Faraday Trans. 2*, **84**, 1263

Botschwina P., 1992, private communication

Botschwina P., Seeger S., Horn M., Fluegge J., Oswald M., Mladenowic M., Hoeper U., Oswald R., Schick E., 1993, Proceedings of the 50'th international meeting of the Division de Chimie Physique de la Societe Francaise de Chimie (Mont Sainte-Odilie), to be published with the American Institute of Physics.

Burberry M.S., Morrell J.A., Albrecht A.C., Swofford R.L., 1979, *J. Chem. Phys.*, **70**, 5522

Carter S., Handy N.C., Mills I.M., 1993, *J. Chem. Phys.*, **99**, 4379

Davidsson J., Gutow J.H., Zare R.N., 1990, *J. Phy. Chem.*, **94**, 4069

Fairbanks W.M., Hänsch T.W., Schawlow A.L., 1975, *J. Opt. Soc. Am.*, **65**, 199

Giver L.P., 1978, *J. Quant. Spectrosc. Radiat. Transfer*, **19**, 311

Herriott D.R., Kogelnik H., Kompfner R., 1964, *Appl. Opt.*, **3**, 523

Herzberg G., 1945, *Infrared and Raman Spectra of Polyatomic Molecule*, Van Nostrand

Kritchman E., Shtrikman S., Slatkine M., 1978, *J. Opt. Soc. Am.*, **68**, 1257

Lovas F.J., Suenram, Ogata T., Yamanoto S., 1992, *Astrophys. J.*, **399**, 325

O'Keefe A., Deacon D.A.G., 1988, *Rev. Sci. Instrum.*, **59**, 2544

O'Keefe A., Scherer J.J., Cooksy A.L., Sheeks R., Heath J, Saykally R.J, 1990, *Chem. Phys. Lett.*, **172**, 214

Powers,D.E., Hansen S.G., Geusic M.E., Puiu A.C., Hopkins J.B., Dietz T.G., Duncan M.A., Langridge-Smith P.R.R., Smalley R.E., 1982, *J. Phys. Chem.*, **78**, 1627

Rempe G., Thompson R.J., Kimble H.J., Lalezari R., 1992, *Optics. Letters*, **17**, 363

Romanini D., Lehmann K.K., 1993, *J. Chem. Phys.*, **99**, 6287

Smith A.M., Lehmann K.K., Klemperer K., 1986, *J. Chem. Phys.*, **85**, 4958

Smith A.M., Jørgensen U.G., Lehmann K.K., 1987, *J. Chem. Phys.*, **87**, 3034

Smith A.M., Coy S.L., Klemperer K., Lehmann K.K., 1989, *J. Mol. Spectrosc.*, **134**, 134

Smith I.W.M., 1981, *J. Chem. Soc. Faraday Trans. 2*, **77**, 2357

Stoeckel F., Atkinson G.H., 1985, *Appl. Optics.*, **24**, 3591

Stoeckel F., Melieres M.-A., Chenevier M., 1982, *J. Chem. Phys.*, **76**, 2191

Welsh H.L., Cummining C., Stansbury E.J., 1951, *J. Opt. Soc. Am.*, **41**, 712

White J.U., 1942, *J. Opt. Soc. Amer.*, **32**, 285

Yariv A., 1989, *Quantum Electronic*, Wiley and Sons

Oscillator Strengths and Excited State Lifetimes in Metallic Hydrides and Oxides

Joël Schamps

Laboratoire de Dynamique Moléculaire et Photonique, U.R.A 779 C.N.R.S, U.F.R de Physique, Université de Lille I, 59655 Villeneuve d'ascq, France

1 Introduction

It is really impressive to see the amount of information astrophysicists have been able to extract from the analysis of that faint glimmering light that stars send to tell us about them. The main piece of this analysis consists in dispersing the light to get the spectrum, that is to say intensity versus wavenumbers. From the measurements of wavenumbers it is possible to identify the atomic and molecular species from comparison with high-resolution laboratory spectroscopic data. Once the constituents have been identified, models can be constructed in order to simulate the metabolism of circumstellar atmospheres. In such models a crucial point is the abundance of the identified atomic and molecular constituents.

In principle abundances can be deduced from intensities since intensities are proportional to them. But actually intensities are also proportional to something that is often poorly known, sometimes not at all, especially in the case of molecules. That "something" reflects the propensity of a given transition to occur: it is the (squared) transition dipole moment (or higher multipole moments when necessary) or, in an initiated language, quantities associated with this moment called oscillator strengths, Einstein coefficients and excited state lifetimes. The purpose of this chapter is to review oscillator strengths and lifetimes for a class of molecules that are important in cool stars, the metallic oxides and hydrides, a class for which much information remains to be gained if one wants to deduce quantitatively reliable values of abundances from intensities. In the present status of knowledge, it is often necessary to assume more or less reasonable values for the unknown oscillator strengths. Certainly we are now gratifyingly far from the drastic assumptions of the first determinations of stellar abundances in which, in the lack of transition probability data, it was assumed as a stopgap that all transitions of all molecules had equal oscillator strengths (Russel 1934). The correct way to proceed requires a knowledge of accurate, or at least approximate, values of molecular transition probabilities. Unfortunately, as examplified in the following, the situation is still far from being yet satisfactory in that matter.

The characteristics of stellar light are linked to the well-known theory of the dynamical behaviour of quantum states of material systems. Briefly it can be said that when photons fall on a material medium, gaseous, liquid or solid, they can be absorbed, i.e. annihilated, by the molecules in this medium if they are resonant with one of the numerous rovibronic transitions between an initially populated state i (for instance the ground state) and an excited (discrete or not) state j of the molecules. To allow for energy conservation, the absorption of optical radiation is accompanied by the promotion of the molecule into this excited state j:

$$M(i) + h\,c\,\nu_{ij} \rightarrow M^*(j) \tag{1}$$

where ν_{ij} is the wavenumber of the i-j transition.

Once the molecule is in the excited state j, three kinds of processes can induce the depopulation of this state. The first one is stimulated emission that is usually small in the visible under LTE conditions. As it is described by the same radiative quantities as absorption, it need not be considered further. The second one is spontaneous emission towards any accessible i′ level (among which the i level) lying lower than the excited state j, with emission of a resonant photon according to:

$$M^*(j) \rightarrow M(i') + h\,c\,\nu_{i'j} \tag{2}$$

The third de-excitation process is non-radiative decay, a denomination that in fact covers many depopulation mechanisms, internal mechanisms, such as dissociation, predissociation, autoionisation, or external mechanisms, i.e. involving other molecules, such as collisional transfer or quenching.

2 Radiative Properties: Definitions

The absorption and emission processes are most interesting to astrophysicists: since direct measurements of stellar properties in the stellar environment are obviously out of the question, practically all the information collected about stars has been provided by achieving spectral analysis of the absorption by photospheres of the continuum emitted by the stellar cores and, to a somewhat lesser extent, of the re-emission of light by these absorbing photospheres.

An absolute intensity is defined as the energy flux emitted per second in emission and as the decrease in the energy flux per second in absorption. It is proportional to the density number of the emitting or absorbing species, to the energy hcν of the emitted or absorbed photons and to the transition probability of the process.

As said above, transition probabilities and, as a consequence, absolute intensities of radiative effects are governed by the value of a central quantity: the square of the transition moment. Since the vast majority of the transitions of interest are dipole-allowed ones we restrict the discussion to the dipole transition moment between states i and j. It is defined as:

$$\mu_{ij} = < i | \sum_{\alpha} q_\alpha \overrightarrow{r_\alpha} | j > \tag{3}$$

where the summation runs over all the charges q_α of the system at positions $\overrightarrow{r_\alpha}$ and $|i>$ and $|j>$ are two rovibronic states. Extension to higher multipole radiation is straightforward though somewhat cumbersome.

Now it often happens that the electronic part of the wavefunction can be separated from the nuclear (rotational and vibrational) part. This is known as the Born-Oppenheimer (B.O.) approximation (1927). In the B.O. approximation, the rovibronic transition moment is given by:

$$\mu_{ij} =< v''J''|\mathcal{R}_e(R)|v'J' > \tag{4}$$

where $\mathcal{R}_e$(R) is the electronic transition dipole moment, a function of the internuclear distance; the $'$ and $''$ signs refer to the upper and lower electronic states, respectively. Moreover, since the rotational part of the wavefunction depends only on angles, μ_{ij} factorizes into

$$\mu_{ij} = \mu_{v'v''} S_{j'j''} \tag{5}$$

in which $S_{j'j''}$ is the rotational strength or Hönl-London factor of the transition, a simple function of the rotational quantum numbers that can be handled by conventional spherical tensor techniques. The $\mu_{v'v''}$ element is the vibronic transition moment:

$$\mu_{v'v''} =< v''|\mathcal{R}_e(R)|v' > \tag{6}$$

The $\mu_{v'v''}$ vibronic transition moment is typical of each transition and is the quantity to be focused on in the following. If the $\mathcal{R}_e$(R) electronic transition moment is constant or if it varies linearly with R over the region where the vibrational wavefunctions are non-negligible, it is possible to simplify further the expression of the vibronic moment that becomes:

$$\mu_{v'v''} = \mathcal{R}_e(\overline{r_{v'v''}}) < v'|v'' > \tag{7}$$

$(\overline{r'_v v''})$ is a particular separation characteristic of the v', v'' couple and called their r-centroid (Nicholls 1969); $< v'|v'' >$ is the vibrational overlap, the square of which is the well-known Franck-Condon factor. This factorization bears the name of Franck-Condon approximation.

In fact, intensities of molecular transitions are seldom characterized by their transition moments. Instead of these moments several inter-related quantities are introduced depending on whether photon or energy flux is detected and whether absorption or emission is considered. Thus for the case of absorption measured by a decrease in photon flux the transition rate is given by:

$$dN_{v'}/dt = -B_{v''v'} u_\nu N_{v'} \tag{8}$$

where u_ν is the differential energy density of the radiation field at wavenumber $\nu_{v'v''}$. The decrease in photon flux is expressed in terms of the (vibronic) Einstein coefficient in absorption:

$$B_{v''v'} = |\mu_{v'v''}|^2 2\pi^2/3h^2\epsilon_0 c \tag{9}$$

i.e, numerically, for μ in Debye units and u_ν in $J/(m^3cm^{-1})$:

$$B_{v''v'}(\text{in}J^{-1}m^3cm^{-1}s^{-1}) = 4.056 \times 10^{17}|\mu_{v'v''}|^2 \tag{10}$$

Note that $B_{v''v'}$ is frequency independent. Note also that for $\Sigma \to \Pi$ transitions and for them only, μ should be replaced by $\sqrt{2}\mu$ in this formula and each time it appears in all others that follow (Whiting et al. 1980). For absorption measurements by a decrease in energy flux, the absolute intensity is proportional to the first power of the wavenumber and it is currently expressed in terms of the absorption oscillator strength $f_{v''v'}$:

$$f_{v''v'} = \nu_{v'v''}|\mu_{v'v''}|^2 16\pi^2 m_e c/3h^2e^2 \tag{11}$$

i.e, numerically, for μ in Debye units and ν in cm^{-1},

$$f_{v''v'} = 4.703 \times 10^{-7}\nu_{v'v''}|\mu_{v'v''}|^2 \tag{12}$$

Note that $f_{v''v'}$ is dimensionless. It can be shown from sum rules that its maximum value is unity, a value that would correspond to a "fully allowed" transition that leaves nothing for others.

Alternatively the absorption lineshape can be described in terms of so-called absorption cross-sections $\sigma_{v'v''}$ that lead, when integrated over wavenumbers, to the integrated cross-section:

$$\sigma^0_{v''v'} = (4\pi^2e^2/m_e\epsilon_0)f_{ij} \tag{13}$$

i.e.numerically,

$$\sigma^0_{v''v'}(\text{in cm}^2) = 4.162 \times 10^{-19}\nu_{v'v''}|\mu_{v'v''}|^2 \tag{14}$$

The microscopic absorption cross-section $\sigma(\nu)$ is simply related to the macroscopic absorption coefficient $K(\nu)$ of the exponential Beer-Lambert law for absorption through the density number $N_{v''}$: $K(\nu)=N_{v''}\sigma(\nu)$

For emission over a single transition, from an excited level down to a lower-lying one, it is customary to introduce the spontaneous emission probability in photon flux detection as another Einstein coefficient:

$$A_{v'v''} = \nu^3_{v'v''}|\mu_{v'v''}|^2 16\pi^2/3h\epsilon_0 \tag{15}$$

i.e. numerically, with the above unit conventions (ν in cm^{-1} and μ in Debye):

$$A_{v'v''}(\text{in s}^{-1}) = 3.137 \times 10^{-7}\nu^3_{v'v''}|\mu_{v'v''}|^2 \tag{16}$$

Note the ν^3 dependence of $A_{v'v''}$. The spontaneous emission absolute intensity of the $v' \to v''$ transition, i.e. the energy flux emitted per unit volume of the source per second on this transition, is therefore proportional to the fourth power of the wavenumber:

$$I_{v'v''}(\text{in Wm}^{-3}) = N_{v'}hc\nu_{v'v''}A_{v'v''} = N_{v'}\nu^4_{v'v''}|\mu_{v'v''}|^2 16\pi^2c/3\epsilon_0 \tag{17}$$

However, the global spontaneous emission from a level v' takes place towards all the lower-lying v'' levels, so that the depopulation rate of this level is a sum over all the possible transitions:

$$dN_{v'}/dt = -(\sum_{v''} A_{v'v''})N_{v''} \tag{18}$$

It follows that the population of the upper state decreases exponentially with a time constant called the radiative lifetime of the excited vibronic level:

$$\tau_{v'} = (\sum_{v''} A_{v'v''})^{-1} \tag{19}$$

Thus, in general, measuring the lifetime of an excited level does not lead to state-to-state transition moments but instead to a global value. However, as discussed later, there may be favourable situations in which rough values of transition moments can be extracted from lifetimes.

Before going further, it should be again recalled that all the constants considered above have been defined for ν standing for wavenumber, not frequency. As a general rule the problem of unities and that of statistical weight factors have led for long to considerable misleading in the problem of conversion between band strengths, oscillator strengths, transition probabilities and radiative lifetimes (Nicholls & Stewart 1962; Tatum 1967; Schadee 1967, 1971, 1978; Brown et al. 1976; Kuzmenko et al. 1979; Kuznetsova et al. 1974, 1980; Whiting 1972; Whiting & Nicholls 1974). Today this problem is definitely overcome since the recommendations given in the central paper by Whiting et al. (1980) seem to have been universally adopted, including for radiative lifetimes (Larsson 1983). The reader should refer to the latter two papers for details.

3 Effects of Perturbations and Predissociations

When the Born-Oppenheimer model holds, i.e when electronic states are isolated from each others, all the rovibronic levels of a given electronic state have nearly identical lifetimes and all oscillator strengths between rovibronic levels of two electronic states follow regular laws: they differ because of different Franck-Condon factors but they correspond to the same (R) electronic transition moment. This is not always so simple, however, and defects to the B.O. approximation frequently appear. In such cases, the actual vibronic state j becomes a mixing of two (or more) quasi-degenerate B.O. states. The coupling that can induce such a mixing may have various origins: for example, kinetic, spin-orbit, Coriolis, or hyperfine interactions. This effect bears the generic name of perturbation. Perturbations are reviewed in full details in an excellent book by Lefebvre-Brion & Field (1986). The name of perturbation is more specifically reserved to the interaction of a bound state with another bound state. For a bound state perturbed by a free state of the continuum, one usually speaks of predissociation when the bound level lies above the dissociation limit of a lower-lying electronic state, or of preionisation when it lies above the ionisation limit of the molecule.

A bound-bound perturbation is experimentally observed as both shifts in the positions of spectroscopic lines and drastic changes in the line intensities and lifetimes compared to some regular evolutions that would be observed in the absence of the perturber. We are concerned here by intensities and lifetimes and from this point of view two cases may be distinguished according to the radiative nature of the perturber. If the perturber is a metastable state $|m>$, i.e. a state that cannot radiate towards lower states in the B.O. approximation because its transition moment towards these states is zero, the squared transition moment of the $|j>$ perturbed state to the $|i>$ state becomes:

$$|<j^{pert}|\mu|i>|^2 = |\{\alpha < j| + (1-\alpha^2)^{1/2} < m|\}|\mu|i>|^2 = \alpha^2|\mu_{ij}|^2 \qquad (20)$$

Since the α coefficient is smaller than unity, the oscillator strengths and the Einstein coefficients of the transition are reduced by a factor α^2. In the same way, assuming a single electronic channel depopulation, that to state i, the lifetime of the j level is lengthened by a factor $1/\alpha^2$ while the perturber, which had normally an "infinite" radiative lifetime, becomes able to radiate. To summarize, perturbation of a vibronic level j by a metastable state produces an intensity borrowing and an increase of lifetime.

Now, if the perturber, denoted $|r>$, is also a radiative state (unperturbed lifetime τ_r),the (squared) transition moment to state i is:

$$\begin{aligned}|<j^{pert}|\mu|i>|^2 &= |\alpha<j|\mu|i> + (1-\alpha^2)^{1/2}<r|\mu|i>|^2 \\ &= \alpha^2|\mu_{ij}|^2 + (1-\alpha^2)|\mu_{ri}|^2 + 2\alpha(1-\alpha^2)\mu_{ji}\mu_{ri} \qquad (21)\end{aligned}$$

The last term can be positive or negative. This gives rise to an interesting quantum interference effect: for a rotational level that is not in general that for which the perturbation lineshift is maximum (except when $|\mu_{ji}| = |\mu_{ri}|$) there is a complete extinction of the line and correlatively an enhancement of the other component: the quantum interference effect results in a complete transfer of intensity.

Let us now consider briefly bound-free perturbations, more precisely let us consider predissociation. In this case the bound B.O. level is quasi-degenerate with the dissociation continuum of another electronic state. This occurs for instance when the potential curve of the bound state is crossed by a dissociative potential curve or sometimes when a bound level lies higher than the dissociation continuum of another stable electronic state. When the two electronic states are coupled by some perturbation element H, the quasi-bound state contains in its wavefunction a more or less pronounced dissociative character. The net effect of this is to widen the corresponding lines in absorption as a consequence of the time-energy Heisenberg relation and to induce disappearance of the expected lines in emission.

Concerning the dynamics of the process, the effect of a predissociation is to add a non-radiative rate constant in the rate equation for the perturbed state:

$$dN/N = -(k_r + k_{nr})dt \tag{22}$$

which leads to:

$$N(t) = N_0 exp[-(k_r + k_{nr})t] \tag{23}$$

The perturbed lifetime is given by:

$$1/\tau_{pert} = 1/\tau_r + 1/\tau_{nr} \tag{24}$$

with $1/\tau_{nr}$ deduced from the Fermi golden rule:

$$1/\tau_{nr} = 4\pi^2/h|He|^2| < v|\chi_E > |^2 \tag{25}$$

From this, the effect of predissociation is seen to decrease the lifetime of the perturbed state.

4 Methods

There are essentially three classes of methods that can be used to determine oscillator strengths in laboratory. Two of them are experimental methods and relate to absolute intensity measurements and to time-resolved spectroscopy, respectively. The third one consists in ab initio calculations of transition moments.

In the absolute intensity measurement method, the wavenumber-integrated absorption coefficient is computed from the equivalent width (Davis et al. 1986) deduced from the exponential (Beer-Lambert law) decrease in the incident flux over a path-length of absorbing gas. The difficulty arises in the passage from the macroscopic data to the microscopic ones. This requires to use the number density in the medium, an information that is in general poorly known for metal compounds that are difficultly volatile; indeed the uncertainty may sometimes turn on several orders of magnitude. Another way to proceed is to use, when possible, another known absolute transition probability of the same molecule as a mean for calibration. In spite of these difficulties the intensity method, when it can be worked out, is quite interesting because it yields directly individual level-to-level oscillator strengths.

The second method for determining oscillator strengths consists in attempting to extract them from lifetime measurements. The simplest method for measuring lifetimes is to compute them from the exponential decay of emission after fast removal of the pulsed dye-laser excitation. The advantage of this pulse-decay method is that it provides directly lifetimes without need for other data. Another way for probing the decay of the population of a state is to photoionize this state and to detect the decay of the number of ejected electrons (Doverstal & Weijnitz 1992). The experiments often involve a supersonic molecular beam, a device that displays the advantage of providing practically collision-free conditions. Another possibility offered by the molecular beam technique is to move the measurement from time to space. Instead of the time constant of the decay one can measure the length over which molecules continue to emit after the excitation point and then divide by the mean velocity of the beam to convert this length into a lifetime. This method is especially suitable for long lifetimes since these correspond to significant lengths. Lifetime measurements can also be carried out in a simple cell at moderately high pressure (a few torrs). However it is then necessary to extrapolate the measured lifetimes to zero-pressure in order to eliminate the non-radiative part of the lifetime due to collisions. The lifetime measurement method has had a lot of variants. Among the most original ones, let us quote the recent lifetime measurements of Hikmet et al. (1992) on copper halide excited states up to high vibrational levels formed in reactive collisions of metastable copper atoms with halogen compounds. In this experiment the removal of excitation of the halide is obtained by stopping abruptly its formation in the excited state owing to a copper laser that depopulates instantaneously the metastable atomic copper state through stimulated emission.

In time-resolved spectroscopy the main problem is to prepare a sufficient concentration of excited molecules. This requires a powerful optical pumping device. The method is therefore restricted to the spectral regions where lasers operate (though other sources can be used in the UV such as synchrotron radiation [Castex et al. 1980]). Because of this, the time-resolved method is frequently called laser-induced decay measurements. Other complications may happen that alter the interpretation of the recorded results, for instance when multi-exponential decays occur due to intense transfers or cascading processes.

Though time-resolved spectroscopy is the ideal way for determining lifetimes, it is not able to yield values of individual state-to-state transition moments and oscillator strengths since it results in principle from depopulation over numerous vibronic channels. However, in a fairly large number of cases, it may happen that one particular electronic depopulation channel is favoured against all others. In such a case, the summation is restricted to the vibrational transitions within this single electronic channel. Owing to the closure condition on Franck-Condon factors, an approximate value of the electronic transition moment $\mathcal{R}_e$ can then be obtained assuming that this moment is constant, and furthermore that all the $\nu_{v'v''}$ factors can be replaced by the ΔT_e electronic spectral term difference. One has then approximately:

$$(\tau_{v'})^{-1} \approx (16\pi^2/3h\epsilon_0)\mathcal{R}_e^2 \Delta T_e^3 \tag{26}$$

Measuring experimentally $\tau_{v'}$ thus yields a rough value of $\mathcal{R}_e^2$ that can be multiplied by relevant Franck-Condon factors to go back to rough values of the $A_{v'v''}$ vibronic transition probabilities. Note, however, that this is nothing but a makeshift, though a useful one because lifetime measurements are often the only experimental way to have access to transition moments of metallic compounds, seeing the difficulty for doing quantitative absorption spectroscopy with them.

The last method, and in the author's opinion the most promising one, for obtaining oscillator strengths and lifetimes is to determine transition moments by means of ab initio calculations (see Langhoff & Bauschlicher, Malmqvist, and

Peyerimhoff, this volume). Some years ago such a statement would probably have been considered as somewhat arrogant. This is no longer so today. Recent state-of-the-art ab initio calculations have begun to reach such a level of sophistication that they are able to provide reliable determinations of oscillator strengths and lifetimes with an accuracy probably comparable to the experimental one, though it remains difficult to assess this accuracy a priori (Langhoff & Chong 1978).

The power of the ab-initio approach is threefold. First, it allows to determine the form of transition moments over any desired range of internal coordinates contrarily to experiment that can only sample a very limited geometry variation. Second, it applies to any transition, all the transitions of a molecule if wanted, even the least intense ones, whereas experiments either in emission or absorption suffer from severe limitations in the number of accessible transitions. Finally, it makes possible a fine understanding of the whole spectrum of a molecule, an identification of "dark" states that are important in dynamics studies, an interpretation of the relative role of orbitals, polarisation, and configuration interaction in the observed intensities. The force of ab initio calculations will be thoroughly illustrated below.

5 Radiative Properties: A Review of Data

The rest of this chapter is devoted to a review, molecule per molecule, of the present status of knowledge about oscillator strengths and lifetimes in the metallic monoxides and monohydrides. It is organized as follows. First the radiative properties of all the hydrides are reviewed following the order of the columns of the periodic classification, the transition and rare-earth series being put at the end. Then the same is done for monoxides.

5.1 Alkali Monohydrides: LiH, NaH, KH, CsH

Lithium monohydride LiH is the lightest molecule containing a metal. This gives it a special position that explains the attraction it has exerted on molecularists in spite of its little interest for astrophysicists even though this little interest probably arises out of certain practical circumstances, namely the fact that its electronic transitions lie far away in the ultraviolet.

Nevertheless the series of papers on LiH (Way & Stwalley 1973; Stwalley & Zemke 1976; Stwalley et al. 1977; Stwalley 1977; Zemke & Stwalley 1978, 1980; Zemke et al. 1978; Vidal & Stwalley 1982, 1984) published between 1973 and 1984 represents a fine example of the application of the ideas developed above. These authors decided to calculate systematically radiative transition probabilities for the three systems interconnecting the $X^1\Sigma^+$, $A^1\Sigma^+$ and $B^1\Pi$ states of LiH. The electronic transition moments $\mathcal{R}_e(R)$ had already been carefully calculated by Docken & Hinze (1972) and these ab initio data were used by Zemke et al. with minor extrapolations. It should be noted that a recent re-calculation of these moments (Gadea 1993) has confirmed the accuracy of Docken's and Hinze's work. Then Zemke et al. constructed full potential curves for the three states using spectroscopic data (see Huber & Herzberg 1979) from which they deduced all the bound vibrational energy levels as well as their wavefunctions. Combining the ab initio data and the experimental potential curves they calculated oscillator strengths and transition probabilities between all bound vibronic levels. They also calculated lifetimes of each v' level by summing the $A_{v'v''}$ Einstein coefficients over all the lower-lying vibrational levels v'' including bound levels and also (less accurately) free levels from the dissociation continuum of the ground state. Thus in the A state the bound-free contribution to the lifetime, completely negligible at low v', was found to become predominant for levels above v'=17. Zemke et al. also estimated the magnitude of the predissociation of the A vibronic levels by the $X^1\Sigma^+$ dissociation continuum and found that it was completely negligible in LiH. The results obtained by Zemke et al. are in excellent agreement with the experimental determinations of lifetimes by Dagdigian (1976), Wine & Melton (1976), Brieger et al. (1983) for the $A^1\Sigma^+$ state and by Von Moers et al. (1987) for the $B^1\Pi$ state.

For NaH, there are also both theoretical and experimental results. Sachs et al. (1975) have carried out ab initio calculations of electronic transition moments similar to those of Docken & Hinze (1972) on LiH but extended to some triplet-triplet transitions in addition to the singlet-singlet ones. They predicted band oscillator strengths and Einstein coefficients for the most intense system, the $A^1\Sigma^+$-$X^1\Sigma^+$ one. Their calculation was later used by Telle (1986) to predict lifetimes of the $A^1\Sigma^+$ vibronic levels of NaH. These semi-theoretical results fit well with the experimental measurements carried out by Baltayan et al. (1976) for $v' = 3-5$, by Dagdigian (1976) for v'=8, by Brieger et al. (1981, 1983) for $v' = 12-15$, and still higher by Nedelec & Giroud (1983) except for the anomalously long lifetime found for the highest level reached in this last study (v'=21) that was attributed to potential curve crossing at long internuclear distance.

Very few results exist for KH. Its fluorescence was observed by dye-laser excitation by Cruse & Zare (1974), Ihle & Wu (1974), and Ennen & Ottinger (1975), but with so high a pressure that relaxation was more rapid than the laser pulse decay. Using a heat-pipe oven, Giroud & Nedelec (1982) were able to measure the lifetime of the $A^1\Sigma^+$ state of KH. For the highest vibrational level they pointed out a sudden decrease of the lifetime at the opposite of their finding for NaH (Nedelec & Giroud 1983).

Such a sudden decrease of the lifetime appears also in CsH and this even in fairly low vibrational levels of the $A^1\Sigma^+$ state (as low as v'=12). This was clearly reported by Ferray et al. (1984) after the first measurements by Tam & Happer (1976) and Hsie et al. (1978) on this molecule. In fact the measured lifetime falls at an approximately constant value that is just equal to the lifetime of the cesium atomic state corresponding to the dissociation. This strongly suggests that this effect is due to predissociation of the highest vibrational levels of CsH ($A^1\Sigma^+$) by the continuum of the ground state. This effect, unobserved in the light LiH and NaH hydrides and suspected at the highest v' in KH, becomes quite clear in CsH. It has been reproduced theoretically by Telle (1984, 1986), though considerably smoothed over, on the basis of the ab initio transition moment calculations of

Laskowski & Stallcop (1981). It would be interesting to measure the lifetimes in RbH to see whether the same predissociation effect is present or not in this heaviest alkali hydride for which no lifetime measurement has yet been reported.

5.2 Alkaline-earth Hydrides: MgH, CaH

MgH has considerable astrophysical importance.Transition probabilities have been calculated by Henneker & Popkie (1971) and Popkie (1971). These ab initio predictions, though less refined than what can be done nowadays, are in good agreement with the values of lifetimes determined later by Nedelec & Dufayard (1978) for the $v'=0$ and $v'=1$ levels of the $A^2\Pi$ states of MgH and MgD (between 40 and 50 ns).

For CaH, Klynning et al. (1982) determined the perturbation-free lifetimes of the $B^2\Sigma^+$ ($v=0$) state to be 57 ns leading to an absorption oscillator strength equal to 0.11.

5.3 Boron and Aluminium Hydrides: BH, AlH

Laboratory measurements of the lifetime of the lowest vibronic level of the $A^1\Pi$ state ($\tau=125\pm5$ns) of BH have been reported by Dufayard & Nedelec (1978). Calculations by Douglas et al. (1989) reproduce fairly well the only experimentally reported lifetimes, thus giving confidence in the other calculated quantities.

Similar experiments using laser-induced fluorescence for the corresponding $A^1\Pi$ states of AlH and AlD have also been carried out for the $v'=0$ and $v'=1$ levels by Baltayan & Nedelec (1979). The lifetime values they found were noticeably shorter than in BH ($\tau_0=66$ns and $\tau_1=83$ns in AlH). An interesting result in AlH is their measurements of the lifetimes of rotational levels in the $v'=1$ state in the region of the predissociation: the first predissociated rotational level, $N'=6$, has a much shorter lifetime (50ns) than the preceeding level, $N'=5$,(83ns), that does not suffer significantly from this predissociation.

Three ab initio calculations have been published relating to potentials and transition moments for the A state of AlH. These were obtained from full-CI calculations by Bauschlicher & Langhoff (1988), from CASSCF calculations by Matos at al. (1987) and by the polarization propagator method by Scuseria et al. (1989). Here also, the agreement with experimental lifetimes is reasonably good and even less sensitive to the shape of the potential curves than in BH. However a good agreement for lifetimes does not definitely ascertain the reliability of the level-to-level quantities that enter in this global value. Thus, Rice et al. (1992) recently determined ratios of the Einstein coefficients and band oscillator strengths involving five bands of the $A^1\Pi - X^1\Sigma^+$ transition of AlH by using relative integrated intensities. Owing to a calibration based on Baltayan's and Nedelec's lifetimes (1979), they derived absolute Einstein coefficients. They found that the so-measured values of these coefficients were considerably smaller than the calculated ones (Matos et al. 1987) whereas the overall summation carried out by these authors to obtain theoretical values of lifetimes were in much better agreement with experiment, probably because of compensating balance effects to be associated with the closure relation on Franck-Condon factors.

5.4 Group IV-A Metallic Hydrides: GeH, SnH, PbH

Only the heavy members of the Group IV-A of the periodic table will be considered here since the others, carbon and silicium, are metalloids and do not enter into the scope of this review.

Three experimental lifetime determinations have been published for GeH. All of them relate to the lowest excited doublet state $A^2\Delta$. Osmundsen et al. (1985) report a rather short lifetime of 12±2ns for the three lowest vibrational levels in the photolysis of GeH_4. The largest and agreeing values of about 80±5ns determined by Erman et al. (1989) and of 93±10ns found in laser-induced fluorescence by Bauer et al. (1989) for the $A^2\Delta$ (v'=0) level seem to be more reliable. These values are about six times smaller than in CH (Bauer et al. 1989). This is explained by the predissociation of the GeH $A^2\Delta$ state (Erman et al. 1989). It would be interesting to determine the τ_{nr} part of this predissociation in the total lifetime by a simple difference between the experimental transition probability and that deduced from the ab initio transition moment calculations performed by Chapman et al. (1988).

For the SnH and PbH molecules, the first and up to now the only known moments come from theoretical calculations, not experiment as is generally the case. Like for GeH, Chapman et al. (1988) determined the transition moments of the $^2\Delta - ^2\Pi$ and $^4\Sigma^- - ^2\Pi$ systems from refined ab initio calculations using a $\omega - \omega$ coupling scheme. In principle, oscillator strengths could be calculated from these data, but the predissociation could probably not be ignored in this case as it could strongly affect the pure radiative values.

5.5 Transition Element Hydrides: ScH, TiH, FeH, MnH, ZnH, CdH, HgH

Transition elements are known to be present in circumstellar atmospheres. Although dissociation equilibrium simulations by Querci (1993) seem to indicate that their hydrides are in general less abundant than the corresponding oxides, their astrophysical importance cannot be denied. Molecules like FeH and TiH for instance have been clearly identified in stellar sources (Wing et al. 1977; Yerle 1979) and NiH, CrH and MnH among others are strongly expected to be present in non-negligible amount. However the structure of their spectra are often not yet fully understood and there has been only a few attempts to measure oscillator strengths and lifetimes for some of these compounds. Most of the information about their dynamics comes from recent careful ab initio calculations.

Thus for ScH and TiH, Anglada et al. (1989, 1990) systematically computed a large number of transition moments starting from several low-lying states. These calculations were meant to identify the most intense transitions. As they were done at a single separation, the R-dependence of the moments is not known and oscillator strengths and lifetimes cannot be deduced from them with accuracy. Anglada et al. (1990) also quote unpublished calculations on VH and YH.

The transition moments of FeH have been calculated by Langhoff & Bauschlicher (1990) who use them to compute Einstein coefficients for the strong $F^6\Delta - X^6\Delta$ band system. They also calculate lifetimes for several states: $e^6\Pi$ (53ns), $g^6\Phi$ (55ns), $F^6\Delta$ (1 μs), $A^4\Pi$ (4 μs) and some electronic oscillator strengths.

Similar calculations on MnH by Langhoff et al. (1989) provide a number of ab initio transition moment functions for the quintet and septet states. From these it has been possible to compute Einstein coefficients for the three strongest transitions of MnH and to derive radiative lifetimes for the four lowest vibronic levels of the $A^7\Pi$ (τ=0.1μs), $b^5\Pi$ ($\tau = 2 - 3\mu$s) and $c^5\Sigma^+$ (τ=0.3μs) states.

The only experimental data about radiative effects in transition metal hydrides are the lifetime measurements of Jourdan et al. (1976) on CdH and CdD, those of Dufayard & Nedelec (1977) on ZnH, ZnD, CdH and CdD and those of Nedelec & Dufayard (1978) on HgH and HgD, relating to the $A^2\Pi$ state. These authors report a lifetime of about 70 ns for the v'=0 and v'=1 levels of the A states of the zinc and cadmium hydrides and a little longer for HgH (101ns) and HgD (134ns).

5.6 Alkali Oxides: LiO, NaO, KO

The observed electronic spectrum of alkali monoxides is limited to a spread progression of vibronic bands that have been found in the red and the near infrared (Woodward et al. 1989; Pfeifer & Gole 1984). The multiple collision conditions necessary for its observation are not favourable to experimental dynamic measurements. This explains why the only paper about the lifetimes of LiO, NaO, and KO is an ab initio one due to Langhoff et al. (1992). These authors have calculated the radiative lifetimes for the $C^2\Pi$ (v'=0) level to be 66, 91 and 315 ns for LiO, NaO and KO, respectively. They have shown that the depopulation channel towards the $X^2\Pi$ ground state is the only significant one. Nothing is known about CsO and RbO spectroscopy.

5.7 Alkaline-earth Monoxides: BeO, MgO, CaO, SrO, BaO

A great deal of literature has been devoted to alkaline-earth monoxide spectroscopy. For BeO, however, no experimental data about lifetimes and oscillator strengths seem to have been published. To our knowledge, the only available information about transition moments in BeO is the full configuration interaction benchmark calculations by Bauschlicher & Langhoff (1988) for selected dipole and quadrupole transitions.

There are many more results for MgO. Here also, several ab initio calculations have been published. The first one dates back to 1974. Huron et al. (1974) performed configuration interaction calculations of ten transition dipole moments in the singlet and in the triplet manifolds. They were led to criticize a previous experimental determination of the $B^1\Sigma^+ - X^1\Sigma^+$ oscillator strength by Main et al. (1967; 1969). The discrepancy between experiment and theory for this system and others was kept going by the shock-tube quantitative absorption data of Evans & Mackie (1977) and of Svyatkin et al. (1980a) that both involved however perilous number density determinations. A better agreement was obtained for lifetime measurements from fluorescence decay. Thus the lifetime for the v=0 level of the $B^1\Sigma^+$ state that depopulates via two electronic channels towards $A^1\Pi$ and $X^1\Sigma^+$ was experimentally measured to be 21.5 ns by Busener et al. (1987) or 32.7 ns by Diffenderfer et al. (1983) who also derived a theoretical value of 24 ns. In the same way, there is good agreement between the experimental value (11.8 ns) reported for the lifetime of the $d^3\Delta$ state (Diffenderfer et al. 1983) and the theoretical one (9.3 ns) calculated by Yarkony (1988).

All the data for the heavier alkaline-earth oxydes are of experimental nature. For CaO, Svyatkin et al. (1980 b,c) determined the $B^1\Pi-X^1\Sigma^+$ and $C^1\Sigma^+-X^1\Sigma^+$ electronic transition strengths using the same shock-tube absorption intensity measurements as in their work on MgO (Svyatkin et al. 1980a). Pasternack & Dagdigian (1978) estimated the overall lifetime of the upper states that produce the orange arc bands to be 32.7 ns and Irvin & Dagdigian (1981) found a lifetime of 155±60ns for the v=6 level of the $A^1\Sigma^+$ state. These lifetime values have been recently confirmed with better accuracy (27.8±1.2 and 149±11 ns, respectively) by Plane & Nien (1991).

For SrO, the only information on the transition probabilities are the works of Svyatkin et al. (1982) that studied the absorption intensities of the B–X and C–X transitions as a continuation of their series on alkaline-earth oxides and of Belyaev et al. (1982) who determined the transition probability of the B–X system from flame absorption spectrometry.

Only laser-induced fluorescence has been used to study BaO. The vibronic lifetimes of several states were determined. Thus Johnson (1972) measured average lifetimes of levels of the $A^1\Sigma^+$ state and found values around 0.35μs. Three years later Pruett & Zare (1975) measured the lifetime of the $A'^1\Pi$ state to be 9μs. They turned to account the difference between the A and A′ lifetimes to separate the weak, long-lifetime A′–X fluorescence by delaying the observation of the A′–X emission until the A–X emission dies away. Hsu et al. (1980) investigated the relaxation of the $C^1\Sigma^+$ state in its three lowest vibrational levels, finding a short lifetime of 10 ns. From relative intensity measurements, they were able to measure fluorescence branching ratios and to show that the value obtained for this lifetime was the result of not less than four competing depopulation electronic channels for which they determined electronic dipole transition moments. Finally, Darrah & Silvers (1985), going back to the $A^1\Sigma^+$ state, studied the effects of perturbations on the rotational level lifetimes and observed lifetime lengthenings by a factor of two within a single vibronic state because of perturbation by components of the metastable $b^3\Pi$ state.

5.8 Boron and Aluminium Oxides: BO_2, BO, AlO

Using a pulsed dye laser, Huie et al. (1978) observed the fluorescence of boron dioxide BO_2 and measured lifetimes of vibrational levels of the $A^2\Pi_u$ state. They report $\tau(000)$=87.2±2.6 ns and $\tau(100)$=76.3±1.4 ns for this triatomic molecule which appears as being the only one involving a metal for which radiative data have been published.

In the same paper, Huie et al. (1978) also determined radiative lifetime values for the v'=1 and v'=2 levels of the $A^2\Pi$ state of boron monoxyde BO. They found τ_1=131±15ns and τ_2=103±6ns, respectively. Two years later, Clyne & Heaven (1980) reported much longer lifetimes for the v'=0 (τ_0=1780±40 ns) and the v'=4 (τ_4=1940±170 ns) of this $A^2\Pi$ state. Configuration interaction calculations of the transition probabilities by Kozlov et al. (1990) seem to be in favour of the short lifetimes given by Huie (1978) though their results depend strongly on the molecular orbital basis used in their computations.

For AlO the determination of the $B^2\Sigma^+-X^2\Sigma^+$ oscillator strengths and of the lifetime of the $B^2\Sigma^+$ state is full of controversy. Here again absolute intensity measurements are difficult because they are subject to large systematic uncertainties in the determination of the molecular concentration. The oscillator strengths determined by this method by Vanpee et al. (1970) from flame emission and by Hooker & Main (1971) from shock-tube absorption are an order of magnitude shorter than the most recent values deduced from laser fluorescence. Other emission intensity measurements by Hebert & Tyte (1964) and by Tyte & Hebert (1964) as well as shock-tube studies by Linton & Nicholls (1969) were re-calibrated by Hebert et al. (1980) using independently determined lifetimes (Johnson et al. 1972; Dagdigian et al. 1975); thus the concentration measurement problem was avoided and more reliable values could be obtained. Fluorescence decay measurements proved again to be the best method in this case. Of the five $B^2\Sigma^+$ lifetime determinations using this method, four give comparable results. These are the measurements by Johnson et al. (1972): τ_0=128±6 ns; by Dagdigian et al. (1975): τ_0=100±7 ns, τ_1=102±7 ns; by Salzberg et al. (1991): τ_1=97±12 ns, and by Campbell et al. (1992): τ_1=102±10 ns. It seems that the value of τ_0=272±14 ns reported by Wentink et al. (1971) in the fifth paper, the oldest one, should be rejected. It would be interesting to compare these results to a careful ab initio determination. The two theoretical studies due to Michels (1972) and to Yoshimine et al. (1973) date back to the early seventies and could be much refined today.

5.9 Transition Metal and Rare-earth Oxides: ScO, YO, TiO, ZrO, FeO, CuO, LaO

Unlike the case of the hydrides, the literature on radiative properties of the transition element oxides is mainly of experimental rather than of theoretical nature. This is due to the large number of electrons to correlate, which makes ab initio calculations somewhat more difficult on oxides than on hydrides, especially for such sensitive quantities as transition moments.

In the course of their study of nascent product state distributions in reactions of Group III-B atoms with various oxiders, Liu & Parson (1977) detected time-resolved fluorescence from which they deduced the lifetimes of molecular excited states in various vibronic levels. These measurements concern the first excited

$^2\Pi$ and $^2\Sigma^+$ electronic states of ScO, YO and LaO. All the measured lifetimes lie in the range 25–40 ns. These values compare well with those measured in the isoelectronic alkaline-earth monohalides by Dagdigian et al. (1974).

For LaO, recent measurements using laser-induced decay fluorescence have been carried out by Carette & Bencheik (1993). This study confirms the lifetime found by Liu & Parson (1977) for the $B^2\Sigma^+$ state (32 ns instead of 34 ns) but not for the $C^2\Pi$ state for which they find a three-times longer lifetime (77 ns instead of 27 ns).

There has been considerable interest in the determination of radiative properties of titanium monoxyde TiO that is known to be one of the main features in the absorption spectra of M-type stars (Kiess 1948; Merrill et al. 1962; Machara & Yamashito 1976). The time-resolved laser pulse-fluorescence decay method has been widely used for TiO. It led to the determination of lifetimes around 30 ns for the v'=0,1,2 vibrational levels of the $C^3\Delta$ state (Steele & Linton 1978). Feinberg & Davis (1976, 1977) also measured a short lifetime of 17.5 ns for the $c^1\Phi$ state. Recently, the same method was used by Simard & Hackett (1991) for the long lifetime of the $E^3\Pi$ (v=0) state (770±40 ns), by Doverstal & Weijnitz (1992) for the v=0 levels of the $A^3\Phi_2$ (τ=102 ns), $B^3\Pi_0$ (τ=55 ns) and $C^3\Delta_1$ (τ=32 ns) states and by Carette & Schamps (1992) for the $B^3\Pi$ (v=1) state (τ=44±2 ns). This last value is in excellent agreement with that determined by Davis et al. (1986) from absorption coefficient measurements that concerned also $A^3\Phi$ (τ=51±9 ns, i.e. half the recent value found from population probing techniques by Doverstal & Weijnitz [1992]), $C^3\Delta$ (τ=18.5±3 ns, smaller than values from time-resolved spectroscopy) and $b^1\Pi$ (τ=192±55 ns, the only experimental determination for this state). Various other determinations of oscillator strengths (leading to lifetime values assuming a single-channel depopulation) have been performed by absorption-coefficient measurements (Golden 1967; Collins & Faÿ 1974) or absolute-intensity studies (Zyrnicki 1975; Price et al. 1971, 1974; Linton & Nicholls 1970). Recently, ab initio calculations for TiO were carried out by Schamps et al. (1992) that reproduced well the lifetimes of short-lived states. This agreement enabled them to predict a value of 42 ns for the as yet unmeasured lifetime of the $f^1\Delta$ state. However, the agreement for the long-lived $E^3\Pi$ and $b^1\Pi$ states was much less satisfactory by factors of more than two as a probable consequence of an insufficient description of electronic correlation in their calculations.

The other transition-element monoxides are much less documented. For ZrO, that is isovalent to TiO, only the lifetime of the J'=17, v'=0 level of the $B^3\Pi$ state has been measured (Hammer & Davis 1979). The analysis of the decay of resonant fluorescence radiation led to a value of 32.5±2 ns for the lifetime of the B state from which transition probabilities and oscillator strengths for the bands of the $B^3\Pi-X^3\Delta$ system were estimated. Fortunately the theoretical study of the spectroscopy of the ZrO molecule by Langhoff & Bauschlicher (1990) counterbalances the lack of experimental data.

FeO was studied by West & Broida (1975) whose photoluminescence and chemiluminescence experiments allowed to determine lifetimes of the order of

450±100 ns for the radiative lifetimes of the $A^5\Sigma^+$ and $B^5\Pi$ upper electronic states of the orange bands.

Finally, two studies should be mentioned for CuO. The first one concerns the pulse-decay determination of lifetimes of four ionic valence states: $A^2\Sigma^-$ (τ=0.65 μs), $C^2\Pi$ (τ=1.3 μs), $D^2\Delta$ (τ=1.8 μs), $A'^2\Sigma^+$ (τ >5 μs). This study made possible a separation of quartet states from doublet ones on the basis of large differences in their lifetime magnitudes (Delaval et al. 1983). The second study is an ab initio calculation on the spectroscopy of CuO (Hoppe & Peyerimhoff 1992). In this study transition moments and lifetimes were systematically computed for numerous valence states in various molecular orbital bases. In spite of the strong dependence of results on the basis used, correspondence between observed and calculated states could be given although the situation is quite complicated in this molecule.

6 Conclusion

The transition element and lanthanide oxide data complete this review that the author wishes to be as exhaustive as possible. Some papers have probably escaped his bibliographic search and he will be grateful to those readers who will be kind enough to mention to him any omission or mistake he could have made.

The major conclusion that arises from this compilation is the urgent need for further experiments and calculations on radiative properties of small molecules. In spite of the abundant literature on the subject, the number of molecules, and for each of these molecules the number of band systems, that have been sampled from the point of view of radiative dynamics is eventually scarce compared to what remains to do. It would be much longer to list all the molecules, if only diatomics, for which radiative properties have not been investigated at all than those for which something is known. For astrophysicists the priority should concern those poorly studied molecules that are yet expected to be significantly present in cool stellar sources, e.g. FeO, LaO, ScO, VO, YO, NiO, CoO, AlO, CaO, NiO_2, YO_2, ScO_2, ZrO_2, VO_2, and the corresponding hydrides.

Yet the situation evolves in the right direction. Whereas the possibilities for getting new experimental data seem to slow down, if not to saturate, the fast development of ab initio calculations is more than encouraging. Computer facilities and program improvements have significantly increased the possibilities of quantum chemistry. Electronic transition moments can now be computed with an accuracy that rivals the best experimental determinations. Even they replace them in the extremely frequent cases when experimental measurements are not feasible. However, one must be cautious about the use of ab initio calculations in predicting accurate radiative properties. For example, small shifts in potential curves may sometimes change drastically the calculated values of oscillator strengths. Also, the possible presence of predissociation effects may decrease, by orders of magnitude, the radiative lifetimes deduced from only the calculated transition moments. Anyway, in the vast majority of cases, ab initio calculations, checked whenever possible by experimental measurements, should be the key for providing a dramatic increase in our knowledge about molecular radiative properties in the coming years.

References

Anglada J., Bruna P.J., Peyerimhoff S.D., 1989, *Molec. Phys.*, **66**, 541
Anglada J., Bruna P.J., Peyerimhoff S.D., 1990, *Molec. Phys.*, **69**, 281
Baltayan P., Jourdan A., Nedelec O., 1976, *Phys. Letters*, **58A**, 443
Baltayan P., Nedelec O., 1979, *J. Chem. Phys.*, **70**, 2399
Bauer W., Engelhardt B., Wiesen P., Becker H., 1989, *Chem. Phys. Letters*, **158**, 321
Bauschlicher C.W., Langhoff S.R., 1988, *J. Chem. Phys.*, **89**, 2116
Bauschlicher C.W., Langhoff S.R., 1988, *Theor. Chem. Acta*, **73**, 43
Belyaev V.N., Lebedeva N.L., Krasnov K.S., Gurvich L.V., 1982, *Opt. Spectrosc.*, **55**, 1075
Born M., Oppenheimer R., 1927, *Ann. Physik*, **84**, 457
Brieger M., Hese A., Penn A., Sodeik A., 1981, *European Conf. on Atomic Physics*, Heidelberg, p.353
Brieger M., Penn A., Sodeik A., Hese A., 1983, *Chem. Phys.*, **75**, 1
Busener H., Heinrich F., Hese A., 1987, *Chem. Phys.*, **112**, 139
Campbell M.L., McClean R.E., Garland N.L., Nelson H.H., 1992, *Chem. Phys. Letters*, **194**, 187
Carette P., Schamps J., 1992, *J. Molec. Spectrosc.*, **154**, 448
Carette P., Bencheikh M., 1993, to be published
Castex M.C., LeCalve J., Haaks D., Jordan B., Zimmerer G., 1980, *Chem. Phys. Letters*, **70**, 106
Chapman D.A., Li J., Balasubramanian K., Lin S.H., 1988, *J. Chem. Phys.*, **88**, 3826
Clyne M.A., Heaven M.C., 1980, *Chem. Phys.*, **51**, 299
Collins J.G., Faÿ T.D., 1974, *J. Quant. Spectrosc. Radiat. Transfer*, **14**, 1259
Cruse J.A., Zare R.N., 1974, *J. Chem. Phys.*, **60**, 1182
Dagdigian P.J., Cruse J.A., Zare R.N., 1974, *J. Chem. Phys.*, **60**, 2330
Dagdigian P.J., Cruse J.A., Zare R.N., 1975, *J. Chem. Phys.*, **62**, 1824
Dagdigian P.J., 1976, *J. Chem. Phys.*, **64**, 2609
Darrah S.D., Silvers S.J., 1985, *Chem. Phys. Letters*, **121**, 377
Davis S.P., Littleton J.E., Phillips J.G., 1986, *Astrophys. J.*, **309**, 449
Delaval J.M., David F., Lefebvre Y., Bernage P., Niay P., Schamps J., 1983, *J. Molec. Spectrosc.*, **101**,358
Diffenderfer R.N., Yarkony D.R., Dagdigian P.J., 1983, *J. Quant. Spectrosc. Radiat. Transfer*, **29**, 329
Docken K.K., Hinze J., 1972, *J. Chem. Phys.*, **57**, 4928
Docken K.K., Hinze J., 1972, *J. Chem. Phys.*, **57**, 4936
Douglas C.H., Nelson H.H., Rice K.J., 1989, *J. Chem. Phys.*, **90**, 6940
Doverstal M., Lindgren B., Sassenberg U., Yu H., 1991, *Physica Scripta*, **42**, 572
Doverstal M., Weijnitz P., 1992, *Molec. Phys.*, **75**, 1357
Dufayard J., Nedelec O., 1977, *J. Phys.*, **38**, 449
Dufayard J., Nedelec O., 1978, *J. Chem. Phys.*, **69**, 4708
Ennen G., Ottinger C., 1975, *Chem. Phys. Letters*, **36**, 16
Erman P., Gustavsson O., Larsson M., 1983, *Physica Scripta*, **27**, 256
Evans P.J., Mackie J.C., 1977, *J. Molec. Spectrosc.*, **65**, 169
Feinberg J., Bilal M.G., Davis S.P., Phillips J.G., 1976, *Appl. Letters*, **17**, 147

Feinberg J., Davis S.P., 1977, *J. Molec.Spectrosc.*, **65**, 264
Feinberg J., Davis S.P., 1978, *J. Molec.Spectrosc.*, **69**, 445
Ferray M., Visticot J.P., Sayer B., Telle H.H., 1984, *J. Chem. Phys.*, **81**, 191
Gadea X., 1993, to be published
Giroud M., Nedelec O., 1982, *J. Chem. Phys.*, **77**, 3998
Golden S.A., 1967, *J. Quant. Spectrosc. Radiat. Transfer*, **7**, 225
Hammer P.D., Davis S.P., 1979, *J. Molec. Spectrosc.*, **78**, 337
Hebert G.R., Tyte D.C., 1964, *Proc. Phys. Soc.*, **83**, 629
Hebert G.R., Nicholls R.W., Linton C., 1980, *J. Quant. Spectrosc. Radiat. Transfer*, **23**, 229
Henneker W.H., Popkie H.E., 1971, *J. Chem. Phys.*, **54**, 1763
Herzberg G.H., 1950, *Spectra of Diatomic Molecules*, Van Nostrand, N.Y.
Hikmet I., Kowalczyk P., Sadeghi N., 1992, *Chem. Phys. Letters*, **188**, 287
Hippe D., Peyerimhoff S.D., 1992, *Molec. Phys.*, **76**, 293
Hooker W.J., Main R.P., 1971, *AlO(A−X)oscillator strengths and collisional relaxation rates*, Final Rep., KMS Techn. Center, Glendale, Cal.
Hsieh Y.K., Yang S.C., Tam A.C., Stwalley W.C., 1978, *J. Chem. Phys.*, **68**, 1448
Hsu Y.C., Hegemann B., Pruett J.G., 1972, *J. Chem. Phys.*, **72**, 6437
Huber K.P., Herzberg G.H., 1979, *Constants of Diatomic Molecules*, Van Nostrand, N.Y.
Huie R.E., Long N.J.T., Thrush B.A., 1978, *Chem. Phys. Letters*, **55**, 404
Huron B., Malrieu J.P., Rancurel P., 1974, *Chem. Phys.*, **3**, 277
Ihle H.R., Wu G.H., 1974, *J. Chem. Phys.*, **63**,1605
Irvin J.A., Dagdigian P.J., 1981, *J. Chem. Phys.*, **74**, 6178
Johnson S.E., 1972, *J. Chem. Phys.*, **56**, 149
Johnson S.E., Capelle G., Broida H.P., 1972, *J. Chem. Phys.*, **56**, 663
Jourdan A., Negre J.M., Dufayard J., Nedelec O., 1976, *J. Phys. Lettres*, **37**, L29
Kiess C.C., 1948, *Publ. Astr. Soc. Pacif.*, **60**, 252
Klynning L., Martin H., Nylen P., Royen P., 1982, *Physica Scripta*, **25**, 362
Kovacs I., 1969, *Rotational Structure in the Spectra of Diatomic Molecules*, Hilger, London
Kozlov V.A., Pupyshev V.I., Stepanov N.F., 1990, *Opt. Spectrosc.*, **68**, 595
Kuzmenko N.E., Kuznetsova L.A., Monyakin A.P., Kuzyakov Y.Y., Plastinin Y.A., 1979, *Sov. Phys. Usp.*, **17**,405
Kuznetsova L.A., Kuzmenko N.E., Kuzyakov Y.Y., Plastinin Y.A., 1980, *Optical Probabilities in Diatomic Molecules*, Nauka, Moscow
Langhoff S.R., Chong D.P., 1978, *J. Chem. Phys.*, **69**, 194
Langhoff S.R., Bauschlicher C.W., Rendell A.P., 1989, *J. Molec. Spectrosc.*, **138**, 108
Langhoff S.R., Bauschlicher C.W., 1990, *J. Molec. Spectrosc.*, **141**, 243
Langhoff S.R., Bauschlicher C.W., 1990, *Astrophys. J.*, **349**, 369
Langhoff S.R., Partridge H., Bauschlicher C.W., 1992, *Chem. Phys.*, **153**, 1
Larsson M., 1983, *Astron. Astroph.*, **128**, 291
Laskowski B., Stallcop J.R., 1981, *J. Chem. Phys.*, **74**, 4883
Lefebvre-Brion H., Field R.W., 1986, *Perturbations in the Spectra of Diatomic Molecules*, Academic Press, N.Y.
Linton C., Nicholls R.W., 1969, *J. Quant. Spectrosc. Radiat. Transfer*, **9**, 1
Linton C., Nicholls R.W., 1970, *J. Quant. Spectrosc. Radiat. Transfer*, **10**, 311
Liu K., Parson J.M., 1977, *J. Chem. Phys.*, **67**, 1814
Machara H., Yamashito Y., 1976, *Publ. Astr. Soc. Japan*, **28**, 135

Main R.P., Carlson D.J., Dupuis R.A., 1967, *J. Quant. Spectrosc. Radiat. Transfer*, **7**, 805
Main R.P., Schadee A., 1969, *J. Quant. Spectrosc. Radiat. Transfer*, **9**, 713
Matos J.M.O., Malmqvist P.A., Roos B.O., 1987, *J. Chem. Phys.*, **86**, 5032
Merril P.J., Deutsch A.J., Keenan P.C., 1962, *Astrophys. J.*, **136**, 121
Michels H.H., 1972, *J. Chem. Phys.*, **56**, 665
Nedelec O., Dufayard J., 1978, *J. Chem. Phys.*, **69**, 1833
Nedelec O., Giroud M., 1983, *J. Chem. Phys.*, **79**, 2121
Nicholls R.W., Stewart A.L., 1962, in *Atomic and Molecular Processes*, ed.D.R.Bates, Academic Press, N.Y.
Nicholls R.W., 1969, in *Physical Chemistry-III*, ed.H.Eyring,D.Henderson&W.Yost, Academic Press, N.Y.
Osmundsen J.F., Abele C.C., Eden J.G., 1985, *J. Chem. Phys.*, **83**, 2159
Pasternack L., Dagdigian P.J., 1978, *Chem. Phys.*, **33**, 1
Pfeifer J., Gole J.L., 1984, *J. Chem. Phys.*, **80**, 565
Plane J.M., Nien C.F., 1991, *J. Chem. Soc. Faraday Trans.*, **87**, 677
Popkie H.E., 1971, *J. Chem. Phys.*, **54**, 4597
Price M.L., Sulzmann K.G., Penner S.S., 1971, *J. Quant. Spectrosc. Radiat. Transfer*, **11**, 427
Price M.L., Sulzmann K.G., Penner S.S., 1974, *J. Quant. Spectrosc. Radiat. Transfer*, **14**, 1273
Pruett J.G., Zare R.N., 1975, *J. Chem. Phys.*, **62**, 2050
Querci F., 1993, unpublished
Rice J.K., Pasternack L., Nelson H.H., 1992, *Chem. Phys. Letters*, **189**, 43
Russell H.N., 1934, *Astrophys. J.*, **79**, 317
Salzberg A.P., Santiago D.I., Asmar F., Sandoval D.N., Weiner B.R.,1991, *Chem. Phys. Letters*, **180**, 161
Sachs E.S., Hinze J., Sabelli N.H., 1975, *J. Chem. Phys.*, **62**, 3384
Schadee A., 1967, *J. Quant. Spectrosc. Radiat. Transfer*, **7**, 169
Schadee A., 1971, *Astron. Astrophys.*, **14**, 401
Schadee A., 1978, *J. Quant. Spectrosc. Radiat. Transfer*, **19**, 451
Schamps J., Sennesal J.M., Carette P., 1992, *J. Quant. Spectrosc. Radiat. Transfer*, **48**, 147
Scuseria G.E., Geertsen J., Oddershede J., 1989, *J. Chem. Phys.*, **90**, 2338
Simard B., Hackett P.A., 1991, *J. Molec. Spectrosc.*, **148**, 127
Steele R.E., Linton C., 1978, *J. Molec.Spectrosc.*, **69**, 66
Stwalley W.C., Zemke W.T., 1976, *Int. J. Quantum Chem. Symp.*, **10**, 223
Stwalley W.C., Zemke W.T., Way K.R., Li K.C., Proctor R., 1977, *J. Chem. Phys.*, **66**, 5412
Stwalley W.C., Zemke W.T., Way K.R., Li K.C., Proctor R., 1977, *J. Chem. Phys.*, **67**, 4785
Stwalley W.C., 1977, *Contemp. Phys.*, **18**, 65
Svyatkin I.A., Kuznetsova L.A., Kuzyakov Y.Y., Leiko I.P., 1980a, *Opt. Spectrosc.*, **48**, 13
Svyatkin I.A., Kuznetsova L.A., Kuzyakov Y.Y., 1980b, *J. Quant. Spectrosc. Radiat. Transfer*, **23**, 307
Svyatkin I.A., Kuznetsova L.A., Kuzyakov Y.Y., 1980c, *J. Quant. Spectrosc. Radiat. Transfer*, **24**, 25
Svyatkin I.A., Kuznetsova L.A., Kuzyakov Y.Y., 1982, *Opt. Spectrosc.*, **53**, 339
Tam A.C., Happer W., 1976, *J. Chem. Phys.*, **64**, 2456

Tatum J.B., 1967, *Astrophys. J. Suppl.*, **14**, 21
Telle H.H., 1984, *J. Chem. Phys.*, **81**, 195
Telle H.H., 1986, *J. Molec. Struct.*, **143**, 565
Tyte D.C., Hebert G.R., 1964, *Proc. Phys. Soc.*, **84**, 830
Vanpee M., Kineyko W.R., Caruso R., 1970, *Combustion and Flame*, **14**, 381
Vidal C.R., Stwalley W.C., 1982, *J. Chem. Phys.*, **77**, 883
Vidal C.R., Stwalley W.C., 1984, *J. Chem. Phys.*, **80**, 2697
Von Moers F., Heitz S., Busener H., Sagner H.J., Hese A., 1987, *Chem. Phys.*, **116**, 215
Way K.R., Stwalley W.C., 1973, *J. Chem. Phys.*, **59**, 5298
Wentink T., Pederson N., Diebold G., 1971, *Radiative Behaviour of Metal Oxides in Laser Blowoff*, DASA Report No 2704
West J.B., Broida H.P., 1975, *J. Chem. Phys.*, **62**, 2566
Whiting E.E., 1972, PhD thesis, York University, Toronto, Canada
Whiting E.E., Nicholls R.W., 1974, *Astrophys. J. Suppl.*, **27**, 1
Whiting E.E., Schadee A., Tatum J.B., Hougen J.T., Nicholls R.W., 1980, *J. Molec. Spectrosc.*, **80**, 249
Wine P.H., Melton L.A., 1976, *J. Chem. Phys.*, **64**, 2692
Wing R.F., Cohen J., Brault W., 1977, *Astrophys. J.*, **216**, 659
Woodward J.R., Hayden J.S., Gole J.L., 1989, *Chem. Phys.*, **134**, 395
Yarkony D.R., 1988, *J. Chem. Phys.*, **89**, 7324
Yerle R., 1979, *Astron. Astrophys.*, **73**, 346
Yoshimine M., McLean A.D., Liu B., 1973, *J. Chem. Phys.*, **58**, 4412
Zemke W.T., Stwalley W.C., 1978, *J. Chem. Phys.*, **68**, 4619
Zemke W.T., Crooks J.B., Stwalley W.C., 1978, *J. Chem. Phys.*, **68**, 4628
Zemke W.T., Way K.R., Stwalley W.C., 1978, *J. Chem. Phys.*, **69**, 402
Zemke W.T., Stwalley W.C., 1978, *J. Chem. Phys.*, **69**, 409
Zemke W.T., Stwalley W.C., 1980, *J. Chem. Phys.*, **73**, 5684
Zyrnicki W., 1975, *J. Quant. Spectrosc. Radiat. Transfer*, **15**, 575

The Berkeley Program on Molecules of Astrophysical Interest

Sumner P. Davis

Department of Physics University of California at Berkeley Berkeley, California 94720, USA

1 Introduction

A systematic program of laboratory analyses of selected molecular spectra of astrophysical interest started in 1958 and continues to the present time. The program includes production of spectral atlases, tabulations of spectral lines, analyses, calculations of excitation energies and molecular parameters, measurements of radiative lifetimes, and determinations of transition strengths. Work has been completed or is in progress on the spectra of ArH^+, C_2, carbon clusters, CN, CS, CaCl, CaH, CaS, FeD, FeH, HgH, HgD, InI, LaO, LaS, OD, OH, SH, Si_2, SiC_2, TiCl, TiO, TiO^+, VO, YS, ZrCl, ZrO, and ZrS. The basic needs for astronomically useful data have not changed, but laboratory and analysis methods have become more sophisticated in order to cope with ever greater demands for consistency, accuracy, and breadth of information. The Fourier transform spectrometer and computer codes for analyses have enhanced our ability to satisfy some of these demands.

The need for better laboratory spectra of astrophysically important molecules is a continuing one, no less important today than it was 35 years ago when the program at Berkeley was started. The genesis of the Berkeley program was at a meeting of the Joint Commission for Spectroscopy held in Columbus in 1956, where Charlotte Moore read a statement about molecular spectra of astrophysical interest. She proposed that a systematic program on the laboratory analyses of selected molecular spectra be planned immediately and carried out as soon as possible. Two years later at the Moscow meeting it was reported that a program directed by F.A. Jenkins and J.G. Phillips was being started in response to the well-defined need. In 1959 I joined the program and became a co-principal investigator in 1961 when Prof. Jenkins passed away (Phillips & Davis 1962). Details of the program up to 1987 are given in Publications of the Astronomical Society of the Pacific (Davis 1987), which should be together with this for a complete picture. I'd like to mention here, a little bit of how we started and what we worked with, and then take you through the process of producing and analyzing a spectrum with our present instrumentation. The description will be abbreviated, but I hope you will end with an increased appreciation of the quality and scope of our results as well as their limitations.

2 Early goals

2.1 Scientific bibliography

Right at the start it was recognized that an up-to-date bibliography and progress report of our work as well as others' were essential for keeping track of spectroscopy of astrophysical interest. The Berkeley Newsletter Analysis of Molecular Spectra was started and continues as a bimonthly publication. It is available to anyone for a modest subscription fee. In this newsletter appear references from about 150 journals, separated into categories of diatomic and small polyatomic species, and general information including theoretical studies. It is currently running about 450 references in each issue. The task of collecting references from current scientific journals is borne entirely by Dr. David M. Eakin, one of the early students, and for many years an award-winning high school science teacher. Cumulative files dating from 1986 are kept by Prof. Phillips, and data for particular molecules can be had for the asking. Complete back issues are not generally available. Fig. 1 is a reproduction of a recent title page.

BERKELEY NEWSLETTER

ANALYSIS OF MOLECULAR SPECTRA

No. 200 July, 1992

You have undoubtedly noticed that this is Issue Number 200 of these bi-monthly Newsletters. It comes as quite a shock to us to realize that we have been distributing the Newsletters for over thirty three years. We are gratified that our 160 recipeints world-wide find it useful. and plan to continue its distribution as long as we are physically able to do so and as long as it proves to be of interest.

BIBLIOGRAPHY

Al_2 Electronic structure calculations of small Al_n (n=2–8) clusters
Al_3 T. Bastug et.al.
Al_4 *Z. f. Phys. D* **22**, 641 (1992)

AlF The infrared emission spectrum of gaseous AlF
H.G. Hedderich and P.F. Bernath
J. Molec. Spectros. **153**, 73 (1992)

AlH The ground-state infrared spectra of aluminum monodeuteride: the mass-independent molecular parameters of group IIIa hydrides
R.D. Urban and H. Jones
Chem. Phys. Lett. **190**, 609 (1992)

Fig. 1. Title page of the Berkeley Newsletter.

2.2 Spectrum Line Lists

The original goal was simply to record an entire spectrum from 400 nm to 1200 nm, analyze it, calculate molecular parameters, and provide a list of all identified lines. The spectrum chosen for the first one was the red system of CN covering the region from 480 nm to 1200 nm, the long wavelength limit of photographic plates. Hundreds of plates were taken, tens of thousands of spectrum lines were measured semiautomatically, reduced by computer, analyzed, and finally tabulated band by band, branch by branch. At that time it was impractical to make a finding list. The final result was a book (Davis & Phillips 1963). Work on the Swan system of C_2 soon followed, and it too was incorporated into a book (Phillips & Davis 1968). Subsequently book publishing became too expensive for long line lists. We settled for publication in the journals so far as possible.

2.3 Atlases

Spectrum atlases – plots of spectral intensity vs. either wavelength or wavenumber – are often more helpful than line lists when trying to identify the presence of particular species in stellar spectra, or to identify unknown spectral lines. We made enlargements of spectra for our own use and for publication, but the production of Gatterer et al.(1957) type of atlas seemed an unwise use of our time.

2.4 Analysis and Calculation of Parameters

Measurements of line positions must be separated into bands and branches and assigned to a specific rotational quantum number before useful parameters can be calculated. The line-by-line picking of branches soon became inadequate to separate complex overlapping structures and a more automated procedure was developed (Phillips 1959). He simply programmed a computer to do what he did by hand, and had it print out enough data for him to determine whether a branch was real or not. After that, parameter calculation was straightforward enough. However, line lists were still essential because using the parameters to reproduce the spectrum was not possible for everyone, since large computers were not readily available.

2.5 Sources and Spectrometers

Early spectroscopists exercised great ingenuity in constructing sources to coax unwary species to radiate and give up their secrets, but we used a few relatively simple ones. Our CN sources, for example, were an arc, a Geissler tube, and a King furnace.

The first spectrometer was a concave diffraction grating in a 6.3 m Paschen mounting which covered the spectrum up to 1200 nm in first order. The first grating was ruled by Rowland, and the second by Bausch and Lomb. The mounting dated back to the early 30's, and was used by R. T. Birge in some of his early work. Later I obtained a nearly perfect original plane grating 13.5 cm wide (it was a generous gift from George Harrison at M.I.T. after he made six replica. He sent them out to silence his critics who scoffed at an interferometrically-controlled continuous-motion ruling engine). It had no ghosts of any significance and produced nearly symmetrical lines, even for those of high intensity. We installed it in a Czerny-Turner mounting with 3-meter focal length mirrors. It served us well for many years, and it proved to be superior in throughput, resolution, line-shape, and dispersion to the concave grating.

2.6 Wavelength Standards

A major concern when making line lists is the accuracy of the wavelengths, both relative and absolute. Iron lines were the standards for many years, but they suffer from being irregularly distributed across the spectrum and are of widely differing intensities. Harrison and others pushed for using thorium lines as standards because they are nearly evenly distributed, their intensities are reasonably uniform, and there are lots of them. As ruled gratings improved, recorded line shapes became more symmetrical, and the standards for accuracy rose even higher. We kept a list of those thorium lines which consistently gave smooth curves for interpolating wavelengths. There was still little hope of measuring line shapes even with the best of gratings – who has ever seen subsidiary maxima of spectrum lines, even though every elementary text explains and illustrates their positions and intensities.

2.7 Intensities and Equivalent Widths

Initially intensities were of secondary importance except on a relative basis. We arbitrarily assigned 100 to the strongest lines in a band and 1 to the weakest, because diatomic molecular line intensities change slowly and predictably, without the extreme variations found in atomic spectra.

The measurement of line intensities and shapes and equivalent widths was tentative at best. On our semi-automatic comparator we had a photomultiplier to register transmitted light. The intensity scale was calibrated with a step weakener as the plates were being exposed. Strong lines and very strong lines and extraordinarily strong atomic lines all produced the same photomultiplier output, which was automatically registered when a line position was measured. Shapes and hence wavelengths changed with intensity because of ghosts, satellites, and "grass" resulting from errors in ruling. A line of maximum strength might be recorded with an intensity of 100, but a line 1000 times stronger was also recorded as 100, although it might have a width many times greater because of saturation and blooming of the emulsion.

2.8 Comments

Work proceeded as above for some years, although improvements in computers made our data reduction much simpler and allowed us to measure and process it at ever increasing rates. By the time tab cards (popularly known as IBM cards) became obsolete, we had processed a million lines, and had that many cards stored in our comparator room to prove it. After a time our research took a dramatic turn for the better because of a combination of developments in instrumentation. It was a change of far greater significance than going from concave gratings to interferometrically ruled plane gratings.

3 Current Practices

With the advent of the Fourier transform spectrometer at the National Solar Observatory at Kitt Peak, there came major improvements in measuring line positions, intensities, shapes, and equivalent widths. Most importantly, the instrumental function is specifically determinable and can be corrected for if necessary. Ghosts, satellites, and grass or their equivalents are negligible for practical purposes. With this change came digital data processing and the recording of spectra as sets of numbers rather than images on a photographic plate or traces on a chart recorder.

3.1 The Spectrometer

A few words about the FTS may be appropriate. A Fourier transform spectrometer is a scanning Michelson interferometer. The input must be light from a stationary source (no changes in spectral composition or amplitude over short time intervals), but the output is an interferogram rather than a frequency or wavelength spectrum. The interferogram must undergo a Fourier transformation in order to display the spectrum. A diffraction grating automatically takes a Fourier transform of the incoming radiation and displays it with a one-to-one correspondence of wavelength and angle of diffraction. Here, the FTS only maps optical frequencies into very much lower spatial frequencies, and we must compute the transform. All data are digitized at an appropriate sampling rate so the mathematics of data reduction is similar to that of continuous functions. With the digitization comes ease of interactive manipulation at the keyboard of a computer. The data and resulting spectrum can be displayed and modified at will, to extract all the inherent information.

The maximum path difference of the Kitt Peak instrument is one meter, making the resolution limit 0.010 cm^{-1} (with this change in instrumentation we switch from wavelengths to wavenumbers, so we speak of cm^{-1} rather than nm, with small intervals specified as millikaysers; 1 mK = 0.001 cm^{-1}. The instrument therefore has a resolving limit of 10 mK).

The wavenumber ranges depend on the available detectors and beam splitters. Practically speaking the observable spectrum extends from 400 cm^{-1} (25 μm) to 40000 cm^{-1} (250 nm) with three sets of beam splitters and detectors. All of our observations to date lie in the region 1800 cm^{-1} (5.5 μm) to 40000 cm^{-1}. The time required for a single scan at maximum resolution is about 15 minutes. A minimum number of 4 scans is desirable because it reduces the noise by a factor of 2.

3.2 Light Sources

All light sources used with the FTS must be steady in intensity and spectral composition, with no sudden changes on a time scale consistent with the spectral range being observed. Just think of the Fourier transform of a sudden change. If the transform has frequencies in the range being observed, then the source is unsatisfactory. Let's consider some examples. A sinusoidal modulation, such as occurs when a power supply is insufficiently filtered or a plasma is oscillating, produces sidebands on every spectral line.

Sharp spikes in intensity, which frequently and irregularly occur in many sources, produce a broad noise spectrum. A dramatic illustration of this is found in the use of a furnace for infrared spectra. With something like ZrO_2 as a charge, when it melts, small hot particles are produced. These fly around inside the furnace until steady-state conditions are reached, and sometimes cling to the top of the bore, where they come loose at irregular intervals for quite some time. As they pass across the bore, they cross the region imaged on the spectrometer entrance aperture and cause large fluctuations in the observed intensity. Hot particles are bright in the infrared! The result is lots of noise in the transformed spectrum from these spikes in intensity.

In spite of these problems, for emission spectra we have had good success with use of several sources: the furnace, discharge tubes both sealed and unsealed and with and without electrodes, hollow cathodes, inductively coupled plasmas, diffusion flames, and a carbon arc (once or twice).

For absorption spectra, the problems usually have nothing to do with the stability of the background source but instead with its spectral composition and emissivity, as will be made clear below with a specific example. The "blackbody" temperature of a lamp may be 2500 K in the visible region of the spectrum, but decrease to 1800 K in the near infrared. When something like the sun is used for background radiation, the spectrum with the laboratory absorber in place must be ratioed with the source spectrum alone.

3.3 Data Reduction and Analysis, Emission Spectra

Raw data are written onto a magnetic tape, and then transformed and placed in a format which is used by the code DECOMP (Brault & Abrams 1989). It is an IBM PC compatible interactive data processing code especially for input from the NSO Kitt Peak FTS (it can, however, be adapted easily for use with spectral data obtained from any kind of instrument). The picking of branches and the subsequent analysis is accomplished with the help of the code ANALYSIS (Pecyner & Davis 1988).

The whole process goes something like this: The transform of the FTS interferogram is a set of numbers representing the signal strength at equally spaced points on the wavenumber axis. In turn we (1) plot them at low resolution to take a first look at the spectrum; (2) subtract the background; (3) make a low resolution atlas to get a perspective of the entire system, especially the large-scale features such as band heads; (4) make atlases at full resolution, both low and high dispersion; (5) pick the lines; (6) fit the lines; (7) prepare a finished line list; (8) pick the branches; (9) calculate the molecular parameters; (10) calculate FC factors and potential curves;

To make this sequence clearer let's follow it through each step with examples from our recent or current work. Let's start with how we treat the data after they are transformed.

Fig. 2 shows a low-resolution plot of the transformed data for FeH, to illustrate the need for background subtraction. The spectrum comes from a carbon tube furnace charged with iron, hydrogen, and helium. The data, consisting of about two million points, were separated into 4096-point sections and the points in each section averaged to make this plot.

Fig. 3 shows a plot of the minima in each of the sections. This curve is smoothed and filtered as necessary, stretched to cover the entire original set of data points, and then subtracted from the data.

Fig. 4 is a low-resolution plot of the background-subtracted data. These three plots are linear in the intensity scale.

Fig. 5 shows a full resolution plot with low dispersion. The intensity scale is linear up to 10, and logarithmic thereafter.

Fig. 6 shows part of a full resolution plot with high dispersion. The scale is linear/log as in the previous Figure.

Fig. 7 shows the same spectral lines, this time on a linear scale, whose positions have been located by our line-finding program. A line is located by finding the slopes at the points where the second derivatives of the data are zero, determining where the tangents cross, and averaging this position with that of the maximum of the data for the line. This position is only a first approximation. Lines must always be fitted in order to obtain accurate measures of their positions, intensities, and widths, because there is only a limited number of sampling points in each line, and a sampling point does not generally coincide with the peak of a line. The lower part of the Figure shows residuals after the lines are fitted and the fitting function subtracted. For the fitting, the model of a Voigt profile is chosen. This profile has gaussian and lorentzian components. The program performs a non-linear least-square fit, and adjusts the intensity, position, width, and ratio of gaussian to lorentzian component, in order to get the best fit. Any of these parameters can be fixed when it is appropriate to do so.

Fig. 8 shows a magnified version of the residuals. It is evident that there is an oscillating component to the spectrum which has not been accounted for. This oscillation is called ringing, and is visible at the feet of sharp spectral lines – lines narrower than the resolving limit of the spectrometer. With proper filtering and fitting, the ringing can be properly removed and the line restored to its original shape as emitted by the source. The process is not the same as simply smoothing the data to hide the oscillations, but has a sound basis in information theory.

Fig. 9 shows the numerical values of the molecular parameters for each line. The intensities are accurate to a few percent, widths and equivalent widths to a few percent, and position to an accuracy depending on the signal-to-noise ratio. In broad terms it has a wavenumber precision as good as the experimental

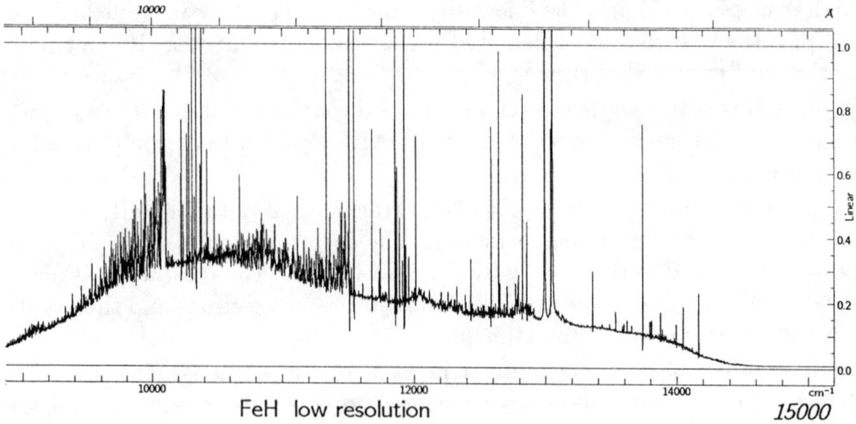

Fig. 2. Low resolution plot of the FeH furnace spectrum from 9000 to 15000 cm^{-1}.

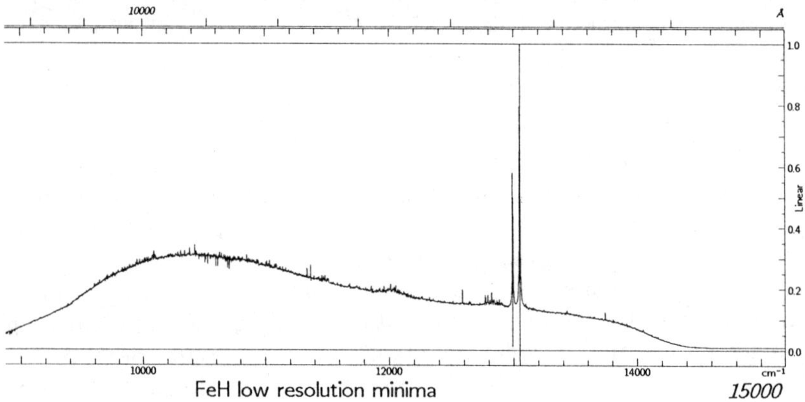

Fig. 3. Plot of minima in the low resolution FeH spectrum.

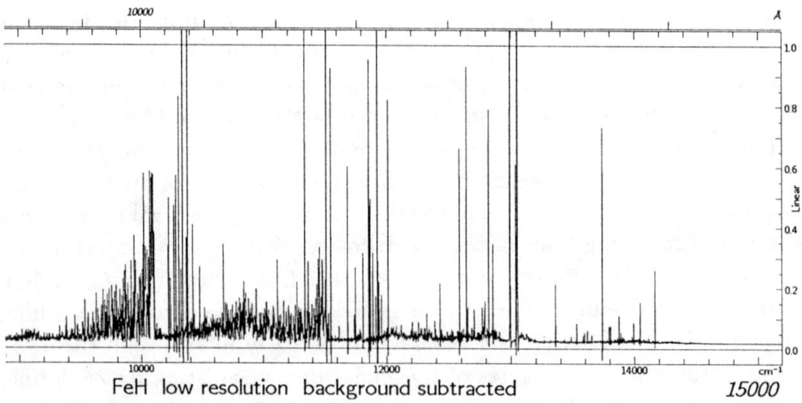

Fig. 4. Background-corrected low resolution FeH spectrum

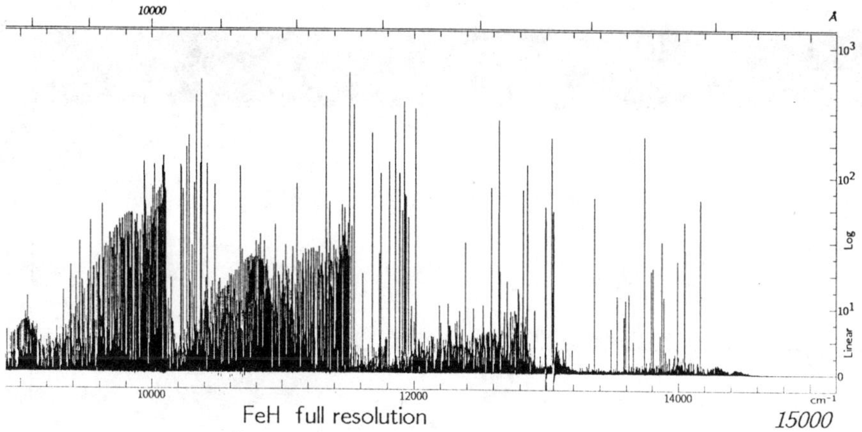

Fig. 5. Full-resolution plot of background-corrected FeH spectrum. Note that weak features are enhanced because the intensity scale is logarithmic above 10

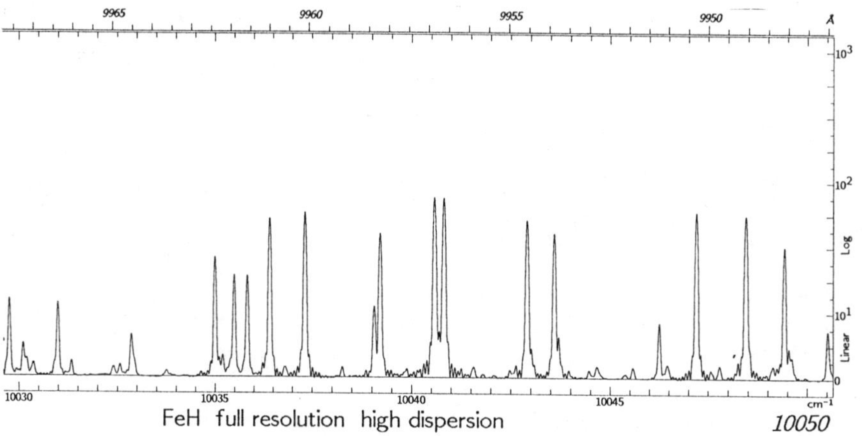

Fig. 6. FeH spectrum plotted at full resolution with high dispersion

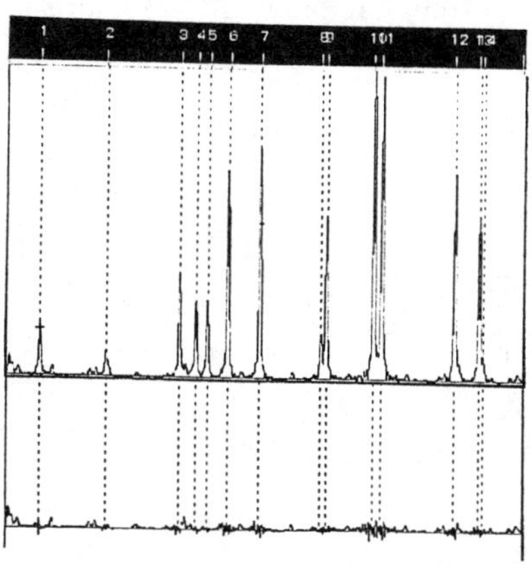

Fig. 7. (upper) FeH spectrum with lines picked by a line-finding program. (lower) Residuals after fitting and subtracting the fitting function (Voigt profile)

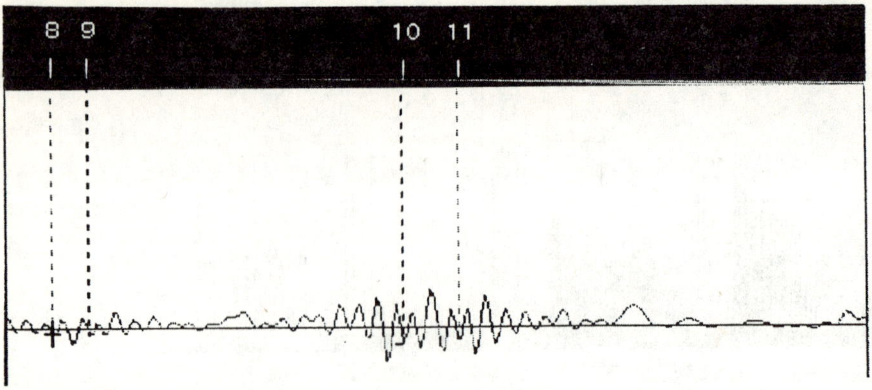

Fig. 8. Magnified residuals showing spectrum line "ringing".

No.	Sigma	Amp	Wid(mK)	Dmp	Itn	H	T	Ew	rms	even	odd
1	10030.9690	.0355	76.4	.664	8	0	L	3.76	.0011	.0007	.0009
2	10032.8438	.0189	99.4	.750	10	0	L	2.70	.0011	.0002	.0011
3	10034.9711	.0798	69.3	.507	6	0	L	7.22	.0023	.0023	.0012
4	10035.4643	.0578	74.0	.572	5	0	L	5.73	.0018	.0010	.0014
5	10035.8026	.0569	74.5	.404	5	0	L	5.31	.0010	.0007	.0007
6	10036.3877	.1586	72.0	.314	4	0	L	13.80	.0027	.0025	.0009
7	10037.2975	.1772	71.6	.299	5	0	L	15.24	.0030	.0029	.0009
8	10039.0365	.0328	81.1	.000	10	0	L	2.84	.0015	.0012	.0009
9	10039.1883	.1214	74.6	.273	9	0	L	10.78	.0017	.0015	.0008
10	10040.5597	.2301	74.8	.314	5	0	L	20.79	.0043	.0040	.0017
11	10040.7993	.2280	74.7	.311	3	0	L	20.57	.0042	.0038	.0016
12	10042.8837	.1550	72.5	.397	7	0	L	14.05	.0031	.0027	.0018
13	10043.5663	.1219	70.8	.303	12	0	L	10.38	.0018	.0017	.0008
14	10043.6822	.0153	55.8	.945	13	0	L	1.32	.0017	.0011	.0013

Fig. 9. Line parameters resulting from the fitting process. "Amp is the amplitude on an arbitrary scale, "Wid(mK) is the width at half maximum in units of 10-3 cm^{-1} (millikayser), "Dmp" is the damping constant – 0 is pure gaussian profile, 1 is pure lorentzian, "Itn" is the number of iterations in the fitting, "H" and "T" are informational columns, "Ew" is the equivalent width (area), "rms", "even", and "odd" are the residuals of the fit

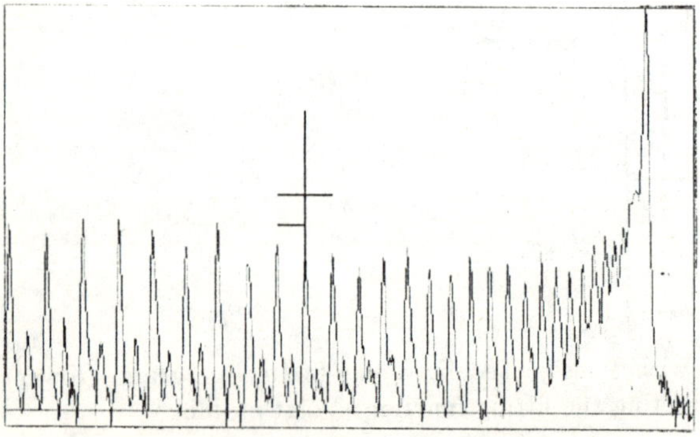

Fig. 10. (0-0) band head of CaS at 15200 cm^{-1}, degraded to the red.

repeatability of the source's optical positioning and its operating conditions. In many cases this is a few tenths of a mK. For the absolute calibration of wavenumbers, a first approximation is taken from the wavenumber of the laser which controls the position of the mirrors. It is temperature stabilized, and in practice the repeatability is a few mK. A few standard lines from argon or CO or other species at any convenient wavenumber are sufficient to set absolute standards everywhere in the spectrum. From this file we construct a line list.

Fig. 10 shows this bandhead, degraded to the red (left). Now we take the linelist and plot it, instead of plotting the full spectrum profile. We obtain a histogram, or "stick" spectrum, as shown in Fig. 11. We can easily see the branches, and assign a few lines; these lines are then extended by extrapolation, more picked, and the process proceeds. The numbers at the top are the coefficients of the fit, and the last one gives an idea of how important the last term is. "Leaf" refers to which branch the line at the cursor belongs; "Number" is the running number m in the polynomial fit, equal to –J for P-branches and J+1 for R-branches; "Sigma" is the wavenumber in cm-1, and "Weight" tells whether the line is included in the fit or not (lines more than 3.2 σ standard deviations).

With a few more keystrokes the molecular parameters are calculated, as shown in Fig. 12. The covariance-correlation matrix is also printed out, as are data on the spectrum lines. Column 1 is the running number m, column 2 is the wavenumber in cm^{-1}, column 3 is the intensity, and column 4 tells whether the line is included in the purged fit.

Finally, calculation of Franck-Condon factors is routine, accomplished with another code (Jarmain & McCallum 1970).

3.4 Data Reduction and Analysis, Absorption Spectra

Data reduction and analysis for absorption spectra is very little different from that for emission spectra, except for initial treatment of the data. The background is no longer subtracted from the spectrum, but ratioed with it. The next step is to take the logarithm to transform the data to an absorbance spectrum. From then on the procedures are identical. Let's look at the solar spectrum to illustrate the ratioing process.

Fig. 13 is a section of the solar spectrum observed from Spacelab 3, from 2000 to 2500 cm^{-1} in the infrared. The plot has been normalized so that the top of the plot is 1.0. The bottom is arbitrarily set to 0.4 The bandheads of a CO sequence are quite evident, as is the fact that the background radiation is far from constant in intensity. As a result, the plot does not truly represent the relative intensities of the lines. Instead, they are given by the ratio of the depth of each peak from the back ground, to the distance from zero intensity to the background, and consequently a background correction by ratioing is required.

Fig. 14 shows the same spectrum with the background straightened. A casual glance might indicate that the atlas is now an accurate representation of widths and intensities. Evidently some of the early atlases were prepared in just this way.

```
COEFF(0)=  .151945E+05    COEFF(1)=  .342713E+00    COEFF(2)= -.990549E-02
COEFF(3)= -.443040E-06    COEFF(4)= -.530684E-08    COEFF(5)=  .614617E-12
STD.RESIDUE=  .0255   ERR. OF LAST COEF.=  .2E-12

15185.000     Leaf= 0   Number=  42.0    Sigma= 15191.323    Weight=1.15200.000
                                15191.32279
```

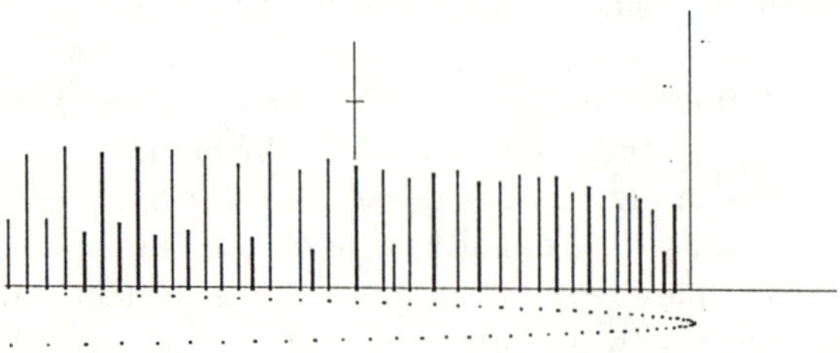

Fig. 11. Histogram of (0-0) band of CaS with P and R branches marked.

```
CaS  (0-0) Band, P and R branches

MODEL NUMBER 1             DEGREE OF FIT= 5

      RESULTS AFTER PURGING

  Vo=       15194.43361 +/-    .00096
  B'=     .16648921E+00 +/-   .98E-05
  D'=     .12594174E-06 +/-   .99E-09
  H'=     .58787490E-12 +/-   .31E-13
  B"=     .17638233E+00 +/-   .98E-05
  D"=     .11989182E-06 +/-   .98E-09

      STANDARD RESIDUE=    .0058
  m         sigma         int      fit   residual

-176.0  14824.63078  .1614      0.    -.71227
-173.0  14836.20041  .1796      0.    -.59507

-125.0  14996.37290  .4831      1.    -.00274
-124.0  14999.21302  .5617      1.    -.00853

 125.0  15080.36917  .4596      0.     .01359
 -91.0  15081.24413  .6392      1.    -.00320
 124.0  15082.56551  .3696      1.     .00310

  25.0  15196.86401  .1537      0.    -.05182
  23.0  15197.05039  .3449      0.     .02815
Head
```

Fig. 12. Molecular parameters. Only a few of the lines are shown, to illustrate the branches: -m = J for P branches, M = J+1 for R branches.

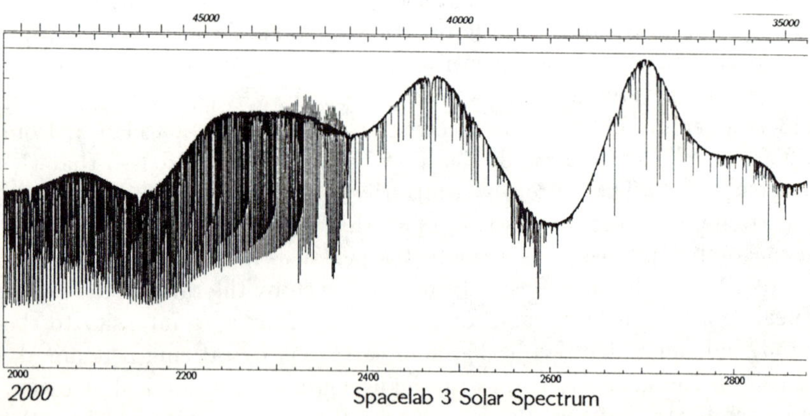

Fig. 13. A portion of the solar spectrum as observed from Spacelab 3. The intensity scale on the vertical axis is linear from .4 to 1.0

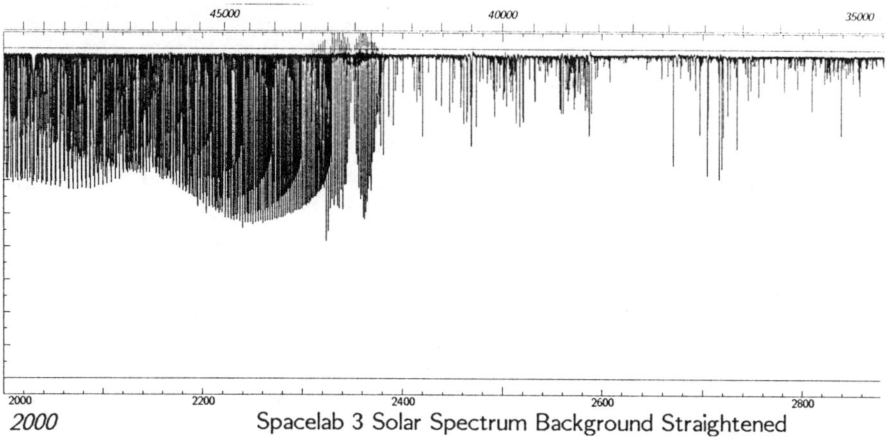

Fig. 14. Spacelab 3 solar spectrum with the background arbitrarily set to 1.0. The intensity scale is linear from .4 to 1.0

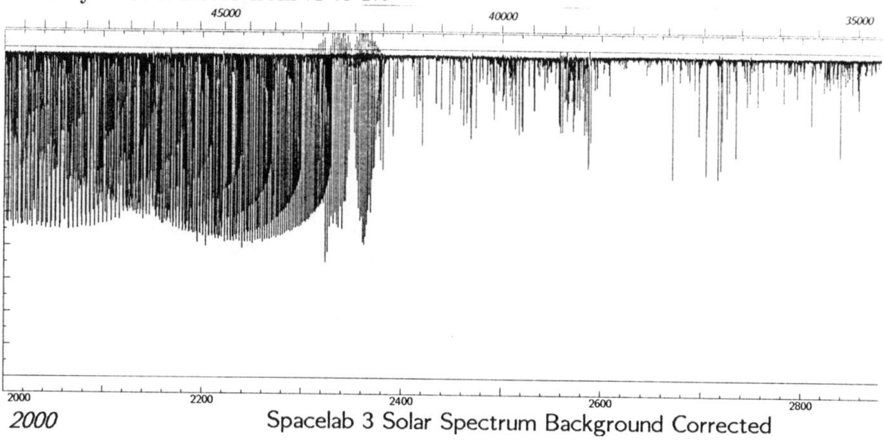

Fig. 15. Spacelab 3 solar spectrum with the background arbitrarily set to 1.0. The intensity scale is linear from .4 to 1.0

However, a look at Fig. 15 which is the correctly ratioed spectrum, shows this is not the case. In the region near 2000 cm^{-1}, for example, true intensities are about one-third greater than the background-straightened spectrum shows. The gradual and monotonic changes in intensities give a clue that the representation is now accurate.

When we take the logarithm of this spectrum, plot it, and pick and fit the lines as before, we then have the true intensities and equivalent widths from which we can make further calculations.

3.5 Line Lists, Analyses, and Atlases

From time-to-time we get requests for line lists. We are happy to provide them when we are able, and will often make up something special. However, it has only been the last several years since the advent of PCs that it is easy to prepare them and write the information on diskettes. Earlier work is occasionally stored on paper, except for the original data which are on magnetic tape. Anything unpublished almost always has to be redone, and of course ever after is then readily available. The same is true for analyses – computation of molecular parameters – but in these cases all our work has been published.

We also get requests for atlases geared to a specific region of a spectrum, with a specified dispersion. These are fun to answer because it means someone has a pressing astronomical problem which needs resolution, and maybe we can help. Let me tell you of one we are working on now. Prof. Jesse Greenstein needs a spectrum of VO to compare with the stellar spectrum of GD165B, "the faintest, coolest known M-dwarf", and the next coolest star PC0025+0447. He suggested that I look at a paper (Kirkpatrick et al. 1993) to compare spectra. On my next trip to Kitt Peak I charged the electric furnace with vanadium oxide and took a spectrum from 8000 to 20000 cm^{-1}, from which I prepared atlas plots to match the spectrum presented in the paper referenced above.

Fig. 16 shows a comparison. Greenstein's comment was, "If I blur my eyes and stand on my head, I would think that it looks about right". I have inverted the emission spectrum in this figure, thereby avoiding the need to stand on one's head. Then he goes on to suggest how I might translate my emission spectrum into an absorption spectrum with saturation, for a possible better match. Other tasks have absorbed my time, but the translation is on my list of things to do.

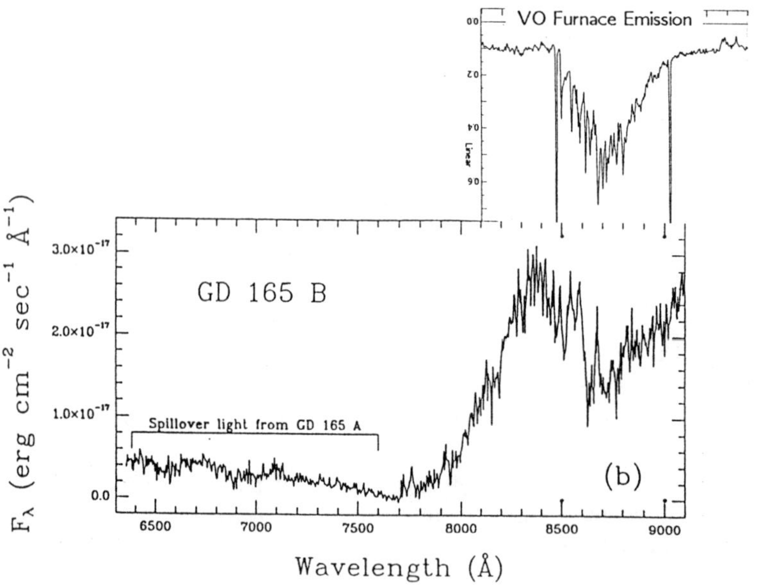

Fig. 16. Absorption spectrum from GD165B (Kirkpatrick et al. 1993) compared with VO furnace emission spectrum (inverted)

3.6 Comments

What's next? Our three most recent publications are transition rates from absorption measurements in ZrO (Littleton et al. 1993), an unidentified band in the infrared spectrum of FeH (Phillips & Davis 1993), and the infrared spectrum of the Meinel system of OH (Abrams et al. 1993). For current work, the complex spectrum of FeH in the green centered on 530 nm (18400 cm^{-1} to 19400 cm^{-1}) is being extensively studied. There are over 1000 lines in this region strong enough to merit assignment to rotational subbands, but there are no obvious subband heads or any obvious rotational series. A program searching for P and R branch pairs has found four groups of red-degraded subbands with heads in the regions 18800, 18970, 19170, and 19360 cm^{-1}. There is no evidence of Q branches. The structure of the branches is quite compatible with an assignment to FeH. If this preliminary assignment proves correct, the carrier is an FeH transition with delta-lambda equal to zero. Only about one-quarter of the 1000 plus lines in this region have been assigned to any branch.

We have an abundance of data for CN from every known source, and have put them in the hands of M.L.P. Rao (Andhra University) for unraveling the perturbations and making a global fit. We are completing our analysis of CaS; we are trying to unravel why f-values in the CaH spectrum are different for P and R branches; and we are thinking about new solar data from the space shuttle.

For the time being LaS, LaO, YS, and ZrS are dormant, although we have data for them all.

References

Abrams M.C., Davis S.P., Rao M.L.P., Engleman R. Jr., Brault J.W., 1993, *Astrophys. J. Suppl.*, submitted

Brault J.W., Abrams M.C., 1989, *Opt. Soc. Am. Tech. Digest*, **6**, 120

Davis S.P., 1987, *Publ. A.S.P.*, **99**, 1105

Davis S.P., Phillips J.G., 1963, *The Red System of the CN Molecule*, (Berkeley: University of California Press)

Gatterer A., Junkes J., Salpeter E.W., Rosen B., 1957, *Molecular Spectra of Metalic Oxides*, Specola Vaticana

Jarmain W.R., McCallum J.C., 1970, *TRAPRB: a computer program for molecular transitions*, (Physics Department, Univ. Western Ontario, Canada)

Kirkpatrick J.D., Henry T.J., Liebert J., 1993, *Astrophys. J.*, **406**, 701

Littleton J.E., Davis S.P., Song M., 1993, *Astrophys. J.*, **404**, 412

Pecyner R., Davis S.P., 1988, *Appl. Opt.* **27**, 3775

Phillips J.G., 1959, *Astrophys. J.*, **130**, 308

Phillips J.G., Davis S.P., 1962, *J. Opt. Soc. Am.*, **52**, 480

Phillips J.G., Davis S.P., 1968, *The Swan System of the C_2 Molecule*, (Berkeley: University of California Press)

Production, Detection and Study in the Infrared of Unstable Molecules and Radicals

Mireille Morillon-Chapey

Laboratoire de Physique Moléculaire et Applications, UPR 136, CNRS/Université Paris-Sud, Bât 350, 91405 ORSAY Cedex, France

1 Introduction

This subject is very wide. I shall present, here, a summary of our works related to radicals created in electric discharges. I shall review the methods and technical means used for studying these species in the infrared. This review will be accompanied with a detailed bibliography in which one will find the description of these methods and the most typical results obtained in our group in Orsay since 1986.

2 Scientific Aim of the Orsay Research Group

It is mainly the spectroscopic characterization of very different media: interstellar matter, atmospheres, plasmas, flames etc...

2.1 Our role is:

To create the molecular species. This implies the set up of various sources (chemical reactions, electric discharges)

To obtain the spectral parameters i.e. to determine line position and line intensity constants.

To determine in the best case, the molecular dipole moment and potential functions.

2.2 Our tools are:

(a) A Fourier Transform Spectrometer (FTS) with the following characteristics:

- home built apparatus (Guelachvili 1978)
- recording mode: step by step
- maximum non-apodized resolution: 0.0027 cm^{-1}
- large spectral domain: from the visible to 500 cm^{-1}

- free spectral range up to 4000 cm^{-1}
- precision of the wavenumber scale: 10^{-4}–10^{-5} cm^{-1}

(b) A Tunable Diode Laser Spectrometer (TDLS) (Spectra Physics) used for real time monitoring of the plasma sources and studying line-broadening effects.

(c) A HP 9000–350 workstation to compute the spectra and to treat the data.

2.3 The problems to solve are of different natures:

- safety problems which need technical solutions
- intensity measurements by FTS which require understanding of complex instrumental effects
- production of molecules:

(a) by electric discharges with optimization of the sources and control of their stability by means of the TDLS. The impossibility of measuring the concentration of the molecules in the source leads to indirect determinations of transition moments.

(b) by chemical reactions: we must obtain samples of very high purity and keep and check this purity during the experiments. Pressure and temperature of the sample must be recorded during the experiments.

In spite of the fact that, at present, many of our works imply the use of chemical reactions in order to study molecules of atmospheric interest like HNO_3, $ClNO_2$, $ClONO_2$ ... in this paper I will only present the techniques and results connected to the electric discharges.

3 Electrical Sources of Radicals

High concentrations of unstable molecules, in the laboratory, are generally difficult to obtain. This is mainly due to their short lifetime relative to their high chemical reactivity. Therefore we set up various experimental devices designed to efficiently produce unstable species. Figure 1 shows a simplified typical experimental configuration. The plasma reactor can be simultaneously observed by the diode-laser and Fourier-transform spectrometers.

The main sources that we use are:

(1) A radiofrequency reactor described by Chollet et al. (1986): It is a stainless tube, 3m long and 22 cm in diameter, which contains a multiple reflection cell permitting optical paths of several tens of meters. The plasma is excited by a RF discharge. The first application of this device was the study of a Silane plasma emission spectrum in which we detected SiH: it was the first detection of this radical in the infrared (Chollet et al. 1986). The results of the SiH study are given by Betrencourt et al. (1986). These experiences showed that in this discharge the vibrational temperatures are higher than the rotational temperatures. The other molecules detected using a Silane plasma are: HNSi (Elhanine et al. 1991,

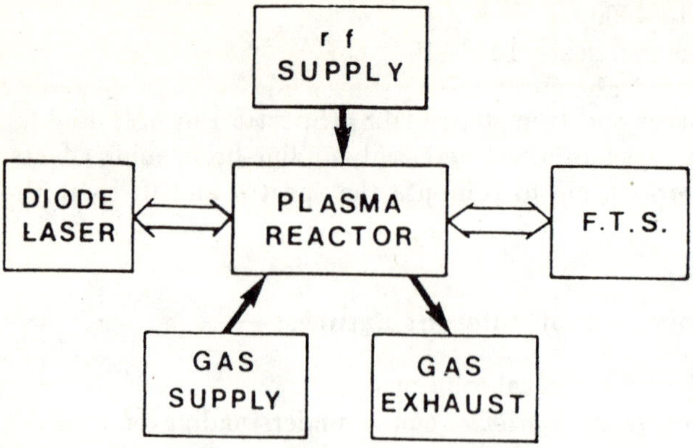

Fig. 1. The figure shows a simplified typical experimental configuration designed to efficiently produce and observe unstable species.

Elhanine et al. 1993) and SiN (Elhanine et al. 1992). Other types of plasma were achieved with the RF source: from a $H_2 + N_2$ mixture we studied the NH radical (Boudjaadar et al. 1986) and from a H_2S plasma we studied SH and the negative ion SH^- (Elhanine et al. 1988).

(2) A microwave source, well known for its advantages: (a) no electrodes and henceforth non impurities (b) wide pressure range (from 10^{-3} to 10 Torr), and (c) large density of generated species (10^{12}–$10^{13}/cm^3$). An important application was the observation of emission spectra of the OH radical from a $H_2O + O_2$ plasma. We obtained simultaneously pure rotational transitions in the v=0 and v=1 states and the rovibrational fundamental band (Benidar 1991). The experiences showed a thermal equilibrium between vibration and rotation.

(3) A D.C. discharge in which we observe the emission of the positive column in the infrared. One characteristic of this often used discharge is that it allows to selectively excite the vibrations. For example, Benidar (1991) shows that for OH this discharge allowed the observation of the four first rovibrational bands with Δv=1. We hope to apply this possibility of highly excited vibrations to the ClO radical for an intensity study.

(4) A hollow-cathode discharge: an infrared emission spectrum of N_2^+ is described by Ferguson et al. (1992).

4 Selective Detection of the Radicals

A crucial problem linked to these sources is to discriminate the usually small signals due to the transient species from the large spectral background of the parent molecule or other stable species which are present at the same time. For this, a property of the species of interest, by which it differs from the precursors, must be chosen and varied in a well defined and periodic manner. In laser spectroscopy, modulation methods have proved to be very successful. However such methods have rarely been applied to high resolution FTS. In our group, new modulation schemes are being developed for the interferometer with the following specificity: the modulation of the output signal of the interferometer, essential to improve all the detections in FTS, is here, produced by the source in a selective manner and not by the interferometer itself. This selective manner depends on the nature of the species, as will be shown in a few examples.

(i) Paramagnetic species: We realized two types of modulation for these very usual species: (a) Zeeman-Modulation: constant and periodic magnetic fields are applied to the molecules in a cell. The effect of this source modulation on the spectrum is that only the lines which are field sensitive are detected as shown on NO (Elhanine et al. 1988) and (Guelachvili 1986). (b) Polarization modulation: based on the polarization property of the Zeeman transitions in a magnetic field. The principle and results are explained by Elhanine et al. (1989).

(ii) Short lived species: We showed that a selective detection by FTS is possible by modulating their production processes on a time scale similar to their lifetime. Thus the long-lived species will not be modulated and therefore not detected. An application to the OH radical detection in emission by concentration modulation is given by Benidar et al. (1991).

(iii) Molecular ions: The most powerful technique is the velocity modulation. It has been very successful in detecting many new ions with tunable infrared lasers, especially in the group of R.J. Saykally at Berkeley. We applied this technique to High-Resolution FTS with its well-known multiplex and wide spectral-range advantages. The principle is given by Martin & Guelachvili (1990) with an application to ArH^+, showing for the first time a velocity modulation in emission.

5 Intensity Measurements

It is well known that rovibrational intensities for radical species are particularly difficult to determine using experimental methods. However, the interaction between vibration and rotation (Herman-Wallis effect) can be exploited to determine the transition moment independently of any information on molecular abundance. The effect of this interaction results in a transfer of intensity in a band, from the P branch to the R one or vice-versa. The first spectacular evidence of this effect was seen on SH produced in the SH_2 radio frequency plasma (Benidar et al. 1991). Then, thanks to this Herman-Wallis interaction, we studied the intensities of: NH (Chackerian et al. 1989), ArH^+ (Martin & Guelachvili 1991) and OH (Benidar 1991).

Let us notice that, by this method we do not need any information upon the population distribution in the levels, i.e. this is valid even if the source is not in thermal equilibrium. This is shown on the OH emission spectra: as seen above we used spectra from two different sources (M.W and D.C. discharges). In the D.C. discharge the vibrational levels are higher and the rotational distribution does not follow a Boltzmann distribution as it does for the microwave discharge. Therefore, Benidar (1991) shows that these two quite different experiments lead to the same Herman-Wallis function and consequently to the same dipole moment function.

References

Benidar A., Farrenq R., Guelachvili G., Chackerian C., 1991, *J. Mol. Spectrosc.*, **147**, 383

Benidar A., 1991, *These Universite Paris-Sud*

Benidar A., Guelachvili G., Martin P.A., 1991, *Chem. Phys. Letters*, **177**, 563

Betrencourt M., Boudjaadar D., Chollet P., Guelachvili G., Morillon-Chapey M., 1986, *J. Chem. Phys.*, **84**, 4121

Boudjaadar D., Brion J., Chollet P., Guelachvili G., Vervloet M., 1986, *J. Mol. Spectrosc.*, **119**, 352

Chackerian C., Guelachvili G., Lopez-Pineiro A., Tipping R.H., 1989, *J. Chem. Phys.*, **90**, 641

Chollet P., Guelachvili G., Morillon-Chapey M., Gressier P., Schmitt J.P.M., 1986, *J. Opt. Soc. Am. B*, **3**,, 687

Elhanine M., Farrenq R., Guelachvili G., 1991, *J. Chem. Phys.*, **94**, 2529

Elhanine M., Hanoune B., Guelachvili G., 1993, *J. Chem. Phys.*, **99**, 5

Elhanine M., Hanoune B., Guelachvili G., Amiot C., 1992, *J. Phys.*, **II 2**, 931

Elhanine M., Farrenq R., Guelachvili G., Morillon-Chapey M., 1988, *J. Mol. Spectrosc.*, **129**, 240

Elhanine M., Farrenq R., Guelachvili G., 1988, *Mikrochim. Acta (Wien)*, **II**, 265

Elhanine M., Farrenq R., Guelachvili G., 1989, *Appl. Opt.*, **28**, 4024

Ferguson D.W., Rao K., Martin P.A., Guelachvili G., 1992, *J. Mol. Spectrosc.*, **153**, 599

Guelachvili G., 1978, *Appl. Opt.*, **17**, 1322

Guelachvili G., 1986, *J. Opt. Soc. Am. B*, **3**, 1718

Martin P.A., Guelachvili G., 1990, *Phys. Rev. Letters*, **65**, 2535

Martin P.A., Guelachvili G., 1991, *Chem. Phys. Letters*, **180**, 344

Laboratory Millimeter Wave Spectroscopy of Small Reactive Species

C. Demuynck, M. Bogey, H. Bolvin, M. Cordonnier, J.L. Destombes, A. Walters

Laboratoire de Spectroscopie Hertzienne, Universite de Lille, F−59655 Villeneuve d'ascq Cedex,France.

1 Introduction

The discovery of a large variety of molecules by radio astronomy has been a very strong motivation for the development of laboratory millimeter wave spectroscopy. Among them, the reactive species, neutral and/or ionic, have been early recognized as playing a very important role in the chemistry of the interstellar and circumstellar medium. While the laboratory spectroscopy of free radicals started relatively early, with the observation of the OH radical by the group of Townes (Dousmanis et al. 1955), the detection of molecular ions proved to be a much more difficult task, and the first millimeter line due to an ion was actually detected by radio astronomy (Buhl & Snyder 1970). It was called "U89.2" until it was tentatively attributed to HCO^+ by Klemperer (1970) on the basis of both considerations on the chemistry of the interstellar medium, and of ab initio calculations for the prediction of the expected line frequency. This identification was later confirmed by more elaborated ab initio calculations (Wahlgren et al. 1973, Kraemers & Diercksen 1976), and by the observation of a transition attributed to $H^{13}CO^+$ (Snyder et al. 1976), but the definite confirmation was the observation of the same transition in a laboratory glow discharge by the group of Woods (Woods et al. 1975).

Due to their key-role in the astro-chemistry (see for example Viala 1986 and references therein), molecular ions are of course of special interest to laboratory spectroscopists, but many other types of molecules are also very important: free radicals and reactive species are relatively abundant in the interstellar medium; small cyclic molecules have been searched for since the discovery of C_3H_2 in a wide variety of sources; refractory molecules have been detected in circumstellar envelopes, and the observation of rare isotopic forms can give information on the mechanisms of isotopic fractionation. In all cases, the detection of these species request a preliminary observation of their spectra in the laboratory.

In addition to its significance for astrophysics, millimeter wave spectroscopy is a very powerful tool for probing the physico-chemistry of low pressure flames and plasmas: High-resolution spectroscopy is actually a very sensitive, non-intrusive technique for detecting stable or unstable species and estimating their temperatures. Moreover, in this frequency range, spectra are usually not too congested, which makes the problem of identifying blended spectra less critical than in the infrared range.

Also, being inherently a very high resolution technique (typically of the order of 100kHz), millimeter wave spectroscopy gives essential information about the energy-level configurations and molecular structure, making possible a detailed analysis of the rotational, fine and hyperfine structures, and leading to an accurate determination of the molecular parameters. The molecular structures determined by this method can be compared with the results of ab initio calculations, and provide reliable tests for the validity of the various levels of calculation. In this view, the reactive species are of special interest, since they widen the possibilities of comparing the bonding properties between atoms in the same column of the Mendeleieff table, such as carbon and silicon.

2 Basic Principles

In its principle, a millimeter wave spectroscopy experiment is extremely simple,as shown in Fig. 1, it is basically an absorption experiment. The source of radiation is a monochromatic, frequency-swept source, usually a klystron, a carcinotron, or, more recently, a gunn diode. The radiation passes through the absorption cell and a sensitive detector (generally a broadband InSb bolometer) is used to detect the absorption resulting from the interaction of the molecules with the electromagnetic field.

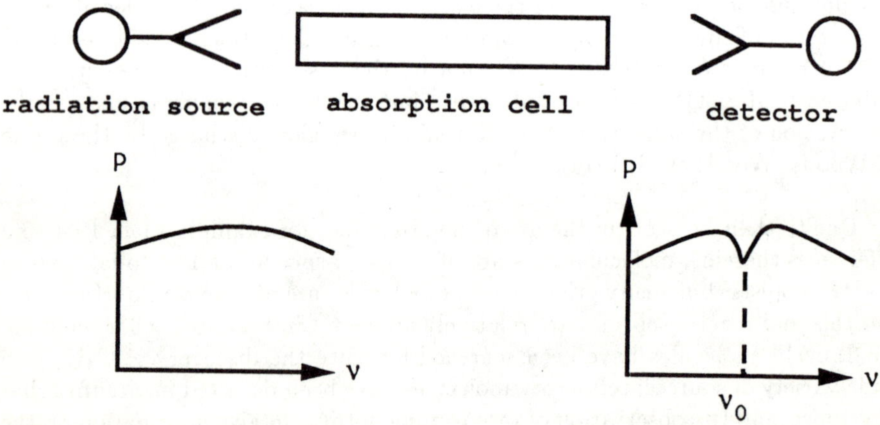

Fig. 1. Basic principle of an experiment of millimeter wave spectroscopy

Since the absorption of radiation by an atom or a molecule is mainly due to its electric dipole moment, only polar molecules can be detected by this method.

A few non-polar molecules have been detected through magnetic dipole transitions, like O_2 (Miller & Townes 1953), but the corresponding absorption is usually weak, and this is a severe limitation in the application of radio astronomy to the detection of interstellar molecules.

In order to avoid collisional broadening of the transitions, the gas in the absorption cell must be kept at a rather low pressure, ranging typically from 1 to 10 P. In this pressure range, the linewidth is of the order of a few hundred kHz, and millimeter wave spectroscopy is characterized by a high resolution (of the order of 50 to 100 kHz).

In the frequency range typical of millimeter wave spectroscopy (currently from 60 to 300 GHz approximatively; up to 400 to 600 GHz in more special cases), molecular transitions occur between rotational levels within a given vibrational state. Since the reactive species of interest are generally small molecules (4 to 5 atoms or less), their spectra are usually not very dense, and the probability of getting blended lines is rather small. This characteristic of millimeter wave spectroscopy is at the origin of the 2 different types of information which can be found in such a spectrum:

The first situation happens when enough lines have been measured and identified (which is not always straightforward!): the spectrum can be fitted using conventional models, and molecular constants are derived. The interpretation of these molecular constants gives information on the electronic and geometrical structure of the molecule. In particular, the rotational constants A, B, C are in first approximation inversely proportional to the inertial moments of the molecules, and, when several isotopic forms of a molecule have been observed, (i.e. several set of rotational constants corresponding to the same geometrical structure, but to different atomic masses), an accurate molecular structure can be determined. Of course, for a discussion of the accuracy on the bond lengths and angles, the validity of the theoretical model must be very carefully checked, but such a discussion is irrelevant in this paper. Moreover, knowing the molecular constants and using the convenient theoretical model, it is possible to calculate with an accuracy comparable to the experimental uncertainty the lines which, sometimes, cannot be directly measured for technical reasons.

The second situation happens when the spectrum of a molecule is already known: it is not necessary to observe the complete spectrum to identify this molecule in a complicated mixture: since the lines are rarely blended, the identification can rely on the observation of only a few characteristic lines, in contrast to infrared spectroscopy, where the identification is much more difficult. This property of millimeter wave spectra is of special importance for radio astronomy.

3 Experimental Problems

Even if the principles of the experiment are very simple, its realization confronts the spectroscopists with many problems: First of all, the frequency range to be swept must be roughly defined, or, in other words, the line frequencies must be predicted. Then, the spectrometer must be very sensitive, easily scanned in frequency, and, as much as possible, automatized in order to save time. The last problem, but not the least, is to produce the reactive species which, in terrestrial conditions, are usually very short-lived.

4 Prediction of the spectra

An ideal case is of course when the line frequency is already known. This condition occurs when the problem is to confirm, by laboratory observation, the assignment of a line observed by radio astronomy, (as mentioned above in the case of HCO^+). HCO^+ is not the only example, other ions, like HCS^+ (Thaddeus et al. 1981; Gudeman et al. 1981) and $HOCO^+$ (Thaddeus et al. 1981; Bogey et al. 1984) were also observed first by radio astronomy and then identified in the laboratory.

Important information can also be derived from other types of spectra, electronic spectra (see for example SiC_2, Michalopoulos et al. 1984; Gottlieb et al. 1989) or vibrational spectra, as in the case of HNSi (Elhaninne et al. 1991; Bogey et al. 1991a) or $C_2H_3^+$ (Bogey et al. 1992b). When no experimental data is available, the prediction of the spectra relies entirely on ab initio calculations. Ab initio calculations provide the potential energy surfaces, the minima of which correspond to the stable forms of the molecules. They also provide the values of the molecular parameters (bond lengths and angles) corresponding to the equilibrium structures of the molecules. From these parameters, the values of the rotational constants can be easily calculated. Due to the high resolution of millimeter wave spectroscopy, the accuracy of these ab initio calculations was for a long time not sufficient to allow an unambiguous identification of a molecule (see the example of HCO^+, Klemperer 1970). Fortunately, the accuracy of ab initio calculations has improved very much during the last years, and, as shown later, the role of quantum chemists is sometimes crucial in the procedure of identification of molecular spectra (see for example Bogey et al. 1987; 1991b; 1992a; Cordonnier et al. 1992).

5 The spectrometer

Since free radicals and molecular ions are very reactive species, their stationary concentration is very low, and the main quality of the spectrometer is a sensitivity as high as possible. A conservative value for the sensitivity of a spectrometer working in the 50–450 GHz range is $\Delta P/P = 10^{-7}$ per cm of absorption path (provided a convenient radiation source is used). Using typical values for the various parameters (dipole moment, linewidth, oscillator strength...), it is possible to estimate the lowest detectable concentration to be 10^8 to 10^7 per cm^3, which makes millimeter wave spectroscopy a competitive technique to detect reactive species, especially for the closed-shell molecules which are difficult to study by visible or UV spectroscopy.

The spectrometer we have built in Lille has been described elsewhere (Bogey et al. 1993). Below 340 GHz, we use harmonic generation from phase-locked klystrons or Gunn oscillators. Above 340 GHz, we use two phase-locked Thomson CSF carcinotrons emitting in the 340-470 GHz frequency range. In this frequency range, the high power emitted by the radiation source allows a very good sensitivity to be reached (about 10^{-8} cm^{-1} for a 10Hz bandwidth detection). The detection is achieved by a QMC Instrument liquid-helium-cooled InSb detector. The radiation is frequency-modulated, and the signal is demodulated at twice the modulation frequency, which provides a second derivative lineshape. A HP 9000-310 micro computer is used to automatically sweep the frequency and simultaneously record the spectrum. It also ensures frequency measurement and, when necessary to improve the signal-to-noise ratio, numerical signal processing.

6 Methods of production

Due to their short lifetime, reactive species can be detected only when they are produced directly in the absorption cell. Depending on the type of molecule of interest, various techniques can be used: Relatively fragile, moderately reactive molecules need soft chemical methods, like vacuum flash thermolysis and/or vacuum gas-solid phase reactions, which were used together to produce the 2H-azirine NCH_2CH, a cyclic isomer of methyl cyanide (Bogey et al. 1986). In the case of very reactive free radicals and molecular ions, the species are most conveniently produced and observed within a plasma excited by a D.C. discharge, and we will restrict ourself to this case in this paper. In a D.C. discharge, two parts of the discharge are of special interest. Generally, the positive column is used when free radicals or reactive molecules are searched for. In normal conditions ($I=50-200$ mA), this part of the discharge fills nearly the whole length of the discharge tube. A very short part of the discharge, called the negative glow and located near the cathode, is more suitable for the production of molecular ions, and especially of protonated molecules, because of the higher electron density. In order to extend this negative glow to the whole length of the cell, an axial magnetic field (typically 200 G), produced by a solenoid coiled round the cell, is applied to the discharge (De Lucia et al. 1983). In this regime, the discharge is characterised by a low discharge current ($1-10$ mA). The magnetic field is also used to discriminate the lines due to an ion from those due to a neutral: The intensity of the lines due to ions is strongly correlated to the strength of the magnetic field, whereas those due to neutral are insensitive to the field. (De Lucia et al. 1983; Bogey et al. 1988). Furthermore, when the positive column is used, the magnetic field is also very useful to check the Zeeman effect on the lines, in order to discriminate paramagnetic species from others.

A scheme of the discharge tube built in Lille is presented in Fig. 2. The cell consists of a 2.5 m long, 5 cm internal diameter Pyrex tube, with a jacket for liquid nitrogen cooling. Condensable gases are introduced through a 2 m long axial glass tube drilled all along its length. Both ends are closed by Teflon windows at Brewster incidence to improve the transmission of the cell.

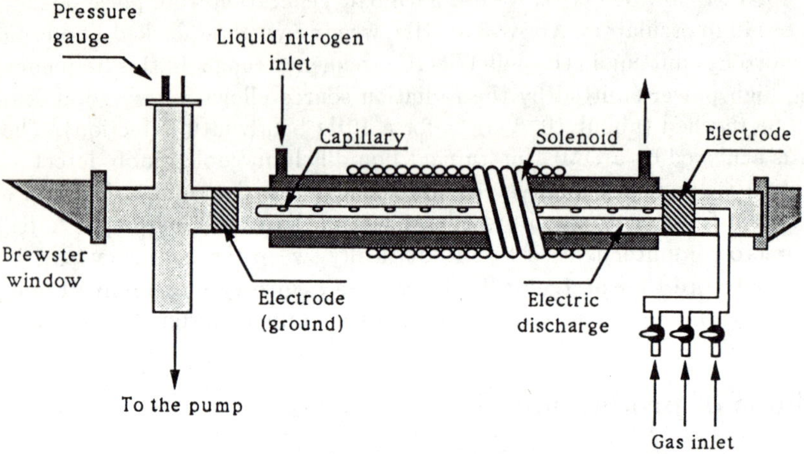

Fig. 2. Schematic diagram of the discharge cell

7 Experimental Procedure

To achieve a detection, measurement, and identification of the millimeter wave spectrum of a reactive species, not only good frequency predictions, a very sensitive spectrometer and an efficient technique of production are necessary: A bit of good luck is also essential, as well as lots of patience.

The main problem is that the physico-chemistry of the electric discharge is extremely mysterious, and in many cases our understanding of what happens in the discharge is nearer cooking recipe than conventional scientific argument. In these conditions, the only procedure is to scan the convenient spectral range (when correctly estimated) in various experimental conditions: chemical mixture, total and partial pressures, temperature and discharge conditions. Of course, sometimes it is successful, sometimes it is not. But the most fascinating is that, in some case, during an unsuccessful search for a given molecule, we found another species, much more interesting than the one we searched for first. In such a case, first the procedure of identification of the spectrum, and then of the molecule itself, and then possibly of its structure, is something like a detective story, as we experienced it first with the weakly bound molecular ion ArH_3^+ (Bogey et al. 1987), and will be described in the next section on other examples.

8 Some Typical Examples

The dibridged disilyne $Si(H_2)Si$: In the course of a search for the molecular ion SiH^+, unidentified lines were observed in a silane/argon plasma, in the 450 GHz frequency range. The lines had a typical behaviour, being maximum in intensity in the abnormal regime of the discharge (very low discharge current), and disappearing nearly completely in the positive column. However their intensity did not show the variation versus the magnetic field strength typical of "ionic" lines. Moreover, they did not exhibit any Zeeman splitting. After a few more chemical tests, we concluded that they were due to a closed-shell, neutral molecule containing only silicon and hydrogen atoms.

Figure 3 presents a stick diagram of a part of the observed spectrum. It shows the very characteristic pattern of a c-type Q branch, and the assignment of this spectrum by conventional methods (Winnewisser et al. 1968) was straightforward. Finally, a total of 87 lines was measured in the 347–472 GHz range, and the spectrum was fitted using a conventional Hamiltonian, with a standard deviation of 20 kHz.

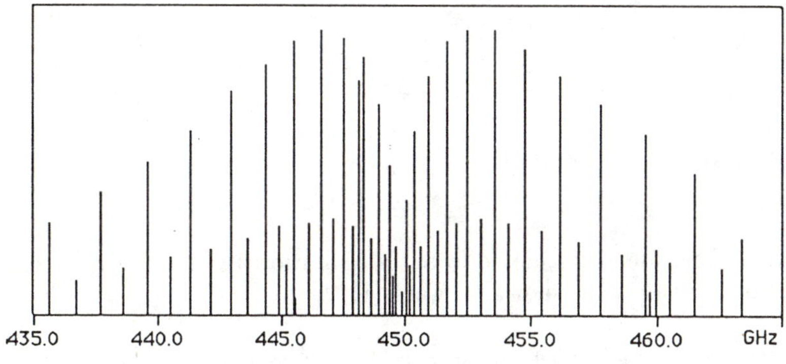

Fig. 3. Stick diagram of a part of the observed spectrum

The identification of this unknown molecule relied on a careful examination of the values of the rotational constants A, B, C, and of the characteristics of the observed spectrum:

(a) The B and C constants (= 7250 MHz) are of the same order of magnitude as those of molecules like S_2 (B = 8.8 GHz) or Si_2 (B = 7.2 GHz), which indicates that the molecule contains 2 silicon atoms.

(b) The relatively high value of A (= 157 GHz) suggests that A is mainly determined by a few H atoms out of the Si-Si axis, and that the number of H atoms do not exceed 3 (by comparison with known molecules).

(c) The remarkable 1:3 alternation in the intensities of adjacent lines, shown on Fig. 3, is characteristic of 2 exchangeable hydrogen atoms. All these observations suggests that the observed molecule is the disilyne Si_2H_2, a fundamental molecule never observed spectroscopically, but theoretically studied by quantum chemistry methods (Colegrove & Schaefer 1990; Grev & Schaefer 1992).

(d) These ab initio calculations showed that the ground state of Si_2H_2 has a singlet non-classical bridged structure (Fig. 4). With such a C_{2v} symmetry, the molecule is polar, and, with the calculated structure, the dipole moment lies along the c-axis, in agreement with the experiment.

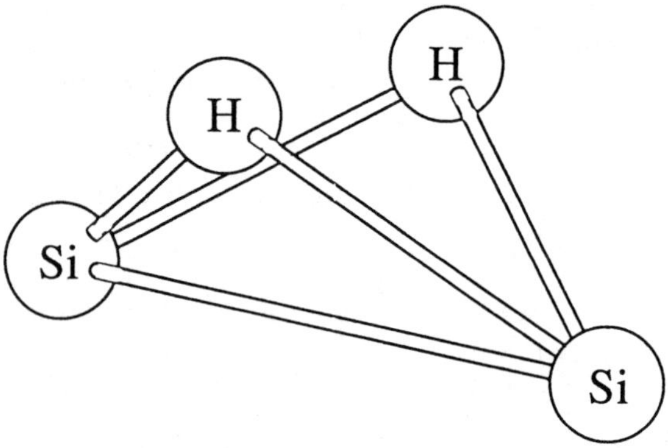

Fig. 4. Molecular structure of the disilyne $Si(H_2)Si$

(e) The A,B,C constants calculated from the ab initio structure are very close to the experimental values, as shown on Table 1.

(f) Other potential molecules can be ruled out in view of spectroscopic arguments, for example: the silasilene polar isomer H_2SiSi would give rise to an a-type spectrum, very different from the observed spectrum. Si_2H_4 has a trans-bent geometry in its ground state, with a center of symmetry, and then no dipole moment.

A definite confirmation of the identification of the molecule came with the observation of the ^{29}Si and ^{30}Si monosubstituted forms: The rotational constants were calculated assuming the structure derived from ab initio calculations, and the spectra were predicted using a conventional hamiltonian. The spectra were observed in natural abundance, and a set of rotational constants was derived for each isomer (Bogey et al. 1991b). Later, the $Si(D_2)Si$ spectrum was also observed

Table 1. Rotational constants of $Si(H_2)Si$ (in MHz). Calculations from Colegrove & Schaefer (1990)

	Experiment	Calculation
A	157198.8	163450.0
B	7281.3	7236.0
C	7199.7	7154.0
B−C	81.6	81.9

by replacing SiH_4 with SiD_4 in the discharge cell, and another set of rotational constants was derived.

Due to the symmetry of the molecule, its structure depends only on 3 geometrical parameters, namely the Si-H and Si-Si distances, and the dihedral angle <HSiSiH>. Using together the 4 sets of rotational constants, it was possible to derive a substitution structure (Kraitchman 1953; Costain 1958) which is a very good approximation of the equilibrium structure. As shown in Table 2, the agreement between experimental and theoretical structures is excellent (Bogey et al. 1993).

Table 2. Molecular structure of $Si(H_2)Si$

	This work	ref(1)	ref(2)
Si - Si (Å)	2.2154	2.216	2.222
Si - H (Å)	1.6680	1.668	1.681
<H SiSi H>(°)	104.22	104.0	106.5

(1) Colegrove & Schaefer 1990, (2) Grev & Schaefer 1992.

9 The Monobridged Form Si(H)SiH

Another case of very interesting unidentified lines occurred during a search for SiH_3. This radical is clearly of astrophysical interest, but there is a large uncertainty on its spectrum because it presents both spin-orbit coupling and inversion internal motion. Several frequency scans were carried out in various experimental conditions: discharge in pure silane or in argon-silane mixtures, at room temperature or with liquid nitrogen cooling of the cell. Again, a lot of unidentified lines were observed, which cannot be ascribed to SiH_3. Clearly, the molecule responsible for these lines was a short-lived molecule, appearing only in the abnormal regime of the discharge, it was not an ion, and it was a closed-shell molecule. Moreover, this mysterious molecule had a chemical behaviour very similar to that of the dibridged disilyne $Si(H_2)Si$. From a spectroscopic point of view, a 6 GHz frequency scan around 210 GHz allowed the observation of a typical a-type pattern, with no indication of spin statistics from the relative intensities of the lines.

The identification of the K-components was straightforward, but several trials were necessary to determine the J-value. Finally, the spectrum was identified and fitted, and line predictions were made in the 420–470 GHz frequency range. It turned out that these predictions allowed us to identify some unassigned lines observed during our preceeding study of $Si(H_2)Si$. A total of 160 lines were finally measured between 180 and 470 GHz. They were fitted using a conventional hamiltonian in order to derive the molecular constants. As shown in Table 3, there is a striking similarity between the B and C constants of $Si(H_2)Si$ and the unknown molecule. It means that, here again, the molecule contains only 2 silicon atoms.

Table 3. Rotational constants (in MHz) of the unknown molecule and $Si(H_2)Si$

	unknown molecule	$Si(H_2)Si$
A	262092.3	157198.8
B	7361.9	7281.3
C	7161.8	7199.7

Here again, the identification of the molecule was greatly aided by using the results of ab initio calculations. Indeed, in the papers where the results concerning $Si(H_2)Si$ were presented (Colegrove & Schaefer 1990; Grev & Schaefer 1992), the authors also characterised the other stationary points on the Si_2H_2 potential surface. In addition to the already known disilavinylidene isomer $SiSiH_2$ (see for example Koseki & Gordon 1989), they discovered another low-lying isomer, the monobridged form Si(H)SiH, which turned out in fact to be lower in energy than the disilavinylidene. They also showed that the energy difference between the di- and mono-bridged forms is low enough (8.7kcal/mol) to make this isomer an attractive candidate for our unknown molecule. Not only were the calculated rotational constants of this new isomer in good agreement with the experimental values (see Table 4), but other arguments were much more in favour of Si(H)SiH than of $SiSiH_2$ too. $SiSiH_2$ has 2 exchangeable H atoms, which is not consistent with our observations. Moreover, it has a rather small a-type dipole moment (0.15D). In contrast to this, Si(H)SiH has no equivalent H atoms, its dipole moment has 2 components, the strongest (0.96D) along the a-axis, and a much weaker one (0.04D) along the b-axis, which is consistent with the typical a-type spectrum observed.

As in the preceeding case, the definite confirmation of the identification of the molecule requested the observation of another isotopomer. The spectra of the ^{29}Si and ^{30}Si monosubstituted forms were too weak to be observed in natural abundance, but the deuterated form was observed by replacing SiH_4 by SiD_4 in

Table 4. Rotational constants (in MHz) of the unknown molecule and the Si_2H_2 isomers (Calculations from Grev & Schaefer 1992)

	A	B	C
Experimental	262092.3	7361.9	7161.8
$SiSiH_2$	168663.0	6708.0	6452.0
Si(H)SiH	275510.0	7375.0	7181.0

the discharge. As expected, the spectrum was observed very near the predicted spectrum, and the observed spectrum was fitted to derive the molecular constants. In the case of this molecule, 5 geometrical parameters are necessary to determine the structure, for example the Si-Si distance, 2 Si-H distances, and 2 angles (see Fig. 5). With 2 isomers, only 4 independent rotational constants are available, since, for a planar molecule, the 3 rotational constants are not independent. It is then impossible to determine completely the molecular structure from the experimental data. In order to determine a preliminary structure, the $<Si_1\text{-}Si_2\text{-}H_1>$ angle was arbitrarily fixed to the ab initio value of 52.5°. It led to the following values for the various structural parameters: $Si_1\text{-}Si_2$ = 2.119 Å; $Si_2\text{-}H_1$ = 1.629 Å; $Si_2\text{-}H_2$ = 1.474 Å; $<Si_1\text{-}Si_2\text{-}H_2>$ = 157.5o (Cordonnier et al. 1992).

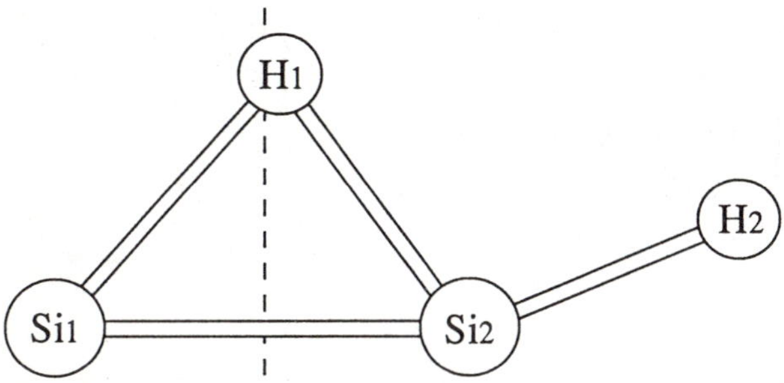

Fig. 5. Molecular structure of the monobridged Si(H))SiH

10 The Protonated Acetylene $C_2H_3^+$

The observation of protonated species was early suggested as an indirect way to detect non polar molecules like acetylene, C_2H_2, in the interstellar medium (Herbst et al. 1977). Moreover, protonated acetylene $C_2H_3^+$ is expected to play an important role in the hydrocarbon chemistry of the interstellar medium (Herbst & Leung 1989), and Glassgold et al. (1992), reexamined the chemistry of this ion, and concluded that with the existing large radiotelescopes some millimeter wave transitions should be observable in interstellar clouds and/or C-rich circumstellar envelopes.

$C_2H_3^+$ has also received considerable attention from the quantum chemists, due to the possible existence of classical and non-classical (bridged) structures in this small carbo-cation (Fig. 6). From the various ab initio calculations, it seems to emerge that the classical structure is a transition state between two equivalent forms of the non-classical isomer (Lindh et al. 1991; see also Crofton et al. 1989 for a review of earlier works). The height of the barrier is estimated to be about 1300 cm^{-1} (Lindh et al. 1991), which is low enough to allow tunneling on a measurable time scale.

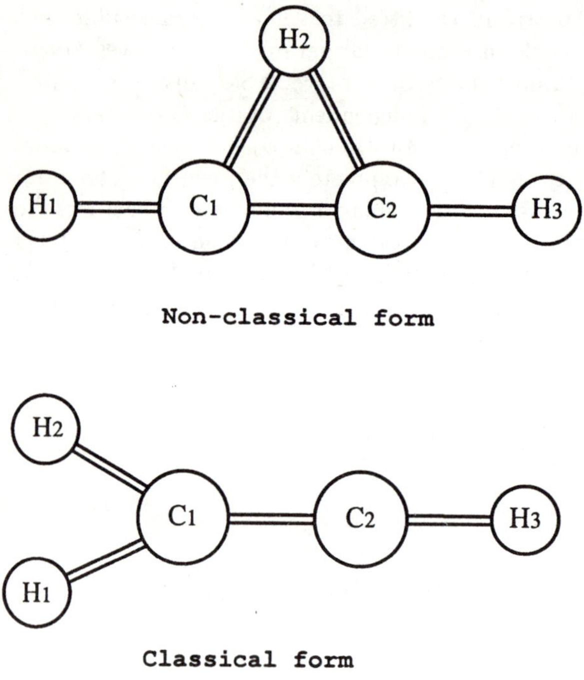

Fig. 6. Molecular structures of the protonated acetylene $C_2H_3^+$

In contrast to these numerous theoretical results, very little was known experimentally on the spectroscopic properties of this ion, until the pioneering work of Oka's group in Chicago. They first observed infrared lines of $C_2H_3{}^+$ in 1985 (Crofton & Oka 1985), and published recently their main conclusions (Crofton et al. 1989): The very rich infrared spectrum observed in the 3.2μm range is due to the bridged form of $C_2H_3{}^+$. Moreover, tunneling splittings expected from the moderate barrier height were experimentally resolved in the excited vibrational state. However, in spite of extensive theoretical works (Hougen 1987; Escribano& Bunker 1987; Escribano et al. 1988; Gomez & Bunker 1990), there is presently no theoretical model to interpret quantitatively this tunneling splitting within experimental accuracy. Consequently, Crofton et al. (1989) used infrared line combination differences to determine the B and C rotational constants of $C_2H_3^+$ in its ground vibrational state, but were unable to determine A, preventing any accurate prediction of the pure rotational b-type spectrum to be made.

However, we decided to carry out a search for this rotational spectrum, since, in view of the reasonably large dipole moment ($\simeq$1.25 D according to Lee & Schaefer 1986), it could be observed at cm and mm wavelengths (Glassgold et al. 1992).

The ions were produced in the negative glow discharge, in a mixture of argon, hydrogen and acetylene. A 5 GHz range was scanned around 432 GHz, the frequency of the intense $9_{18}-9_{09}$ transition predicted using the molecular constants of Crofton et al. (1989), and several lines were observed. Among them, only one exhibited the intensity evolution versus magnetic field characteristic of an ion. Several chemical tests were performed to make sure that the ion contains only C and H atoms. Assuming that this line was indeed the expected transition, its frequency was included in a fit together with the combination differences of Crofton et al. (1989), leading to a preliminary determination of the rotational constant A, and to a prediction of the frequencies of other transitions, which were readily observed.

A total of 17 b-type transitions were finally measured and fitted using a standard hamiltonian, leading to the determination of the 3 rotational constants and 4 centrifugal distortion constants (Bogey et al. 1992b). In order to confirm the identification of the molecule, we compared the molecular constants obtained by fitting (a) only the combination differences of Crofton et al. (1989); (b) both the combination differences and the pure rotational transitions; (c) only the rotational transitions. The results are presented in Table 5. Their consistency is remarkable and confirms that the molecular ion observed by Oka's group and by us is the same carbo-ion, namely $C_2H_3{}^+$ in its non-classical, bridged form. Moreover, all ab initio calculated rotational constants are in agreement with these results.

Table 5. Rotational constants of $C_2H_3{}^+$ (in MHz)

	IR data	IR + mmw data	mmw data
A	396925.(fixed)	399956.00(17)	399955.983(18)
B	34234.8(20)	34237.532(58)	34237.5395(66)
C	31371.8(20)	31371.74(12)	31371.760(14)

One of the most exciting problems concerning this molecule is related to the internal motion of the protons. In spite of an extensive search, no splitting of the lines resulting from this tunneling motion was observed in the pure rotational spectrum. This result indicates that the barrier is probably higher than the calculated value (Lindh et al. 1991), and this problem should be re-investigated in the future, since the tunneling splitting is expected to be very sensitive to the height of the barrier (Escribano 1993).

Finally, the frequencies of some (for technical reasons) non observed lines of potential astrophysical interest were predicted using our set of molecular constants (Bogey et al. 1992b).

Recently, this molecular ion was searched for in the interstellar medium. The detection was negative, but an upper limit for its abundance was determined (Guelin 1993).

11 Conclusion

Recent improvements in the methods of production as well as in the performances of the spectrometers (sensitivity, frequency coverage, automatisation) have made possible the spectacular developments in the topic of high resolution spectroscopy of transient species. From the examples presented above, it is clear that quantum chemistry plays often a key-role in the procedure of identification of a new molecule.

The necessary interplay between theory, laboratory spectroscopy and radio astronomy has proved to be particularly successful, and many exciting discoveries are still expected in this field, as examplified by the recent laboratory and radio astronomical detection of MgNC, the first magnesium-containing interstellar molecule (Kawaguchi et al. 1993). It is also noteworth that radio astronomy is not only a strong motivation for laboratory spectroscopists, it can be also, indirectly, at the origin of the accidental discovery of fascinating molecules in the laboratory, as illustrated by ArH_3^+ (Bogey et al. 1988), $Si(H_2)Si$ (Bogey et al. 1991b), and Si(H)SiH (Cordonnier et al. 1992).

Acknowledgement. This research was partially supported by the European Communities (project 892 001 59/OP1: "Structure and Dynamics of Molecular Ions") and by the C.N.R.S. (Groupe De Recherche "Physico-Chimie des Molecules et Grains Interstellaires").

References

Bogey M., Demuynck C., Destombes J.L., 1984, *Astron. Astrophys.*, **138**, L11
Bogey M., Destombes J.L., Denis J.M., Guillemin J.C., 1986, *J. Mol. Spectr.*, **115** , 1
Bogey M., Bolvin H., Demuynck C., Destombes J.L., 1987, *Phys. Rev. Lett.*, **58**, 988

Bogey M., Demuynck C., Destombes J.L., Walters A., 1991a, *Astron. Astrophys.*, **244**, L47

Bogey M., Bolvin H., Demuynck C., Destombes J.L., 1991b, *Phys. Rev. Lett.*, **66**, 413

Bogey M., Cordonnier M., Demuynck C., Destombes J.L., 1992a,*J. Mol. Spectr.*, **155**, 217

Bogey M., Cordonnier M., Demuynck C., Destombes J.L., 1992b, *Astrophys. J.*, **399**, L103

Bogey M., Bolvin H., Cordonnier M., Demuynck C., Destombes J.L., Csaszar A.G., 1993, in preparation

Buhl D., Snyder L.E., 1970,*Nature*, **228**, 267

Colegrove B.T., Schaefer III H.F., 1990, *J. Phys. Chem.*, **94**, 5593

Cordonnier M., Bogey M., Demuynck C., Destombes J.L., 1992, *J. Chem. Phys.*, **97**, 7984

Costain C.C., 1958, *J. Chem. Phys.*, **29**, 864

Crofton M.W., Oka T., 1985, *40th Symposium on Molecular Spectroscopy, Columbus, Ohio*

Crofton M.W., Jagod M.F., Rehfuss B.D., Oka T., 1989, *J. Chem. Phys.*, **91**, 5139

De Lucia F.C., Herbst E., Plummer G.M., Blake G.A., 1983, *J. Chem. Phys.*, **78**, 2312

Destombes J.L., Demuynck C., Bogey M., 1988, *Phil. Trans. R. Soc. Lond. A*, **324**, 147

Dousmanis G.C., Sanders T.M. Jr, Townes C.H., 1955, *Phys. Rev.*, **100**, 1735

Elhanine M., Farrenq R., Guelachvili G., 1991, *J. Chem. Phys.*, **94**, 2529

Escribano R., Bunker P.R., 1987, *J. Mol. Spectr.*, **122**, 325

Escribano R., Bunker P.R., Gomez P.C.,1988, *Chem. Phys. Lett.*, **150**, 60

Escribano R., 1993, *private communication*

Glassgold A.E., Omont A., Guelin M., 1992, *Astrophys. J.*, **396**, 115

Gomez P.C., Bunker P.R., 1990, *Chem. Phys. Lett.*, **165**, 351

Gottlieb C.A., Vrtilek J.M., Thaddeus P., 1989, *Astrophys. J.*, **343**, L29

Grev R.S., Schaefer III H.F., 1992, *J. Chem. Phys.*, **97**, 7990

Gudeman C.S., Haese N.N., Piltch N.D., Woods R.C., 1981, *Astrophys. J.*, **246**, L47

Guelin M., 1992, *private communication*

Herbst E., Green S., Thaddeus P., Klemperer W., 1977, *Astrophys. J.*, **215**, 503

Herbst E., Leung C.M., 1989, *Astrophys. J. Suppl. Ser.*, **69**, 271

Hougen J.T., 1987, *J. Mol. Spectr.*, **123**, 197

Kawaguchi K., Kagi E., Hirano T., Takano S., Saito S., 1993, *Astrophys. J.*, **406**, L39

Klemperer W., 1970, *Nature*, **227**, 267

Koseki S., Gordon M.S., 1989, *J. Phys. Chem.*, **93**, 118

Kraemers W.P., Diercksen G.H.F., 1976, *Astrophys. J.* , **205**, L97

Kraitchmann J., 1953, *Am. J. Phys.*, **21**, 17

Lee T.J., Schaefer III H.F., 1986, *J. Chem. Phys.*, **85**, 3437

Lindh R., Rice J.E., Lee T.J., 1991, *J. Chem. Phys.*, **94**,8008

Michapopoulos D.L., Geusic M.E., Langridge-Smith P.R.R., Smalley R.E., 1984, *J. Chem. Phys.*, **80**, 3556

Miller S.L., Townes C.H., 1953, *Phys. Rev.*, **90**, 537

Snyder L.E., Hillis J.M., Lovas F.J., Ulich B.L., 196, *Astrophys. J.*, **209**, 67

Thaddeus P., Guelin M., Linke R.A., 1981, *Astrophys. J.*, **246**, L41

Viala Y.P., 1986, *Astron. Astrophys. Suppl. Ser.*, **64**, 391

Wahlgreen U., Liu B., Pearson P.K., Schaefer III H.F., 1973, *Nature*, **246**, 4

Winnewisser G., Winnewisser M., Gordy W., 1968, *J. Chem. Phys.*, **49**, 3465

Woods R.C., Dixon T.A., Saykally R.J., Szanto P.G., 1975, *Phys. Rev. Lett.*, **35**, 1269

Table of Contents for "Poster Session Proceedings of IAU Colloquium 146: Molecular Opacities in the Stellar Environment"

Edited by Peter Thejll and Uffe Gråe Jørgensen

The present book, "Molecules in the stellar environment", is based on the invited talks of IAU Colloquium 146 held in Copenhagen, May 1993. The poster papers of the same conference are issued as a special poster-edition printed by Copenhagen University. The poster session proceedings is available by writing to the editors (address: Niels Bohr institute, Blegdamsvej 17, DK 2100 Copenhagen, Denmark; email addresses: thejll@nbivax.nbi.dk or uffegj@nbivax.nbi.dk). The table of contents is given below.

Posters:

Index

Springer-Verlag and the Environment

We at Springer-Verlag firmly believe that an international science publisher has a special obligation to the environment, and our corporate policies consistently reflect this conviction.

We also expect our business partners – paper mills, printers, packaging manufacturers, etc. – to commit themselves to using environmentally friendly materials and production processes.

The paper in this book is made from low- or no-chlorine pulp and is acid free, in conformance with international standards for paper permanency.

Lecture Notes in Physics

For information about Vols. 1–389
please contact your bookseller or Springer-Verlag

Vol. 390: J. Heidmann, M. J. Klein (Eds.), Bioastronomy - The Search for Extraterrestrial Life. Proceedings, 1990. XVII, 413 pages. 1991.

Vol. 391: A. Zdziarski, M. Sikora (Eds.), Ralativistic Hadrons in Cosmic Compact Objects. Proceedings, 1990. XII, 182 pages. 1991.

Vol. 392: J.-D. Fournier, P.-L. Sulem (Eds.), Large-Scale Structures in Nonlinear Physics. Proceedings. VIII, 353 pages. 1991.

Vol. 393: M. Remoissenet, M.Peyrard (Eds.), Nonlinear Coherent Structures in Physics and Biology. Proceedings. XII, 398 pages. 1991.

Vol. 394: M. R. J. Hoch, R. H. Lemmer (Eds.), Low Temperature Physics. Proceedings. X,374 pages. 1991.

Vol. 395: H. E. Trease, M. J. Fritts, W. P. Crowley (Eds.), Advances in the Free-Lagrange Method. Proceedings, 1990. XI, 327 pages. 1991.

Vol. 396: H. Mitter, H. Gausterer (Eds.), Recent Aspects of Quantum Fields. Proceedings. XIII, 332 pages. 1991.

Vol. 398: T. M. M. Verheggen (Ed.), Numerical Methods for the Simulation of Multi-Phase and Complex Flow. Proceedings, 1990. VI, 153 pages. 1992.

Vol. 399: Z. Švestka, B. V. Jackson, M. E. Machedo (Eds.), Eruptive Solar Flares. Proceedings, 1991. XIV, 409 pages. 1992.

Vol. 400: M. Dienes, M. Month, S. Turner (Eds.), Frontiers of Particle Beams: Intensity Limitations. Proceedings, 1990. IX, 610 pages. 1992.

Vol. 401: U. Heber, C. S. Jeffery (Eds.), The Atmospheres of Early-Type Stars. Proceedings, 1991. XIX, 450 pages. 1992.

Vol. 402: L. Boi, D. Flament, J.-M. Salanskis (Eds.), 1830-1930: A Century of Geometry. VIII, 304 pages. 1992.

Vol. 403: E. Balslev (Ed.), Schrödinger Operators. Proceedings, 1991. VIII, 264 pages. 1992.

Vol. 404: R. Schmidt, H. O. Lutz, R. Dreizler (Eds.), Nuclear Physics Concepts in the Study of Atomic Cluster Physics. Proceedings, 1991. XVIII, 363 pages. 1992.

Vol. 405: W. Hollik, R. Rückl, J. Wess (Eds.), Phenomenological Aspects of Supersymmetry. VII, 329 pages. 1992.

Vol. 406: R. Kayser, T. Schramm, L. Nieser (Eds.), Gravitational Lenses. Proceedings, 1991. XXII, 399 pages. 1992.

Vol. 407: P. L. Smith, W. L. Wiese (Eds.), Atomic and Molecular Data for Space Astronomy. VII, 158 pages. 1992.

Vol. 408: V. J. Martínez, M. Portilla, D. Sàez (Eds.), New Insights into the Universe. Proceedings, 1991. XI, 298 pages. 1992.

Vol. 409: H. Gausterer, C. B. Lang (Eds.), Computational Methods in Field Theory. Proceedings, 1992. XII, 274 pages. 1992.

Vol. 410: J. Ehlers, G. Schäfer (Eds.), Relativistic Gravity Research. Proceedings, VIII, 409 pages. 1992.

Vol. 411: W. Dieter Heiss (Ed.), Chaos and Quantum Chaos. Proceedings, XIV, 330 pages. 1992.

Vol. 412: A. W. Clegg, G. E. Nedoluha (Eds.), Astrophysical Masers. Proceedings, 1992. XX, 480 pages. 1993.

Vol. 413: Aa. Sandqvist, T. P. Ray (Eds.); Central Activity in Galaxies. From Observational Data to Astrophysical Diagnostics. XIII, 235 pages. 1993.

Vol. 414: M. Napolitano, F. Sabetta (Eds.), Thirteenth International Conference on Numerical Methods in Fluid Dynamics. Proceedings, 1992. XIV, 541 pages. 1993.

Vol. 415: L. Garrido (Ed.), Complex Fluids. Proceedings, 1992. XIII, 413 pages. 1993.

Vol. 416: B. Baschek, G. Klare, J. Lequeux (Eds.), New Aspects of Magellanic Cloud Research. Proceedings, 1992. XIII, 494 pages. 1993.

Vol. 417: K. Goeke P. Kroll, H.-R. Petry (Eds.), Quark Cluster Dynamics. Proceedings, 1992. XI, 297 pages. 1993.

Vol. 418: J. van Paradijs, H. M. Maitzen (Eds.), Galactic High-Energy Astrophysics. XIII, 293 pages. 1993.

Vol. 419: K. H. Ploog, L. Tapfer (Eds.), Physics and Technology of Semiconductor Quantum Devices. Proceedings, 1992. VIII, 212 pages. 1993.

Vol. 420: F. Ehlotzky (Ed.), Fundamentals of Quantum Optics III. Proceedings, 1993. XII, 346 pages. 1993.

Vol. 421: H.-J. Röser, K. Meisenheimer (Eds.), Jets in Extragalactic Radio Sources. XX, 301 pages. 1993.

Vol. 422: L. Päivärinta, E. Somersalo (Eds.), Inverse Problems in Mathematical Physics. Proceedings, 1992. XVIII, 256 pages. 1993.

Vol. 423: F. J. Chinea, L. M. González-Romero (Eds.), Rotating Objects and Relativistic Physics. Proceedings, 1992. XII, 304 pages. 1993.

Vol. 424: G. F. Helminck (Ed.), Geometric and Quantum Aspects of Integrable Systems. Proceedings, 1992. IX, 224 pages. 1993.

Vol. 425: M. Dienes, M. Month, B. Strasser, S. Turner (Eds.), Frontiers of Particle Beams: Factories with $e^+ e^-$ Rings. Proceedings, 1992. IX, 414 pages. 1994.

Vol. 426: L. Mathelitsch, W. Plessas (Eds.), Substructures of Matter as Revealed with Electroweak Probes. Proceedings, 1993. XIV, 441 pages. 1994

Vol. 427: H. V. von Geramb (Ed.), Quantum Inversion Theory and Applications. Proceedings, 1993. VIII, 481 pages. 1994.

Vol. 428: U. G. Jørgensen (Ed.), Molecules in the Stellar Environment. Proceedings, 1993. VIII, 440 pages. 1994.

Vol. 429: J. L. Sanz, E. Martínez-González, L. Cayón (Eds.), Present and Future of the Cosmic Microwave Background. Proceedings, 1993. VIII, 233 pages. 1994.

New Series m: Monographs

Vol. m 1: H. Hora, Plasmas at High Temperature and Density. VIII, 442 pages. 1991.

Vol. m 2: P. Busch, P. J. Lahti, P. Mittelstaedt, The Quantum Theory of Measurement. XIII, 165 pages. 1991.

Vol. m 3: A. Heck, J. M. Perdang (Eds.), Applying Fractals in Astronomy. IX, 210 pages. 1991.

Vol. m 4: R. K. Zeytounian, Mécanique des fluides fondamentale. XV, 615 pages, 1991.

Vol. m 5: R. K. Zeytounian, Meteorological Fluid Dynamics. XI, 346 pages. 1991.

Vol. m 6: N. M. J. Woodhouse, Special Relativity. VIII, 86 pages. 1992.

Vol. m 7: G. Morandi, The Role of Topology in Classical and Quantum Physics. XIII, 239 pages. 1992.

Vol. m 8: D. Funaro, Polynomial Approximation of Differential Equations. X, 305 pages. 1992.

Vol. m 9: M. Namiki, Stochastic Quantization. X, 217 pages. 1992.

Vol. m 10: J. Hoppe, Lectures on Integrable Systems. VII, 111 pages. 1992.

Vol. m 11: A. D. Yaghjian, Relativistic Dynamics of a Charged Sphere. XII, 115 pages. 1992.

Vol. m 12: G. Esposito, Quantum Gravity, Quantum Cosmology and Lorentzian Geometries. Second Corrected and Enlarged Edition. XVIII, 349 pages. 1994.

Vol. m 13: M. Klein, A. Knauf, Classical Planar Scattering by Coulombic Potentials. V, 142 pages. 1992.

Vol. m 14: A. Lerda, Anyons. XI, 138 pages. 1992.

Vol. m 15: N. Peters, B. Rogg (Eds.), Reduced Kinetic Mechanisms for Applications in Combustion Systems. X, 360 pages. 1993.

Vol. m 16: P. Christe, M. Henkel, Introduction to Conformal Invariance and Its Applications to Critical Phenomena. XV, 260 pages. 1993.

Vol. m 17: M. Schoen, Computer Simulation of Condensed Phases in Complex Geometries. X, 136 pages. 1993.

Vol. m 18: H. Carmichael, An Open Systems Approach to Quantum Optics. X, 179 pages. 1993.

Vol. m 19: S. D. Bogan, M. K. Hinders, Interface Effects in Elastic Wave Scattering. XII, 182 pages. 1994.

Vol. m 20: E. Abdalla, M. C. B. Abdalla, D. Dalmazi, A. Zadra, 2D-Gravity in Non-Critical Strings. IX, 319 pages. 1994.